Clean Energy Production Technologies

Series Editors

Neha Srivastava, Department of Chemical Engineering and Technology, IIT (BHU) Varanasi, Varanasi, Uttar Pradesh, India

P. K. Mishra, Department of Chemical Engineering and Technology, IIT (BHU) Varanasi, Varanasi, Uttar Pradesh, India

The consumption of fossil fuels has been continuously increasing around the globe and simultaneously becoming the primary cause of global warming as well as environmental pollution. Due to limited life span of fossil fuels and limited alternate energy options, energy crises is important concern faced by the world. Amidst these complex environmental and economic scenarios, renewable energy alternates such as biodiesel, hydrogen, wind, solar and bioenergy sources, which can produce energy with zero carbon residue are emerging as excellent clean energy source. For maximizing the efficiency and productivity of clean fuels via green & renewable methods, it's crucial to understand the configuration, sustainability and techno-economic feasibility of these promising energy alternates. The book series presents a comprehensive coverage combining the domains of exploring clean sources of energy and ensuring its production in an economical as well as ecologically feasible fashion. Series involves renowned experts and academicians as volume-editors and authors, from all the regions of the world. Series brings forth latest research, approaches and perspectives on clean energy production from both developed and developing parts of world under one umbrella. It is curated and developed by authoritative institutions and experts to serves global readership on this theme.

More information about this series at http://www.springer.com/series/16486

Jitendra Kumar Saini • Rajesh K Sani
Editors

Microbial Biotechnology for Renewable and Sustainable Energy

Springer

Editors
Jitendra Kumar Saini
Department of Microbiology
Central University of Haryana
Mahendergarh, Haryana, India

Rajesh K Sani
Department of Chemical and Biological Engineering
South Dakota School of Mines and Technology
Rapid City, South Dakota, USA

ISSN 2662-6861 ISSN 2662-687X (electronic)
Clean Energy Production Technologies
ISBN 978-981-16-3851-0 ISBN 978-981-16-3852-7 (eBook)
https://doi.org/10.1007/978-981-16-3852-7

This Springer imprint is published by the registered company Springer Nature Singapore Pte Ltd.
The registered company address is: 152 Beach Road, #21-01/04 Gateway East, Singapore 189721, Singapore

To my late mother

—Jitendra Kumar Saini

Preface

The microorganisms and microbial technologies play a significant role in sustainable development of agriculture, health, environment, and energy. The major areas where microorganisms are important in renewable energy research and development are production of lignocellulolytic enzymes and cellulolytic cocktail development for biomass saccharification for cellulosic ethanol, consolidated bioprocessing for biofuels and biochemicals, bioethanol from plant and algal biomass through fermentation, co-digestion of wastes for biogas, algal-based biofuel, and biohydrogen. In view of the importance of the microbial technologies in sustainable development of renewable energy, biofuels, and bioenergy, this book project was undertaken. This book includes discussion on varied aspects of microbial technologies to produce bioenergy and biofuels from agricultural and forestry wastes. The book also covers techno-economic as well as life cycle assessments of microbial processes to ensure optimal sustainability and profitability. The information to improve the production efficiency and capacity of bio-based products has also been incorporated. Metabolic engineering and synthetic biology aspects to improve microbial production of biofuels and value-added products from biomass have also been touched upon in this book.

After numerous deliberations, we came up with the idea to explore the possibility of developing a book on microbial technologies for renewable energy generation. Luckily, we were able to convince and engage a variety of researchers from all over the globe to contribute the chapters for this book.

We are thankful to our family, friends, students, teachers, and mentors who acted as a source of inspiration to us. At this point, we must appreciate the kind gesture of the entire Springer nature team and the series editors who gave us support and showed trust in our capabilities. They generously extended the book timelines in the hard times of COVID-19 pandemic and kept supporting us continuously due to which we could complete this book in its present form.

This has been our maiden effort to produce a book on microbial technologies in sustainable development of renewable energy to help students, teachers, and researchers. The primary audiences of this book will be the science and engineering researchers, biorefinery professionals, environmentalists, agricultural researchers, scientists, biorefinery and bioeconomy analysts, and policy makers. The secondary audience will be the students as well as faculties, working in the area of biofuels and biorefineries, renewable and clean energy. We hope that the book will act as a knowledge base for students and faculties engaged in biofuel/bioenergy and biorefinery-based bioeconomy teaching and research. We are open to criticism, suggestions, and recommendations from esteemed readers of this book that will help us to explore other aspects of microbial biotechnologies for biofuels and bioenergy.

We dedicate this book to all the persons who are directly or indirectly serving the people affected by COVID-19 pandemic.

Mahendergarh, Haryana, India — Jitendra Kumar Saini
Rapid City, South Dakota, USA — Rajesh K Sani
April 2021

Contents

Editors and Contributors

About the Editors

Jitendra Kumar Saini received his B.Sc. (Industrial Microbiology) and M.Sc. (Microbiology) from Gurukula Kangri University, Haridwar. He obtained his Ph. D. in Microbiology from Govind Ballabh Pant University of Agriculture and Technology, Pantnagar in 2010, after which he worked as a postdoctoral associate at GADVASU, Ludhiana in a World Bank-funded NAIP project on Rumen Microbiology. Later, he joined DBT-IOC Centre for Advanced Bioenergy Research, Indian Oil Corporation Ltd., Research and Development Centre, Faridabad as a Scientific Officer, where he led team on enzyme development for advanced biofuels. He joined Department of Microbiology, Central University of Haryana, Mahendergarh in 2016 as an Assistant Professor and teaches Industrial Microbiology, and Food and Dairy Microbiology to PG students. His current research focuses on enzyme and microbial technologies for sustainable development of energy and environment. Dr. Saini is a recipient of Early Carrier Research grant from Science and Engineering Research Board, Department of Science and Technology, Government of India and a twinning grant from Department of Biotechnology, Government of India. He is currently supervising 2 doctoral, 5 PG dissertation students, and has supervised 15 PG dissertations in the past, besides co-supervising a postdoc. He has filed 1 US patent, is an author of more than 40 articles, is an editor of 2 books, and is an active reviewer for many reputed journals in biofuel and bioenergy research. As a course coordinator, Dr. Saini conducted one-week Global Initiative of Academic Networks (GIAN) course on "Integrated Lignocellulosic Biorefineries for Sustainable Development." Recently, he organized an International Conference AMI-2019 entitled "Microbial technologies in sustainable development of Energy, Environment, Agriculture & Health" as an organizing secretary. Dr. Saini is a Life member of Association of Microbiologists of India (AMI) and Asian Federation of Biotechnology (AFOB).

Rajesh K Sani is a Professor in the Department of Chemical and Biological Engineering and the Department of Applied Biological Sciences at South Dakota School of Mines and Technology. His research expertise includes Rules of Life in Biofilms grown on 2D materials, Extremophilic Bioprocessing of Solid Wastes to Biofuels and Value-added Products (Molecular Biology, Biotechnology, and Metabolic Engineering), Space Biology, Biogas to Liquid fuels (BioGTL, Genome Editing), Biocatalysis (Protein Engineering, Simulations/Modeling, and Bioinformatics), and Biomaterials/Biopolymers (EPSs and PHAs: Biomedical applications). Over the past 15 years, he has acted as the PI or co-PI on over $49.93 million in funded research. He has 2 patent, 11 invention disclosures, published over 105 peer-reviewed articles in high impact factor journals, and has contributed to several book chapters. In addition, he has edited nine books and one Proceeding for various international publishers. Dr Sani has also been leading a research consortium funded by the NSF with the aid of 69 scientists and engineers.

Contributors

P. Abdeshahian Department of Biotechnology, Engineering School of Lorena, University of Sao Paulo, Lorena, Brazil

Cristóbal N. Aguilar Bioprocesses and Bioproducts Research Group (BBG-DIA), Food Research Department, School of Chemistry, Universidad Autónoma de Coahuila, Saltillo, México

Jorge Angulo-López Bioprocesses and Bioproducts Research Group (BBG-DIA), Food Research Department, School of Chemistry, Universidad Autónoma de Coahuila, Saltillo, México

F. A. F. Antunes Department of Biotechnology, Engineering School of Lorena, University of Sao Paulo, Lorena, Brazil

Ankur Choudhary Department of Civil Engineering, Jaypee University of Information Technology, Solan, Himachal Pradesh, India

Debabrata Das Department of Biotechnology, Indian Institute of Technology, Kharagpur, West Bengal, India

K. J. Dussán Department of Engineering, Physics and Mathematics, Institute of Chemistry, São Paulo State University (UNESP), Araraquara, SP, Brazil

D. L. Flumignan Department of Biochemistry and Organic Chemistry, Institute of Chemistry, São Paulo State University (UNESP), Araraquara, SP, Brazil

Tanvi Govil Department of Chemical and Biological Engineering, South Dakota Mines, Rapid City, SD, USA

Composite and Nanocomposite Advanced Manufacturing—Biomaterials Center, Rapid City, SD, USA

Hemansi Department of Microbiology, Central University of Haryana, Mahendergarh, Haryana, India

A. Hernandez-Perez Department of Biotechnology, Engineering School of Lorena, University of Sao Paulo, Lorena, Brazil

Zachary Hogan Department of Mechanical Engineering, South Dakota Mines, Rapid City, SD, USA

A. P. Ingle Department of Biotechnology, Engineering School of Lorena, University of Sao Paulo, Lorena, Brazil

Muniraj Iniyakumar Kumaraguru Institute of Agriculture, Erode, Tamil Nadu, India

J. Beslin Joshi Department of Plant Biotechnology, Centre for Plant Molecular Biology and Biotechnology, Tamil Nadu Agricultural University, Coimbatore, Tamil Nadu, India

Subburamu Karthikeyan Department of Renewable Energy Engineering, Tamil Nadu Agricultural University, Coimbatore, Tamil Nadu, India

Lakshmi Kasirajan Indian Council of Agricultural Research-Sugarcane Breeding Institute, Coimbatore, India

Ashish Kumar Department of Civil Engineering, Jaypee University of Information Technology, Solan, Himachal Pradesh, India

Raubins Kumar Microbial Engineering Group, International Centre for Genetic Engineering and Biotechnology, Aruna Asaf Ali Marg, New Delhi, India

Sudhir Kumar Department of Biotechnology and Bioinformatics, Jaypee University of Information Technology, Solan, Himachal Pradesh, India

Chandan Mahata Advanced Technology Development Centre, Indian Institute of Technology, Kharagpur, West Bengal, India

S. E. Martiniano Department of Biotechnology, Engineering School of Lorena, University of Sao Paulo, Lorena, Brazil

Julie A. Maupin-Furlow Department of Microbiology and Cell Science, Institute of Food and Agricultural Sciences, University of Florida, Gainesville, FL, USA
Genetics Institute, University of Florida, Gainesville, FL, USA

S. Sánchez-Muñoz Department of Biotechnology, Engineering School of Lorena, University of Sao Paulo (USP), Lorena, Brazil

Lata Nain Division of Microbiology, ICAR-Indian Agricultural Research Institute, New Delhi, India

Anju Mayadevi Nair Microbial Engineering Group, International Centre for Genetic Engineering and Biotechnology, Aruna Asaf Ali Marg, New Delhi, India

Adhithya S. Narayanan School of Chemistry and Biotechnology, SASTRA University, Thanjavur, Tamil Nadu, India

E. M. D. Oliveira Department of Biochemistry and Organic Chemistry, Institute of Chemistry, São Paulo State University (UNESP), Araraquara, SP, Brazil

Sandra Pacios Bioprocesses and Bioproducts Research Group (BBG-DIA), Food Research Department, School of Chemistry, Universidad Autónoma de Coahuila, Saltillo, México

Erick M. Peña-Lucio Bioprocesses and Bioproducts Research Group (BBG-DIA), Food Research Department, School of Chemistry, Universidad Autónoma de Coahuila, Saltillo, México

R. R. Philippini Department of Biotechnology, Engineering School of Lorena, University of Sao Paulo (USP), Lorena, Brazil

Ghasideh Pourhashem Department of Coatings and Polymeric Materials, North Dakota State University, Fargo, ND, USA

C. A. Pradro Department of Biotechnology, Engineering School of Lorena, University of Sao Paulo (USP), Lorena, Brazil

R. Priyadharshini Biocatalysts Laboratory, Department of Agricultural Microbiology, Directorate of Natural Resource Management, Tamil Nadu Agricultural University, Coimbatore, Tamil Nadu, India

Pratap Pullammanappallil Department of Agricultural and Biological Engineering, University of Florida, Gainesville, FL, USA

Suchitra Rakesh Department of Microbiology, Central University of Tamil Nadu, Thiruvarur, Tamil Nadu, India

Desikan Ramesh Department of Vegetable Science, Horticultural College and Research Institute for Women, Tiruchirapalli, Tamil Nadu, India

Nathiely Ramírez-Guzmán Center for Interdisciplinary Studies and Research (CEII-UAdeC), Universidad Autónoma de Coahuila, Saltillo, México

D. R. Ribeaux Department of Biotechnology, Engineering School of Lorena, University of Sao Paulo, Lorena, Brazil

T. M. Rocha Department of Biotechnology, Engineering School of Lorena, University of Sao Paulo, Lorena, Brazil

Laihsa Rodriguez Bioprocesses and Bioproducts Research Group (BBG-DIA), Food Research Department, School of Chemistry, Universidad Autónoma de Coahuila, Saltillo, México

Orlando de la Rosa Bioprocesses and Bioproducts Research Group (BBG-DIA), Food Research Department, School of Chemistry, Universidad Autónoma de Coahuila, Saltillo, México

S. Sagia Division of Microbiology, ICAR-Indian Agricultural Research Institute, New Delhi, India

Jitendra Kumar Saini Department of Microbiology, Central University of Haryana, Mahendergarh, Haryana, India

Salvador Saldaña-Mendoza Bioprocesses and Bioproducts Research Group (BBG-DIA), Food Research Department, School of Chemistry, Universidad Autónoma de Coahuila, Saltillo, México

David R. Salem Department of Chemical and Biological Engineering, South Dakota Mines, Rapid City, SD, USA
Composite and Nanocomposite Advanced Manufacturing—Biomaterials Center, Rapid City, SD, USA
Department of Materials and Metallurgical Engineering, South Dakota Mines, Rapid City, SD, USA

Rajesh K Sani Department of Chemical and Biological Engineering, South Dakota Mines, Rapid City, SD, USA
Composite and Nanocomposite Advanced Manufacturing—Biomaterials Center, Rapid City, SD, USA
Department of Chemistry, Biology, and Health Sciences, South Dakota Mines, Rapid City, SD, USA
BuG ReMeDEE Consortium, Rapid City, SD, USA

J. C. Santos Department of Biotechnology, Engineering School of Lorena, University of Sao Paulo, Lorena, Brazil

L. K. Santos Department of Biochemistry and Organic Chemistry, Institute of Chemistry, São Paulo State University (UNESP), Araraquara, SP, Brazil

Karimangalam Murugesan Shivakumar Department of Agricultural Economics, Tamil Nadu Agricultural University, Coimbatore, Tamil Nadu, India

Rishikesh Shukla Department of Biotechnology, Atmiya University, Rajkot, Gujarat, India

Shraddha Shukla Department of Microbiology, Atmiya University, Rajkot, Gujarat, India

D. D. V. Silva Department of Engineering, Physics and Mathematics, Institute of Chemistry, São Paulo State University (UNESP), Araraquara, SP, Brazil

G. M. M. Silva Department of Biotechnology, Engineering School of Lorena, University of Sao Paulo, Lorena, Brazil

S. S. da Silva Department of Biotechnology, Engineering School of Lorena, University of Sao Paulo, Lorena, Brazil

Neha Singh Department of Molecular and Structural Biology, CSIR-CDRI, Lucknow, Uttar Pradesh, India

Surender Singh Division of Microbiology, ICAR-Indian Agricultural Research Institute, New Delhi, India
Department of Microbiology, Central University of Haryana, Mahendargarh, Haryana, India

Saveena Solanki Department of Biotechnology and Bioinformatics, Jaypee University of Information Technology, Solan, Himachal Pradesh, India

R. Terán-Hilares Department of Biotechnology, Engineering School of Lorena, University of Sao Paulo, Lorena, Brazil

Kalyanasundaram Geetha Thanuja Department of Agricultural Microbiology, Tamil Nadu Agricultural University, Coimbatore, Tamil Nadu, India

Sivakumar Uthandi Biocatalysts Laboratory, Department of Agricultural Microbiology, Directorate of Natural Resource Management, Tamil Nadu Agricultural University, Coimbatore, Tamil Nadu, India

Leidy Johana Valencia-Hernández Bioprocesses and Bioproducts Research Group (BBG-DIA), Food Research Department, School of Chemistry, Universidad Autónoma de Coahuila, Saltillo, México

Magan Vaughn Department of Chemical and Biological Engineering, South Dakota Mines, Rapid City, SD, USA

A. V. Paula Department of Biochemistry and Organic Chemistry, Institute of Chemistry, São Paulo State University (UNESP), Araraquara, SP, Brazil

Na Wu Department of Coatings and Polymeric Materials, North Dakota State University, Fargo, ND, USA
Department of Agricultural and Biological Engineering, University of Florida, Gainesville, FL, USA

Shunchang Yang Department of Agricultural and Biological Engineering, University of Florida, Gainesville, FL, USA

Syed Shams Yazdani Microbial Engineering Group, International Centre for Genetic Engineering and Biotechnology, Aruna Asaf Ali Marg, New Delhi, India

Chapter 1
Integrated Waste Biorefinery for Biofuels and Biochemicals

Kalyanasundaram GeethaThanuja, Desikan Ramesh, Muniraj Iniyakumar, Suchitra Rakesh, Karimangalam Murugesan Shivakumar, and Subburamu Karthikeyan

Abstract This chapter summarizes the recent advances in the processing of waste resources to produce biofuels and platform chemicals. There is a growing concern globally on clean energy and environmental sustainability, which is impelling the search for biofuel sources and other platform chemicals. This chapter examines the prospects provided by organic waste materials and waste water and considers their suitability for alternative fuel and fine chemical production, their sources, residue management, conversion and refining technologies, and the circular economy. In addition, the applied aspects of waste conversion by several thermal, chemical, and biological technologies are discussed.

Keywords Waste biorefinery · Biofuels · Biochemicals · Lignocellulose · Clean energy

---

K. GeethaThanuja
Department of Agricultural Microbiology, Tamil Nadu Agricultural University, Coimbatore, Tamil Nadu, India

D. Ramesh
Department of Vegetable Science, Horticultural College and Research Institute for Women, Tiruchirapalli, Tamil Nadu, India

M. Iniyakumar
Kumaraguru Institute of Agriculture, Erode, Tamil Nadu, India

S. Rakesh
Department of Microbiology, Central University of Tamil Nadu, Thiruvarur, Tamil Nadu, India

K. M. Shivakumar
Department of Agricultural Economics, Tamil Nadu Agricultural University, Coimbatore, Tamil Nadu, India

S. Karthikeyan (✉)
Department of Renewable Energy Engineering, Tamil Nadu Agricultural University, Coimbatore, Tamil Nadu, India
e-mail: skarthy@tnau.ac.in

J. K. Saini, R. K. Sani (eds.), *Microbial Biotechnology for Renewable and Sustainable Energy*, Clean Energy Production Technologies,
https://doi.org/10.1007/978-981-16-3852-7_1

1.1 Introduction

The utilization of wastes and agricultural residues is of concern globally in response to confronting climate change and the worldwide endeavor to curtail greenhouse gas (GHG) emissions. Biodegradable wastes and agro-residues have previously been recognized for the production of energy and find a place in the developing biorefinery concepts. India has tremendous waste biomass as a potential feedstock that satisfies most energy needs. The annual consumption of liquid fuels in India is about 50 million metric tonnes. However, it should be borne in mind that the biomass potential and its broad utilization in India can adept to produce double the amount annually. For India, the estimated energy usage for domestic, transport, and industrial sectors are 40%, 40%, and 20%, respectively. The required stake of crude oil and gases has a significant share (90%) for the primary and transport sectors, and the remaining 10% is utilized for the production of industrial chemicals. The escalating prices of crude oil and energy security issues have forced developing countries to search for alternative and cheap energy sources to fulfill their rising energy demand. The Indian energy transition has a long route to go. The wastes must contribute to the bio-economy that uses material and energy in biorefineries and interpose for the energy transition in hybrid technologies, i.e., combing other renewable energies. India foresees installing 175 GW of renewable energy capacity by 2022 with the contribution of solar (57%), wind (34%), biomass (6%), and small hydropower (3%).

1.2 Integrated Waste Biorefinery

The traditional petroleum-based refineries employ fractional partitioning on a raw feedstock to obtain various components. In analogy, biorefinery involves the association of various biomass treatments processed under one umbrella, resulting in the production of different components of commercial use. Subsequently, the entire chain becomes more viable and reduces the waste generated. Biorefineries are envisaged as viable platforms for transforming to a biobased circular economy capable of utilizing a variety of biofuels and platform chemicals. The full-scale biorefinery will also attain sustainability if the basic frameworks are built up. Since the very rationality of biorefinery hands holds sustainability goals, the second generation biorefineries become the main targets. Endowed with huge biomass potential and abundance of lignocellulosic wastes, immense prospects exist in India for the development of 2G biorefineries. These agro-residues reportedly are varied and are available throughout the year in required amounts. However, large agricultural residues, specifically the paddy straw, are burnt in the field due to lack of awareness, policies and, poor valorization. This chapter examines the prospects provided by organic waste materials and wastewaters to consider their suitability for alternative fuel and fine chemical production, their sources, residue management, conversion and refining technologies, and the circular economy. Besides, the applied aspects of waste conversion by several thermal, chemical and biological technologies will be discussed. In summary, the present chapter offers comprehensive and illustrative descriptions of major processing technologies, waste valorization for fuels and chemicals, supply value chain and logistics, techno-economic analysis, and life-cycle assessment, and the circular bio-economy.

Biorefineries aid in the maximum utilization of optimum energy potential of organic wastes and resolve the issues on waste management and GHGs emissions. Wastes can be converted into either gaseous or liquid fuels by suitable enzymatic/chemical treatment. The pretreatment processes involved in biorefining generate products such as paper-pulp, high fructose corn syrups, solvents, acetate, resins, laminates, adhesives, flavor chemicals, activated carbon, fuel enhancers, and undigested sugars. These sources remain generally untapped in the conventional processes. The efficiency and appropriateness of the process rely on their ability to use a wide range of biomass resources obtained from animal or plant materials. The concept of the biorefinery is still in the budding stages in most places of the world due to several factors such as availability of raw material, product supply chain viability, and model flexibility that hamper the progress to commercial scales. Being in a burgeon holds the solution to the optimum utilization of wastes and natural resources that the mankind has always tried to achieve. The onus now lies on governments and corporate organizations to incentivize or finance the research and development in this field.

In the context of increasing global demand for more environmentally friendly sources of energy, biofuels and biochemicals stand on the fore to make different products. Few companies have already explored the production of platform chemicals from these renewable resources. For example, Cargill and Virent Inc. had collaborated to utilize corn dextrose as a feedstock for the production of drop-in low-carbon biofuels and biochemicals. The BioForming® technology of Virent Inc. facilitates the use of plant-derived sugars as feedstocks for renewable drop-in gasoline, lower carbon biochemicals, and jet fuel. Furthermore, bioparaxylene can be produced and used to produce recyclable biopolyester. Comparably Chempolis Ltd., Fortum, and Numaligarh Refinery Ltd., India focused on the bamboo biorefinery concept to convert 300,000 tons into bioethanol, furfural, acetic acid, and biocoal. Biocoal is used as fuel in the combined heat and power (CHP) plant located in Assam, India.

1.2.1 Solid Waste-Based Biorefineries

The substantial growth in population with developed living standards has enhanced the energy demands along with waste generation. The depletion of resources at a faster pace has created innumerable impacts on the environment leading to climate change and global warming. To overcome these drawbacks, various efforts have been made to develop sustainable strategies. A biorefinery is a boon to several industries based on polluting and finite fossil resources and is commercially convenient for the production of biofuels and biopower from biomass. Biorefinery is the sustainable processing of biomass into a spectrum of marketable products and energy (IEA 2008). The concept of biorefinery comprehends a broad range of technologies for the conversion of biomass resources into value-added products integrating biomass conversion processes and equipment. Integrated biorefinery systems aim to optimize the energy use and materials in the total chains from biomass plantation to end-product to ultimate product use, by that the economic viability and sustainability of biorefineries gets developed. Accordingly, tight integration is essential in the integration of platforms, waste and product exchange, application of efficient conversion routes, and optimizing biomass supply chain (Budzianowski and Postawa 2016).

1.2.2 Solid Waste Value Chain and Logistics

In practice, solid waste management begins at the household at the micro-level, firms at the macro level, thereby resulting in a new form of waste. The value chain linkages of solid waste will have distinct dissimilarities with the main manufactured product. While the value chain of the main product will symbolize the value enhancement along its chain, it can be termed as a positive chain, and sometimes, the waste value chain will also exhibit negative value. But the economic performance of waste value chains can be improved by different strategies, such as industrial integration, economies of scale and size, and reducing feedstock logistics. The solid waste value chain holds a significant role in the circular economies to mitigate the challenges of environmental issues. The value chain of several agro-based wastes is depicted in Fig. 1.1.

In developing economies, poor institutional governance, financial crunch, resources shortage, and political matters are some of the important issues in the management of solid waste effectively. Lack of coordination in addressing solid waste management requires a holistic environmental approach to focus upon 3 R: reduce, reuse, and recycle. It could also create employment opportunities, thereby helping economic development. The Thailand waste management experience revealed that different technologies are used for solid waste management (Thiengburanathum et al. 2012). Similarly, a research study on Nigerian experience in handling solid waste and logistics in the performance of the Lagos State Waste Management Authority helped in developing metrics to analyze the efficiency of management (Ayantoyinbo and Adepoju 2018). In Brazil, a reverse logistics network was deployed for the management of solid waste by the Brazilian Waste Management Policy (Ferri et al. 2015).

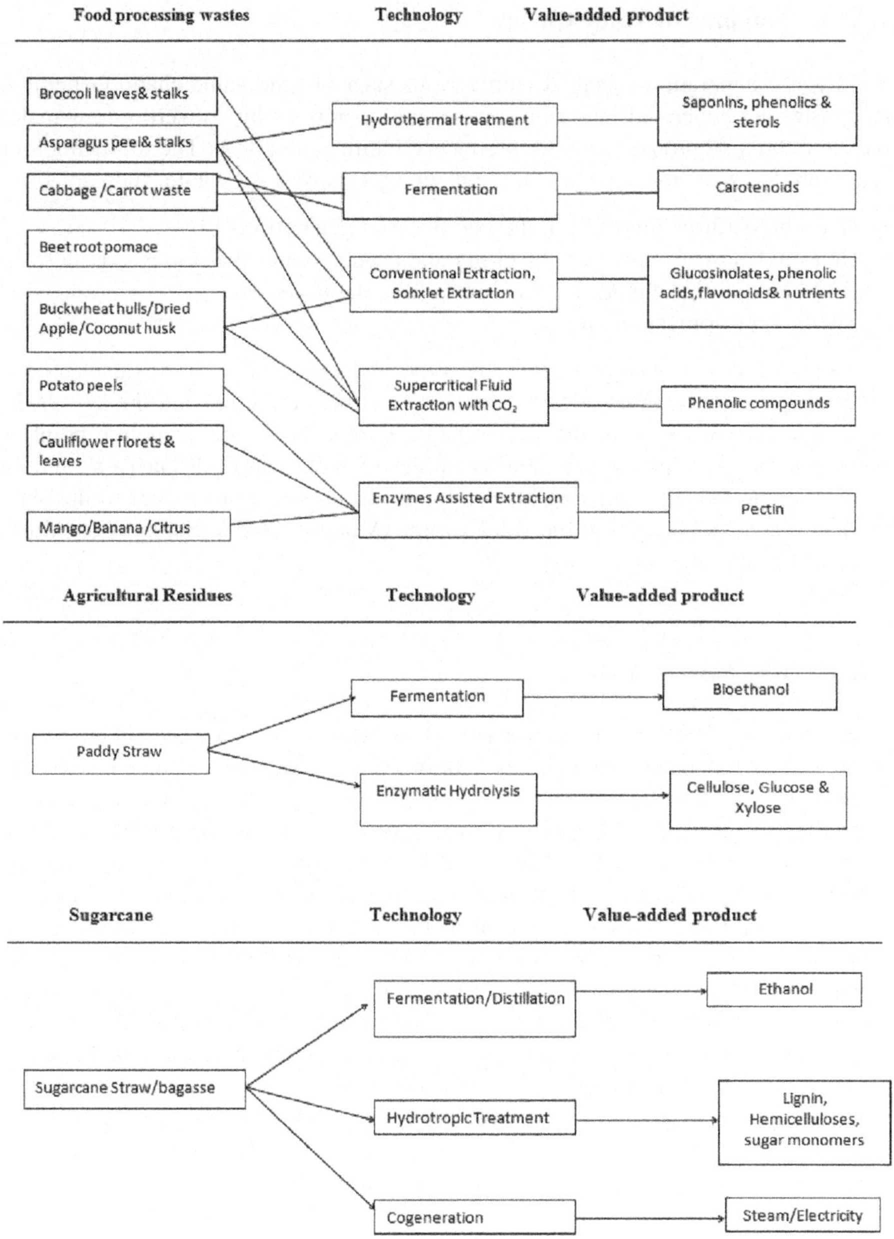

Fig. 1.1 Value chain of various agro-based wastes

1.2.2.1 Mapping of Value Chains

Solid waste generation is linked with various stages of the value chain. Generally, the wastes are generated in various processing operations by converting raw materials into the final products and by-products (Hakim et al. 2017). The following four stages are the sequential process in the mapping of any waste value chains.

- Households/firms involved in the raw material generation,
- Households/firms with manufacturing and production as the prime activities,
- Households/firms engaged in the involved in the logistics,
- The final customers/end users.

While treating the solid wastes, the type and availability of waste materials, processing methods, products recovered, and market conditions are the key determining factors in mapping the value chain process. From the economic point of view, wastes containing appreciable quantities of high-value substances should be converted into valuable building blocks. It also decides the optimal size of the plant to be operated in processing the solid wastes (Muntoni 2019). Some of the waste value chains are presented below.

1.2.2.2 Key Players/Actors

Value chain concept in product/services will be always seen as a value-added stream, where each activity will subsequently adding value. In waste management, there are systems with a negative value in the chain. The households, institutions, markets, and infrastructure are the key players who ultimately decide the efficiency, sustenance, and performance of the waste value chain and logistics. But, in a market economy, the economic agents of this value chain ecosystem compete against each other for individual welfare, resulting in the addition of negativity to the chain. Identifying the systemic constraints that inhibit the performance of these value chains is the foremost concern of the researchers/policy makers. Incorporating a value chain network in solid waste management aims at minimizing the environmental impact of the production process and recycling and minimizing GHG emissions resulting from logistics. To achieve this, the incentives for the key players/actors, opportunities and constraints for the individual players on the value chain, quantification of positive and negative externalities at each node of the waste vale chain should be properly defined. Generating wealth from the waste and identifying existing gaps in solid waste management will lead to fulfilling some of the sustainable development goals of the United Nations.

1.2.2.3 Logistics for Waste Management

One of the new trends in logistics for waste management is effective recycling or reuse of products/services along with transport with minimal damage to the ecosystem. Designing and implementing proper logistics will address most of the environmental and economic aspects of solid waste management.

The conceptual framework for the logistics of solid waste can be as follows:

- Identification and determination of the waste locations;
- Determination of intensity of vehicles and their capacity;
- Estimation of any environmental damage while performing logistics;
- Understanding the demand for solid wastes;
- Determination of costs of operations;
- Preparation of budgets for fleet management;
- Evaluating the effectiveness of the logistical method;
- Final feedback system.

1.2.3 2G and 3G Biofuels

The first-generation biofuels were manufactured from edible feedstocks which appear to be unjustifiable and the controversies on food-versus-fuel appear to be untenable for the commercial use of first-generation biomass. They have significant financial, natural, and political concerns as extensive manufacturing of feedstock demand arable agricultural lands and contributes to ecological degradation. The debate on the food security for the use of biofuels was expunged with the use of second-generation (2G) biomass, which includes the nonedible plant materials, including different feedstocks, i.e., lignocellulosic biomass feedstocks to municipal solid wastes. The variety of biomass feedstocks used for 2G biofuels are wood, organic wastes, food wastes, and specific crop residues. 2G biomass requires to undergo a series of pre-treatment to recover the fermentable sugars embedded in the fibers of the plant. Further, they should undergo fermentation/gasification/pyrolysis to yield ethanol/syngas/ biochar, respectively. The operational cost and the additional steps for the processing of biomass hinder the efficiency. Consequently, third-generation biofuels (3G) gained interest in the field of sustainable energy. They are produced from algal biomass and renders favorable advantages for producing biofuels as they can accumulate large cell lipid content (20–77%) (Jin et al. 2015). The short harvesting cycle and high growth rate of microalgae serve huge potential compared to other biomass. Their low lignin content eliminates the need for pretreatment and increases the production of fuel by transesterification.

1.3 Sources of Wastes

1.3.1 Potential Economic Utilization of Biomass Waste as Feed Stock

In India, the potentiality of lignocellulosic biomass for renewable energy production using the latest technologies has been analyzed for the past decades (Mandade et al. 2016). The exigency on energy endowed with global warming has spurred the world to hunt for alternatives. One of the key alternatives is the production of biofuels and biomaterial building blocks from agro wastes, agro-processing industrial wastes, food waste, biomass feedstocks, and liquid wastes. Generally, the pretreatment process was used for different biomass feedstocks to produce liquid or gaseous biofuels such as biomethane, biohydrogen, bioethanol, and biodiesel (Liu and Wu 2016).

The viable characteristic of biorefinery in reducing the processing cost of biofuels can be encouraged and applied for economic sustenance. An estimate shows that the cost of petroleum fuels is still two to three folds lower than that of second-generation biofuels based on energy equivalent aspects (Carriquiry et al. 2011). In the context of reducing the production cost, there are numerous challenges endowed in the production of biofuels and biochemicals from biomass (Hoekman 2009; Luo et al. 2010; Menon and Rao 2012) that need to be addressed. These challenges are in the areas of biomass production and its logistics, development of energy-efficient biofuel production technologies (pretreatment, enzymatic hydrolysis, and fermentation), co-product production, standards for bioproducts, biofuel supply chain network, societal acceptance, and life cycle assessment (LCA) and environmental impact of biofuel production technologies. These challenges necessitate the need for experts from different areas of research starting from crop cultivation to final product production.

Biofuels and biochemicals produced from lignocellulosic biomass feedstocks offer quite a few welfares to the society, such as (1) renewable and sustainable feedstocks, (2) carbon-neutral, (3) local economic growth and rural employment, (4) alternate eco-friendly solutions for air pollution from in situ biomass burning and biomass rotting in fields, (5) supporting bioeconomy concept and energy security for countries and also reduce the oil imports, (6) new employment opportunities (Greenwell et al. 2013). Apart from this, the current potential uses of biomass residues include animal fodder, mulching, thatching, and fuel in different industries and biomass power plants.

1.3.2 Wastewater as Feed Stocks

A wastewater-based biorefinery integrates the concept of the biorefinery to wastewater treatment. This biorefinery generates valorized products to direct an economically viable process that enhance resource productivity and simultaneously treats wastewater to acceptable standards. It is focused on bioresource recovery in converting the major organic nutrients and trace elements in the wastewater stream to value-added byproducts and concurrently offering clean water as a product. This system contributes to a potential circular bio-based economy to promote the energy and industrial sectors. Thus, integrating the biorefinery system into a wastewater treatment system will promote an exemplar transference that can enhance the system's profitability and reduce environmental pollution. This system also facilitates a linkage between the end-users of water and those who control the wastewater management and can end in resource recovery in closed-loop cycles that fabricate a circular economy.

One of the major impediments in biofuel generation from algae is their high nutrients requirement and higher downstream processing costs. Spanning algal biomass generation with wastewater treatment will resolve these issues. Algal biomass is a capable alternative feedstock in biorefineries, owing to their higher photosynthetic efficiency (Singh et al. 2011), biomass productivity (Bhola et al. 2011) oil content (Mutanda et al. 2011), and the possibility of daily harvesting of algal biomass (Rosenberg et al. 2011). The algal biomass does not compete with food crops and its cultivable area. Algal biofuels are referred to as third-generation biofuels (Gressel 2008). Hitherto, microalgae are a potential resource for liquid biofuel production due to higher biomass productivity (175 tons/ha/year), possibility to cultivate in wastewaters (Jena et al. 2011), mixotrophic growth (Nagajothilakshmi et al. 2016), and cocultivation of algae (Rakesh and Karthikeyan 2019). Rinna et al. (2017) reported that *Botryococcus braunii* has higher nitrogen and phosphorus removal efficiency in wastewater. Simultaneously, this process generates lipid-rich biomass and algal lipid can be utilized as a potential feedstock for biodiesel production. Generally, about 20% of agro-industrial food wastes are utilized as animal feed and the remaining waste may be disposed of through incineration, composting, or landfilling. Nowadays, agro-industries are facing an increase in their growth around the globe. These agro-industrial wastes are inexpensive, abundant, and micronutrient-rich, but they have disposal problems. These wastes could be potential as a substrate for alternate carbon sources for biofuels and biochemicals production. Increasing global demand for biofuels and biochemicals via utilization of waste biomass resources has driven research toward stable, inexpensive resources with concern over global climate change (Hu and Ragauskas 2012). Vijayanand et al. (2017) used different inexpensive and abundantly available agro-industrial wastes for biobutanol production. They confirmed that pretreatment and glucose supplement enhanced the biobutanol production by *Clostridium beijerinckii*. The distinct feature of algal biomass is the coexistence of mixed or multiple species contributing to an array of products from nutraceuticals such as omega-3 fatty acids to complex recombinant proteins, thereby making a valuable biorefinery processing system (Subhadra 2010). *Nannochloropsis* sp. is a potential biomass candidate for lipid resource to produce biodiesel, biohydrogen, and high added-value compounds. The algal biomass cake after the extraction of oils and pigments can produce hydrogen through the fermentation process (Nobre et al. 2013). Agar is obtained from the pulp of marine red seaweed *Gracilaria verrucosa*, and its process residues may be used as bioethanol feedstock (Kumar et al. 2013). However, there is restricted information available for research on selective microorganisms (bacteria/yeast) to exploit for high value-added products and biofuel production (da Silva et al. 2014).

A microbial fuel cell (MFC) is an electrochemical system for converting the chemical energy of organic materials into electrical energy via redox reaction under anoxic conditions (Ledezma et al. 2015). Several industries utilize a huge amount of fresh water and energy for processing and generate a large quantity of wastewater. Generally, this wastewater is directly discharged to land, resulting in environmental problems such as water and soil pollution. Hence, recently the emphasis has shifted to utilize various industrial effluents for MFC cell feed (Sahu 2019). MFC have the potential to overcome wastewater management issues. It is an ideal technique to use industrial wastes material in wastewater to fuel and hence obtaining electrical energy as the end-product (Pant et al. 2013). The efficiency of microbial fuel cells usually depends on the suitable cathode, an anode (Bi et al. 2018), and cation exchange capacity of the material used to treat wastewater (Rahimnejad et al. 2015). Recently, many industrial wastewaters such as starch processing, brewery, palm oil, paper, and sewage were treated with the MFC concept (Baranitharan et al. 2015; Radha and Kanmani 2017). The complex chemical composition of agro and food processing wastes is a very reliable feedstock in MFCs (ElMekawy et al. 2015). Wastewater treatment sludge consists of a desirable source of microorganisms for microbial fuel cells treating liquid wastes, whereas endogenous microflora can be utilized for MFCs with solid organic waste (Mohan and Chandrasekhar 2011).

1.3.3 Biomass Harvest and Yield

Wastewater-based biorefinery offers new opportunities for both algal cultivation and multiple products generation aspects (Khoo et al. 2019). Harvesting of microalgae is one of the major obstacles to microalgae processing for multiproducts due to its higher initial investment, low biomass concentration, and sedimentation rate (Rakesh et al. 2020). Various methods applied for microalgae harvesting are sedimentation, centrifugation, flotation, and flocculation. Sometimes a combination of two or more methods is used for ideal harvesting (Chutia et al. 2017). Pahl et al. (2013) examined various centrifuges for microalgae harvesting and reported that disc stack centrifuges are extensively used for high-value product recovery from algae in industries. Flocculation of microalgal cells via flocculating agents is one of the desirable methods of harvesting microalgae. The selection of a suitable flocculating agent is an essential condition for this process, i.e., it should be easily available, non-toxic, inexpensive, and should be effective at low concentrations (Branyikova et al. 2018). Rakesh et al. (2014) used multivalent metal salts to initiate flocculation in the microalgal cell suspension.

Jiang et al. (2020) reported that co-flocculation of *Chlorella pyrenoidosa* and *Citrobacter freundii* in the ratio of 1:1.6 showed maximum flocculation efficiency of 97.45%. Autoflocculation is a species-dependent harvesting process that involves interaction between surface molecules of microalgae with the surrounding medium or among themselves. Matter et al. (2019) showed that *Scenedesmus obliquus* autoflocculation efficiency improved from 10.4 to 33.2% when pH increased from 7 to 10. Pandey et al. (2020) evaluated the harvesting of *Scenedesmus* sp. using electro-coagulation-flocculation showed effective harvesting efficiency (>99%) under optimal conditions. Autoflocculation and bioflocculation are found to be inexpensive and effective dewatering techniques for algal harvesting. Autoflocculation has a high sedimentation rate without any addition of the flocculants. The autoflocculation can be enhanced by a high aeration rate, CO_2 concentration, and nitrogen levels. Bioflocculation is also an efficient, eco-friendly, and cost-effective algal harvesting method.

1.4 Industrial Waste Biorefineries

Huge industrialization across the globe has well served to the generation of industrial wastes and harmful environmental pollutants menacing mankind. A waste biorefinery aims at plausible utilization of wastes into a wide spectrum of bio-based products, thereby providing energy security and pollution control with societal development. The biomass waste accumulation from industries and storage systems is crucial for further processing. In the modern era rather than waste disposal methods like incineration and landfill, reuse and recycle are indispensable. Various industries, including cassava, brewing, wood, and sugarcane industries, contribute to starch residues in either liquid or solid waste. The concept of circular economy is being increasingly adopted in both developing and developed countries not only to reduce, reuse, and recycle the wastes but also to produce a plethora of products such as food, feed, fuels, and chemicals through multiple technologies of valorization. This concept of biorefineries (producing various products from one feedstock or mixed feedstock) is developing at a fast pace to meet the socio, economic, environmental, and geopolitical factors of different countries.

Several wastes such as agricultural wastes, forestry wastes, municipal wastes, industrial wastes, food wastes, and animal wastes are suitable for biorefineries. All the wastes have high potential in terms of processing and getting high-value products (Takkellapati et al. 2018). Among the aforementioned industrial wastes belong to the following categories:

Olive oil wastes (including olive oil crop residues and mill wastewater);

Pulp and paper industry wastes (including lignin-rich waste streams, kraft lignin derivatives, etc);

Sugar industry wastes (press mud, bagasse, molasses distillery spent wash, sugarcane tops, etc.);

Coffee industry wastes.

1.4.1 *Waste Refinery Based on Sugar and Syngas Platforms*

This refinery process falls into a category of two platform biorefinery according to National Renewable Energy Laboratory. The two platforms are (1) The sugar platform in which the wastes are biochemically converted to produce sugars, and (2) The syngas platform where wastes are put into the gasifier to produce syngas. The sugar platform uses biochemical methods such as pre-treatment, hydrolysis, and fermentation to produce sugars. The syngas platform uses thermochemical methods to generate syngas from wastes (Yadav et al. 2019).

1.4.1.1 Sugar Platform

As discussed, the sugar platform involves biochemical steps such as pretreatment, hydrolysis, and fermentation or biological processes into various biofuels and biochemicals. As an example of sugar biorefinery, bioethanol is the major end product produced. Bioethanol can be a renewable resource for various other platform chemical production such as ethylene, propylene, and butadiene and also other chemicals of commercial utility such as acetaldehyde and acetic acid. For example, acetaldehyde and acetic acid are value-added chemicals generated from bioethanol in the sugar biorefinery concept.

1.4.1.2 Syngas Platform

The biomass conversion by the thermochemical process is quite complex and utilizes several component configurations and operating conditions for transforming biomass into synthesis gas or oil. High energy gas production by partial oxidation of industrial wastes at 500–800 °C is referred to as syngas. Initially, the wastes are pretreated to remove unwanted materials, then gasification proceeds with partial oxidation, leading to syngas production. Syngas primarily consists of carbon monoxide (CO) and hydrogen (H_2). The gas composition of syngas depends on the components of biomass feedstocks, the gasifier operational parameters, and gasifier types (Puigjaner 2011). Unpurified syngas also contains small amounts of impurities such as tar, CO_2, and other gases. Hence, most of the syngas platforms use cleaning as the third step in cleaning and purifying syngas to remove impurities. Syngas can produce multiple products such as ammonia, methanol, ethanol, methane petrol, diesel, and chemicals. This can be achieved through different processes including syngas fermentation, Fischer-Tropsch synthesis, methanol synthesis, and ammonia synthesis synthetic natural gas production. Syngas can be a renewable feedstock for the generation of bioethanol by both biochemical and thermochemical routes. The biochemical route involves using microorganisms. For example, *Clostridium autoethanogenum* and *Rhodospirillum rubrum* convert syngas into bioethanol and biohydrogen, respectively. The added advantages of the syngas fermentation method over conventional fermentation are that it needs no pretreatment, utilizes entire biomass; in it reactions occur at ambient conditions, ethanol yield is higher, and no costly enzymes are used. However, poor mass transfer properties of the syngas and low ethanol yield of biocatalysts are the major hurdles for the commercialization of this technology (Munasinghe and Khanal 2010).

1.4.2 Waste Refinery on Cellulosic/Starch-Based Biofuels

The lignocellulosic biomass has candidacy to be transformed into energy-rich hydrocarbon and fine chemicals through thermo-chemical and biochemical pathways. For the industrial wastes considered, the basic waste bio-refinery may consist of a biodigester. Crops with copious quantities of starch such as corn, wheat, and cassava can be employed for enzymatic hydrolysis to yield a sugar solution, which can subsequently ferment and be processed into biofuels and biochemicals. On the other side, the by-products from the processing of starch-rich crops are animal feed with rich proteins. If appropriate technology is applied for sweet sorghum stems, liquid biofuels (e.g., bioethanol, biobutanol), and wood-plastic composites can be generated (Yu et al. 2012). Generally, the plant oils contain fatty acids with 8–24 carbon length chains (Octave and Thomas 2009). Oilseeds can be a rich resource for alternative petroleum products (fuels, chemicals, lubricants, and detergents), which can produce biofuels and high-value fatty acids. Oils of soybean, palm fruits, rapeseeds, and canola seeds are popular to produce biodiesel (Demirbas 2007). Bouaid et al. (2010) scrutinized an integrated process for producing low and high-molecular-weight methyl ester fractions from coconut oil for the production of biodiesel/biolubricants/bio solvents. Rincón et al. (2014) developed an integrated approach for producing biodiesel by transesterification of palm oil, palm wastes, and crude glycerol or methanol from syngas.

1.4.3 Conversion of Sugars from Waste to Hydrocarbon Chemicals

Hydrocarbons are long-chain containing alkanes formed by condensation or head-to-head condensation of fatty acids involving various steps as depicted in Fig. 1.2. They are similar to high octane jet fuel. Sugars produced from wastes can be used to produce hydrocarbons (Ladygina et al. 2006). Sugar-based biorefineries are applicable for different sugar crops such as sugarcane, sugar beet, or sweet sorghum. It is a simple way to extract the saccharose from sugar crops, and it is further processed to produce bioethanol and biochemicals using appropriate technologies. In Brazil, the biorefinery was applied to sugarcane crops to produce bioethanol and biopower using sugar juice and sugarcane bagasse (Mariano et al. 2013). In India, Godavari

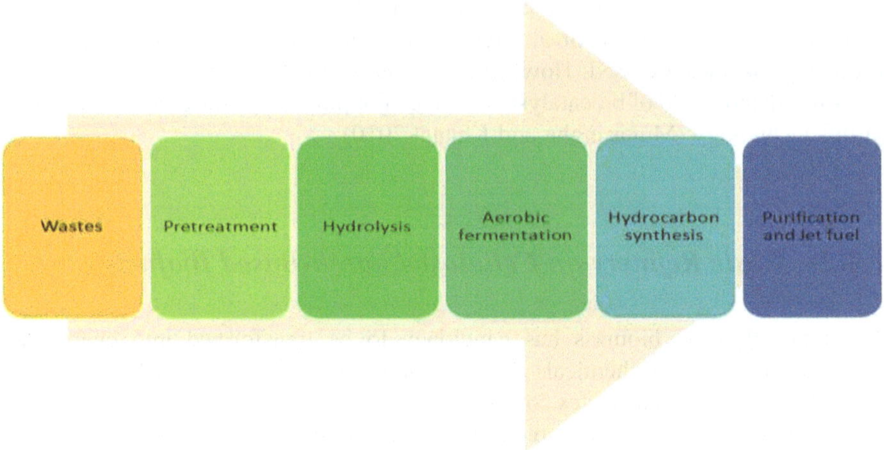

Fig. 1.2 Steps involved in hydrocarbon production

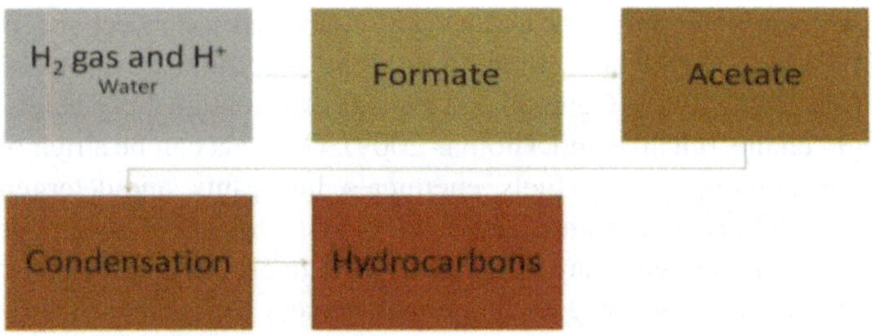

Fig. 1.3 Pathway of hydrocarbon production (Adapted from Ladygina et al. 2006)

Biorefineries Ltd. operates two sugar refineries to produce sugar, bioethanol, and electricity from sugarcane coupled with more than 20 renewable feedstocks. In Colombia, sugarcane biorefineries operate to produce sugar, bioethanol, and electricity from cane juice, molasses, and bagasse, respectively. It helps in establishing a profitable and sustainable biorefinery and offers several benefits such as acceptable GHG emissions, low stillage effluent production, waste minimization, and new job opportunities for both rural and educated people (Moncada et al. 2013).

In the developed nations, the bioethanol pilot plants are used with a small modification to produce hydrocarbons. Initially, the biomass is processed and pre-treated with dilute sulfuric acid. The pretreated biomass is subjected to enzymatic hydrolysis produced onsite and the pathway of hydrocarbon production is depicted in Fig. 1.3. The difference between bioethanol and hydrocarbon is an important aerobic process. The reactor is supplemented with aerators to increase the mass balance ratio of the medium. Another difference is the removal of solids in hydrocarbons production. The majority of microorganisms viz., Cyanobacteria *N. muscorum, Anacystis nidulans*, Gram-negative anaerobic sulfate-reducing bacteria *Desulfovibrio desulfuricans,* Gram-positive aerobic bacteria (eubacteria) *Bacillus* sp., and Yeast *Saccharomyces, Penicillium* sp. are capable of accumulating intracellular and extracellular hydrocarbons.

1.5 Food Industry Waste Biorefinery

The exponential growth of the global population poses threat to finite resources and also surges the sum of waste generated. Among the most generated biowastes, food waste (FW) is of global concern. The ineffective waste management strategies lay a step for a waste generation along the food supply chain accounting for 1.3 billion tons of waste. It is approximately equivalent to one-third of edible parts of food for human consumption (FAO 2019) and total waste generation is projected to increase by 44% by 2025. FW, being rich in moisture content and nutrients (proteins, carbohydrates, and lipids) putrefies upon accumulation, thereby serving as a ground for disease-causing organisms and poses serious environmental threats contributing to 10% of greenhouse gas emissions (IPCC 2019). Given the collective challenges of food wastage with demand for green energy and diminishing fossil fuels, a sustainable biorefinery strategy is the need of an hour for the utilization of this potential feedstock toward assorted product production. An appropriate biorefinery of FW can increase the efficiency of the food supply chain and obtains value-added products by various means such as extraction, biological/chemical conversion, and synthesis.

1.5.1 Energy Recovery and Waste Treatment

The lignocellulosic nature of FW has attracted interest among renewable energy scientists for its conversion into commercially important products. The organic fraction composition of FW endows the high bio-degradability that reduces the need for pre-treatment methods. Energy recovery from the wastes can be employed by either of the processes including combustion, pyrolysis, anaerobic digestion (AD), and gasification. These processes involve the conversion of wastes into energy which may in the form of heat, fuel, or electricity. Various energy recovery processes of FW are depicted in Fig. 1.4. AD serves as a key for reduction, stabilization, and biogas production from FW (Algapani et al. 2017). AD of FW has less environmental impact than incineration and landfilling. However, AD involves complex processes and relies on important parameters such as nutrient contents, particle size, inhibitory compounds, and process parameters like pH, temperature, retention time, organic loading rate, agitation, and inoculum, while various innovations are being developed to enhance and optimize product yield. Two-stage anaerobic digestion of food waste was studied by De Gioannis et al. (2017), which resulted in enhanced methane production as well as associated H_2 production. Co-digestion of FW with sewage sludge has been gaining interest to increase the efficiency of AD. FW sludge

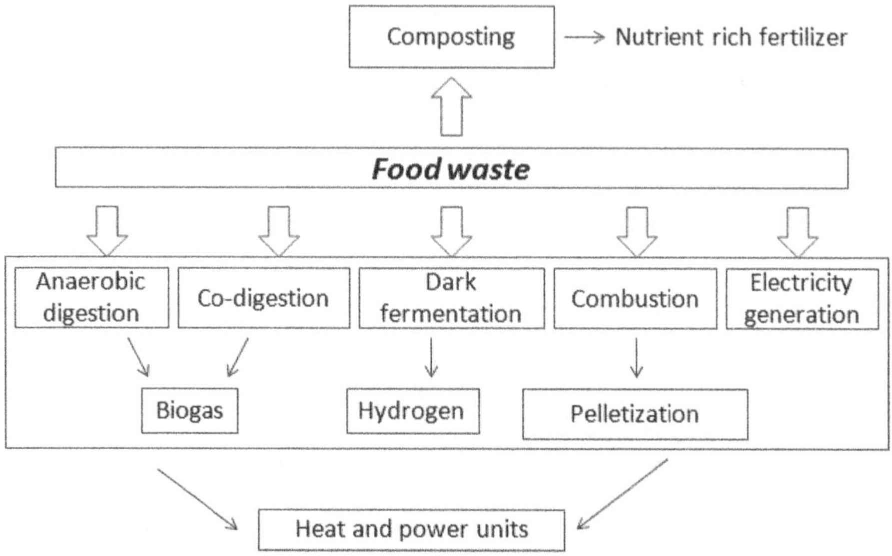

Fig. 1.4 Energy recovery from food wastes

co-digestion with chemically enhanced pre-treated sludge was found to improve methane (CH_4) recovery (Chakraborty et al. 2018). The energy balance of wastewater treatment plants can be enhanced by co-treatment of municipal wastewater with FW associated with increased methane yield (Guven et al. 2019). Co-digestion with animal manure or sewage sludge supplies the needed alkalinity and micronutrients required for the AD. Another common type of co-substrate being used is carbon-to-nitrogen (C/N) ratio rich lignocellulosic biomass, which also helps to prevent rapid acidification in AD using food waste as a single substrate. Co-substrate should have a good C/N ratio, total solids, and enough buffering capacity to sustain pH drop in methanogenesis of FW under dry conditions.

Micronutrient availability plays a pivotal role in the performance and stability of food waste digesters. Improving the design of the digester and operating strategies solves the issue of rapid acidification of FW and the inhibition by methanogens. Appropriately, two-stage systems have been anticipated in which CH_4 production and acid production are divided into two reactors to prevent pH inhibition (Grimberg et al. 2015). Various modifications were applied in a two-stage system to diminish digestion time (Fuess et al. 2017), reduce hydrogen sulfide and CO_2 content, increase methane content (Li et al. 2017), biohydrogen production, or sulfate removal (Yun et al. 2017). Further, a study on three stages of anaerobic digestion of FW with horse manure accelerated the solubilization of organic matters and volatile fatty acid formation with a 23% increase in methane yield (Zhang et al. 2017).

Food waste pre-treatment for AD aims to: (1) improve the lipids/protein digestibility in short retention time, (2) reduce the rapid acidification rate, (3) modifies the physicochemical characteristics of FW to eliminate process inhibition, while the strategy for dark fermentation aims to (1) solubilize complex carbohydrates and make easy access for hydrogen-producing bacteria (HPB), (2) inactivation of hydrogen consuming and non-hydrogen producing microbial communities, and (3) selective enrichment of HPB (Parthiba Karthikeyan et al. 2018).

1.5.2 Food Waste Processing for Platform Chemicals

Management of huge FW seems to be critical for many countries worldwide. It is estimated that supply chain waste alone contributes to 40 percent of waste in food processing (Dahiya et al. 2018). This problem is more aggravated in low-income countries where the infrastructure is not proper. Currently, anaerobic digestion of food wastes is practiced in many parts of the world. However, a high organic load with more fatty acid content reduces the methane yield. Hence more value-added chemical production through biorefinery (tabulated in Table 1.1) is an important option. The large volume and unstable nature of FW pose more risks in the valorization of FW. For instance, fruits and vegetable waste during processing constitute the largest part of the food waste they can be an excellent source for the production of pectin and phenols and gelling agents. Similarly, kefiran, an exopolysaccharide rich in glucose and galactose, can be produced from milk industry wastes. Proteins extracted out of meat and the meat processing industry have a high market value. Platform chemicals are the prime feedstock for the production of secondary chemicals, intermediates, and final products.

1.6 Agroindustry Waste Biorefineries

The agricultural strength in the country provides a huge amount of biomass which is used as feedstock in agro-industries. The diverse variety of lignocellulosic biomass available around the year provides an opportunity for multidrop biorefineries for different bioproducts production.

1.6.1 Problems with Agro-Residues

There are two types of agro-residues viz., crop residues and agro-industrial residues. Crop residues are non-edible parts of the plant collected in the field after the harvest of the main crop. Agro-industrial residues are engendered from different unit operations used in the post-harvest processes. For example, waste residues from wood and food processing industries (Mande et al. 2005). The comprehensive statistical data on the availability of agricultural residues is a must for developing

Table 1.1 Valorization of FW into various value-added products and platform chemicals

Products	Substrate	Pre-treatment/conversion process	Inoculum	Product yield	References
Biofuel	Instant noodle waste	Simultaneous saccharification and fermentation, and chemical trans-esterification	*Saccharomyces cerevisiae* K35	61.1 g/L	Yang et al. (2014)
	Waste cake	Grinding, hydrolysis and centrifugation	*Saccharomyces cerevisiae*	46.6 g/L	Han et al. (2019)
Hydrogen	Canteen waste	Solid-state fermentation (SSF) and dark fermentation (DF)	*Biohydrogenbacterium* R3	52.4 mL H_2/g	Han et al. (2015)
	Canteen waste	Potassium ferrate pretreatment + DF	Municipal solid waste	173.5 mL/g	Kuang et al. (2020)
Butanol	Pea pod waste	Saccharification with acid pre-treatment (1.3% H_2SO_4)	*Clostridium acetobutylicum* NRRL B-527	6 g/L	Nimbalkar et al. (2018)
	Orange peel waste	Steam explosion	*C. acetobutylicum* NCIM 2877	19.5 g/L	Joshi et al. (2015)
	Cassava waste residue	Acid hydrolysis	*C. bifermentans* PNAS- 1	3.36 g/L	Johnravindar et al. (2017)
Ethanol	Citrus peel waste	Simultaneous saccharification and fermentation	*Saccharomyces cerevisiae*	39.6 g/L	Wilkins et al. (2007)
Protease	Bread waste	Autoclaved	*Rhizopus oryzae*	2400 U/g	Benabda et al. (2019)
	Brewery waste	Centrifugation	*B.subtilis*	9.77 U/mL	Blanco et al. (2016)
α-Amylase	Potato peel		*Bacillus subtilis*	600 U/mL	Shukla and Kar (2006)
	Orange peel	Dried and pulverized	*B.amyloliquefaciens*	220 U/mL	Uygut and Tanyildizi (2018)
Bioplastics	Canteen waste		*Serratia ureilytica*	54 ± 3% dry cell weight	Reddy et al. (2015)
	Pineapple waste	Acid hydrolysis	*Ralstonia eutropha* ATCC-17679	88 mg L/L	Vega-Castro et al. (2016)
Lactic acid	Pineapple waste	Ground and filtered	*Lactobacillus delbreuckii*	0.82 gg^{-1} sugar	Idris and Suzana (2006)

	Apple pomace	Dried, milled and autoclaved	*L.rhamnosus*	0.88 gg^{-1} sugar	Gullón et al. (2007)
Citric acid	Pomegranate peel waste	Dried, pulverized followed by SSF	*Aspergillus niger*	278.5 g/kg dry peel	Roukas and Kotzekidou (2020)
	Banana peels	Untreated	*Enterococcus faecium*	15.9 L^{-1}	Abdel-Rahman et al. (2019)
Succinic acid	Fruit and vegetable waste	Hydrolysis by enzyme and SSF	*A. niger* and *Rhizopus oryzae*	27.03 g/L	Dessie et al. (2018)
	Wheat bran	SSF	*Actinobacillus succinogenes*	0.88 gg^{-1} sugar	Du et al. (2008)
	Bread waste	SSF		47.3 gL^{-1}	Leung et al. (2012)

any management strategy. In most of the cases, it is estimated as product yield of crops and residue to crop ratio. The agriculture sector generates billions of tons of non-edible residues every year. These residues create high environmental pollution, management, and economic problems due to improper handling and untapped potentials. Hence, the usage of agricultural residue as a source of high-value products is highly encouraged.

The lignocellulosic agro-industrial wastes are generated in tons every year by most of the developing and under-developing countries (Bhatia et al. 2012). However, various feedstocks with different characteristics pose challenges in collection, transportation, and handling. These wastes are usually burnt freely, which not only causes loss of agricultural biomass but also creates environmental pollution. The major environmental issues may occur due to the poor logistics and mismanagement of the wastes. Therefore, the same can be utilized to produce a variety of bioproducts through proper biomass conversion technologies (Ramesh et al. 2019b). Lignocellulosic agro-industrial wastes mainly comprise cellulose, hemicellulose, and lignin. The cellulose and hemicellulose can be easily converted into fermentable sugar and further fermented to produce bioethanol. The lignin acts as a physical barrier hindering the fermentation for bioethanol production (Ramesh et al. 2018).

1.6.2 Pretreatment of Agro-Residues for Biofuels

Pretreatment of lignocellulosic biomass involves the conversion of complex lignin structures into simple sugars to remove lignin, preserve hemicellulose and reduce the cellulose crystallinity. The pretreatment choice for the biomass depends on the composition and desired products as a result of pretreatment. There are various methods of pretreatment as shown in Fig. 1.5 and aim to attain the formation of sugars by hydrolysis, avoid the loss of fermentable sugars, control the excess inhibitory compounds production, reduce energy consumption, and minimize biofuel production cost.

1.6.3 Agroresidues- Sources, Availability, and Collection

Generally, the cultivation of crops yields not only farm produces but also agro residues or crop residues. There are two types of wastes generated: field and crop processing residues. The field residues mean wastes collected after the harvesting of crop/farm produce. Stem, stalks, leaves, trashes, and straws fall into this category. In crop processing residues, wastes are generated during the processing of farm produce to get the final product or value-added product. The quantities of these agro wastes vary from crop to crop and climatic conditions. These major compositions of these wastes are similar to other lignocellulosic feedstocks, such as cellulose, hemicellulose, and lignin. According to the National Policy for Management of

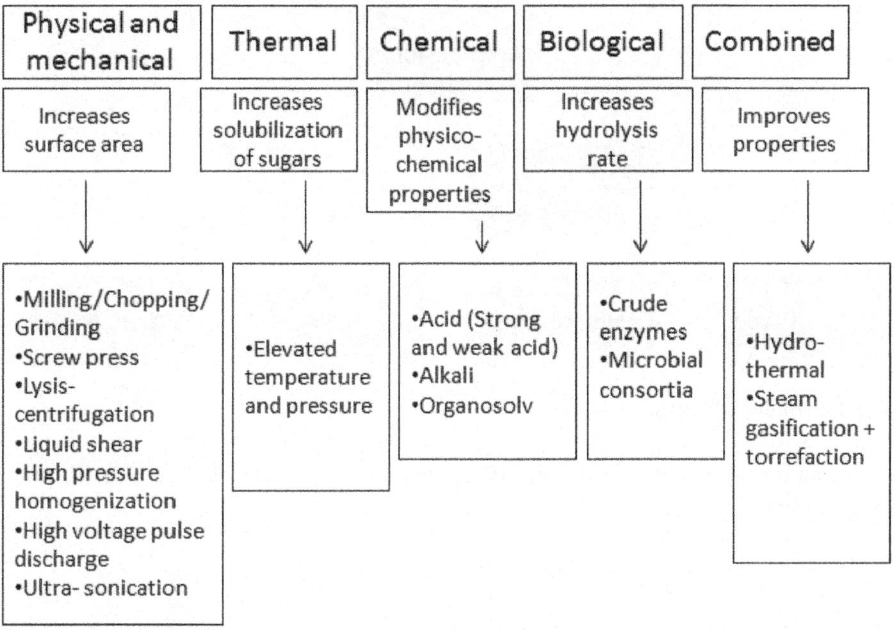

Fig. 1.5 Different types of pre-treatment methods for the production of bio-fuel

Crop Residues (NPMCR) report, the first three states contributing agro-residue generation are Uttar Pradesh (60 Mt) Punjab (51 Mt), and Maharashtra (46 Mt). Among them, 70% of residues were contributed by rice and wheat crops, and out of 500 Mt, 92 Mt of crop residues were burnt in a year (NPMCR n.d.). The collection of agricultural wastes depends on the type of residues generated. Earlier, the residues were collected manually, which is a labor-intensive process. Due to the low bulk density of agro wastes, transportation cost is higher, and sometimes its cost more than the price of residues. The wastes collected from the agricultural field could be achieved through different types of machinery. In the case of straw collection, the baling machines are used to make square or circular-shaped bales of straw. This step can reduce transportation charges due to the higher bulk density of balers. Size reduction types of machinery are commercially available for easy handling of agro residues. Shredder/chopper can be used to reduce the plant materials (e.g., oil palm fronds) into smaller sizes.

1.6.4 Biofuel from Agro-Residues

Agro residues are the most abundant renewable resources on earth. They consist of cellulose 50%, hemicellulose 30%, and lignin 20%. The carbohydrates in the biomass can be biochemically processed through pre-treatment, hydrolysis, and

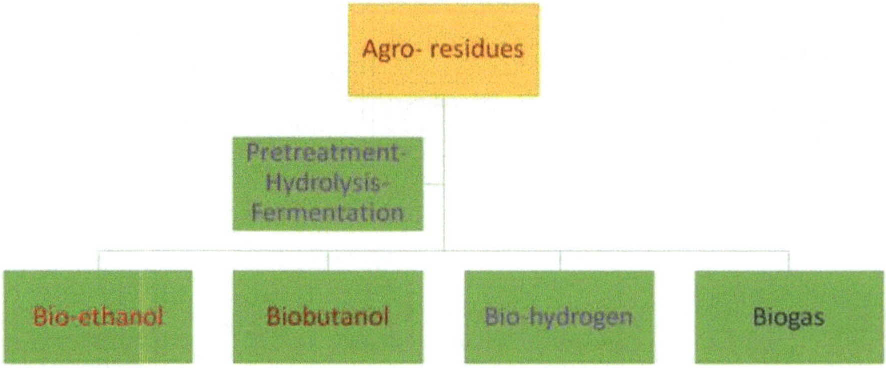

Fig. 1.6 Biofuels from agro-residues

fermentation, resulting in different types of fuels as depicted in Fig. 1.6 (Ramesh et al. 2019a).

1.6.5 Case Study with Paddy Straw

Paddy straw is one of the residues produced in rice cultivation. It can be used for several purposes ranging from animal feed to building blocks. Generally, 1 kg of paddy grain can produce 1.0–1.5 kg of straw (Maiorella 1985). Therefore, larger quantities of straw are generated. However, the length of the straw depends on harvesting methods. Abraham et al. (2016) proposed the paddy straw biorefinery concept based on the thermochemical and biochemical platforms. Thermochemical platforms use any one of the processes such as pyrolysis, hydrothermal liquefaction, gasification, and combustion to get the final product as bio-oil/syngas/ heat/electricity. They also suggested biochemical conversion pathways for biogas gasoline, aromatics, phenolic and liquid biofuels, 5- hydroxymethylfurfural (HMF), and other furfurals. The pretreatment is a pivotal production step for the biochemical conversion technology used for lignocellulosic biomass for biofuel production. It is obligatory to break the biomass structure to make cellulose more accessible to the enzymes, which helps the conversion of carbohydrate polymers to fermentable sugars. Sreekumar et al. (2020) studied straw biorefinery for bioethanol production and heat generation and calculated that production of 1 L of bioethanol requires 3.37 kg of rice straw-based overall mass balance approach.

1.6.6 Case Study with Sugarcane Trash

Sugarcane is one of the important cash crops produced globally. Harvesting of sugarcane leaves enormous quantities of residues in the field itself. For instance, a study conducted by TIFAC India in collaboration with CSIR-NIST states that sugarcane tops are the major residues generated in the county with an annual production of more than 100 MMT. Often, its potential is not realized and it is left in the field for low-value products of compost or many times burnt directly in situ causing serious environmental threats.

One case study was conducted in Brazil on sugarcane trash utilization for fuel and compost production. Dried sugarcane leaves had more nutrients than tops were found by simple enzymatic means. The high moisture content of 82.3% and heating value of cane trash, and heating value make trash an excellent source for biofuel production.

A case study was conducted in India to evaluate the alternative utilization of sugarcane trash. The results indicated that rash utilization reduced the ethanol break-even selling price (BESP). The scientists also studied the percent of retention in soil and their contribution to BESP. The results showed that 50% retention of trash could be beneficial as it doesn't linearly increase the ethanol ESP. More than 50% retention reduced the BESP of Ethanol. As the trash is added to the soil, it reduces the fertilizer requirement for the next crop and increases the crop yield. Reduction in GHG emission was also correlated and transportation due to GHG was calculated. It was revealed that fertilizer saving has more GHG reduction than transportation. However, the study did not include the benefits of the environment, irrigation saving Life cycle analysis. Overall trash utilization can have a beneficial effect on ethanol price and increase soil fertility was reported (Vikash et al. 2018).

1.6.7 Agro-Industry Waste and Sustainable Rural Development

The biorefinery mode operates for converting agro-wastes into a spectrum range of products such as biofuels, biohydrogen, biochemicals, etc., through a cascade of advanced approaches such as pyrolysis, gasification, and other catalytic processes. Such development helps in stabilizing the economy of rural areas by conferring clean energy by the replacement of fossil fuels. Due to the lack of awareness and knowledge on the management of surplus agro-residues, these wastes are frequently ruined on a mass scale for waste management instead of being used in other productive ways (Hiloidhari et al. 2020) Many of the rural areas are equipped with biogas (methane gas) with agro-industrial waste as substrate via anaerobic digestion for various purposes like water heating, broiler operation, drying of grains, etc. (Obi et al. 2016). Methane production via anaerobic digestion makes disposal and treatment of a huge quantity of agro-industrial wastes easier and also reduces the foul smell problem.

Agro-industrial wastes are nutritionally rich in composition and comprise many of the bioactive compounds, which can be utilized as raw material for the production of value-added products viz., biogas, mushroom, biofuel, etc. Many of the valuable products are generated through solid-state fermentation with the help of suitable microbial growth on agricultural residue (Sadh et al. 2018). Gowda and Manvi (2019) utilized agro-residues as a substrate in mushroom cultivation and are also developed simple and low-cost pasteurization equipment for small-scale rural mushroom growers. Vazquez-Olivo et al. (2019) converted lignocellulosic agro-industrial waste to value-added products such as bioactive molecules, phenolic compounds, antioxidants, etc. with a zero-waste process (Um et al. 2017). Hence, the agro-based biorefinery approach will not only produce value-added products but also help in the sustainable development of rural India efficiently.

1.7 Cost Economics of Waste Biorefineries

Cost economics is a must to evaluate the sustainability and financial feasibility of any industry. As the global population swells every year, the rate of increase in waste generation poses major environmental threats and resource crunch. If there are proper directives for waste management, wealth can be generated through recycling or converting them into value-embedded products. The biological source of wastes accumulated in low-income economies is about 50% higher than that in well-developed economies. Hence, for the economic development of emerging economies, waste-based biorefineries are very crucial. Solid waste biorefineries, if properly integrated, would result in generating new entrepreneurs, creation of job opportunities, reduced cost in waste management, waste to value-added products, and lower emissions. Besides capital intensive, most of these biorefinery technologies are energy-centric; hence, there may be scope for emission acting as a negative externality.

1.7.1 Cost of Biomass

Considering the agricultural sector, waste generated from energy crops depends upon the land extent under each crop, its yield potential, production cost, logistics, handling, and proximity to the nearest biorefinery. Mostly, the sources of biomass can be broadly classified into three categories: They are crop residues, energy crops, wastes of industrial origin. Energy crops are those dedicated crops serving as stock materials for biorefineries. The resultant products are the first-generation (1G) biofuels. Corn, soybean, cassava, sweet potato, sugarcane, barley and palm oil are the energy crops used for these purposes around the globe. The demand and supply for these crops are dwindling, according to the market and nonmarket factors.

Crop residues are the organic wastes generated as byproducts obtained during post-harvest processing of field crops which are again classified into primary and secondary residues. The primary residues are the ones obtained on the production site which have alternate applications and the secondary residues are mainly the byproducts obtained while processing. Secondary residues are much cheaper to serve as feedstock in waste biorefineries as they find no alternative applications. Developing economies generate vast industrial wastes. Huge amounts of wastewater from households and industries, wastes from processing industries, animal wastes can be a source of feedstock in the biorefineries.

1.7.2 Cost of Logistics

The commercial viability of waste biorefineries is much dependent upon the location of the site from the biorefineries, harvesting and collection of biomass, transportation mode, time duration of transportation, and processing of biomass. The bulkiness of low energy content of biomass creates logistics much difficult. Cost economics of waste logistics is much dependent on distance, time of travel, the density of biomass, etc. Travel time influences the cost involved in hiring manpower and wear/tear of the vehicles. Biomass density is another prime concern as the requirement of low-density biomass but its huge volume, in turn, falls heavy on the cost of logistics. Figure 1.7 depicts several costs incurred while processing the waste biorefineries.

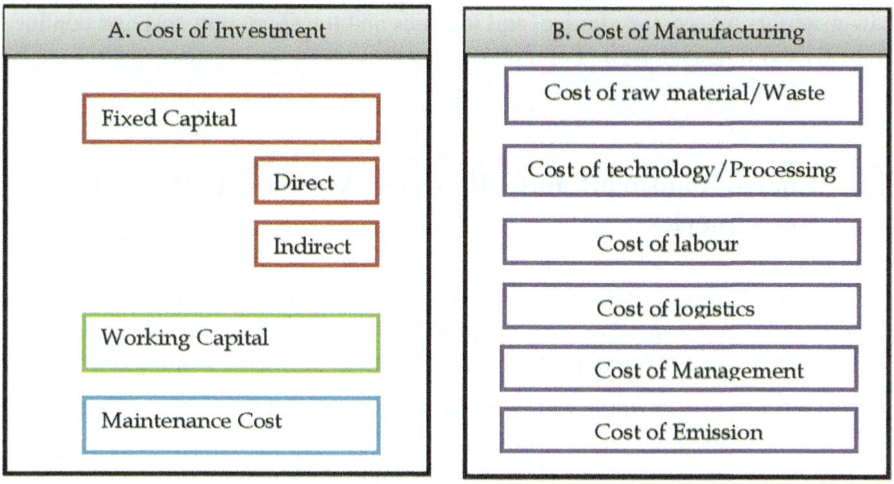

Fig. 1.7 Various costs incurred upon waste biorefinery processing

1.7.3 Economic Assessment of Waste Biorefineries

The economic assessment is an important criterion for evaluating the quality of the waste biorefineries. The required parameters for the economic evaluation of any waste biorefineries on cost streams are investment cost, maintenance cost, interest, taxes, insurance, material cost, logistics, labor cost, management cost, emission cost, etc. and on the revenue side, the quantum and sale of energy products from waste biorefineries. Markets for bio-based products are characterized by high-value innovative chemicals and materials with high energy security with low processing costs and also with a minimal environmental cost. The commercial viability is much dependent on technical, commercial, and sustainable issues.

The estimation of food waste generation at the manufacturing stage for various products through biorefineries and a techno-economic and profitability analysis of routes for their valorization indicated that the markets for the energy products, processing, logistics, and the prices of competing fossil fuel-based products are the key determinants for the commercial viability of the waste biorefineries (Cristóbal et al. 2018). The technical, economic, and environmental assessment case studies conducted by the International Energy Agency (2019) revealed that all the case studies of sugars to lignin, biogas, lipids, and pulp to lignin depicted the potential environmental benefits accrued from developing biobased products through biorefinery processes.

Commercialization, creation of markets, and the economic feasibility of these biobased products are still under investigation as the lower cost of competing for fossil counterparts in the energy markets. The commercial feasibility and the economic viability of lactic acid with biogas through an integrated biorefinery process are more efficient than as a single process. The integrated biorefinery resulted in a minimal amount of waste generated and increased value-added products (Demichelis et al. 2018). The cost of biomass is determined by the selling price, raw material cost, cost of storage, and logistics and transportation cost and confined to place and time (Thorsell et al. 2004).

1.8 Energy Footprint and Life Cycle Assessment of Waste Biorefineries

The economic, as well as the environmental benefits of using agro/industrial wastes, would be further enhanced by the joint production of chemicals and energy products. To convert wastes into wealth, the following are the necessary conditions.

- The product mix should have the highest economic value;
- It should yield the highest benefit; and
- The feedstock requirements should not be bulky to handle.

Generally, the Life Cycle Assessment (LCA) is meant to analyze the resource use pattern and the environmental impacts of the process involved in a production cycle to obtain the final products from raw materials. To estimate the associated energy footprints, carbon dioxide emissions from various feedstock sources are in corporate in the LCA framework. Then, the emissions from various components are added up to decide the aggregated to decide the footprint of the entire system. Direct and indirect emissions are reported separately to improve the system boundaries in terms of energy use and education in emission rate.

1.8.1 Key Issues in Life-Cycle Assessment of Waste Biorefineries

Increased energy consumption owing to rapid urbanization resulted in more GHG emissions leading to unpredictable climate change. Urbanization triggers the enhanced energy consumption and accumulation of more solid and liquid wastes. The amount of waste generated is alarmingly increasing affecting the ecosystem. Hence, it is necessary to identify efficient strategies to reduce ever-increasing environmental hazards. Waste biorefineries have created aspirations aimed toward integrating various conversion technologies for waste management to generate an array of energy products resulting in circular and low-carbon bioeconomy.

1.8.2 Case Studies in Life-Cycle Assessment of Waste Biorefineries

Increased energy consumption owing to rapid urbanization resulted in more greenhouse gas (GHG) emissions and climate change. Urbanization is one of the chief criteria leading to enlarged energy consumption and the accumulation of more solid and liquid wastes. The amount of waste generated is alarmingly increasing at a faster rate affecting the ecosystem. Hence, efficient sustainable waste management strategies are the need of an hour to reduce ever-growing environmental hazards.

The life cycle assessment for the calculation of the greenhouse gas emissions in the organic livestock production systems of Spain has concluded that organic livestock farming is a feasible strategy for reducing GHGs (Horrillo et al. 2021). The results of the ecosystem model for agricultural carbon footprint used for the US Western Corn Belt region showed an enlarging negative carbon footprint due to crop land expansion and associated carbon cost of grain production (Lu et al. 2018). Wang et al. (2015) studied the excessive use of nitrogen fertilizer and its impact on agriculture, and they found that several parameters such as grain yield, input energy, greenhouse gas emission, and carbon footprint were increased with an increase in nitrogen rate.

The GHG emissions of crop production from a life-cycle assessment perception concluded that intensive crop production aiming at economic optimum nitrogen supply helped to mitigate GHG emissions (Torres-Dorante et al. 2009).

1.9 Conclusion and Future Perspectives

With the current availability of 500 million metric tonnes of biomass, India has a potential of about 18 GW of energy from biomass and constitutes 32% of the total primary energy used. Higher than 70% of the country's population relies on biomass for energy needs. The estimated surplus biomass availability per year in India is about 120–150 million metric tonnes of agricultural and forestry residues (equivalent to 18,000 MW). Backward and forward integration at different levels should be considered to advance the overall efficiency of multi-product integrated portfolios. The economic and environmental performance of the biorefinery systems stands at the fore of the evaluation.

Overall, the efficient biorefinery system could provide energy generation, land saving, new business with employment generation, landfills cost savings, reduction of GHG emission, and savings of natural resources. Waste biorefineries are not only the way forward to sustainability but also generate crucial environmental benefits.

References

Abdel-Rahman MA, Hassan SE-D, Roushdy MM, Azab MS, Gaber MA (2019) Free-nutrient supply and thermo-alkaline conditions for direct lactic acid production from mixed lignocellulosic and food waste materials. Bioresour Technol Rep 7:100256

Abraham A, Mathew AK, Sindhu R, Pandey A, Binod P (2016) Potential of rice straw for bio-refining: an overview. Bioresour Technol 215:29–36

Algapani DE, Wang J, Qiao W, Su M, Goglio A, Wandera SM, Jiang M, Pan X, Adani F, Dong R (2017) Improving methane production and anaerobic digestion stability of food waste by extracting lipids and mixing it with sewage sludge. Bioresour Technol 244:996–1005

Ayantoyinbo BB, Adepoju OO (2018) Analysis of solid waste management logistics and its attendant challenges in Lagos metropolis. Logistics 2(2):11

Baranitharan E, Khan MR, Prasad DMR, Teo WFA, Tan GYA, Jose R (2015) Effect of biofilm formation on the performance of microbial fuel cell for the treatment of palm oil mill effluent. Bioprocess Biosyst Eng 38(1):15–24

Benabda O, M'hir S, Kasmi M, Mnif W, Hamdi M (2019) Optimization of protease and amylase production by *Rhizopus oryzae* cultivated on bread waste using solid-state fermentation. J Chem 2019:3738181

Bhatia L, Johri S, Ahmad R (2012) An economic and ecological perspective of ethanol production from renewable agro waste: a review. AMB Express 2(1):65

Bhola V, Desikan R, Santosh SK, Subburamu K, Sanniyasi E, Bux F (2011) Effects of parameters affecting biomass yield and thermal behaviour of *Chlorella vulgaris*. J Biosci Bioeng 111 (3):377–382

Bi L, Ci S, Cai P, Li H, Wen Z (2018) One-step pyrolysis route to three dimensional nitrogen-doped porous carbon as anode materials for microbial fuel cells. Appl Surf Sci 427:10–16

Blanco AS, Durive OP, Pérez SB, Montes ZD, Guerra NP (2016) Simultaneous production of amylases and proteases by *Bacillus subtilis* in brewery wastes. Braz J Microbiol 47(3):665–674

Bouaid A, Martínez M, Aracil J (2010) Biorefinery approach for coconut oil valorisation: a statistical study. Bioresour Technol 101(11):4006–4012

Branyikova I, Prochazkova G, Potocar T, Jezkova Z, Branyik T (2018) Harvesting of microalgae by flocculation. Fermentation 4(4):93

Budzianowski WM, Postawa K (2016) Total chain integration of sustainable biorefinery systems. Appl Energy 184:1432–1446

Carriquiry MA, Du X, Timilsina GR (2011) Second generation biofuels: economics and policies. Energy Policy 39(7):4222–4234

Chakraborty D, Karthikeyan OP, Selvam A, Wong JWC (2018) Co-digestion of food waste and chemically enhanced primary treated sludge in a continuous stirred tank reactor. Biomass Bioenergy 111:232–240

Chutia S, Gohain M, Deka D, Kakoty NM (2017) A review on the harvesting techniques of algae for algal based biofuel production. J Energy Res Environ Technol 4(1):58–62

Cristóbal J, Caldeira C, Corrado S, Sala S (2018) Techno-economic and profitability analysis of food waste biorefineries at European level. Bioresour Technol 259:244–252

da Silva TL, Gouveia L, Reis A (2014) Integrated microbial processes for biofuels and high value-added products: the way to improve the cost-effectiveness of biofuel production. Appl Microbiol Biotechnol 98(3):1043–1053

Dahiya S, Kumar AN, Shanthi Sravan J, Chatterjee S, Sarkar O, Mohan SV (2018) Food waste biorefinery: sustainable strategy for circular bioeconomy. Bioresour Technol 248:2–12

De Gioannis G, Muntoni A, Polettini A, Pomi R, Spiga D (2017) Energy recovery from one- and two-stage anaerobic digestion of food waste. Waste Manag 68:595–602

Demichelis F, Fiore S, Pleissner D, Venus J (2018) Technical and economic assessment of food waste valorization through a biorefinery chain. Renew Sust Energ Rev 94:38–48

Demirbas A (2007) Importance of biodiesel as a transportation fuel. Energy Policy 35 (9):4661–4670

Dessie W, Zhang W, Xin F, Dong W, Zhang M, Ma J, Jiang M (2018) Succinic acid production from fruit and vegetable wastes hydrolyzed by on-site enzyme mixtures through solid-state fermentation. Bioresour Technol 247:1177–1180

Du C, Lin SKC, Koutinas A, Wang R, Dorado P, Webb C (2008) A wheat biorefining strategy based on solid-state fermentation for fermentative production of succinic acid. Bioresour Technol 99(17):8310–8315

ElMekawy A, Srikanth S, Bajracharya S, Hegab HM, Nigam PS, Singh A, Mohan SV, Pant D (2015) Food and agricultural wastes as substrates for bioelectrochemical system (BES): the synchronized recovery of sustainable energy and waste treatment. Food Res Int 73:213–225

FAO (2019) Food loss and food waste. http://www.fao.org/food-loss-and-food-waste/en/454. Accessed 25 Sept 2019

Ferri GL, de Lorena Diniz Chaves G, Ribeiro GM (2015) Reverse logistics network for municipal solid waste management: the inclusion of waste pickers as a Brazilian legal requirement. Waste Manag 40:173–191

Fuess LT, Kiyuna LSM, Ferraz ADN, Persinoti GF, Squina FM, Garcia ML, Zaiat M (2017) Thermophilic two-phase anaerobic digestion using an innovative fixed-bed reactor for enhanced organic matter removal and bioenergy recovery from sugarcane vinasse. Appl Energy 189:480–491

Gowda N, Manvi D (2019) Agro-residues disinfection methods for mushroom cultivation. Agric Rev 40(2):93–103

Greenwell HC, Lloyd-Evans M, Wenner C (2013) Biofuels, science and society. vol. 3. The Royal Society

Gressel J (2008) Transgenics are imperative for biofuel crops. Plant Sci 174(3):246–263

Grimberg SJ, Hilderbrandt D, Kinnunen M, Rogers S (2015) Anaerobic digestion of food waste through the operation of a mesophilic two-phase pilot scale digester—assessment of variable loadings on system performance. Bioresour Technol 178:226–229

Gullón B, Garrote G, Alonso JL, Parajó JC (2007) Production of L-lactic acid and oligomeric compounds from apple pomace by simultaneous saccharification and fermentation: a response surface methodology assessment. J Agric Food Chem 55(14):5580–5587

Guven H, Ersahin ME, Dereli RK, Ozgun H, Isik I, Ozturk I (2019) Energy recovery potential of anaerobic digestion of excess sludge from high-rate activated sludge systems co-treating municipal wastewater and food waste. Energy 172:1027–1036. https://doi.org/10.1016/j.energy.2019.01.150

Hakim RA, Taib CAB, Hossain MTB, Naim MJ (2017) Social responsibility in global supply chain: research trend from 1999 to 2014. Int J Supply Chain Manag 6(4):146

Han W, Ye M, Zhu AJ, Zhao HT, Li YF (2015) Batch dark fermentation from enzymatic hydrolyzed food waste for hydrogen production. Bioresour Technol 191:24–29

Han W, Xu X, Gao Y, He H, Chen L, Tian X, Hou P (2019) Utilization of waste cake for fermentative ethanol production. Sci Total Environ 673:378–383

Hiloidhari M, Bhuyan N, Gogoi N, Seth D, Garg A, Singh A, Prasad S, Kataki R (2020) Chapter 16—agroindustry wastes: biofuels and biomaterials feedstocks for sustainable rural development. In: Kumar RP, Gnansounou E, Raman JK, Baskar G (eds) Refining biomass residues for sustainable energy and bioproducts. Academic, pp 357–388

Hoekman SK (2009) Biofuels in the U.S.—challenges and opportunities. Renew Energy 34 (1):14–22

Horrillo A, Gaspar P, Díaz-Caro C, Escribano M (2021) A scenario-based analysis of the effect of carbon pricing on organic livestock farm performance: a case study of Spanish dehesas and rangelands. Sci Total Environ 751:141675

Hu F, Ragauskas A (2012) Pretreatment and lignocellulosic chemistry. Bioenergy Res 5 (4):1043–1066

Idris A, Suzana W (2006) Effect of sodium alginate concentration, bead diameter, initial pH and temperature on lactic acid production from pineapple waste using immobilized *Lactobacillus delbrueckii*. Process Biochem 41(5):1117–1123

IEA (2008) IEA bioenergy task 42 on biorefineries: co-production of fuels, chemicals, power and materials from biomass. In: Minutes of the third task meeting, Copenhagen, Denmark, 25–26 March 2007

IPCC (2019) Climate change and land

Jena U, Vaidyanathan N, Chinnasamy S, Das KC (2011) Evaluation of microalgae cultivation using recovered aqueous co-product from thermochemical liquefaction of algal biomass. Bioresour Technol 102(3):3380–3387

Jiang J, Jin W, Tu R, Han S, Ji Y, Zhou X (2020) Harvesting of microalgae Chlorella pyrenoidosa by bio-flocculation with bacteria and filamentous fungi. Waste Biomass Valoriz 12(1):145–154

Jin M, Slininger PJ, Dien BS, Waghmode S, Moser BR, Orjuela A, da Costa Sousa L, Balan V (2015) Microbial lipid-based lignocellulosic biorefinery: feasibility and challenges. Trends Biotechnol 33(1):43–54

Johnravindar D, Murugesan K, Wong JWC, Elangovan N (2017) Waste-to-biofuel: production of biobutanol from sago waste residues. Environ Technol 38(13–14):1725–1734

Joshi SM, Waghmare JS, Sonawane KD, Waghmare SR (2015) Bio-ethanol and bio-butanol production from orange peel waste. Biofuels 6(1–2):55–61

Khoo CG, Dasan YK, Lam MK, Lee KT (2019) Algae biorefinery: review on a broad spectrum of downstream processes and products. Bioresour Technol 292:121964

Kuang Y, Zhao J, Gao Y, Lu C, Luo S, Sun Y, Zhang D (2020) Enhanced hydrogen production from food waste dark fermentation by potassium ferrate pretreatment. Environ Sci Pollut Res 27 (15):18145–18156

Kumar S, Gupta R, Kumar G, Sahoo D, Kuhad RC (2013) Bioethanol production from *Gracilaria verrucosa*, a red alga, in a biorefinery approach. Bioresour Technol 135:150–156

Ladygina N, Dedyukhina E, Vainshtein M (2006) A review on microbial synthesis of hydrocarbons. Process Biochem 41(5):1001–1014

Ledezma P, Kuntke P, Buisman CJN, Keller J, Freguia S (2015) Source-separated urine opens golden opportunities for microbial electrochemical technologies. Trends Biotechnol 33 (4):214–220

Leung CCJ, Cheung ASY, Zhang AY-Z, Lam KF, Lin CSK (2012) Utilisation of waste bread for fermentative succinic acid production. Biochem Eng J 65:10–15

Li Y, Liu H, Yan F, Su D, Wang Y, Zhou H (2017) High-calorific biogas production from anaerobic digestion of food waste using a two-phase pressurized biofilm (TPPB) system. Bioresour Technol 224:56–62

Liu C-M, Wu S-Y (2016) From biomass waste to biofuels and biomaterial building blocks. Renew Energy 96:1056–1062

Lu C, Yu Z, Tian H, Hennessy DA, Feng H, Al-Kaisi M, Zhou Y, Sauer T, Arritt R (2018) Increasing carbon footprint of grain crop production in the US Western Corn Belt. Environ Res Lett 13(12):124007

Luo L, van der Voet E, Huppes G (2010) Biorefining of lignocellulosic feedstock—technical, economic and environmental considerations. Bioresour Technol 101(13):5023–5032

Maiorella B (1985) Ethanol. Comprehen Biotechnol 3:861–914

Mandade P, Bakshi BR, Yadav GD (2016) Ethanol from Indian agro-industrial lignocellulosic biomass: an emerging evaluation. Clean Techn Environ Policy 18(8):2625–2634

Mande S, Agarwal AK, Bae J-H, Bardiya N, Reddy MP, Johri R, Rajeshwari K, Pandey S, Dadhich PK, Lal B (2005) Biomass gasifier-based power plants: potential, problems, and research needs for decentralized rural electrification. In: Building the future we want: trends and technologies

Mariano AP, Dias MOS, Junqueira TL, Cunha MP, Bonomi A, Filho RM (2013) Utilization of pentoses from sugarcane biomass: techno-economics of biogas vs. butanol production. Bioresour Technol 142:390–399

Matter IA, Bui VKH, Jung M, Seo JY, Kim Y-E, Lee Y-C, Oh Y-K (2019) Flocculation harvesting techniques for microalgae: a review. Appl Sci 9(15):3069

Menon V, Rao M (2012) Trends in bioconversion of lignocellulose: biofuels, platform chemicals & biorefinery concept. Prog Energy Combust Sci 38(4):522–550

Mohan SV, Chandrasekhar K (2011) Solid phase microbial fuel cell (SMFC) for harnessing bioelectricity from composite food waste fermentation: influence of electrode assembly and buffering capacity. Bioresour Technol 102(14):7077–7085

Moncada J, El-Halwagi MM, Cardona CA (2013) Techno-economic analysis for a sugarcane biorefinery: Colombian case. Bioresour Technol 135:533–543

Munasinghe PC, Khanal SK (2010) Biomass-derived syngas fermentation into biofuels: opportunities and challenges. Bioresour Technol 101(13):5013–5022

Muntoni A (2019) Waste biorefineries: opportunities and perspectives

Mutanda T, Ramesh D, Karthikeyan S, Kumari S, Anandraj A, Bux F (2011) Bioprospecting for hyper-lipid producing microalgal strains for sustainable biofuel production. Bioresour Technol 102(1):57–70

Nagajothilakshmi K, Ramesh D, Karthikeyan S (2016) Mixotrophic growth on wastewater for sustainable algal biomass production. In: 5th International conference on sustainable utilization of tropical plant biomass: bioproducts, biocatalysts and biorefinery, p 188

Nimbalkar PR, Khedkar MA, Chavan PV, Bankar SB (2018) Biobutanol production using pea pod waste as substrate: impact of drying on saccharification and fermentation. Renew Energy 117:520–529

Nobre BP, Villalobos F, Barragan BE, Oliveira AC, Batista AP, Marques PASS et al (2013) A biorefinery from Nannochloropsis sp. microalga–extraction of oils and pigments. Production of biohydrogen from the leftover biomass. Bioresour Technol 135:128–136

NPMCR (n.d.). http://agricoop.nic.in/sites/default/files/NPMCR_1.pdf

Obi F, Ugwuishiwu B, Nwakaire J (2016) Agricultural waste concept, generation, utilization and management. Niger J Technol 35(4):957–964

Octave S, Thomas D (2009) Biorefinery: toward an industrial metabolism. Biochimie 91 (6):659–664

Pahl SL, Lee AK, Kalaitzidis T, Ashman PJ, Sathe S, Lewis DM (2013) Harvesting, thickening and dewatering microalgae biomass. In: Borowitzka MA, Moheimani NR (eds) Algae for biofuels and energy. Springer Netherlands, Dordrecht, pp 165–185

Pandey A, Shah R, Yadav P, Verma R, Srivastava S (2020) Harvesting of freshwater microalgae Scenedesmus sp. by electro–coagulation–flocculation for biofuel production: effects on spent medium recycling and lipid extraction. Environ Sci Pollut Res 27(3):3497–3507

Pant D, Arslan D, Van Bogaert G, Gallego YA, De Wever H, Diels L, Vanbroekhoven K (2013) Integrated conversion of food waste diluted with sewage into volatile fatty acids through fermentation and electricity through a fuel cell. Environ Technol 34(13–14):1935–1945

Parthiba Karthikeyan O, Trably E, Mehariya S, Bernet N, Wong JWC, Carrere H (2018) Pretreatment of food waste for methane and hydrogen recovery: a review. Bioresour Technol 249:1025–1039

Puigjaner L (2011) Syngas from waste: emerging technologies. Springer

Radha M, Kanmani S (2017) Evaluation of paper recycling wastewater treatment accompanied by power generation using microbial fuel cell. Global NEST J 19(4):682–686

Rahimnejad M, Adhami A, Darvari S, Zirepour A, Oh S-E (2015) Microbial fuel cell as new technology for bioelectricity generation: a review. Alex Eng J 54(3):745–756

Rakesh S, Karthikeyan S (2019) Co-cultivation of microalgae with oleaginous yeast for economical biofuel production. J Farm Sci 32(2):1–6

Rakesh S, Saxena S, Dhar DW, Prasanna R, Saxena AK (2014) Comparative evaluation of inorganic and organic amendments for their flocculation efficiency of selected microalgae. J Appl Phycol 26(1):399–406

Rakesh S, TharunKumar J, Gudelli B, Karuppaiyan J, Karthikeyan S (2020) Sustainable cost-effective microalgae harvesting strategies for the production of biofuel and oleochemicals. In: Highlights in bioscience, p 3

Ramesh D, Muniraj IK, Thangavelu K, Karthikeyan S (2018) Pretreatment of lignocellulosic biomass feedstocks for biofuel production. In: Advanced solid catalysts for renewable energy production. IGI Global, pp 33–60

Ramesh D, Muniraj IK, Thangavelu K, Karthikeyan S (2019a) Chemicals and fuels production from agro residues: a biorefinery approach. In: Sustainable approaches for biofuels production technologies. Springer, pp 47–71

Ramesh D, Muniraj IK, Thangavelu K, Karthikeyan S (2019b) Chemicals and fuels production from agro residues: a biorefinery approach. In: Srivastava N, Srivastava M, Mishra PK, Upadhyay SN, Ramteke PW, Gupta VK (eds) Sustainable approaches for biofuels production technologies: from current status to practical implementation. Springer International Publishing, Cham, pp 47–71

Reddy MV, Mohan SV (2015) Polyhydroxyalkanoates production by newly isolated bacteria Serratia ureilytica using volatile fatty acids as substrate: bio-electro kinetic analysis. J Microb Biochem Technol 7:026–032

Rincón LE, Jaramillo JJ, Cardona CA (2014) Comparison of feedstocks and technologies for biodiesel production: an environmental and techno-economic evaluation. Renew Energy 69:479–487

Rinna F, Buono S, Cabanelas ITD, Nascimento IA, Sansone G, Barone CMA (2017) Wastewater treatment by microalgae can generate high quality biodiesel feedstock. J Water Process Eng 18:144–149

Rosenberg JN, Mathias A, Korth K, Betenbaugh MJ, Oyler GA (2011) Microalgal biomass production and carbon dioxide sequestration from an integrated ethanol biorefinery in Iowa: a technical appraisal and economic feasibility evaluation. Biomass Bioenergy 35(9):3865–3876

Roukas T, Kotzekidou P (2020) Pomegranate peel waste: a new substrate for citric acid production by *Aspergillus niger* in solid-state fermentation under non-aseptic conditions. Environ Sci Pollut Res 27(12):13105–13113
Sadh PK, Duhan S, Duhan JS (2018) Agro-industrial wastes and their utilization using solid state fermentation: a review. Bioresour Bioprocess 5(1):1
Sahu O (2019) Sustainable and clean treatment of industrial wastewater with microbial fuel cell. Results Eng 4:100053
Shukla J, Kar R (2006) Potato peel as a solid state substrate for thermostable α-amylase production by thermophilic Bacillus isolates. World J Microbiol Biotechnol 22(5):417–422
Singh A, Nigam PS, Murphy JD (2011) Mechanism and challenges in commercialisation of algal biofuels. Bioresour Technol 102(1):26–34
Sreekumar A, Shastri Y, Wadekar P, Patil M, Lali A (2020) Life cycle assessment of ethanol production in a rice-straw-based biorefinery in India. Clean Techn Environ Policy 22 (2):409–422
Subhadra BG (2010) Sustainability of algal biofuel production using integrated renewable energy park (IREP) and algal biorefinery approach. Energy Policy 38(10):5892–5901
Takkellapati S, Li T, Gonzalez MA (2018) An overview of biorefinery-derived platform chemicals from a cellulose and hemicellulose biorefinery. Clean Techn Environ Policy 20(7):1615–1630
Thiengburanathum P, Theingburanathum P, Madhyamapurush C (2012) Value chain optimization framework for solid waste management in Thailand: a case study of Samui Island
Thorsell S, Epplin FM, Huhnke RL, Taliaferro CM (2004) Economics of a coordinated biorefinery feedstock harvest system: lignocellulosic biomass harvest cost. Biomass Bioenergy 27 (4):327–337
Torres-Dorante LO, Lammel J, Kuhlmann H (2009) Use of a layered double hydroxide (LDH) to buffer nitrate in soil: long-term nitrate exchange properties under cropping and fallow conditions. Plant Soil 315(1):257–272
Um M, Shin G-J, Lee J-W (2017) Extraction of total phenolic compounds from yellow poplar hydrolysate and evaluation of their antioxidant activities. Ind Crop Prod 97:574–581
Uygut MA, Tanyildizi MŞ (2018) Optimization of alpha-amylase production by Bacillus amyloliquefaciens grown on Orange peels. Iran J Sci Technol Trans A Sci 42(2):443–449
Vazquez-Olivo G, López-Martínez LX, Contreras-Angulo L, Heredia JB (2019) Antioxidant capacity of lignin and phenolic compounds from Corn Stover. Waste Biomass Valorization 10(1):95–102
Vega-Castro O, Contreras-Calderon J, León E, Segura A, Arias M, Pérez L, Sobral PJA (2016) Characterization of a polyhydroxyalkanoate obtained from pineapple peel waste using *Ralsthonia eutropha*. J Biotechnol 231:232–238
Vijayanand C, Kamaraj S, Sriramajayam S, Karthikeyan S, Surendrakumar A, Ramesh D (2017) A probe into the biphasic nature of the acetone butanol ethanol (ABE) fermentation by Clostridium acetobutylicum ATCC 824. Int J Curr Microbiol Appl Sci 6(4):2179–2186
Vikash PV, Mandade PV, Shastri Y (2018) Assessment of bagasse and trash utilization for ethanol production: a case study in India. Environ Prog Sustain Energy 37(6):2165–2174
Wang Z-B, Wen X-Y, Zhang H-L, Lu X-H, Chen F (2015) Net energy yield and carbon footprint of summer corn under different N fertilizer rates in the North China plain. J Integr Agric 14 (8):1534–1541
Wilkins MR, Widmer WW, Grohmann K (2007) Simultaneous saccharification and fermentation of citrus peel waste by *Saccharomyces cerevisiae* to produce ethanol. Process Biochem 42 (12):1614–1619
Yadav G, Dash SK, Sen R (2019) A biorefinery for valorization of industrial waste-water and flue gas by microalgae for waste mitigation, carbon-dioxide sequestration and algal biomass production. Sci Total Environ 688:129–135

Yang X, Lee SJ, Yoo HY, Choi HS, Park C, Kim SW (2014) Biorefinery of instant noodle waste to biofuels. Bioresour Technol 159:17–23

Yu J, Zhang T, Zhong J, Zhang X, Tan T (2012) Biorefinery of sweet sorghum stem. Biotechnol Adv 30(4):811–816

Yun Y-M, Sung S, Shin H-S, Han J-I, Kim H-W, Kim D-H (2017) Producing desulfurized biogas through removal of sulfate in the first-stage of a two-stage anaerobic digestion. Biotechnol Bioeng 114(5):970–979

Zhang J, Loh K-C, Lee J, Wang C-H, Dai Y, Wah Tong Y (2017) Three-stage anaerobic co-digestion of food waste and horse manure. Sci Rep 7(1):1269

Chapter 2
Progress in Consolidated Bioprocessing of Lignocellulosic Biomass for Biofuels and Biochemicals

Tanvi Govil, Adhithya S. Narayanan, David R. Salem, and Rajesh K Sani

Abstract Currently, the cost of lignocellulose pretreatment is a weighty obstructing aspect in the economical production of biofuels and value-added biochemicals that make up for a significant percentage of the overall cost of production. For simplicity, process integration that can eliminate pre-treatment is a critical necessity for the process's overall economization. Even though cellulosic biorefining processes currently under commercial development use various process configurations, perhaps

T. Govil
Department of Chemical and Biological Engineering, South Dakota Mines, Rapid City, SD, USA

Composite and Nanocomposite Advanced Manufacturing - Biomaterials Center, Rapid City, SD, USA

A. S. Narayanan
School of Chemistry and Biotechnology, SASTRA University, Thanjavur, Tamil Nadu, India

D. R. Salem (✉)
Department of Chemical and Biological Engineering, South Dakota Mines, Rapid City, SD, USA

Composite and Nanocomposite Advanced Manufacturing - Biomaterials Center, Rapid City, SD, USA

Department of Materials and Metallurgical Engineering, South Dakota Mines, Rapid City, SD, USA
e-mail: david.salem@sdsmt.edu

R. K. Sani (✉)
Department of Chemical and Biological Engineering, South Dakota Mines, Rapid City, SD, USA

Composite and Nanocomposite Advanced Manufacturing - Biomaterials Center, Rapid City, SD, USA

Department of Chemistry, Biology, and Health Sciences, South Dakota Mines, Rapid City, SD, USA

BuG ReMeDEE Consortium, Rapid City, SD, USA
e-mail: rajesh.sani@sdsmt.edu

J. K. Saini, R. K. Sani (eds.), *Microbial Biotechnology for Renewable and Sustainable Energy*, Clean Energy Production Technologies,
https://doi.org/10.1007/978-981-16-3852-7_2

the concept that has gained most interest in the research community is "consolidated bioprocessing," also referred to as "CBP." CBP offers improvement of the lignocellulose depolymerization, effective mass, and energy balances to produce biofuels from agri-substrates at lower cost. This results from the overall simplification of the process operation, subdued costs of utilities/vessels associated with separate hydrolysis, saccharification, and fermentation and elimination of the additional costs associated with the use of the enzymes, using CBP. This book chapter will concentrate on the recent developments and progress in processing lignocellulosic biomass to produce biofuels with more focus on ethanol, butanol, and hydrogen.

Keywords Biobutanol · Bioethanol · Biohydrogen · Bioprocessing · Consolidated · Lignocellulosic

2.1 Introduction

The unified consolidated bioprocessing (CBP) of lignocellulosic biomass (LCB) to produce biofuels and biochemicals is a budding, sustainable, and energy-efficient approach that can facilitate the cost-competitive production of these carbon-neutral platform biofuels on a commercial scale. In the concept of CBP, after the deconstruction and depolymerization of cellulose and hemicellulose via pre-treatments, the next two biologically mediated processing steps for LCB degradation, i.e., saccharification of polymers into monomeric sugars and, finally, fermentation of hexose and pentose to end products, are integrated into a single operation by the action of microorganisms grown solo or as co-cultures (Fig. 2.1). Essentially, to be a qualified CBP host, an organism must possess the genome machinery to synthesize and secrete a multitude of untreated LCB depolymerizing enzymes, assimilate the

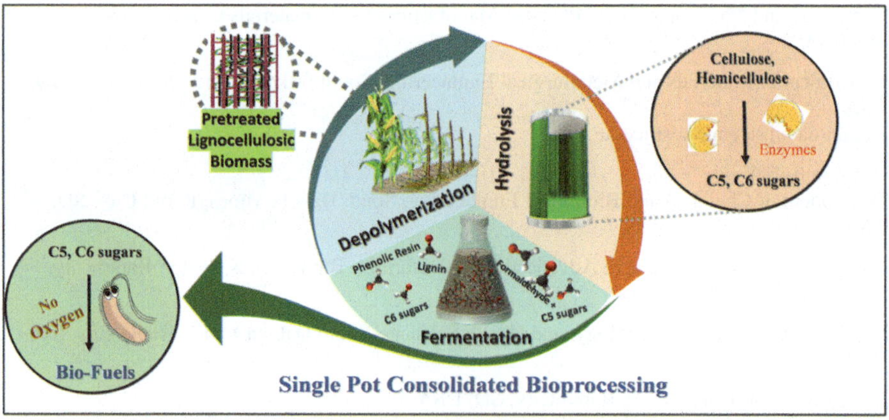

Fig. 2.1 Schematic scheme of consolidated bioprocessing for single step bioprocessing of lignocellulosic biomasses to biofuels

released C5 and C6 monomeric sugars, and finally metabolize these sugars to produce the desired biofuels (Linger and Darzins 2013). Otherwise, synthetic consortia can be utilized to divide the labor for the requisite metabolic functions between distinct specialized microorganisms, as will be reviewed in this book chapter for the explicit synthesis of various biofuels from lignocellulosic biomass.

In principle, CBP presents the prospective for lower cost of biofuels' production due to more straightforward and efficient feedstock usage, reduced energy consumption, and superior biomass conversion efficacies than separate hydrolysis and fermentation practices. When contemplating investment, feedstocks, utilities, and productivity deficit expenditures, a relative cost analysis performed on ethanol production ensued in a prediction of USD0.04/gal for CBP, though saccharification and co-fermentation were estimated at USD0.19/gal (Mbaneme-Smith and Chinn 2015). It is a sustainable eco-friendly method that can decrease costs associated with biomass processing, and it is for this reason that CBP is being cited as an economical option for "next-generation" lignocellulosic biofuel production (Govil et al. 2020; Levin et al. 2015).

In the past years, a multitude of natural or recombinant strategies have been implemented for engineering attractive CBP biocatalysts. The native approach includes improving the productivity and titer of specific biofuels by genome editing and modulation of genetic modification of cellulolytic microorganisms' metabolisms. The recombinant plan includes the heterologous expression of ligninolytic enzymes in host organisms that can biosynthesize the target biofuels (Liu et al. 2020). The biomass degradation and hydrolysis are higher as well as faster at elevated temperatures (Govil et al. 2020). Hence, thermophilic microbes remain the unsurpassed preference as a CBP host because of their ability for concurrent lignocellulose hydrolysis and biofuel production in a "one-pot" process (Liu et al. 2020). While several monocultures of cellulolytic bacteria have been explored, they have their shortcomings. Consequently, tapping co-cultures for novel biorefining processes have become an extremely dynamic field of research for biofuels' production related to the production of value-added co-products.

In the early stages of biofuel production, only ethanol and biodiesel were produced as fuel on an industrial scale (Antoni et al. 2007). However, as new processing technologies emerged, many other biofuels with important commercial potential, such as butanol, hydrogen, and methane, caught researchers' interest. Other than biofuels, many industrial solvents such as acetone, isopropanol, and organic acids (such as lactic acid and butyric acid) produced from lignocellulosic biomass were commercialized as value-added products. Various developments were made in the processing technologies, including the use of genetically modified organisms and/or consortia of organisms to improve biofuels' yield and productivity. In this chapter, the latest developments are discussed for increasing product titers, improving hydrolysis of biomass, and overcoming the limitations in implementing the above-mentioned mechanisms for production of biofuels, including bioethanol, biobutanol, and biohydrogen. In addition, other value-added products from lignocellulosic biomass through consolidated bioprocessing are discussed.

2.2 Biofuels from Lignocellulosic Feedstocks

2.2.1 Bioethanol Production

Bioethanol is an established fuel, being produced from sugar or starchy food crops since 1826 and dominating the biofuel world market with a total market size of USD 43.2 billion in 2019, that is predicted to become USD 64.8 billion by 2025, growing at a CAGR of 14% between 2020 and 2025 (GlobeNewsWire 2020). In the last years, attempts have been made to bio-synthesize bioethanol from second-generation lignocellulosic feedstocks (Fig. 2.2) by various processing techniques such as single pot biorefineries, and combining bioprocessing bioenergy systems with carbon capture (Toor et al. 2020). Consolidated bioprocessing (CBP) advancements to produce bioethanol have the advantage of avoiding the use of enzymes for cellulose and hemicellulose hydrolysis and hence circumvent this cost-increasing item from the production of the second-generation bioethanol. However, finding the right organism that produces enzymes for hydrolysis and fermentation of cellulose and hemicellulose into C5 and C6 sugars into ethanol is most important for CBP's success.

Until now, the anaerobic thermophilic bacterium, *Clostridium thermocellum*, is one of the most studied model candidates for CBP due to its capability to biosynthesize ethanol from cellulosic substrates. *C. thermocellum* strains have been proven to possess a supramolecular ligninolytic enzyme "cellulosome"

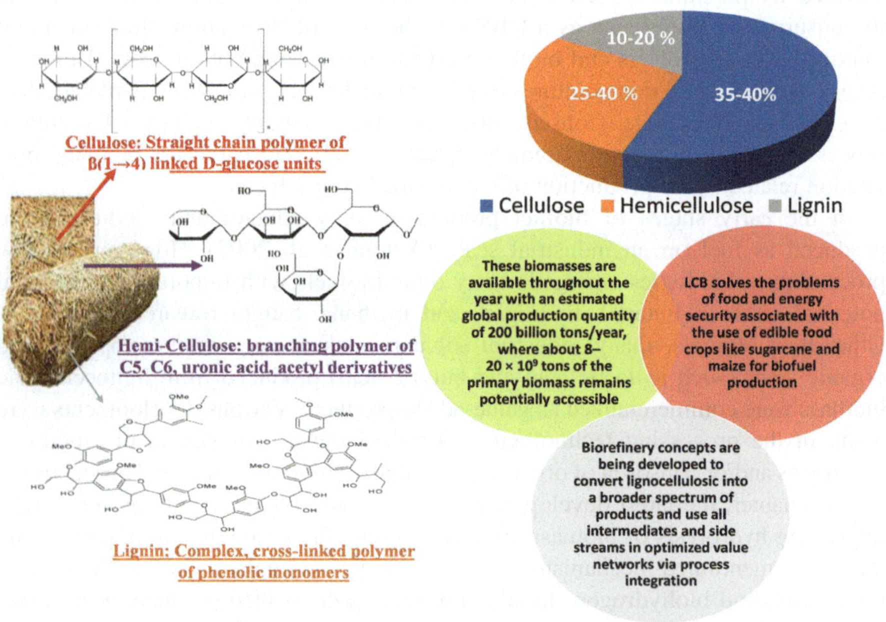

Fig. 2.2 Agro-industrial lignocellulosic biomass scope in second-generation biorefineries

complex on its cell wall that synergistically degrade crystalline cellulose with enormous efficiency, contrasted to free cellulases (Hirano et al. 2016). Some *C. thermocellum* strains such as ATCC 27405^{T} and DSM 1313 (previously known as LQ8) are well established as ethanol-producing strain (Akinosho et al. 2014). Under controlled fermentation conditions, these wild-type strains on average yield 0.15 g ethanol/g cellulose consumed, which encompasses about 10% of the theoretical maximum (Tian et al. 2016). In recent years, with an endeavor to improve the yield as well as the titer for cost-competitive synthesis of ethanol from lignocellulosic biomass, attempts toward finding new strains or their engineered versions have been made.

For example, in 2016, a new strain, *Clostridium thermocellum* ATCC 31924, was studied for superior ethanol production from crystalline cellulose (Singh et al. 2018a). The strain yielded 0.30 g ethanol/g cellulose consumed, with a cellulose conversion of 95.32%. Furthermore, it was found that the addition of a fairly small amount of acetate to the medium pushed the carbon flux away from lactic acid, with an increase in ethanol production by 20% (Singh et al. 2018a). Likewise, in 2018, another thermophilic clostridial strain, DBT-IOC, was isolated and studied for ethanol production from untreated rice straw biomass by CBP (Singh et al. 2018b). The strain DBT-IOC showed a wide substrate consumption spectrum for monomeric (both hexoses and pentoses), polymeric carbohydrates related to LCBs (cellobiose, microcrystalline cellulose, and xylan) as well as untreated lignocellulosic (rice straw) feedstock. This strain thus endorsed the suitability of the new thermophilic (70 °C) *C. thermocellum* strain as a CBP host through its ability to efficiently bio-convert unprocessed LCB to ethanol ((19.48 mM, 0.897 g/L) with 28 g/L of untreated rice straw), while much of the ethanol studies conducted with *Clostridium thermocellum* as a CBP host is confined to cellulose and its derivatives (Singh et al. 2018b).

More recently, in 2020, the suitability of biogenic municipal solid waste (BMSW) was studied as a carbon source for the growth of *C. thermocellum* ATCC-27405 and its ethanol production (Althuri and Mohan 2020). Therein, the authors used the strategy of sequential CBP to enhance ethanol titer where, due to the inability of the strain to digest pentose sugar, the leftover C5-rich broth generated after *C. thermocellum* ATCC-27405 culturing was inoculated with pentose-fermenting yeast, *Pichia stipitis* NCIM-3498, intended for added ethanol production (Althuri and Mohan 2020). The sequential CBP showed a 1.32-fold increase in ethanol production than the CBP on its own, from 18.10 ± 0.01 g/L to 23.99 ± 0.10 g/L. A further increase in ethanol titer to 36.90 g/L was observed when exogenous xylanases were added to the medium (Althuri and Mohan 2020). These reports promote the use of a microbial consortium to advance the processing technologies for viable bioethanol production in a single bioreactor.

In yet another strategy, aimed toward enhanced ethanol production, the engineering of the native host *C. thermocellum* strains was conducted. In their study, Tian et al. (2016) blocked the strain DSM 1313's carbon flux to pathways other than the ethanol pathway. The mutated strain showed an ethanol yield of 0.39 g ethanol/g cellulose (75% of the theoretical maximum), and ethanol titer (25 g/L) was obtained, which is the highest ethanol yield reported to date for a cellulolytic thermophilic anaerobic *C. thermocellum* (Tian et al. 2016).

Other than clostridial strains, thermophiles such as *Geobacillus* sp. and *Thermoanaerobacter* sp. are also reported to produce ethanol. Species in the genus *Thermoanaerobacter* can hydrolyze and utilize a wide variety of amylo- and xylanolytic substrates but are unable to utilize cellulose due to the absence of endoglucanases (Scully and Orlygsson 2015). *Geobacillus thermoglucosidasius* was employed to produce ethanol from 10% (w/v) blended food waste in a CBP setup at 60 °C, and an ethanol titer of 13.5 g/L with a productivity of 0.1 g/L/h has been reported (Bibra et al. 2020). A further rise in the ethanol titer up to 16.1 g/L was observed when *Geobacillus thermoglucosidasius* was sequentially cultivated with *Thermoanaerobacter ethanolicus* in a 1 L bioreactor. Further scaling up the reaction to a 40 L reactor produced a titer of 18.4 g/L of ethanol, accompanied by increase in productivity to 0.15 g/L/h. It was stated that the consortium was able to produce 70.1 L of gasoline-equivalent ethanol per ton of dry food waste, thus providing an efficient way to convert waste biomass to second-generation biofuels (Bibra et al. 2020). Other illustrative models of thermophilic ethanologenic bacteria have been noted within the genus *Caldicellulosiruptor*. An extremely thermophilic cellulolytic bacterium *Caldicellulosiruptor bescii* was reported as a prospective CBP candidate for ethanol production (Williams-Rhaesa et al. 2018).

Even though CBP is an encouraging technology for bioethanol production, particular shortcomings are present, for instance, the recalcitrant nature of the LCBs, leading to longer fermentation cycles, and the period required by the hosts to adapt to the conditions (Rajak and Banerjee 2020). Addition of an optimal laccase, cellulase, and xylanase cocktail for depolymerization of lignin and holocellulose solubilization in Kans grass, followed by fermentation by *S. cerevisiae* in a single vessel, generated 59.96 g/L of ethanol per day (Rajak and Banerjee 2020). Concepts such as these are still under exploration and provide potential for successful development and demonstration of one-step process designs that can be utilized for ethanol production at commercial scales.

2.2.2 Biobutanol Production

In recent times, biobutanol has garnered the interests of the biotechnologists as a prospective liquid biofuel. It has properties matching those of gasoline, which allows butanol to be used in contemporary delivery pipelines and car engines with no modification (Ibrahim et al. 2018). Biobutanol is also an industrial solvent in chemical industries and extractant in pharmaceutical industries, making it a high-demand chemical. Due to its high market value, mass production of butanol from a renewable source became an important goal, motivating researchers to investigate lignocellulosic biomass as a potential feedstock and CBP as the preferred technology. However, as in the production of bioethanol, there is no ideal CBP microorganism capable of executing the single-step bioconversion of lignocellulosic biomass into biobutanol efficiently and at the desired yield.

As part of the microbial selection, numerous cellulolytic bacterial species belonging to the genus *Clostridia* have been employed for biobutanol production following its acetone-butanol-ethanol (ABE) fermentation metabolism from lignocellulosic biomass. Among them, *C. acetobutylicum* has been regarded as an efficient bacterium together with some of its other species, notably, *C. saccharobutylicum* and *C. saccharoperbutylicum* (Keis et al. 2001). Furthermore, its genetically modified strains are the most commonly studied, and are the most employed hosts for biobutanol production (Ibrahim et al. 2018). The first industrial operation of the acetone-butanol-ethanol (ABE) fermentation with the *Clostridium acetobutylicum* strain (initially isolated in 1912 by Claim Weizmann) was carried out in 1916 by using conventional renewable substrates such as maize and molasses via batch fermentation (Jones and Woods 1986). Underlying limitations associated with the process include high cost of the sugar substrates, low solvent yields, and undesirable ratios of acetone and ethanol in the extracted solvent, with the necessity to maintain axenic conditions during the duration of the process. Research to overcome some of these limitations has been initiated, including the use of alternative, cheaper lignocellulosic substrates. *Clostridium acetobutylicum* is a bacterium that naturally digests lignocellulose, but due to the inherent inefficiency of the *C. acetobutylicum* strains toward the consumption of pentose sugars (xylose and arabinose), lower solvent yields are usually obtained.

Much research work is being conducted on increasing the yield and titer of butanol from lignocellulosic biomass via co-cultivation methods using single, co-cultures, or mixed cultures of microbes that possess enzymes efficient in hydrolyzing cellulose and hemicellulose. In this direction, *Clostridium* spp. (*C. beijerinckii*, *C. saccharobutylicum*, *C. saccharoperbutylacetonicum*, *C. butyricum*, and *C. pasteurianum*) have extensively been used (Cao et al. 2016; Ibrahim et al. 2018; Jones and Woods 1986). Also, metabolic engineering has been broadly adopted, which entails (a) overexpression of rate-limiting enzymes, (b) modulation of central metabolic pathways, (c) blocking the competing carbon metabolisms, and (d) cofactor balancing to improve the butanol yields (Ferreira et al. 2020).

The bio-conversion of cellulose to butanol by a mesophilic co-culture of a cellulolytic *C. cellulolyticum* H10 strain with *C. acetobutylicum* was shown in early 1983 (Fond et al. 1983). This mixed cultivation on cellulose resulted in the production of primarily butyric acid (14 g/L) with minor quantities of acetic acid (4 g/L), ethanol (3 g/L), and butanol (1 g/L). The authors of this study reported that a sufficiently high glucose supply rate in the medium is essential to push the metabolism from butyric acid toward the production of ABE solvents. Yu et al. (1985) investigated a sequential co-culture approach with *C. thermocellum* and *C. acetobutylicum* for the conversion of solka floc and aspen wood xylan to solvents. The simultaneous growth of the two anaerobic bacteria together led to an effective hydrolysis of all the cellulose and hemicellulose in the lignocellulose substrate. This resulted in a 1.7–2.6 fold rise in the aggregate solvents produced, which were mostly acids (Yu et al. 1985). These findings were similar to those in a study by Fond et al. (Fond et al. 1983), which implies that the C5 and C6 sugar levels alone did not regulate butanol titers. Previous studies have signaled that a decrease in butyric acid concentrations may be have a considerable effect in sparking butanol production (Jones and Woods 1986). Hence, combinations have been investigated to increase butanol production by including a butyric acid utilizing organism in the co-culture. Notably, by co-culturing *C. pasteurianum*, a butyric acid-metabolizing bacterium with *C. beijerinckii*, 20% higher butanol concentrations were obtained than when using *C. beijerinckii* in pure culture (Jones and Woods 1986). Thus, consortia engineering of cellulolytic, butyric acid utilizing, and butanol-producing *Clostridia* has posed itself as an evolving approach for the production of butanol by CBP, where the participating strains can divide the metabolic needs and adapt to the external environment by adjusting the abundance of each strain (Wen et al. 2020c). Even the co-culturing of *Clostridia* members with strains belonging to a different genus has long been explored for constructing a cross-genus cellulosic butanol system.

In 2018, a *Thermoanaerobacterium* sp. M5 that may well promptly produce 1.17 g/L of butanol from polysaccharide xylan through consolidated bioprocessing under thermophilic conditions (55 °C) was isolated. Concurrent growth of *Thermoanaerobacterium* sp. M5 with *Clostridium acetobutylicum* NJ4 (a butanol-synthesizing bacterium) improved the butanol productivity from xylan in a CBP fermenter to 8.34 g/L (Jiang et al. 2018a). In a further example of co-cultivation, an anaerobic cross-kingdom consortium was established between *C. saccharoperbutylacetonicum* and the white-rot fungus *Phlebia* sp. MG-60-P2 transformant line KO77 (Tri and Kamei 2020). The strain KO77 was created by knocking out of the pyruvate decarboxylase (pdc) gene from host strain MG-60-P2, and this resulted in the fungal KO77 strain, accumulating glucose from cellulose rather than fermenting it to ethanol. In the study, the butanol-producing *C. saccharoperbutylacetonicum* was separately co-cultured with P2 and KO77 fungal strains, with unbleached hardwood kraft pulp (UHKP), a paper industry waste, as the feedstock. The KO77 strain produced a higher amount of butanol (3.2 g/L) than that produced from the P2 strain (2.5 g/L butanol) as well as enhanced saccharification, showing the synergistic effects of the consortium on the production of butanol by CBP (Tri and Kamei 2020).

Furthermore, many symbiotically associated microorganisms have been isolated from the environment that can produce butanol by utilizing lignocellulose (Jiang et al. 2018b). Since it continues to be challenging to modulate microbes in a consortium owing to their intricate build-up and the constraints of accessible genetic tools, the nonmatching incubation temperature of the contributing strains can result in a long, non-isothermal fermentation and, consequently, reduced productivity. Hence, it is crucial to choose fitting hosts for the development of a consortium; and it can be productive to upgrade the performance of synthetic consortia via metabolic modeling, genetic engineering, and optimization of operational factors.

Studies are being conducted to create the mutant strains with engineered butanol metabolic networks. For instance, *C. cellulolytic* was metabolically engineered to introduce the CoA-dependent metabolic pathway from *C. acetobutylicum*, to produce n-butanol with crystalline cellulose as a substrate. However, the productivity was very low (<0.12 g/L) (Gaida et al. 2016). Higashide et al. (2012) diverted the 2-keto acid intermediates of a valine pathway toward isobutanol production in *C. cellulolyticum,* but still yielded only 0.66 g/L of isobutanol (Higashide et al. 2012). Lin et al. (2015) expressed vital genes of an isobutanol pathway under distinct promoters into *C. thermocellum,* and the most efficient strain amongst those engineered produced 5.4 g/L of isobutanol from crystalline cellulose, analogous to 41% of theoretical yield (Lin et al. 2015). Likewise, *Clostridium cellulovorans* DSM 743B was engineered by introducing a coenzyme A (CoA)-dependent acetone-butanol-ethanol (ABE) pathway from yet another *Clostridium strain (C. acetobutylicum* ATCC 824) into it. Next, the adaptive laboratory evolution (ALE) methodology was integrated to improve the production of *n*-butanol *by C. cellulovorans* from deshelled corn cobs (AECC), with a final *n*-butanol production of 3.47 g/L (Wen et al. 2019). Nevertheless, this approach still required optimization, as residuals of butyrate and xylose in good enough amounts were left unutilized in the medium after 84 h of fermentation. Ideally, any successful fermentation process needs butyrate and xylose to be converted to the final product, i.e., n-butanol with highest of the conversion efficiency.

Indeed, the same strain of *C. cellulovorans* was cultivated in a microbial consortium with *C. beijerinckii*, where during co-cultivation, 11.5 g/L of butanol productivity was obtained from 83.2 g/L of deshelled corn cobs extracted with alkali (AECC) (Wen et al. 2017). However, because *C. cellulovorans* was not able to grow in the required pH range (pH 4.5–5.5), the authors of the study used a scheme to adjust the pH in two stages. Therein, initially the pH was maintained at 7.0 to stimulate the growth and secretion of cellulolytic enzymes by *C. cellulovorans*, and then, the pH was kept below 7.0 (preferring acid conditions) to promote ABE fermentation for butanol production by *C. beijerinckii* (Zhiqiang Wen et al. 2017). To avoid this complex CBP implementation, the authors recently engineered *Clostridium cellulovorans* DSM 743B to make it tolerant to acidic pH (below 6.0) by inactivating cell wall lyase genes (Wen et al. 2020b). Furthermore, in the same study, Wen et al. also transferred an alcohol aldehyde dehydrogenase gene (adgE1) from *C. acetobutylicum* ATCC 824 into *C. cellulovorans* DSM 743B to render DSM 743B an ability to produce butanol under acidic pH. This was a strategy that enabled joint production of n-butanol minus any pH adjustment, from the consortium of *C. cellulovorans* DSM 743B and *C. beijerinckii*, and the reported titer of butanol at 3.94 g/L was found to be five times the butanol production attained by their wild-type versions under the same conditions (Wen et al. 2020b). This research group then went a step further, with another innovative metabolic strategy, wherein the carbon flux was shifted from acetyl-CoA toward butanol synthesis pathways, yielding a butanol optimum at 4.96 g/L of from alkali extracted corn cobs (AECC) (Wen et al. 2020a).

Bao et al. (2019) simply overexpressed the alcohol dehydrogenase gene in *C. cellulovorans* and were able to achieve a butanol productivity of 4.0 g/L (Bao et al. 2019), while Wang et al. (2020) overexpressed *Clostridium acetobutylicum's* indigenous xylanase gene (xynB). The overexpression of this gene increased the exogenous xylanase activity 88 fold, from 0.09 to 7.93 U/mL with hemicellulose as the substrate, and yielded 4.03 g/L of butanol. These yields of approximately 4.0 g/L of butanol obtained in studies by Bao et al. (2019) and Wang et al. (2020) are the highest titers of *n*-butanol reported from a monoculture within a CBP setup to date (Bao et al. 2019; Wang et al. 2020).

Thus, butanol production as a biofuel and an industrial solvent from lignocellulosic biomass is developing continuously as new processing approaches are followed to improve the yield and productivity. *Clostridial* species are majorly exploited in butanol production, with co-cultivation and metabolic engineering methods being used to improve the processing technique towards large scale production as a second-generation biofuel. The advancement of the omics technologies has strengthened the existing knowledge of researchers with respect to the genome and phenomes of the potential n-butanol-producing *Clostridium* strains, and much benefit can be gained from the availability of advanced genomic toolkits for genetic manipulation of the microbes for n-butanol/isobutanol production from LCB by CBP. Some of the recently published reviews on the topic, by Wen et al. (2020a, b, c) and Ferreira et al. (2020), summarize the improvement in the field of metabolic modeling and other strategies to butanol production from varied lignocellulosic feedstocks by CBP. Nevertheless, given that there has only been partial realization of generating n-butanol at commercially significant volumes, substantial headway is needed to achieve an economical cell factory for this compound, in which yield and cost-effective production can compete favorably with petrochemical production.

2.2.3 Biohydrogen Production

Hydrogen is a clean, zero-emission fuel that when burnt, in fuel cells or internal combustion engines with oxygen, produces only water. Hydrogen is an energy carrier, 2.2 lb of which holds the equivalent quantity of energy as a gallon or 6.2 lb of gasoline (Nagarajan et al. 2019). This has made hydrogen an attractive alternative for fossil fuels, and the most sustainable energy and carbon source for hydrogen production are the biological feedstocks (Nagarajan et al. 2019). Their bacterial degradation under anaerobic conditions can produce biohydrogen via direct or indirect bio-photolysis, photo-fermentation, and dark fermentation, of which only the latter do not need the input of light energy (Rittmann and Herwig 2012). Under aerobic conditions, oxygen functions as an electron acceptor releasing water. Theoretically, biohydrogen yield from 1 mole of glucose via dark fermentation is limited to only 2 mol of hydrogen from facultative anaerobes vs. 4 mols of hydrogen from obligate thermophilic anaerobes (Nagarajan et al. 2019), but the maximum hydrogen yield or substrate conversion efficiency, expressed as ($Y(H_2/S)$ accomplished in dark fermentation employing complex carbon polysaccharides, is in the range of 1.1–2.6 mol H_2 per mole substrate, which signifies nearly 30–33% of the theoretical upper limit. The metabolic constraints of cellular physiology dictate these reduced yields, and statistically based evidence in a study by Rittmann and Herwig (2012) shows that for a similar carbon source, the $Y(H_2/S)$ is superior at thermophilic specifications in contrast to that under mesophilic conditions. According to the study, the average levels of hydrogen generated by thermophiles (Thermoanaerobacterales), obligate anaerobes (Clostridiaceae), and facultative anaerobes (Enterobacteriaceae) are estimated to be 2.92, 1.87, and 1.15 mol H_2 per mole glucose, respectively (Rittmann and Herwig 2012). However, because of the higher hydrogen evolution rate/volumetric productivity [(HER); mmol/L/h] as well as superior biological production capacity [qH_2, mmol g^{-1} h^{-1}] associated with mesophilic hydrogen production, it is the mesophilic systems that are being mostly studied and are currently exploited for the synthesis.

To date, it is the anaerobic members from the genus *Clostridium, Pseudomonas, Aeromonas,* and *Bacillus* that are the most successful hydrogen producers under mesophilic conditions. On the other hand, *Thermoanaerobacter* sp., *Caldicellulosiruptor sp., Thermotoga* sp., and a number of thermophilic *Clostridium* strains such as *C. thermocellum* are successful thermophilic hydrogen-fermenting bacteria (Nagarajan et al. 2019; Saleem et al. 2020). Furthermore, other than pure cultures, mixed cultures acquired from environmental reserves, such as anaerobic sludge of various wastewater treatment plants like sewage sludge and palm oil mill effluent (POME) sludge, have also been explored for hydrogen production, with these samples providing a plethora of indigenous anaerobic fermentative bacteria (Kumar et al. 2016).

Recently, a study was conducted to isolate two anaerobic bacterial strains, RTUA and RTUB, capable of hydrolyzing cellulose and hemicellulose with simultaneous hydrogen production (Saleem et al. 2020), and the efficacy of these strains to valorize complex substrates of rotten fruits and vegetables and wheat straw was tested. Initially, during growth on monomeric sugars, the strain RTUA showed higher hydrogen yield with glucose (2.69 H_2 mol/mol substrate) and RTUB with fructose (1.23 H_2 mol/mol substrate). When cultivated with complex substrates, RTUA and RTUB provided comparable values with rotten vegetables (~0.70 H_2 mol/mol substrate), rotten fruits (~0.40 H_2 mol/mol substrate), and untreated wheat straw (0.39 H_2 mol/mol substrate). When the wheat straw biomass was pretreated with *Trichoderma,* RTUA produced a 1.16 H_2 mol/g wheat straw, while RTUB produced 1.05 H_2 mol/mol of the substrate. With NaOH-pretreated biomass, maximum yields were seen, with 2.04 H_2 mol/mol substrate and 1.42 H_2 mol/g substrate for RTUB. RTUA produced maximum hydrogen with NaOH-treated wheat straw as a substrate (156.4 mL H2/g VS) (Saleem et al. 2020). This study defined the concept of utilizing diverse groupings of complex polysaccharides and food wastes for biohydrogen production under the concept of CBP.

In the production of biohydrogen from LCB, thermophiles are generally considered suitable modes for CBP because of their ability to hydrolyze lignocellulose and to produce biofuels simultaneously and efficiently. One such extreme thermophile, *Caldicellulosiruptor saccharolyticus* strain DSM 8903 (temperature: 70 °C), has been reported to produce hydrogen on various agricultural wastes such as wheat straw, switchgrass, bagasse, and maize leaves, without a prior physical, chemical, or biological pre-treatment (Talluri et al. 2013). In their study, DSM 8903 generated 4 moles of H_2 from 1 mole of pure glucose. The reported yield of hydrogen per gram of unprocessed switchgrass and microcrystalline cellulose was 11.2 and 9.4 mmol, respectively, with the significant by-product being acetate in both the cases (Talluri et al. 2013). Moreover, another hyperthermophilic bacteria belonging to the *Caldicellulosiruptor,* i.e., *C. bescii, has been demonstrated to economically consume biosolids as the only available carbon source. A* H_2 yield of 4.40 mmol/g volatile solid added was achieved in this study, and the authors claimed this yield to be the highest dark fermentative H_2 yield achieved from biosolids to date (Yilmazel et al. 2015).

In another paper, a thermophilic *Thermoanaerobacterium sp.* strain F6 was also able to produce hydrogen by direct utilization of various hemicellulose- and cellulose-rich substrates along with complex lignocellulosic biomass such as corn cob and sugarcane bagasse (Jiang et al. 2019). When F6 strain was cultivated with xylan, a maximum of 370.70 ± 1.59 mmol/l of hydrogen from 60 g/L substrate was reported by the authors. When corncob was investigated as a substrate, 66.71 ± 1.80 mmol/L H_2 was shown to be accumulated from 60 g/L of the corncob. To further investigate the ability of the strain to utilize complex LCBs, a sugarcane bagasse was used as a substrate, but the reported yield of hydrogen was much lower, at 30.24 ± 1.65 mmol/L from 30 g/L of the substrate (Jiang et al. 2019). Cao and co-workers obtained yields of 3.47 mmol, 3.53 mmol, and 3.23 mmol of hydrogen from each gram of corn stalk, wheat straw, and corn stalk, respectively, with *Thermoanaerobacterium thermosaccharolyticum* M18 in a single pot consolidated bioprocessing (Cao et al. 2014).

In the perspective of biohydrogen generation, another group of naive thermophilic cellulolytic bacteria can produce hydrogen as a by-product of their anaerobic metabolism, belonging to the genus *Clostridia. Clostridium thermocellum* ATCC 27405 was studied for batch synthesis of hydrogen by utilizing distinct cellulosic sources (α-cellulose, shredded filter paper, cellobiose, and delignified wood fibers (DLWs)) (Levin et al. 2006). In this study, DLW was shown to be an effective carbon source delivering an average yield of 1.6 mol H_2/mol glucose, with acetate, ethanol, lactate, and formate as the major fermentation end products. *This same* thermophilic bacterium strain, *Clostridium thermocellum* ATCC 27405, was later shown to grow on waste date seeds to produce hydrogen via consolidated processing (Rambabu et al. 2020). The effects of various surfactants and buffering agents were tested on hydrogen production. As a surfactant, the supplementation of Triton X-100 at an optimum dose of 0.75% w/v in the medium has been shown to enhance the hydrogen production by 30%. This increased to 33.2% when the authors added 15 mM of sodium carbonate in the test medium. Together, the addition of Triton X-100 and sodium carbonate at their most optimal concentrations augmented the maximum achieved titer of hydrogen to 146.19 mmol/l (productivity of 0.443 mmol/g/h) from waste date seeds, representing a synergistic increase of 40.6% (Rambabu et al. 2020). An analogous observation was made in a study by Tian et al. (2015), where the authors used *Clostridium thermocellum* ATCC 27405 to degrade unprocessed sugarcane bagasse (SCB) for hydrogen production, and it was reported that hydrogen yield was appreciably improved by complementing the medium with $CaCO_3$ (Tian et al. 2015). Thus, both these papers (Tian et al. 2015; Rambabu et al. 2020) provided some evidence that optimal dosage of additives can enhance the biohydrogen production from a lignocellulosic precursor, and such studies provide a unique approach to improve biohydrogen production from LCB.

As LCB is generally challenging to hydrolyze, co-cultures were also tested to boost lignocellulose conversion and increase the hydrogen yield. For instance, an aerobic bacterium, *Geobacillus* sp. strain WSUCF1, was sequentially cultured with an anaerobic consortium (*Thermoanaerobacterium*-98.19%, *Clostridia*, and *Geobacillus*) to increase the biohydrogen yield in the CBP system (Bibra et al. 2018). The anaerobic consortium alone, when cultured with 2% w/v prairie cordgrass (PCG) yielded 2.2 mmol of hydrogen per gram of PCG. However, this consortium's sequential growth with strain WSUCF1 yielded 3.74 mmol hydrogen/g PCG, making it a suitable system for low-cost biohydrogen production (Bibra et al. 2018). de Vrije et al. (2009) even obtained enhanced hydrogen yield during the growth of the cellulolytic bacterium *Caldicellulosiruptor saccharolyticus* with the moderately thermophilic halophilic bacterium *Thermotoga neapolitana* from *Miscanthus*, a perennial C_4 grass (de Vrije et al. 2009).

The utility of thermophilic strains to produce biohydrogen with human waste stimulants as a substrate was demonstrated in 2018 in an innovative approach (Wang et al. 2018). In this experiment, four thermophilic consortia were developed by handling the samples from hot springs (C1), wastewater treatment plants (C2), and landfill compost (C3). The third consortium, mainly dominated by *Thermoanaerobacterium sp.,* followed by *Caloribacterium,* produced the maximum amount of H_2, 88.18 ± 3.79 mL with a production rate of 0.30 ± 0.03 mmol H_2/L/h with space crew human waste stimulants as the substrate. The consortium C3 was further studied for optimum conditions, and the authors reported a yield of 3.99 mmol/g of H_2 at pH 7.0 and temperature 60 °C (Wang et al. 2018). This research can aid in creating a self-sustainable and economic system for long-term space missions. Overall, these studies prove the effectiveness of using sequential cultivation techniques to produce biohydrogen with high yields and low production rates from unprocessed lignocelluloses by CBP and also establish the importance of thermophiles to produce hydrogen from LCB in a single step.

Nevertheless, even some mesophilic strains have been tested for the ability to produce hydrogen from agri-based substrates. *Clostridium* strain BOH3, a mesophilic strain, was tested for the concurrent production of butanol (13.50 ± 0.12 g/L) and hydrogen (4.41 ± 0.04 L/L) from rice bran and sesame oil cake (Rajagopalan et al. 2016). In the latest research, the same bacterium, *Clostridium* sp. strain BOH3, was studied for biohydrogen production only with fruit wastes as a carbon source (Mahato et al. 2020). These studies confirmed the ability of *Clostridium* strain BOH3 as a CBP model and established clostridial strains as highly suitable for large-scale hydrogen production by CBP.

Since cost-competitive hydrogen production with environment-friendly methods is essential to promote hydrogen as fuel in the transport sector, consolidated bioprocessing is considered one of the lowest-cost methods to reuse the agricultural, municipal, and food wastes as lignocellulosic feedstock to produce hydrogen. Furthermore, from this section, it is evident that rapid developments and innovative approaches are being made to increase CBP's biohydrogen production. With these continuing efforts, we can expect a shift to a zero-carbon economy in which biohydrogen is a key component.

2.2.4 *Other Products*

Lignocellulosic feedstocks can be utilized to biosynthesize a sizable number of chemicals. This section will investigate some examples of organic chemicals (isopropanol, butyric acid, lactic acid, diacids, and fatty acids) that can be produced from lignocellulosic biomass, either individually or as co-products with biofuels.

2.2.4.1 Isopropanol

Isopropanol is a significant chemical used for formulating superior octane gasoline and diesel oil, and it is also one of the significant precursors used for propylene production, making it an essential chemical in the energy field and other industries (Liu et al. 2019). Xin et al. (2017) utilized *Clostridium* sp. strain NJP7 for improved butanol (2.06 g/L) as well as isopropanol (0.54 g/L) production via a fermentative acetone-isopropanol-butanol pathway from birchwood xylan (Xin et al. 2017). In a study conducted by Liu et al. (2019), 7 g/L of isopropanol was achieved from a bacterial consortium, EMSD5 (composed of members from genus *Bacillus, Clostridium, Escherichia,* and *Lysinibacillus*), grown with corncob as the carbon source (Liu et al. 2019). This research demonstrated the capability of the consortium to produce isopropanol utilizing lignocellulosic biomass.

2.2.4.2 Butyric Acid

Butyric acid is a valuable chemical used in the chemical, textile, food, pharmaceutical, and bioplastic industry and is produced from lignocellulosic biomass. Delignified rice straw was reported as a feedstock for the production of 33.9 g/L of butyric acid by co-culture of *Clostridium thermocellum* ATCC 27405 and *C. thermobutyricum* ATCC 49875 (Chi et al. 2018). By examining the co-cultures' metabolism pathways, it was found that during this twin-clostridial growth, there was a push of carbon flux in *C. thermobutyricum* ATCC 49875 towards butyric acid formation (Chi et al. 2018). In another study, Ai et al. (2016) have reported a productivity of 16.2 g/L of butyric acid from rice straw pretreated with sodium hydroxide, when grown with an unidentified cellulose degrading yet butyrate-producing microbial community (Ai et al. 2016). These studies indicate that CBP is also an effective process for the low-cost production of secondary metabolites from lignocellulosic feedstock without supplementary cellulolytic enzymes.

2.2.4.3 Lactic Acid

Another organic acid, lactic acid, is another in-demand chemical used in the food and chemical industries. An artificial cross-kingdom consortium consisting of an aerobic fungus *Trichoderma reesei* (that secretes cellulolytic enzymes) with *Lactobacilli* was studied for bioconversion of LCB to lactic acid in a CBP setup (Shahab et al. 2018). The authors of the study achieved a productivity of 34.7 g/L of lactic acid, when the fermentation experiments with microcrystalline cellulose as a substrate were conducted. During further experiments with nondetoxified beech wood that was pretreated with steam, the study reported lactic acids productivity of 19.8 g/L (Shahab et al. 2018). This is the only study that establishes the value of a consortium-based CBP technique to produce lactic acid inexpensively from lignocellulosic substrate. The rest of the published studies to produce lactic acid from lignocelluloses have used a simultaneous saccharification and fermentation (SSF) methodology that needs supplementation of expensive cellulolytic enzymes.

2.2.4.4 4-Carbon Acids

Other than mono acids, four carbon diacids, such as malic and succinic acids, are also essential chemicals, especially in biopolymer industries. In one study, *Myceliophthora thermophila*, a cellulolytic thermophilic fungus, was metabolically engineered to convert lignocellulosic biomass to C4 diacids (Li et al. 2020). The genes Aopyc (pyruvate carboxylase), Aomdh (malate dehydrogenase), and Aomae (malate transporter) from *Aspergillus oryzae* were engineered into the fungus, thus constructing the reductive tricarboxylic acid (rTCA) pathway for diacids' production. In initial studies with crystalline cellulose (Avicel) as the substrate, the productivity of C4-diacids in batch fermentation reported in the study was 72.6 g/L (65.4 g/L malic acid and 7.2 g/L succinic acid). The corresponding output of malic acid and succinic acid reached 181 g/L, and 19.7 g/L, respectively (total diacid titer 200.7 g/L), when the fermentation of crystalline cellulose by *Myceliophthora thermophila was carried out in a* 5-L fed-batch fermentation vessel (Li et al. 2020). Furthermore, when corncob was used as a substrate, the study reports that 110.4 g/L C4-diacids was produced (105 g/L malic acid and 5.4 g/L succinic acid) with a productivity of 0.40 g/g corncob. Thus, it was established that CBP techniques could also be used to produce bulk chemicals that can be beneficial in various chemical and biological industries (Li et al. 2020).

2.2.4.5 By-Products

Sometimes essential chemicals are produced as by-products along with biofuels by a single fermentation system. For instance, an integrated CBP process was developed with a cross-kingdom consortium comprising *C. thermocellum* ATCC 31924 strain and a marine microalgal strain *Schizochytrium sp.* DT3 for bioethanol as well as omega-3 fatty acids production simultaneously, with pretreated rice straw biomass as a substrate (Singh et al. 2020). In the two-step bioprocess adopted in the study, (1) a pretreated rice straw biomass was fermented anaerobically with *C. thermocellum* ATCC 31924 to generate 1.8 g/L bioethanol, accompanied by 29.40% solubilization of rice straw biomass, and (2) *Schizochytrium sp.* DT3 was grown aerobically on the spent rice straw-derived sugars, yielding docosahexaenoic acid, docosapentaenoic acid, palmitic acid, eicosapentaenoic acid, and stearic acid, at 44%, 16.12%, 13.95%, 7.24%, and 5.07% of total fatty acid, respectively (Singh et al. 2020). This process is not yet commercialized, but the development of such processes has significant commercial potential in the biorefinery industry.

The competent production of ABE solvents (acetone, butanol, and ethanol) via fermentation from alkali-extracted, deshelled corn cobs (AECC) was achieved by developing a CBP comprising of a consortium of metabolically engineered *Clostridium cellulovorans* DSM 743B and *Clostridium beijerinckii* NCIMB 8052 (Wen et al. 2017). An MMME (multivariate modular metabolic engineering) strategy was used in this study to engineer the twin-clostridial consortium, where (1) the strain *C. cellulovorans* DSM 743B was engineered to suppress the lactic acid pathway and diverge the carbon flux for butyric acid production, and (2) bacterium *C. beijerinckii* NCIMB 8052 was modified to enhance ethanol production via enhanced assimilation of organic acids and pentose sugars produced during the process. When this engineered consortium was grown with AECC, 22.1 g/L of solvents, represented by acetone, butanol, and ethanol at 4.25 g/L,11.5 g/L, and 6.37 g/L, respectively, was produced from 83.2 g/L of substrate (Wen et al. 2017). Thus, the developed consortium qualified as a model for effective ABE fermentation simultaneously producing acetone, ethanol, and butanol from lignocellulosic feedstock by CBP. In another study, a recently isolated *Clostridium* sp. strain NJP7 capable of ABI fermentation from hemicellulose was characterized (Xin et al. 2017). When cultivated with xylan as the primary substrate, the *Clostridium* sp. strain NJP7 showed the ability to degrade xylan and ferment sugars to solvents. It produced 5.84 g/L of total solvents, comprising acetone, butanol, and isopropanol (ABI) at 3.24 g/L, 2.06 g/L, and 0.54 g/L, respectively. The authors claim these concentrations to be the highest ABI biosynthesis levels from lignocellulose reported to date (Xin et al. 2017). It can be inferred that other beneficial chemicals may be produced from lignocellulosic biomass through consolidated bioprocessing, either as the primary product or as a biofuel bioproduct.

2.3 Conclusion and Perspectives

Second-generation biofuels from lignocellulosic biomass have a high potential to overcome dependency on fossil fuels and lead to a cleaner environment with greenhouse gas reduction. These fuels can be synthesized by thermochemical conversion and microbial fermentation, with the latter taking center stage with various new techniques and advancements. Even though there are multiple microbial fermentation methods, consolidated bioprocessing (CBP) is the lowest cost and most effective way to produce second-generation biofuels from lignocellulosic biomasses. For CBP to work, simultaneous saccharification and fermentation of the biomass are required. From recently reported data, it can be inferred that thermophiles have gathered more interest in the quest of increased biofuels yield and productivity because of their ability to both degrade lignocellulosic biomass and produce biofuels efficiently. Much interest is also developing in the concept of co-culture engineering for efficient single pot conversion, where different single microbial strains are selected for their specialized efficiency in one of these functions, while lacking capability in the other function. Thus, at least one of the participating strains is devoted to the production of LCB depolymerizing enzymes, and the other microorganism converts the released sugars to the final product. Moreover, the use of genetic engineering to modify the CBP strains' metabolism, after detailed study of the associated pathways and its metabolic flux modeling, will result in microbes that can deliver unique natural biofuels on a large scale. In this chapter, several examples were provided which support the feasibility of these strategies for improving product titer, yield, and productivity. To enable large-scale application and commercialization, this technology requires further development to achieve higher efficiency and robustness, but the current high pace of progress strongly suggests that this technology will dominate the biofuel industry in the coming decades.

Acknowledgments We acknowledge the support from the CNAM-Bio Center, supported by the South Dakota Governor's Office of Economic Development. The authors also gratefully acknowledge the support from the National Science Foundation (Award #1736255, #1849206, and #1920954) and the Department of Chemical and Biological Engineering at the South Dakota School of Mines and Technology.

Conflicts of Interest The authors declare no conflict of interest.

References

Ai B, Chi X, Meng J, Sheng Z, Zheng L, Zheng X, Li J (2016) Consolidated bioprocessing for butyric acid production from rice straw with undefined mixed culture. Front Microbiol 7:1648

Akinosho H, Yee K, Close D, Ragauskas A (2014) The emergence of *Clostridium thermocellum* as a high utility candidate for consolidated bioprocessing applications. Front Chem 2:66–66

Althuri A, Mohan SV (2020) Sequential and consolidated bioprocessing of biogenic municipal solid waste: a strategic pairing of thermophilic anaerobe and mesophilic microaerobe for ethanol production. Bioresour Technol 308:123260

Antoni D, Zverlov VV, Schwarz WH (2007) Biofuels from microbes. Appl Microbiol Biotechnol 77(1):23–35

Bao T, Zhao J, Li J, Liu X, Yang S-T (2019) n-Butanol and ethanol production from cellulose by *Clostridium cellulovorans* overexpressing heterologous aldehyde/alcohol dehydrogenases. Bioresour Technol 285:121316

Bibra M, Kumar S, Wang J, Bhalla A, Salem DR, Sani RK (2018) Single pot bioconversion of prairie cordgrass into biohydrogen by thermophiles. Bioresour Technol 266:232–241

Bibra M, Rathinam NK, Johnson GR, Sani RK (2020) Single pot biovalorization of food waste to ethanol by *Geobacillus* and *Thermoanaerobacter* spp. Renew Energy 155:1032–1041

Cao G-L, Zhao L, Wang A-J, Wang Z-Y, Ren N-Q (2014) Single-step bioconversion of lignocellulose to hydrogen using novel moderately thermophilic bacteria. Biotechnol Biofuels 7(1):82

Cao G, Sheng Y, Zhang L, Song J, Cong H, Zhang J (2016) Biobutanol production from lignocellulosic biomass: prospective and challenges. J Bioremed Biodegr 7(4):363

Chi X, Li J, Wang X, Zhang Y, Antwi P (2018) Hyper-production of butyric acid from delignified rice straw by a novel consolidated bioprocess. Bioresour Technol 254:115–120

de Vrije T, Bakker RR, Budde MAW, Lai MH, Mars AE, Claassen PAM (2009) Efficient hydrogen production from the lignocellulosic energy crop Miscanthus by the extreme thermophilic bacteria *Caldicellulosiruptor saccharolyticus* and *Thermotoga neapolitana*. Biotechnol Biofuels 2(1):12

Ferreira S, Pereira R, Wahl SA, Rocha I (2020) Metabolic engineering strategies for butanol production in Escherichia coli. Biotechnol Bioeng 117(8):2571–2587

Fond O, Petitdemange E, Petitdemange H, Engasser J (1983) Cellulose fermentation by a coculture of a mesophilic cellulolytic *Clostridium* and *Clostridium acetobutylicum*

Gaida SM, Liedtke A, Jentges AHW, Engels B, Jennewein S (2016) Metabolic engineering of *Clostridium cellulolyticum* for the production of n-butanol from crystalline cellulose. Microb Cell Factories 15(1):6

GlobeNewsWire (2020) Global bioethanol market (2020 to 2025)—rising demand from the pharmaceutical industry presents opportunities. In: Research and markets

Govil T, Wang J, Samanta D, David A, Tripathi A, Rauniyar S, Salem DR, Sani RK (2020) Lignocellulosic feedstock: a review of a sustainable platform for cleaner production of nature's plastics. J Clean Prod 270:122521

Higashide W, Li Y, Yang Y, Liao JC (2012) Metabolic engineering of *Clostridium cellulolyticum* for production of isobutanol from cellulose. Appl Environ Microbiol 78(19):7171–7171

Hirano K, Kurosaki M, Nihei S, Hasegawa H, Shinoda S, Haruki M, Hirano N (2016) Enzymatic diversity of the *Clostridium thermocellum* cellulosome is crucial for the degradation of crystalline cellulose and plant biomass. Sci Rep 6(1):35709

Ibrahim MF, Kim SW, Abd-Aziz S (2018) Advanced bioprocessing strategies for biobutanol production from biomass. Renew Sust Energ Rev 91:1192–1204

Jiang Y, Guo D, Lu J, Dürre P, Dong W, Yan W, Zhang W, Ma J, Jiang M, Xin F (2018a) Consolidated bioprocessing of butanol production from xylan by a thermophilic and butanologenic Thermoanaerobacterium sp. M5. Biotechnol Biofuels 11(1):89

Jiang Y, Zhang T, Lu J, Dürre P, Zhang W, Dong W, Zhou J, Jiang M, Xin F (2018b) Microbial co-culturing systems: butanol production from organic wastes through consolidated bioprocessing. Appl Microbiol Biotechnol 102(13):5419–5425

Jiang Y, Lu J, Lv Y, Wu R, Dong W, Zhou J, Jiang M, Xin F (2019) Efficient hydrogen production from lignocellulosic feedstocks by a newly isolated thermophlic *Thermoanaerobacterium* sp. strain F6. Int J Hydrog Energy 44(28):14380–14386

Jones DT, Woods DR (1986) Acetone-butanol fermentation revisited. Microbiol Rev 50 (4):484–524

Keis S, Shaheen R, Jones DT (2001) Emended descriptions of *Clostridium acetobutylicum* and *Clostridium beijerinckii*, and descriptions of *Clostridium saccharoperbutylacetonicum* sp. nov. and *Clostridium saccharobutylicum* sp. nov. Int J Syst Evol Microbiol 51(Pt 6):2095–2103

Kumar G, Mudhoo A, Sivagurunathan P, Nagarajan D, Ghimire A, Lay C-H, Lin C-Y, Lee D-J, Chang J-S (2016) Recent insights into the cell immobilization technology applied for dark fermentative hydrogen production. Bioresour Technol 219:725–737

Levin DB, Islam R, Cicek N, Sparling R (2006) Hydrogen production by Clostridium thermocellum 27405 from cellulosic biomass substrates. Int J Hydrog Energy 31(11):1496–1503

Levin DB, Verbeke TJ, Munir R, Islam R, Ramachandran U, Lal S, Schellenberg J, Sparling R (2015) Chapter 17—omics approaches for designing biofuel producing cocultures for enhanced microbial conversion of lignocellulosic substrates. In: Himmel ME (ed) Direct microbial conversion of biomass to advanced biofuels. Elsevier, Amsterdam, pp 335–363

Li J, Lin L, Sun T, Xu J, Ji J, Liu Q, Tian C (2020) Direct production of commodity chemicals from lignocellulose using *Myceliophthora thermophila*. Metab Eng 61:416–426

Lin PP, Mi L, Morioka AH, Yoshino KM, Konishi S, Xu SC, Papanek BA, Riley LA, Guss AM, Liao JC (2015) Consolidated bioprocessing of cellulose to isobutanol using *Clostridium thermocellum*. Metab Eng 31:44–52

Linger JG, Darzins A (2013) Consolidated bioprocessing. In: Lee JW (ed) Advanced biofuels and bioproducts. Springer, New York, pp 267–280

Liu L, Yang J, Yang Y, Luo L, Wang R, Zhang Y, Yuan H (2019) Consolidated bioprocessing performance of bacterial consortium EMSD5 on hemicellulose for isopropanol production. Bioresour Technol 292:121965

Liu Y-J, Li B, Feng Y, Cui Q (2020) Consolidated bio-saccharification: leading lignocellulose bioconversion into the real world. Biotechnol Adv 40:107535

Mahato RK, Kumar D, Rajagopalan G (2020) Biohydrogen production from fruit waste by Clostridium strain BOH3. Renew Energy 153:1368–1377

Mbaneme-Smith V, Chinn MS (2015) Consolidated bioprocessing for biofuel production: recent advances. Energy Emission Control Technol 3:23

Nagarajan D, Lee D-J, Chang J-S (2019) Recent insights into consolidated bioprocessing for lignocellulosic biohydrogen production. Int J Hydrog Energy 44(28):14362–14379

Rajagopalan G, He J, Yang K-L (2016) One-pot fermentation of agricultural residues to produce butanol and hydrogen by Clostridium strain BOH3. Renew Energy 85:1127–1134

Rajak RC, Banerjee R (2020) An innovative approach of mixed enzymatic venture for 2G ethanol production from lignocellulosic feedstock. Energy Convers Manag 207:112504

Rambabu K, Show P-L, Bharath G, Banat F, Naushad M, Chang J-S (2020) Enhanced biohydrogen production from date seeds by *Clostridium thermocellum* ATCC 27405. Int J Hydrog Energy 45:22271–22280

Rittmann S, Herwig C (2012) A comprehensive and quantitative review of dark fermentative biohydrogen production. Microb Cell Factories 11(1):115

Saleem A, Umar H, Shah TA, Tabassum R (2020) Fermentation of simple and complex substrates to biohydrogen using pure *Bacillus cereus* strains. Environ Technol Innov 18:100704

Scully SM, Orlygsson J (2015) Recent advances in second generation ethanol production by thermophilic Bacteria. Energies 8(1):1–30

Shahab RL, Luterbacher JS, Brethauer S, Studer MH (2018) Consolidated bioprocessing of lignocellulosic biomass to lactic acid by a synthetic fungal-bacterial consortium. Biotechnol Bioeng 115(5):1207–1215

Singh N, Mathur AS, Gupta RP, Barrow CJ, Tuli D, Puri M (2018a) Enhanced cellulosic ethanol production via consolidated bioprocessing by *Clostridium thermocellum* ATCC 31924☆. Bioresour Technol 250:860–867

Singh, N., Puri, M., Tuli, D.K., Gupta, R.P., Barrow, C.J., and , Mathur, A.S. 2018b. Bioethanol production by a xylan fermenting thermophilic isolate *Clostridium* strain DBT-IOC-DC21. Anaerobe, 51, 89–98

Singh N, Gupta A, Mathur AS, Barrow C, Puri M (2020) Integrated consolidated bioprocessing for simultaneous production of omega-3 fatty acids and bioethanol. Biomass Bioenergy 137:105555

Talluri S, Raj SM, Christopher LP (2013) Consolidated bioprocessing of untreated switchgrass to hydrogen by the extreme thermophile *Caldicellulosiruptor saccharolyticus* DSM 8903. Bioresour Technol 139:272–279

Tian Q-Q, Liang L, Zhu M-J (2015) Enhanced biohydrogen production from sugarcane bagasse by *Clostridium thermocellum* supplemented with CaCO3. Bioresour Technol 197:422–428

Tian L, Papanek B, Olson DG, Rydzak T, Holwerda EK, Zheng T, Zhou J, Maloney M, Jiang N, Giannone RJ, Hettich RL, Guss AM, Lynd LR (2016) Simultaneous achievement of high ethanol yield and titer in *Clostridium thermocellum*. Biotechnol Biofuels 9(1):116

Toor M, Kumar SS, Malyan SK, Bishnoi NR, Mathimani T, Rajendran K, Pugazhendhi A (2020) An overview on bioethanol production from lignocellulosic feedstocks. Chemosphere 242:125080

Tri CL, Kamei I (2020) Butanol production from cellulosic material by anaerobic co-culture of white-rot fungus Phlebia and bacterium Clostridium in consolidated bioprocessing. Bioresour Technol 305:123065

Wang J, Bibra M, Venkateswaran K, Salem DR, Rathinam NK, Gadhamshetty V, Sani RK (2018) Biohydrogen production from space crew's waste simulants using thermophilic consolidated bioprocessing. Bioresour Technol 255:349–353

Wang Z, Cao X, Li N, Yang Z, Lei M, Zhao Y, Wang L, Li Z, Liu D, Niu H, Ying H (2020) Production of butanol directly from hemicellulose through secretory expression of a xylanase in *Clostridium acetobutylicum*. Energy Fuel 34(3):3376–3382

Wen Z, Minton NP, Zhang Y, Li Q, Liu J, Jiang Y, Yang S (2017) Enhanced solvent production by metabolic engineering of a twin-clostridial consortium. Metab Eng 39:38–48

Wen Z, Ledesma-Amaro R, Lin J, Jiang Y, Yang S (2019) Improved n-butanol production from *Clostridium cellulovorans* by integrated metabolic and evolutionary engineering. Appl Environ Microbiol 85(7):e02560–e02618

Wen Z, Ledesma-Amaro R, Lu M, Jin M, Yang S (2020a) Metabolic engineering of *Clostridium cellulovorans* to improve butanol production by consolidated bioprocessing. In: ACS synthetic biology

Wen Z, Ledesma-Amaro R, Lu M, Jiang Y, Gao S, Jin M, Yang S (2020b) Combined evolutionary engineering and genetic manipulation improve low pH tolerance and butanol production in a synthetic microbial *Clostridium* community. Biotechnol Bioeng 117:2008–2022

Wen Z, Li Q, Liu J, Jin M, Yang S (2020c) Consolidated bioprocessing for butanol production of cellulolytic Clostridia: development and optimization. Microb Biotechnol 13(2):410–422

Williams-Rhaesa AM, Rubinstein GM, Scott IM, Lipscomb GL, Poole IIFL, Kelly RM, Adams MWW (2018) Engineering redox-balanced ethanol production in the cellulolytic and extremely thermophilic bacterium, *Caldicellulosiruptor bescii*. Metab Eng Commun 7:e00073

Xin F, Chen T, Jiang Y, Dong W, Zhang W, Zhang M, Wu H, Ma J, Jiang M (2017) Strategies for improved isopropanol–butanol production by a *Clostridium* strain from glucose and hemicellulose through consolidated bioprocessing. Biotechnol Biofuels 10(1):118

Yilmazel YD, Johnston D, Duran M (2015) Hyperthermophilic hydrogen production from wastewater biosolids by *Caldicellulosiruptor bescii*. Int J Hydrog Energy 40(36):12177–12186

Yu EKC, Chan MKH, Saddler JN (1985) Butanol production from cellulosic substrates by sequential co-culture of *Clostridium thermocellum* and *C. acetobutylicum*. Biotechnol Lett 7 (7):509–514

Chapter 3
Next Generation Biofuels from Macroalgae: Prospects and Challenges

Shraddha Shukla, Rishikesh Shukla, Neha Singh, Hemansi, and Jitendra Kumar Saini

Abstract Algae are unicellular as well as multicellular organisms and have been called as micro and macroalgae, respectively. Algae offer multiple potential benefits to address the today's need of the renewable feedstocks and energy source. Based on the source of feedstock, biofuels have been classified into first generation, second generation, and latest third/fourth generation biofuels. The latest technologies and scientific studies on algae have discovered the third and fourth generations of biofuels. Currently, researchers' focus is on biofuels, and they have shown that marine macroalgae have substantial stamina to replace the first- and second-generation biofuels as they are eco-friendly and utilize carbon neutral energy. Algal biomass is a sustainable resource of energy and therefore could offer an economic, environment-friendly, and industrially growing area in biofuel research. Algal biofuel is produced from the lipid stored in the algal cells. Algae have very high carbon dioxide fixation rate, low land requirement, and require significantly less area for their cultivation and mass production. In addition, they certainly have high photosynthetic efficiency per area. The challenges in the algal biofuel production lies in the economic large-scale production of the microalgae lipid, which can be modulated by enhancing the lipid content without trailing the growth rate of the strains.

Keywords Algae · Bioethanol · Biofuel · Renewable · Third-generation bioethanol

S. Shukla
Department of Microbiology, Atmiya University, Rajkot, Gujarat, India

R. Shukla
Department of Biotechnology, Atmiya University, Rajkot, Gujarat, India

N. Singh
Department of Molecular and Structural Biology, CSIR-CDRI, Lucknow, Uttar Pradesh, India

Hemansi (✉) · J. K. Saini
Department of Microbiology, Central University of Haryana, Mahendergarh, Haryana, India
e-mail: hemansi9042@cuh.ac.in

J. K. Saini, R. K. Sani (eds.), *Microbial Biotechnology for Renewable and Sustainable Energy*, Clean Energy Production Technologies,
https://doi.org/10.1007/978-981-16-3852-7_3

3.1 Introduction

Algae are small, single celled, or multicellular. Algae are plant-like organisms, but lack true leaves and root system. Algae have been divided into microalgae and macroalgae based on their cellular morphology. Microalgae are commonly called as phytoplanktons. Approximately, 50,000 species of microalgae are known till date, which are classified based on their ultra-structure, biochemical composition, and life cycle. Algae have been classified into five groups and are believed to be both eukaryotic and prokaryotic (Table 3.1). Major groups are as follows: green algae (Chlorophyceae), cyanobacteria (Cyanophyceae), diatoms (Bacillariophyceae), golden algae (Chrysophyceae), yellow-green algae (Xanthophyceae), red algae (Rhodophyceae), brownalgae (Phaeophyceae), dinoflagellates (Dinophyceae) and 'pico-plankton' (Prasinophyceae and Eustigmatophyceae). Blue-green algae also known as cyanobacteria are prokaryotic photosynthetic unicellular microalgae. These algae conduct photosynthesis directly in the cytoplasm as they lack true cell organization. Photosynthetic cyanobacteria are mostly found in water systems such as streams, rivers, lakes, and oceans and carry out carbon assimilation, energy production and rapid polar and nonpolar lipid accumulation in algal biomass in the presence of sunlight. Macroalgae (generally called seaweeds) are multicellular algae that can grow in both fresh and salt water. These are fast-growing species and can reach sizes up to 60 m in length and are broadly classified into three distinct groups on the basis of their pigmentation: Phaeophyceae (brown seaweed), Rhodophyceae (red seaweed), and Chlorophyceae (green seaweed). While considering the general

Table 3.1 Different classes and common characteristics of the different algal groups

S. no.	Class	Common characteristics	Chlorophyll	Storage compound
1.	Bacillariophyceae (golden brown algae)	Diatoms, unicellular/ colonial, silicate cell wall, fucoxanthin	Chl a, c	Triglycerides (TAGs) and carbohydrates
2.	Chrysophyceae (brown algae)	Unicellular, fucoxanthin, cellulose and pectin in cell wall	Chl a, c	Oil droplets and carbohydrates
3.	Eustigmatophyceae (yellow green algae) promising for biofuel production	Eukaryotic, unicellular, presence of polysaccha-rides in cell wall	Chl a	Large amount of polyunsaturated fatty acids
4.	Chlorophyceae (green algae)	Unicellular, colonial or filamentous, eukaryotic	Chl a, b	Starch and oil droplets
5.	Prymnesiophyceae (golden brown algae)	Mostly marine, fucoxanthin		Chrysolaminarin: a carbohydrate reservoir
6.	Cyanophyceae	Prokaryotic, unicellular, multicellular or colonial, phycocyanin, can fix nitrogen	Chl a	Low level of lipid reserve material

term algae, the number of species has been estimated to be between one and ten million, most of which are microalgae.

Other algae belong to eukaryotes and contain a well-defined nucleus enclosed within a nuclear membrane. All eukaryotic algae contain intracellular chloroplast that comprises chlorophyll inside photosynthetic lamellae, which is the site for photosynthesis. Algal chloroplasts are of various shapes with diverse types of chlorophyll molecules. Different types of microalgae have characteristic color based on the presence of different types of chlorophyll and pigments. Apart from the chloroplast, well developed endoplasmic reticulum is also present in all eukaryotic algae. Pyrenoides are the site for all the enzymatic reactions leading to the glucose conversion to starch. These pyrenoides are generally present within the chloroplast of golden, red, and green algae. Vacuole also exists as a lipid storage organelle within the algal cells.

Chapman classified the algae by a modern approach on the basis of (1) existence of different nucleus, (2) photosynthetic pigments amount and chemical composition, (3) food reservoir type, (4) composition of cell wall, (5) existence, type, number, orientation of flagella, and (6) reproduction mode. Remarkably, all algal species are not likewise important for the production of biofuels. Microalgae are of prime importance for producing biofuels. Amongst the microalgae, the Chlorophyceae taxonomic group is the most potential species for biofuels. They are capable of performing photosynthesis, important for life on the Earth and produce approximately half of the atmospheric oxygen. They also use the greenhouse gas (GHG) carbon dioxide simultaneously to grow photoautotrophically. They represent an almost unexploited resource on the Earth and hence can be utilized immensely for producing third generation bioethanol.

3.2 Third Generation Biofuels

Biofuels, as the name suggests, are referred to the fuels that are obtained from the biomass. Depending on the means they have been utilized to turn as a fuel source, they are grouped in two broad categories—primary and secondary biofuels; the primary being those that can be directly used as a fuel without processing and the secondary ones have to undergo processing to be utilized as biofuels. In the present scenario, where an emerging requirement of fuel sources has to be met, biofuels are the hope of the hour, and thus, the focus on the secondary biofuels is much needed. Biofuels like biodiesel and bioethanol generated from the biomass are processed and then used in vehicles and industries. The secondary biofuels are further categorized into three generations: first, second, and third generation biofuels on the basis of different parameters such as type of processing technology, type of feedstock, or their level of development (Fig. 3.1).

First generation biofuels have certain limitations because it is derived from crops and thus creates a competition for agricultural land (Schenk et al. 2008); and therefore, second generation of biofuels emerged that are produced from

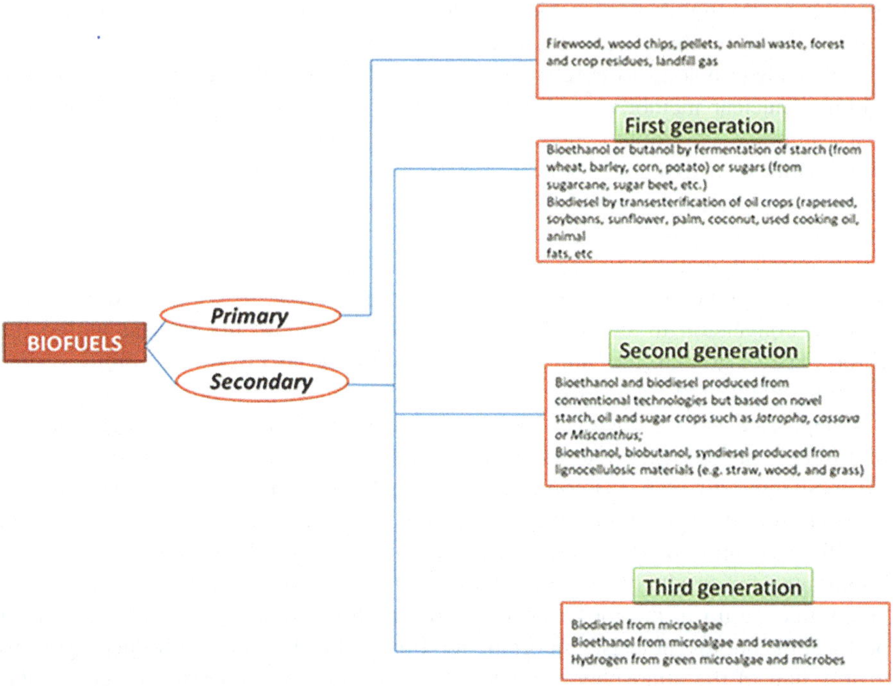

Fig. 3.1 Classification of biofuels

lignocellulosic biomass. Lignocellulosic biomass is the plant residues and hence does not compete with the food production category. Nevertheless, the disadvantage of the second generation is that it is not cost-effective. It requires expensive techniques for processing the lignocellulosic biomass to biofuels, and hence, large-scale production of second-generation biofuels is still noneconomic (Brennan and Owende 2010). Therefore, third generation biofuels that are microalgae derivatives emerged as a feasible alternative energy resource. Third generation biofuels is a recent term that is used for algae derived biofuels. It surpasses the major drawbacks en route with first- and second-generation biofuels (Nigam and Singh 2011; Chisti 2007). Microalgae are capable of producing approximately 15- to 300-fold higher amounts of oil for biodiesel production than the traditional crops on the basis of area. Another major benefit of utilizing microalgae is the shorter harvesting cycle than the conventional crops. The harvesting cycle of microalgae is ≈1–10 days (depending on the process), with significantly higher yields upon harvesting. Other characteristic properties such as algae can be genetically manipulated and oils produced by them can be converted directly in different types of fuels make them a potential source for biofuel production.

3.3 Potential of Macro/Microalgae: Advantages over Traditional Feedstocks

Algae are the potential source of next generation biofuel. The advantages of algal use as biofuels are manifolds. Algae grow very rapidly and produce biomass and oils which can be utilized for the biofuel production. Nearly 70% of the earth surface is covered with water, which can be utilized for the economic cultivation of algae up to five harvests per year. The generation time is less; therefore, algal biomass can be doubled in every few hours and biomass can be harvested. Algae being autotrophs, utilize direct sunlight and CO_2 for their growth. Photosynthetic algae are capable of utilizing almost 2 kg of CO_2/kg of biomass as algae produce about half of the oxygen present in the atmosphere (Williams and Laurens 2010). In contrast to terrestrial crops, microalgae do not require any fertile soil and freshwater supply and can tolerate extreme seasonal conditions (Williams and Laurens 2010). Thus, the cost of biofuel can be highly reduced. These microalgae do not interfere with the growth of terrestrial crops. The use of algae excludes the negative impact of food security by offering high sugar content and low lignin content than the terrestrial plants (Rajkumar et al. 2016). Few classes of algae can utilize heavy metals present in the ocean and accumulate high sugar content as reserve polymers which can easily be converted into biofuels. Finally, the efficacy of algae for photoconversion up to 5% is an added advantage in the direction of third generation biofuel production (Fig. 3.2).

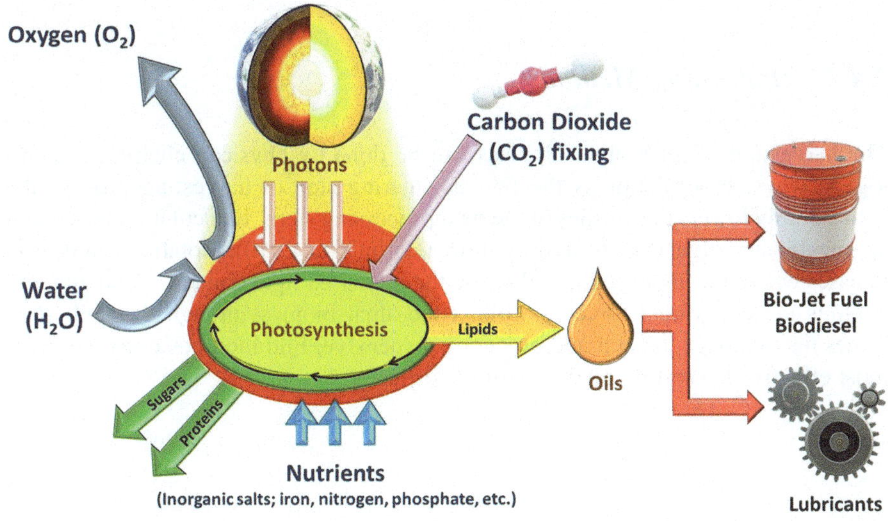

Fig. 3.2 Advantages of macroalgae for biofuel generation

3.4 Bioethanol Production Process from Algae

Microalgal biomass has emerged as the promising source for biofuels and is capable of meeting the demands without compromising either with food products or agricultural land for food production (Chisti 2007). The three main components of the algal biomass are carbohydrates, proteins, and lipids (Um and Kim 2009).

An integrated production of biofuels from microalgae consists of microalgal cultivation, separation of the cells from the growth medium, and subsequent lipid extraction for the biodiesel production through transesterification. Once the oils are obtained, amylolytic enzymes are used that promote fermentable sugar production; then, these are fermented and then distilled using conventional distillation technologies.

3.4.1 Microalgae Cultivation

First, the desired strain has to be selected to obtain the product of interest. The design of bioreactors is a crucial step for commercialization of the product. The cultivation system used for the microalgae should have the following characteristic properties which include high area productivity, high volumetric productivity, cost-effectiveness, and simplified control of the parameters (oxygen, temperature, pH, and turbulence) (Olaizola 2003). Depending on the purpose, the microalgae strain, and the final product, the cultivation system can be selected keeping in mind the various advantages and disadvantages of each kind.

3.4.2 Harvesting Methods

The harvesting of microalgal biomass can be done by physical, chemical, or biological ways. Flocculation is the first dewatering step of harvesting process that considerably lowers the hurdles of the next processing steps. Flocculation is meant to aggregate microalgal cells from the culture (Harun et al. 2010). Flotation can be used in cases where the algae begin to float by the increased lipid content. Centrifugation exploits the centrifugal forces to collect the algal biomass though the high shear forces may damage the cell and process is expensive. Filtration has emerged as the most effective harvesting method with its different forms.

3.4.3 *Production of Bioethanol from Algal Biomass*

The extraction can be achieved through oil press, liquid/liquid extraction, supercritical fluid extraction (SFE), and ultrasound systems (Harun et al. 2010). Solvent extraction has turned out to be one of the most useful methods, while ultrasound is an another promising way. After extraction, the obtained lipids can be transformed to biofuels via transesterification. The harvested algal biomass almost has 90% of the water content, and further processed using dehydration and extraction procedures, which leads to the formation of bioethanol.

3.4.4 *Dehydration of Algal Biomass*

Although there are many different methods for reducing the water content, each method has its own set of advantages and disadvantages. The common methods used are sun drying, spring drying, and freeze drying. Sun drying is cost effective, but extended time duration and area requirements makes it less worthy; spring drying is much worthy but not cost-effective. Spring drying is the most effective, but it is quite costly and hence becomes difficult for a large-scale operation (McKendry 2002; Prakash et al. 1997).

3.4.5 *Extraction of Bioethanol*

After dehydration, the biomass needs to be fermented in order to obtain bioethanol; for this purpose, the microalgal biomass is crushed and the obtained starch is transformed into sugars, which is then mixed with yeast and water in fermenters (Singh et al. 2011). The fermented bioproduct is then processed by distillation to eliminate the other extra matter from the thinned alcohol product. Different methods may be used for the extraction of biofuels as discussed earlier.

3.5 Lipid and Biomass Enhancement Strategies

Today, the population blast is the major concern for the modern generation. The world population is increasing day by day, and as per the current research, it is estimated that, by 2050, the existing population will increase by 1.5% of the current population. At present, the huge demand for energy and fuels is being met mainly through utilization of fossil fuels: oil, natural gas and coal—finite resources (Sajjadia et al. 2018). This could lead to the extinction of all fossil fuel reservoirs and aids in the environmental pollution owing to the heavy release of CO_2. This problem has driven a shift in the research direction toward algae as the novel lipid and biomass resource.

3.5.1 Lipid Content in Algae

Algae are wide lipid producers which cover a variety of lipids, viz. polar and neutral lipids, esters, sterols, and different modified versions of carotenoids. Polar lipids play a vital structural role in cell membrane and are called as high quantity of Poly Unsaturated Fatty Acids (PUFAs). Sterols and phospholipids are key membrane lipids, and some polar lipids act as signal tranducers in algae. Lipids are stored in algae in their late stationary phase in form of triglycerides (TAGs), which include mostly saturated fatty acids (FAs) and a few unsaturated fats. These forms of fats are stored in the cytosolic lipid bodies and can be easily catabolized into energy (Sharma et al. 2012). The fat content in macroalgae can range from 30 to 80% of total biomass; however, as the fat content is increased, the growth rate of algae is dropped rapidly. *Schizochytrium* sp. is known to produce up to 80% of the total mass as lipid which can be extracted and transformed into the next generation biofuels (Deng et al. 2009). In general, Chlorophyceae is the most promising species for algal oil extraction.

3.5.2 Lipid Composition and Suitability of Algae

Suitability of algae in the research field of biofuel is primarily determined by their lipid composition of saturated and unsaturated lipids. Unsaturated fatty acids such as palmitic acid (16:1), oleic acid (18:1), linoleic acid (18:2), and linolenic acid (18:3) are the most important components of biofuels. Very little information is available for the use of saturated fatty acids (SFA) in biofuel production. Some microalgae synthesize lipids largely consisting of polyunsaturated fatty acids such as C22:6 (42%) in *Aurantiochytrium* sp., C22:5 + C22:6 (39.4%) in *Schizochytrium limacinum*, and C20:5 (25%) in *Porphyridium cruentum* (Sajjadia et al. 2018). Biofuels obtained from microalgae largely depend on the composition of fatty acids produced by the algal species.

Algal species producing fats predominantly of saturated fatty acids are the most suitable to be used in low temperature. Long chain SFA increases the pouring ability of biodiesel (Sajjadia et al. 2018). In contrast to this, the biofuels produced mainly by algal species producing high unsaturated fatty acids are easily oxidized, which may lead to inadequate engine performance due to the deposition of insoluble particles. For this issue, the proper selection of appropriate algal strain that may produce high yield of lipid along with quality composition of saturated and unsaturated fatty acid is highly desired. Moreover, the study of the impact of extreme environmental conditions on the production and composition of fatty acids by algal strain needs to be considered.

3.5.3 Lipid Content Enhancement Strategies

Metabolic activities influence cellular composition, growth rate, and lipid composition in algae (Fig. 3.3). Nitrogen, phosphorus, and silicate are the most vital nutrient contents in culture medium which are crucial for the growth and metabolic production of lipid in microalgae. Carbon source, salt concentration, intensity of sunlight, and atmospheric temperature are among the other effective factors. The common effective ratio of C:H:O:N:P for optimal algal growth needs to match $C_{106}H_{181}O_{45}N_{16}P$, theoretically so as to sustain the effective algal biomass. N:P ratio is also very crucial, as low N:P ratio (5:1) reflects nitrogen is limiting, whereas a high N:P ratio (30:1) suggests P is a limiting factor.

The research on algae limitation for nutrients is carried out by manipulating the nutrient contents in three different approaches. In starvation, firstly, the algae are allowed to grow in nutrient-rich conditions, and then the culture is transferred to the same nutrient limitation condition. This starvation triggers a sharp biological shock and results in the storage of high energy compounds. During the study of impact of nutrient limitation, metabolic shift in algae can be carried out by growing the algae under nutrient limitation in continuous culture. In nutrient-limited growth, the algae are grown in a medium in which all the macro and micronutrients are abundant except for the one limiting nutrient. This limiting nutrient can limit the yield of algal biomass due to the physiological reaction toward limiting nutrient, a phenomenon known as "law of the minimum" in microbiology. The effects of nutrient depletion

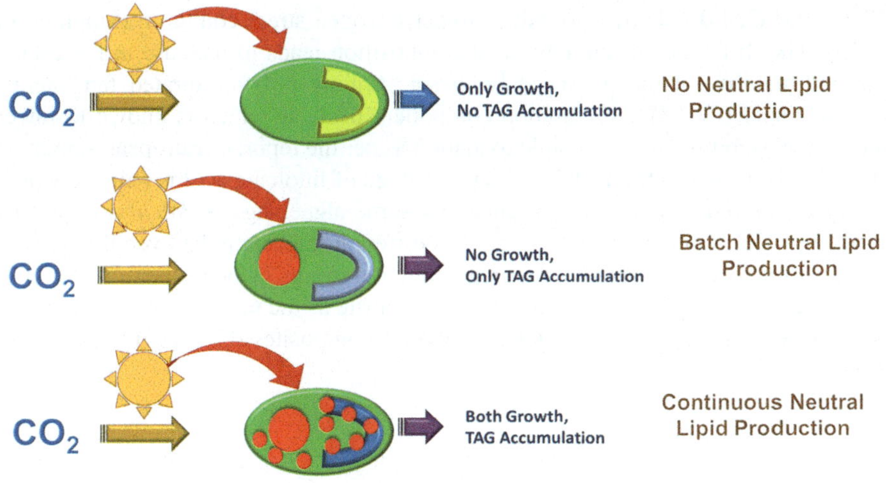

Fig. 3.3 Modulation of lipid content from microalgae for biodiesel production

can be studied by monitoring the growth of algae in small batches under nutrient depletion condition. The depleted nutrient is believed to specify the biomass yield. In this study, the algal cells are first grown in an environment which is enriched with required nutrients. Following the rise in biomass of algae due to growth, the nutrient depletion occurs. Due to this nutrient depletion, the growth rate and photosynthesis rate are reduced and the metabolic processes are shifted so as to adjust the organism to this nutrient depleted state.

3.6 Enhancement Parameters on Lipid Productivity

3.6.1 Nitrogen and Algal Lipid Content

Nitrogen is required for the biosynthesis of the major macromolecule, i.e., protein. It is also a common component for the buildup of structural and functional processes of the cell. Nitrogen is a crucial factor for lipid content and algal growth promotion. Nitrogen limitation causes enhancement in lipid and/or carbohydrate content and reduction in protein, efficiency of photosynthesis, and microalgae growth rate (Jiang et al. 2012). During nitrogen starvation, the C fixation increases, and the extra carbon is converted into lipids and carbohydrates as storage compounds. As the organic nitrogen pool in cell starts to decrease, the lipid production is induced in algae. Nitrogen concentration has a significant quantitative and qualitative impact on the lipid production and is very crucial for the biofuel production. The overall major fat profile in algae remains same in control and nitrogen starved media, but there is a significant difference in proportion of fatty acids (saturated and unsaturated). Under nitrogen stress condition, the ratio of saturated fatty acids (i.e., C16:0 and C18:0) is decreased. In *Chlorella* and *Ankistrodesmus falcatus*, saturated fatty acid content of C23:0 and C24:0 is known to reduce under nitrogen stress condition (Singh et al. 2015). The difference in the fatty acid composition leads to increase in the cetane no. and stability of the produced biodiesel. Among polyunsaturated fatty acids, linolenic acid (C18:3) is a common component of algae which is known to lower the biofuel stability due to possible oxidation as per the report of European standards (EN14214). Under nutrient stress, the percentage of linolenic acid is reduced which increases the biodiesel quality produced by some algae, e.g., *A. falcatus* (Converti et al. 2009). Different nitrogen sources has different impact on the overall fatty acid composition in algae. For instance, urea utilization results in the generation of higher PUFA, e.g., C18:4, C20:5, C22:6; nitrate and nitrite as the nitrogen source increases the ratio of saturated (C14:0) and unsaturated composites (C16:1) (Fidalgo et al. 1998).

3.6.2 Phosphorus and Algal Lipid Content

Phosphorus has a key role in various metabolisms such as signal transduction, transfer of energy, and photosynthesis. When phosphorus supply is abundant, the algal cells assimilate the excess phosphorus inside as polyphosphate granules. When phosphorus is in limited amount, the algal membrane phospholipid components are enriched by nonphosphorus glycolipids and sulpholipids as a P conservation mechanism as reported in *Chlorella* sp. (Liang et al. 2013). During phosphorus depleted state, cell division is reduced, which results in the accumulation of carbon in the form of TGA rich in saturated and monounsaturated fatty acids. It causes alterations in the biosynthesis process and enhances the stored amount of lipid in *Dunaliella parva*, *Chlorella* sp., *Scenedesmus* sp., and many other species. Unsaturated fatty acids which make the highest proportion of lipids in green algae *Chlamydomonas acidophila* (*C. acidophila*), *Chlorella* sp. (Spijkerman and Wacker 2011), and *Phaeodactylum tricornutum* (*P. tricornutum*) significantly increase under P starving condition (El-Sheek and Rady 1995). Similar results have been reported in *Scenedesmus quadricauda* (Ahlgren et al. 1998) which showed a lower percentage of PUFA 18:3(9,12,15) under P-limited conditions.

3.6.3 Effects of Carbon

Almost 50% of the algal biomass has been made up of carbon, and CO_2 is the main supply of the carbon source. The effects of CO_2 levels on total lipid contents, biomass (dw), and lipid profile of the microalgae has been reported in a few studies as summarized in Table 3.2. Carbon has a very pronounced impact on the fatty acid composition of algae. It is known that a high carbon monoxide concentration

Table 3.2 The effect of metal ions on metabolism, biomass, and lipid content in algae

Metal	Metabolic role	Effect on biomass	Effect on lipid content
Fe concentrations up to almost 2×10^{-3} g/L	Fundamental enzymatic reactions of photosynthesis	Increased	Can't promote lipid accumulation
Mg^{2+} 2×10^{-3}–8×10^{-3} g/L	Promote acetyl-CoA carboxylase (ACCase) in vivo activity	Increased	Increased
Ca^{2+} 5×10^{-4}–5×10^{-3} g/L	Signal transduction of environmental stimuli	–	Increased
Heavy metals (cadmium, copper 31.4 mg/L—Approximately 0.49 mM and zinc)	Alter the lipid metabolism	–	Enhancement of lipid content
Silica deficiency	Most common stress in diatoms	–	Promoting storage lipid

enhances the ratio of C18:1 and C18:2, whereas it decreases the polyunsaturated fatty acids (C18:3). In contrast to CO effect, the studies report that low CO_2 concentration (less than 2%) increases the unsaturated fatty acids (C18:1, C18:2) and high carbon dioxide (2–10%) favours the biosynthesis of the saturated fatty acids (104). It can be concluded that the high CO_2 concentration tends to reduce the unsaturation.

3.6.4 Effect of Metals

Several metal ions such as iron, magnesium, calcium, chromium, and other heavy metals (e.g., Cd, Cu, Zn) have a very promising impact on the lipid content of algae. The metabolic effect of various metal ions and their subsequent impact on the algal biomass and lipid content have been summarized in Table 3.2. It is observed that an increase in iron concentration has a positive influence on increasing the saturated: unsaturated fatty acid in algae. Rocchetta et al. (2006) demonstrated the increased saturated fatty acids production in the treated algal cells with the higher metal (e.g., Cr) concentrations. Under increased metal concentrations, the algal cells increase the carbon assimilation which leads to the production of C14:0, C16:0, and C18: 0 enriched fatty acids with a decreased rate of production for PUFA. When iron is abundantly present at a higher CO_2 atmosphere, it increases the fatty acid production of longer carbon chain (Takagi et al. 2006; Ghasemi et al. 2012). Therefore, 2% CO_2 can be effectively used in combination with high metal ion concentrations for the production of long-chain saturated fatty acids containing C12:0 and C16:0.

3.7 Effective Process Conditions

Several studies have been carried out to recognize the effects of various parameters on the growth and productivity of algae. Light, temperature, nutrients, pH, salinity, and dissolved oxygen are examples of the important key process conditioning factors that can be optimized for maximizing the third-generation biofuel production.

3.7.1 Light

Light is the most important parameter for photosynthesis steps by algae. Microalgal cells experience three light zones based on the intensity of light (Fig. 3.3). According to Beer–Lamberts Law, the light intensity is continuously lessened as it penetrates any surface (Bernard 2011; Yuan et al. 2014; Blanken et al. 2016). Hydrodynamics of the culture, i.e., how cells move in media, their closeness towards light source, and also other factors, play a vital role in the study. Algae are photosynthetic organisms and utilize a particular suitable wavelength (400–700 nm) of light. High light intensity leads to the inhibition of pigment system II of photosynthetic apparatus by inhibiting the crucial electron transfer proteins during photosynthesis (Moheimani and Parlevliet 2013; Quaas et al. 2015). A moderate light (optimum range) is required for the optimal level of photosynthesis; high intensity leads to photoinhibition of early growth and low light intensity cannot support the cell growth for successive generations. Day and night length also plays a crucial role in algal growth by altering its biochemical composition. Wahidin et al. (2013) reported that the lipid content and growth rate of *Nannochloropsis* sp. was enhanced when the L–D cycle was changed from 12:12–18:6 h. It has been observed that the red light generally promotes increased biomass production, whereas the blue and far red-light wavelength positively impacts the lipid and carotenoids accumulation. Therefore, light needs to be optimized accordingly to obtain a balance between photoprotection, photosynthesis, and biochemical composition of algae.

3.7.2 Salinity

Biosynthesis of lipid is a known resistance mechanism toward salt stress. Environmental salt concentration has a sharp effect on the lipid productivity, fatty acid profile, and growth rate. It is reported that as the salt content is increases, the ratio of monounsaturated fatty acids between palmitic acid C16:1) and oleic acids (C18:1)) enhances along with decreased proportion of polyunsaturated FAs (PUFA), which in turn favors the production of good quality biofuels. These changes in the fatty acid profiling in *Chlamydomonas mexicana*, *Scenedesmus obliquus* (Salama et al. 2014), *Botryococcus braunii* (Rao et al. 2007), *Cladophora vagabunda* (Elenkov et al. 1996) play a vital role in the direction of keeping the membrane fluidity and its destruction. These changes in fatty acids are consistent until the algae are grown within the optimum salinity level. Cao et al. (2014) stated that the PUFA increases abruptly in fractions of lipid till the optimum concentration of sodium chloride is attained.

3.7.3 Effects of Temperature

The temperature fluctuation has substantial impact on lipid production, lipid profiling, and biomass yield in algae. Both high and low temperature severely affect the organism's growth. High temperature has more severe negative impact on algal growth due to the denaturation of proteins and enzymes. The temperature effect on algal growth can be understood by a bell-shaped growth curve. Plasma membrane fluidity needs to be maintained at such temperature fluctuations. Therefore, higher and lower temperature alters the biosynthetic pathway of lipid resulting in variable expression of relevant lipids, so that the algae gets adjusted with the changing temperature. In several algae, i.e., *Chlorella vulgaris* (*C. vulgaris*), exposure to high temperature (38 °C) leads to the decolourization of algal biomass from green to brown, reduced biomass, and ultimately death (Converti et al. 2009). Temperature range of 15–30 °C is ideal for the growth and photosynthesis of most of the algal species.

To survive the imbalance between energy supply and consumption, algal cells might change their size and shape as it hampers the photosynthetic ability by modulating the RUBISCO enzyme efficiency (Atkinson et al. 2003). The relationship between environmental temperature fluctuation and alteration in lipid, carbohydrate, and protein biosynthesis varies among different groups of algae. However, it has been observed that, low temperature stress mostly enhances the lipid accumulation in algal cell with an increased biosynthesis of saturated fatty acids (Renaud et al. 2002). This effect can be attributed to the increase in the glycolysis intermediates which can be directed towards high pyruvic acid biosynthesis and therefore, an increase in the overall lipid biosynthesis (Wang et al. 2016). Increasing temperature increases the monounsaturated fatty acids while downregulating the PUFA biosynthesis. According to the report, the low temperature activates the fatty acid desaturase enzyme leading to the conversion of oleic acid (18:1) to linoleic acid (18: 2) and linolenic acid (18:3). This shift in the expression of polyunsaturated fatty acids enables the algae to maintain its membrane fluidity (Renaud et al. 2002; Wang et al. 2016). However, the overall lipid profile response toward increased/decreased temperature fluctuation is highly dependent on algal species.

3.7.4 Effects of pH

pH affects the algal growth by altering the mineral absorption capacity, i.e., iron and carbon. Most of the algae grow well between pH 7 and 9, and the pH of the algal culture increases sharply during the day time because of photosynthesis, whereas at night, the respiration leads to decrease in pH. Algal culture can be maintained at an optimal pH range by supplementing CO_2 or mineral acids. Higher pH conditions inhibit the growth of algae, whereas low pH supports the algal growth and lipid accumulation (Cao et al. 2014). The lipid accumulation in algae also alters dramatically with pH alteration. pH range of 7.0–9.5 is known to promote significant lipid accumulation in the algal species. Under nitrogen limitation and increased medium pH, *Chlorella* produces higher triacylglycerol. Algae grown under pH stress, e.g., pH 7.6 and 9.5, show increased lipid accumulation, where the alkali pH stress favors higher saturated lipid production and reduced glycolipid and polar lipid biosynthesis.

3.8 Scope and Challenges of Bioethanol Production

Nowadays, algal biomass for biofuel production hasemerged as a solution to many problems that arises due to an increase in industrialization. Water ecosystem management for cultivation offers a comparatively cheap and environment friendly option. Sustainable, biofriendly practices are the main reasons that make biofuels beneficial. There are still many hurdles in the path for sustainable algal biomass production; the production cost is still much higher that needs to be reduced. Developing new and genetically modified strains that may lead to higher yields and thus lower the production cost is the need of the hour (Chia et al. 2018). Current techniques are still much expensive, and we need more focus and research on the development of more efficient and cost effective harvesting methods and reduction of production cost (Wu et al. 2005; Zhu et al. 2014; and Ambati et al. 2014). In addition to these, the biomass cultivation systems (i.e., PBR system and open pond system) should also be made more economical and of course much sustainable.

3.9 Biotechnological Engineering of Microalgae

Algal biofuels are known to be a sustainable alternative to traditional fuels, though they need to overcome several hindrances for competing in the fuel market. Studies on microalgal biotechnology have gained attention in the past few decades, and the industry is expanding into new areas. However, it is remarkable to comment here that the microalgae are still not a well-studied group if biotechnology is the parameter. Among thousands of species that are supposed to exist, only a few hundred are investigated at molecular and biochemical levels and some are cultured industrially. The major bottlenecks in making algal biofuels as an alternative source of energy are the lack of two characteristics—high content of lipid and faster growth rate in current microalgal species (Ghosh et al. 2016; Chen et al. 2017). Also, the lack of sufficient light harvesting capacity in natural growth environment is one of the major obstacles (Stephenson et al. 2011). Studying the molecular complexities of metabolism, biosynthesis of TAG, its regulation, and metabolic flux, possibly will assist in strain improvement, and hence, maximize the biofuel production in microalgae in a cost-effective manner. There are many promising ways in which these unicellular organisms may be engineered, and all these methods finally lead to increase in the valuable content of these algae. Augmenting synthesis of oil in microalgae mainly hinge on the enzymes manipulation which are convoluted in biosynthesis of lipid or other competitive pathway intended to sidetrack the carbon and equivalent fluxes in the direction of lipid biosynthesis. Manipulating the expression of enzymes such as PEP, pyruvate dehydrogenase, acetyl-CoA synthetase, NAD(H) kinase, etc. has been proved to considerably enhance the lipid content without adversely affecting the cell growth. The most commonly used technique is the manipulation of genes involved in the metabolic pathway; however, this strategy has seen mixed response (Bajhaiya et al. 2017). In recent times, the transcriptional regulation of biosynthesis of oil has become prevalent for controlling the expression of multiple constituents of a metabolic pathway at the same time (Courchesne et al. 2009). Moreover, efforts to manipulate numerous other targets like improving light using efficiency, cell dormancy control, refining carbon sequestration, etc. have gained attention; these approaches influence the lipid content indirectly. The enhanced biomass yield is crucial for the overall energy output (Barry et al. 2015). In microalgae, the biomass productivity is governed by CO_2 fixation rate, abiotic stress, and light utilization efficiency (Chu 2017). Genetic engineering in combination with omics analysis facilitates the recognition of the major transcription regulators, enzymes, and promoters for stress response. This could be beneficial and facilitates future molecular studies.

Engineering metabolic pathway for achieving enhanced productivity of the algal strains by gene editing propounds a robust mechanism to overcome genetic shortfalls (Ng et al. 2017). Recently, CRISPR/Cas9, a genome editing technique, has emerged. The utility of this technique in improving microalgae traits for biofuels and nutraceutical applications has a significant scope. Numerous advanced studies led to success in improving the microalgae species, which endorse the technology for its efficiency in producing targeted mutants. The main advantage of CRISPR technology in algae is it provides the ease of multiplexing and manipulating of metabolic pathways unlike traditional knock out approaches. One of the successful examples is engineering of lipid in oleaginous microalgal strains by blocking the metabolic routes such as generating starch, degrading lipid, etc., which are competitive to lipid production. There are certain drawbacks that has to be always kept in mind whenever genetic engineering is considered, e.g., they should always be environment friendly, the newly edited gene is not that easily thrown out of the transformants, etc. In short, the genetically engineered variants should be complementing the finding new ones and those already known and not substitute them (Pulz O and Gross W 2004).

3.10 Sustainable Economy and Industrial Growth

Macroalgae are promising source for the commercial utilization. This approach where the high value coproducts are formed from the algal biomass along with the substantial biofuel production is known as a "biorefinery approach". Many attractive valuable biochemicals are yet to be discovered from the micro and macroalgae for their maximum utilization in industries. In biorefinery approach, the collaborative simulation of bioprocessing and environment friendly technologies occurs in an economic manner. Micro and macroalgae have vast industrial applications including high cost bioactive compounds, health foods, and natural pigments (Fig. 3.4). Algal biomass has immense applications, including i. wastewater treatment and CO_2 mitigation, ii. food and nutrition, iii. Aanimal and aquatic feed, iv. cosmetics,

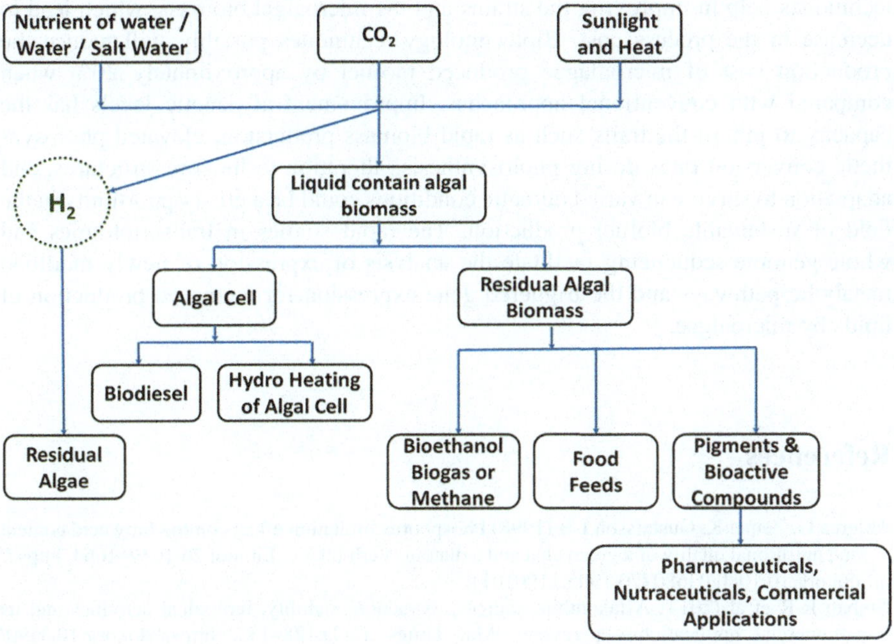

Fig. 3.4 Coproducts from algal biomass along with substantial biofuel production

v. fatty acid production, vi. synthesis of pigments, and vii. Bbiofertilizers. Therefore, the strategical bioproduction of biofuels from algae with maximizing the industrially important bioproducts should be carried out to utilize the algal biomass at its maximum ease. In this approach, the algae are cultivated in an algal farming facility, e.g., CO_2 mitigation and waste water treatment plant and algal biomass is harvested, following which, the bioactive products are effectively extracted, and finally, the biofuel can be produced by thermal processes (pyrolysis, liquefaction, or gasification).

3.11 Conclusion and Future Prospects

Regardless of the fascinating studies in the past few decades, microalgal cultivation is still challenging for industrialization. Despite the fact that the commercialization of third generation bioethanol is far from reality because of its high cost, microalgae could be a potential substitute of energy in the terms of biofuel and its production system. It does not raise the debate of food vs. fuel and set pressure on agricultural lands and forests. Various technologies such as metabolic engineering and gene editing could be promising for making this technology economically feasible. These techniques help in improving the strains and the microalgal biomass, which lead to decrease in the process cost. Biotechnology techniques possibly will reduce the production cost of microalagae produced biofuel by approximately 20% when compared with conventional approaches. Improvement at genetic levels has the capacity to inherit the traits such as rapid biomass production, elevated photosynthetic conversion rates during photosynthesis, alteration to its core structures, and adaptation to survive in varied climatic conditions could be a great opportunity in the field of sustainable biofuel production. The rapid studies in transcriptomics and whole genome sequencing facilitate the analysis of expression of newly modified metabolic pathways and the triggered gene expression for increased production of lipids by microalgae.

References

Ahlgren G, Zeipel K, Gustafsson I-B (1998) Phosphorus limitation effects on the fatty acid content and nutritional quality of a green alga and a diatom. Verh Int Ver Limnol 26:1659–1664. https://doi.org/10.1080/03680770.1995.11901011

Ambati R-R et al (2014) Astaxanthin: sources, extraction, stability, biological activities and its commercial applications—a review. Mar Drugs 12(1):128–152. https://doi.org/10.3390/md12010128

Atkinson D, Ciotti BJ, Montagnes DJ (2003) Protists decrease in size linearly with temperature: ca. 2.5% degrees C(−1). Proc Biol Sci 270:2605–2611. https://doi.org/10.1098/rspb.2003.2538

Bajhaiya A-K, Moreira J-Z, Pittman J-K (2017) Transcriptional engineering of microalgae: prospects for high-value chemicals. Trends Biotechnol 35:95–99. https://doi.org/10.1016/j.tibtech.2016.06.001

Barry A-N, Starkenburg S-R, Sayre R-T (2015) Strategies for optimizing algal biology for enhanced biomass production. Front Energy Res 3:1. https://doi.org/10.3389/fenrg.2015.00001

Bernard O (2011) Hurdles and challenges for modelling and control of microalgae for CO_2 mitigation and biofuel production. J Process Control 21:1378–1389. https://doi.org/10.1016/j.jprocont.2011.07.012

Blanken W, Postma P-R, De Winter L, Wijffels R-H, Janssen M (2016) Predicting microalgae growth. Algal Res 14:28–38. https://doi.org/10.1016/j.algal.2015.12.020

Brennan L, Owende P (2010) Biofuels from microalgae—a review of technologies for production, processing, and extractions of biofuels and co-products. Renew Sust Energ Rev 14:557–577. https://doi.org/10.1016/j.rser.2009.10.009

Cao J, Yuan H, Li B, Yang J (2014) Significance evaluation of the effects of environmental factors on the lipid accumulation of *Chlorella minutissima* UTEX 2341 under lownutrition heterotrophic condition. Bioresour Technol 152:177–184. https://doi.org/10.1016/j.biortech.2013.10.084. Epub 2013 Nov 5

Chen B, Wan C, Mehmood M-A, Chang J-S, Bai F, Zhao X (2017) Manipulating environmental stresses and stress tolerance of microalgae for enhanced production of lipids and value-added products—a review. Bioresour Technol 244:1198–1206. https://doi.org/10.1016/j.biortech.2017.05.170

Chia S-R et al (2018) Sustainable approaches for algae utilization in bioenergy production. Renew Energy 129:838–852. https://doi.org/10.1016/j.renene.2017.04.001

Chisti Y (2007) Biodiesel from microalgae. Biotechnol Adv 25(3):294–306. https://doi.org/10.1016/j.biotechadv.2007.02.001

Chu W-L (2017) Strategies to enhance production of microalgal biomass and lipids for biofuel feedstock. Eur J Phycol 52:419–437. https://doi.org/10.1080/09670262.2017.1379100

Converti A, Casazza AA, Ortiz EY, Perego P, Del Borghi M (2009) Effect of temperature and nitrogen concentration on the growth and lipid content of *Nannochloropsis oculata* and *Chlorella vulgaris* for biodiesel production. Chem Eng Process Process Intensif 48:1146–1151. https://doi.org/10.1016/j.cep.2009.03.006

Courchesne N-M-D, Parisien A, Wang B, Lan C-Q (2009) Enhancement of lipid production using biochemical, genetic and transcription factor engineering approaches. J Biotechnol 141:31–41. https://doi.org/10.1016/j.jbiotec.2009.02.018

Deng X, Li Y, Fei X (2009) Microalgae: a promising feedstock for biodiesel. Afr J Microbiol Res 3: 1008–1014

Elenkov I, Stefanov K, Dimitrova-Konaklieva S, Popov S (1996) Effect of salinity on lipid composition of *Cladophora vagabunda*. Phytochemistry 42:39–44. https://doi.org/10.1016/0031-9422(95)00857-8

El-Sheek MM, Rady AA (1995) Effect of phosphorus starvation on growth, photosynthesis and some metabolic processes in the unicellular green alga *Chlorella kessleri*. Phyton 35:139–151

Fidalgo J-P, Cid A, Torres E, Sukenik A, Herrero C (1998) Effects of nitrogen source and growth phase on proximate biochemical composition, lipid classes and fatty acid profile of the marine microalga *Isochrysis galbana*. Aquaculture 166:105–116. https://doi.org/10.1016/S0044-8486(98)00278-6

Ghasemi Y, Rasoul-Amini S, Naseri AT, Montazeri-Najafabady N, Mobasher MA, Dabbagh F (2012) Microalgae biofuel potentials. Appl Biochem Microbiol 48:126–144

Ghosh A, Khanra S, Mondal M, Halder G, Tiwari O-N, Saini S et al (2016) Progress toward isolation of strains and genetically engineered strains of microalgae for production of biofuel and other value added chemicals: a review. Energy Convers Manag 113:104–118. https://doi.org/10.1016/j.enconman.2016.01.050

Harun R et al (2010) Bioprocess engineering of microalgae to produce a variety of consumer products. Renew Sustain Energy Rev 14(3):1037–1047. https://doi.org/10.1016/j.rser.2009.11.004

Jiang Y, Yoshida T, Quigg A (2012) Photosynthetic performance, lipid production and biomass composition in response to nitrogen limitation in marine microalgae. Plant Physiol Biochem 54: 70–77. https://doi.org/10.1016/j.plaphy.2012.02.012

Liang K, Zhang Q, Gu M, Cong W (2013) Effect of phosphorus on lipid accumulation in freshwater microalga *Chlorella* sp. J Appl Phycol 25:311–318

McKendry P (2002) Energy production from biomass (part 2): conversion technologies. Bioresour Technol 83(1):47–54. https://doi.org/10.1016/S0960-8524(01)00119-5

Moheimani NR, Parlevliet D (2013) Sustainable solar energy conversion to chemical and electrical energy. Renew Sust Energ Rev 27:494–504. https://doi.org/10.1016/j.rser.2013.07.006

Ng I-S, Tan S-I, Kao P-H, Chang Y-K, Chang J-S (2017) Recent developments on genetic engineering of microalgae for biofuels and bio-based chemicals. Biotechnol J 12:1600644. https://doi.org/10.1002/biot.201600644

Nigam P-S, Singh A (2011) Production of liquid biofuels from renewable resources. Prog Energy Combust Sci 37(1):52–68. https://doi.org/10.1016/j.pecs.2010.01.003

Olaizola M (2003) Commercial development of microalgal biotechnology: from the test tube to the marketplace. Biomol Eng 20(4–6):459–466. https://doi.org/10.1016/S1389-0344(03)00076-5

Prakash J et al (1997) Microalgal biomass drying by a simple solar device. Int J Solar Energy 18(4):303–311. https://doi.org/10.1080/01425919708914325

Quaas T, Berteotti S, Ballottari M, Flieger K, Bassi R, Wilhelm C et al (2015) Non-photochemical quenching and xanthophyll cycle activities in six green algal species suggest mechanistic differences in the process of excess energy dissipation. J Plant Physiol 172:92–103. https://doi.org/10.1016/j.jplph.2014.07.023. Epub 2014 Sep 3

Rajkumar R et al (2016) Potential of macro and microalgae for biofuel production. Bioresour Technol 9(1):606–1633. https://doi.org/10.15376/biores.9.1.1606-1633

Rao AR, Dayananda C, Sarada R, Shamala TR, Ravishankar GA (2007) Effect of salinity on growth of green alga *Botryococcus braunii* and its constituents. Bioresour Technol 98:560–564. https://doi.org/10.1016/j.biortech.2006.02.007. Epub 2006 Jun 19

Renaud SM, Thinh L-V, Lambrinidis G, Parry DL (2002) Effect of temperature on growth, chemical composition and fatty acid composition of tropical Australian microalgae grown in batch cultures. Aquaculture 211:195–214

Rocchetta I, Mazzuca M, Conforti V, Ruiz L, Balzaretti V, del Carmen Ríos de Molina M (2006) Effect of chromium on the fatty acid composition of two strains of *Euglena gracilis*. Environ Pollut 141:353–358. https://doi.org/10.1016/j.envpol.2005.08.035. Epub 2005 Oct 4

Sajjadia B, Chena W-Y, Abdul A, Raman A, Ibrahim S (2018) Microalgae lipid and biomass for biofuel production: a comprehensive review on lipid enhancement strategies and their effects on fatty acid composition. Renew Sust Energ Rev 97:200–232. https://doi.org/10.1016/j.rser.2018.07.050

Salama ES, Abou-Shanaba RA, Kim JR, Lee S, Kim SH, Oh SE et al (2014) The effects of salinity on the growth and biochemical properties of *Chlamydomonas mexicana* GU732420 cultivated in municipal wastewater. Environ Technol 35:1491–1498. https://doi.org/10.1080/09593330.2013.871350

Schenk P, Thomas-Hall S, Stephens E, Marx U, Mussgnug J, Posten C, Kruse O, Hankamer B (2008) Second generation biofuels: high efficiency microalgae for biodiesel production. Bioenergy Res 1:20–43. https://doi.org/10.1007/s12155-008-9008-8

Sharma K-K, Schuhmann H, Schenk P-M (2012) High lipid induction in microalgae for biodiesel production. Energies 5:1532. https://doi.org/10.3390/en5051532

Singh A et al (2011) Renewable fuels from algae: an answer to debatable land-based fuels. Bioresour Technol 102(1):10–16. https://doi.org/10.1016/j.biortech.2010.06.032

Singh P, Guldhe A, Kumari S, Rawat I, Bux F (2015) Investigation of combined effect of nitrogen, phosphorus and iron on lipid productivity of microalgae *Ankistrodesmus falcatus* KJ671624 using response surface methodology. Biochem Eng J 94:22–29. https://doi.org/10.1016/j.bej.2014.10.019

Spijkerman E, Wacker A (2011) Interactions between P-limitation and different C conditions on the fatty acid composition of an extremophile microalga. Extremophiles 15:597–609. https://doi.org/10.1007/s00792-011-0390-3

Stephenson P-G, Moore C-M, Terry M-J, Zubkov M-V, Bibby T-S (2011) Improving photosynthesis for algal biofuels: toward a green revolution. Trends Biotechnol 29:615–623. https://doi.org/10.1016/j.tibtech.2011.06.005

Takagi M, Karseno, Yoshida T (2006) Effect of salt concentration on intracellular accumulation of lipids and triacylglyceride in marine microalgae Dunaliella cells. J Biosci Bioeng 101:223–226. https://doi.org/10.1263/jbb.101.223

Um BH, Kim YS (2009) A chance for Korea to advance algal-biodiesel technology. J Ind Eng Chem 15(1):1–7. https://doi.org/10.1016/j.jiec.2008.08.002

Wahidin S, Idris A, Shaleh SRM (2013) The influence of light intensity and photoperiod on the growth and lipid content of microalgae Nannochloropsis sp. Bioresour Technol 129:7–11. https://doi.org/10.1016/j.biortech.2012.11.032. Epub 2012 Nov 16

Wang Y, He B, Sun Z, Chen Y-F (2016) Chemically enhanced lipid production from microalgae under low sub-optimal temperature. Algal Res 16:20–27

Williams P-J-l-B, Laurens L-M-L (2010) Microalgae as biodiesel & biomass feedstocks: review & analysis of the biochemistry, energetics & economics. Energy Environ Sci 3:554–590. https://doi.org/10.1039/B924978H
Wu L-C et al (2005) Antioxidant and antiproliferative activities of Spirulina and Chlorella water extracts. J Agric Food Chem 53(10):4207–4212. https://doi.org/10.1021/jf0479517
Yuan S, Zhou X, Chen R, Song B (2014) Study on modelling microalgae growth in nitrogen-limited culture system for estimating biomass productivity. Renew Sust Energ Rev 34:525–535. https://doi.org/10.1016/j.rser.2014.03.032
Zhu L-D et al (2014) Microalgal biofuels: flexible bioenergies for sustainable development. Renew Sust Energy Rev 30:1035–1046. https://doi.org/10.1016/j.rser.2013.11.003

Chapter 4
Co-Digestion of Lignocellulosic Wastes with Food Waste for Sustainable Biogas Production

Ankur Choudhary, Ashish Kumar, and Sudhir Kumar

Abstract Lignocellulosic waste (LW) and food waste (FW) have great potential for biomethane production. The recalcitrant nature of LW limits its use in anaerobic processes. Without pretreatment, it is difficult to utilize LW at higher organic loading rates (OLRs) through an anaerobic process. Therefore, pretreatment is a compulsory step for efficient utilization (i.e., at higher OLRs) of LW. There are various techniques of pretreatment of LW, and every method has its own advantages and disadvantages and generally makes the process expansive. Pretreatment of LW can change the biomass structure by removing lignin, increasing the surface area, and decreasing the crystalline nature and length of the polymer chain. On the other hand, FW is readily digestible biomass and can be utilized at a comparatively higher organic loading rate than that of LW. Although mono-anaerobic digestion of FW at higher organic loading rates leads to the accumulation of higher volatile fatty acids, the process becomes unstable. This chapter provides state-of-the-art knowledge on the current status of mono-anaerobic digestion of LW and FW for sustainable biogas production, limitations such as pretreatment, low organic loading rates, higher hydraulic retention time, low buffering capacity, and higher accumulation of volatile fatty acids. Besides, the advantages of co-anaerobic digestion of LW with FW over mono-digestion of LW and FW, the technological advancements being made are also discussed.

Keywords Lignocellulosic waste · Food waste · Pretreatment · Mono-anaerobic digestion · Co-digestion

---

A. Choudhary · A. Kumar (✉)
Department of Civil Engineering, Jaypee University of Information Technology, Solan, Himachal Pradesh, India

S. Kumar
Department of Biotechnology and Bioinformatics, Jaypee University of Information Technology, Solan, Himachal Pradesh, India

J. K. Saini, R. K. Sani (eds.), *Microbial Biotechnology for Renewable and Sustainable Energy*, Clean Energy Production Technologies,
https://doi.org/10.1007/978-981-16-3852-7_4

4.1 Process of Sustainable Production of Biogas

4.1.1 Anaerobic Digestion

Biodegradable matter is decomposed by microorganisms in the presence or absence of oxygen. The process of decomposition of any organic substance in the presence of oxygen is known as aerobic digestion, whereas if the decomposing takes place in the absence of oxygen, it is generally known as anaerobic digestion. A general process of anaerobic digestion process is presented in Fig. 4.1. Biogas is one of the types of renewable energy which is generally an outcome of the anaerobic digestion process. Amongst the various advantages, one of the main advantages of anaerobic digestion is that a diversity of biodegradable matters can be utilized via this process (Dolan et al. 2011).

Biogas generally constitutes methane (CH_4), carbon dioxide (CO_2), hydrogen sulfide (H_2S), hydrogen, nitrogen, ammonia, oxygen, and water. However, primarily CH_4 and CO_2 comprise approximately 90% (Naik et al. 2010; Choudhary et al. 2020a). Apart from CH_4, one of the very important outcomes of this process is digestate. Digestate is generally a slurry that produces after the biochemical reaction during the anaerobic digestion process. Dominantly, it is water and a very small quantity of solids. This is generally very rich in nutrients and widely used as a fertilizer during farming (Tampio et al. 2016). Therefore, anaerobic digestion diminishes the demand for fossil fuels which otherwise would be required during the generation of conventional chemical fertilizers. Anaerobic digestion has other several advantages as well, such as it reduces the reliance on the usages of fossil fuels and hence indirectly helps in the curtailment of greenhouse gases into the environment which generally takes place during the burning of the conventional fossil fuel.

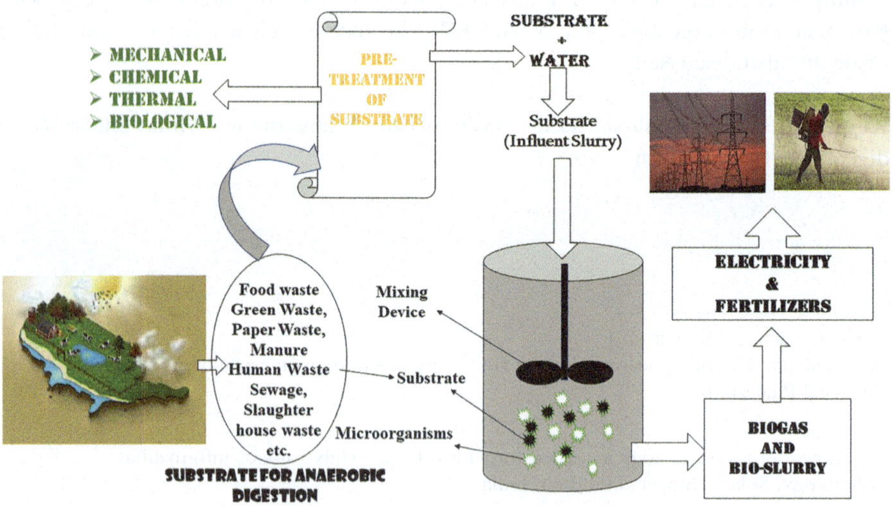

Fig. 4.1 Process of anaerobic digestion

Due to all these advantages, currently anaerobic digestion is becoming a very popular waste management technology across the world. It is also being practiced and becoming very popular amongst the industries in developing countries due to its simple design, working, and requirement of low-capital investment (Börjesson and Mattiasson 2008; Forgács 2012).

Under optimum system variables such as organic loading rate (OLR) and temperature, the process may yield high usage of biomass, i.e., the ratio of energy output/input ratio of 28 MJ/kg (Kabir et al. 2015).

The best feature about this technology is that small-scale and pilot-scale anaerobic reactors can be developed at the local level and be fed with materials available at the regional level. Due to this reason, a huge number of small-scale (household) anaerobic digesters can be found in developing countries. Based on some studies available, approximately 30 million, 3.8 million, and approximately 200,000 anaerobic digesters are running in China, India, and Nepal, respectively (Jiang et al. 2011; Rajendran et al. 2012). However, in African nations, this technology has not been established much, and only very few anaerobic digesters are running at the small-scale level (Amigun et al. 2008). On the other hand, the scenario of anaerobic digesters is opposite in European countries and America. In these regions, the anaerobic digesters are larger than small-scale household digesters when compared with developing countries. In Europe, various waste materials such as sludge, energy crops, and different animal dungs are utilized in anaerobic digestion, and approximately 10,000 anaerobic reactors are working here. According to a study, the anaerobic digesters will be increased by five-fold in Europe in future, whereas a study reported that the number of biogas plants will reach 200 million by 2020 (Deublein and Steinhauser 2008).

4.1.2 Biochemistry of Anaerobic Digestion

Anaerobic digestion is a process in which various biological and chemical processes occur simultaneously. During this process, the biodegradable matter is degraded by a variety of microorganisms and biogas is a major output of this process. Hydrolysis, acidogenesis, acetogenesis, and methanogenesis are the biochemical process which simultaneously takes place within the reactor (Fig. 4.2). In these processes, a variety of bacteria and archaea are involved which have a syntrophic relationship with each other (Deublein and Steinhauser 2008).

In anaerobic digestion, hydrolysis is the primary step. During hydrolysis, hydrolytic bacteria are involved which hydrolyze the biodegradable matter. Generally, in this process, large molecules are broken down into smaller ones, i.e., polymers are degraded into soluble monomers and oligomers. Biochemistry of anaerobic digestion of FW is illustrated in Fig. 4.3.

Cellulases, hemicellulases, lipases, amylases, and proteases are the enzymes associated in this phase (Taherzadeh and Karimi 2008). Almost, all types of biodegradable matters can be decomposed, and all the abovementioned enzymes are

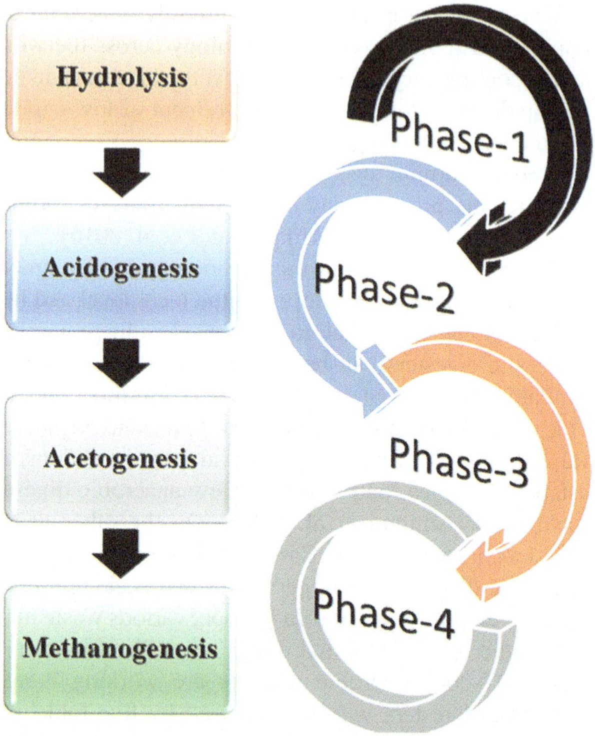

Fig. 4.2 Various phases in the anaerobic digestion process

involved in this. However, it has been noticed that during the anaerobic digestion of any lignocellulosic biomass, lignin does not decompose (Fernandes et al. 2009).

Hydrolysis is a time-dependent process, and it decisively depends upon the nature of biodegradable matter, i.e., complex or rapidly digestible. For example, the hydrolysis of food waste is rapid (Choudhary et al. 2020b) when compared with any lignocellulosic biomass. Nevertheless, in the case of complex biodegradable matter, rapid hydrolysis can be accomplished if suitable enzymes are generated by the microorganisms and suitable contact between biodegradable matter and enzyme is achieved (Taherzadeh and Karimi 2008). Even though, the complex biodegradable matter may take a few weeks to decompose (Deublein and Steinhauser 2008). Therefore, the hydrolysis phase in a complex biodegradable matter such as lignocellulosic wastes is measured as the rate-limiting step (Taherzadeh and Karimi 2008).

Acidogenesis is an acid-forming phase in the anaerobic digestion process. The by-products of the hydrolysis phase are utilized in this phase and are further converted into volatile fatty acids (VFAs) with the help of obligate and facultative anaerobes. Propionic acid, valeric acid, formic acid, acetic acid, and butyric acid are amongst the main VFAs produced during this phase. Apart from the VFAs, alcohols and hydrogen are other by-products in this process. Hydrogen has a major role during this process; it regulates the expected by-products in this phase.

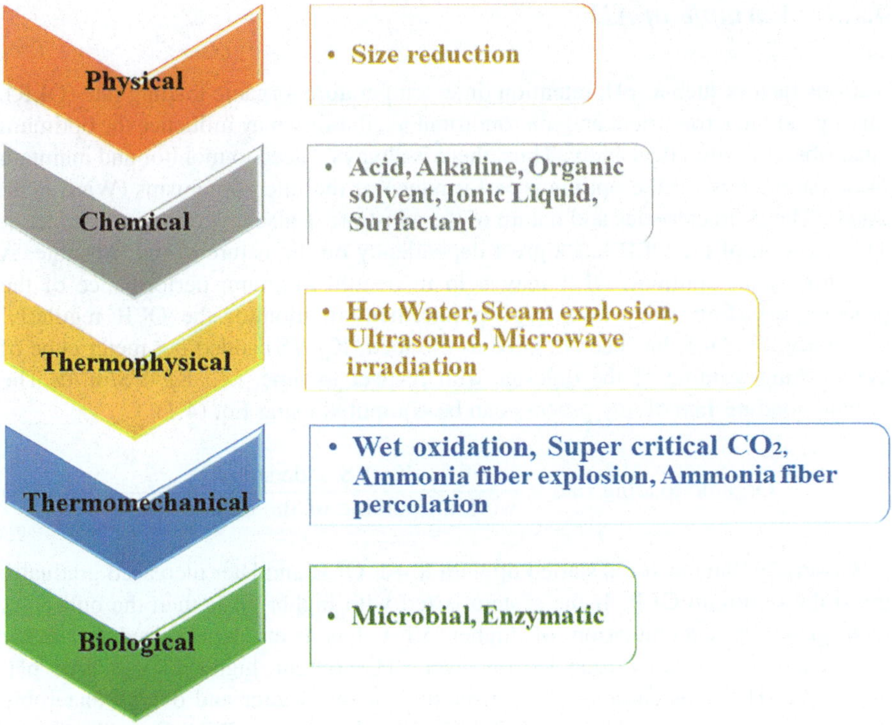

Fig. 4.3 Biochemistry of the anaerobic digestion process of food waste

A low partial hydrogen pressure results in hydrogen, carbon dioxide, and acetate. On the contrary, if partial pressure exerted by the hydrogen is high, then the formation of alcohols and VFAs takes place. Therefore, during the process, the condition should be optimal, i.e., to avoid high partial pressure, otherwise accumulation of VFA takes place which may result in the inhibition of the whole process. These products are furthermore decomposed under the optimal conditions for the production of biogas (Schink 1997).

The by-products of the acidogenesis phase are subjected to two different pathways. In the first pathway, hydrogen, acetate, and carbon dioxide can straightforwardly be used by methanogens for the production of methane. Hence, alcohols carrying greater than one carbon atoms and VFAs carrying greater than two carbon atoms are further degraded to hydrogen and acetate in this phase by obligatory hydrogen-producing bacteria (Bryant 1979; Schink 1997).

During the methanogenesis phase, anaerobic archaea convert acetate or hydrogen to methane and carbon dioxide.

4.1.3 Variable of AD

Various factors such as pH, retention time, temperature, organic loading rate (OLR), mixing, and macronutrient and micronutrient availability may influence the optimum anaerobic digestion decisively. Thus, there is always a need to monitor and maintain these parameters for the optimum performance of the microorganisms (Ward et al. 2008). The characteristics and nature of the substrate is also a very important factor. The decision of the OLR has a great dependency on the nature of the substrate. A decision of an optimum OLR may help in overall optimum performance of the process; therefore, there is a huge requirement to monitor the OLR regularly. Generally, the OLR has been defined as the feed (Kg VS) added per meter cube of the working volume of the digester with respect to time, i.e., Kg VS/m^3/d. The organic loading rate of any process can be computed using Eq. (4.1).

$$\text{Organic loading rate} = \frac{\text{Kg VS added/day}}{\text{working volume of the digester (m}^3\text{)}} \tag{4.1}$$

Generally, the reactor is started up with lower OLR and later increased gradually up to the optimum OLR. If the system is fed with higher OLR than the optimum, then generally accumulation of higher VFA has been experienced by many researchers and which further led to lower CH_4 content, higher CO_2, higher pH, and higher H_2S concentration, i.e., instability of the reactor and overall anaerobic digestion process (Mata-Alvarez et al. 2000; Bouallagui et al. 2004; Choudhary et al. 2020b). Nonetheless, reactors running over extremely low OLR have no techno-economic feasibilities because the real capacity of the reactor has not been utilized.

Apart from that, another crucial parameter is retention time and generally reported as hydraulic retention time (HRT). The HRT is the time for which liquid sludge exist in the digester. It is also generally called solid retention time (SRT), which indicates the duration spent by a solid particle within the reactor or with microorganisms (Appels et al. 2008). The HRT can be calculated using Eq. (4.2).

$$\text{HRT (days)} = \frac{V\ (\text{m}^3)}{Q(\text{m}^3/\text{day})} \tag{4.2}$$

HRT is calculated based on the following formula, where *V* is the working volume of the digester in m^3, *Q* is the flow rate of the sludge (m^3/day).

Generally, if the feed is complex to digest, then HRT is more significant; on the other hand, if the feed is easily digestible, then SRT is more important (Speece 2008). To increase the efficiency of the process, a short retention time is generally favorable. A shorter retention time may reduce the overall capital investment of the project(Chandra et al. 2012). There is always necessity of managing OLR and HRT in such a manner that optimum anaerobic digestion can occur. This means that while running the digester at higher OLR, the HRT should be appropriately higher to ensure an adequate interaction between the substrate and the microorganisms

(Demirer and Chen 2005). For continuously and semi-continuous anaerobic reactors, HRT and SRT are equal. Nonetheless, these reactors are not subjected to re-circulation; in the case of re-circulation, the HRT and SRT will increase.

Temperature is amongst the most important parameters which can affect the whole anaerobic digestion process decisively. Generally, temperature fluctuations during the anaerobic digestion process may be favorable for a certain group of microorganisms, but may not be favorable for the other groups. The process of methanogenesis is most affected by any fluctuation in the temperature during the anaerobic digestion process. Anaerobic digestion is performed at three temperature ranges, i.e., thermophilic, mesophilic, and psychrophilic. The growth optimal is around 10, 37, and 50 °C for thermophilic, mesophilic, and psychrophilic, respectively (Kashyap et al. 2003; Wiegel et al. 2007; Coelho et al. 2011).

pH is also an important parameter during the anaerobic digestion process. During the process of hydrolysis, acidogenesis, acetogenesis, and methanogenesis, there is a requirement of a wide range of pH (Mittal 1996). A significant number of microorganisms prefer the pH in the range of 7.0–8.5 (Kanokwan 2006). There are, however, microorganisms that can survive in an acidic and basic range of pH. The acidogenic microorganisms can survive in the acid range, i.e., near pH 5.0. Nevertheless, in accordance with the microorganisms involved in all the processes, the pH of the anaerobic process needs to be maintained in the range of 6.6–7.3 (Babel et al. 2004; Sitorus et al. 2013; Zhou et al. 2016). The pH beyond this range may affect the overall process, more specifically during the methanogenesis process (Kim et al. 2004; Taconi et al. 2008). An optimum OLR is highly recommended for the optimum pH during the whole process, which is attributed to the fact that if the anaerobic digester is subjected to the higher OLR, it may affect the intermediate by-products such as VFA which may further affect the pH adversely. It is recommended that the concentration of VFA should always be less than 2000 mg/L for an optimum anaerobic digestion process (Jain and Mattiasson 1998).

However, if the concentration of the VFA is higher in the system, it can be lowered by adding or maintaining the alkalinity in the system. The alkalinity can be defined as the buffering capacity of the system and generally measured in terms of mg/L as $CaCO_3$. If there is adequate alkalinity present in the system, then it can neutralize the high VFA concentration, and hence, this will lead the system towards the optimum pH, i.e., near to 7.0. Few substrates have the alkalinity by default such as animal dungs. On the other hand, few substrates have low-buffering capacities such as food wastes and lignocellulosic wastes (Banks and Humphreys 1998). Therefore, monitoring of alkalinity and VFA in such systems becomes mandatory. A great indicator of the stability of the anaerobic digestion process is the ratio of VFA to alkalinity. It is suggested that for optimum anaerobic digestion, this ratio should be nearly about 0.3–0.4. However, it should never exceed 0.8. The VFA/alkalinity ratio exceeding 0.8 shows system instability, and hence, process inhibition, or less efficiency of the process (Wang et al. 2012).

Nutrients are significantly important for any anaerobic digestion process. There must always be an abundance of nutrients in the system, and even a small shortage of them can cause process inhibition. The synthesis and growth of enzymes are associated with biochemical and metabolic pathways of the process's microorganisms. Generally, nutrients can be categorized into two types, i.e., micronutrients and macronutrients (Mara and Horan 2003).

The much known fundamental macronutrients are nitrogen (N), phosphorus (P), carbon (C), and sulfur (S). These are very important for the multiplication and growth of microorganisms. During the anaerobic digestion process, carbon and nitrogen levels play even more critical role. Nitrogen is very important for the overall development and growth of the microorganisms, whereas carbon acts as food for the microorganisms. Deficiency of nitrogen in any system may lead to unsatisfactory consumption of the carbon, or in other words, it will prohibit the growth of the microorganisms (Resch et al. 2011). As a result of that, the overall biogas production will be reduced. Therefore, the ratio of C to N is always a decisive parameter during the anaerobic digestion process (Hobsen et al. 1981; Chandra et al. 2012), and it can be adjusted/optimized by adjusting the ratios of substrates during the design of OLR.

4.2 Types of Lignocellulosic and Food Wastes

4.2.1 Lignocellulosic Material as a Substrate

As we have discussed in the previous section of this chapter, for the production of biogas, various substrates such as animal dungs, organic fraction of municipal solid waste, wastewater, sewage sludge, and agricultural residues can be used (Koniuszewska et al. 2020; Ferdeş et al. 2020; Atelge et al. 2020; Choudhary et al. 2020a, b, c). Amongst these substrates, several are lignocellulosic. There is ample availability of lignocellulosic substrates across the world. The carbohydrate content present in the lignocellulosic wastes makes it more attractive for the production of biogas via anaerobic digestion. Generally, lignocellulosic substrates can be divided into two categories, i.e., lignocellulosic residuals and cultivated feedstocks, known as energy crops. The major drawback with the lignocellulosic residuals is that they have a high percentage of lignin and therefore are less suitable for the utilization in anaerobic digestion. Due to this reason, only lignocellulosic residues as substrate (without pretreatment and co-digestion) have relatively low methane yield (Kainthola et al. 2019a). On the other hand, energy crops have a smaller fraction of lignin when compared to lignocellulosic residuals. Energy crops primarily consist of cellulose and hemicellulose (Kabir et al. 2015; Chen et al. 2018). Moreover, along with cellulose and hemicellulose, the energy crops' residues consist of various non-structural carbohydrates such as fructose, fructans, pectins, glucose, sucrose, and extractives (Kabir et al. 2014). The utilization of lignocellulosic wastes such as giant reed stems, wheat straw (Dell'Omo and Spena 2020), rice straw (Liu et al. 2019), corn stover (You et al. 2019), and Napier grass (Phuttaro et al. 2019) is common across the world (Kainthola et al. 2019a).

4.2.2 *Food Waste as a Substrate*

Food waste (FW) is a great substrate for anaerobic digestion, and it has a huge potential for producing biomethane (Pramanik et al. 2019; Choudhary et al. 2020b). FW generally consists of complex and organic material. There are various types of FW, such as vegetable and fruit waste, brewery waste, kitchen waste, and dairy waste (Xu et al. 2018). The composition and characteristics of FW vary with the geographical area (Meng et al. 2015; Xu et al. 2018). FW consists of carbohydrates, fats, protein, and sugar. FW is generally acidic and has less alkalinity. Fisgativa et al. (2016) studied various types of food waste and reported that the average pH of FW was 5.1; the C/N was reported at 18.5%. Also, they have reported carbohydrates, protein, and fat fraction in the FW as 57.2%, 62.2%, and 15%, respectively (Fisgativa et al. 2016). Generally, carbohydrates and protein have a rapid hydrolysis rate when compared to lipids.

Vegetable and fruit waste have low lipid and comparatively higher cellulose content. Due to the presence of animal fat and vegetable oil, the kitchen waste carries high lipid content (Bong et al. 2018). The lipid content may vary in the range of 11.8–33.22% in the case of fruit and vegetable and kitchen waste, respectively (Wang et al. 2014; Yong et al. 2015). FW with higher lipid content can produce significantly higher biomethane when compared to protein and carbohydrates (Li et al. 2017). Nevertheless, very high lipid content may inhibit the process as well because of the formation of a high concentration of long-chain fatty acids (Leung and Wang 2016; Li et al. 2017). FW carrying significantly higher carbohydrate may decisively affect the C/N ratio. This is attributed to the fact that high carbohydrate content may increase carbon content, and hence, quick acidification may occur due to heavy loading of carbon into the system (Li et al. 2017).

The total solid may fall in the range of 10.7–41% in any type of food waste which indicates significantly higher moisture content, i.e., about 60–90%. Due to the presence of higher moisture content, FW is also considered as a rapidly digestible substrate for the anaerobic digestion (Zhang et al. 2014).

The C/N of the FW may vary in the range of 12.7–28.84. The pH generally falls in the acidic range, i.e., 4.1–6.5. The biomethane potential of every variety of FW may vary in the range of 346–551.4 mL/gVS, which is comparatively higher than animal dungs and various other wastes (Lehtomäki et al. 2007).

4.3 Mono-digestion and its Limitations

When only one substrate is fed into the digester, such process is referred to as mono-digestion. Mono-digestion of lignocellulosic waste and FW has several limitations that will be discussed further.

During mono-digestion of FW and lignocellulosic substrates when the anaerobic digester runs at comparatively higher OLRs, the accumulation of VFA is a major limitation of the anaerobic mono-digestion process. Due to this reason, the process faces several challenges such as instability, ammonia inhibition, insufficient alkalinity, production of H_2S, and less ultimate biomethane potential.

On the other hand, if the digester runs at lower OLRs, then the process becomes economically unfeasible. Secondly, to enhance the biomethane yield of the process, often various pretreatments are suggested, which again makes the process less environment friendly and economically less attractive (Mata-Alvarez et al. 2011; Nghiem et al. 2017).

In the case of lignocellulosic substrates, the C/N ratio is significantly high, which creates nitrogen deficiency during the anaerobic digestion process. Therefore, the risk of production of inhibitors such as furfural and hydroxymethylfurfural becomes very high. Moreover, due to the presence of lignin, hydrolysis occurs at a relatively slower pace and consequently the HRT of the process increases significantly (Kabir et al. 2013; Yong et al. 2015; Achinas et al. 2017). Apart from this, the low C/N ratio is also a major limitation of the mono-digestion of FW (David et al. 2018).

4.4 Pretreatment Technologies

4.4.1 Pretreatment of FW

FW generally consists of a rapidly digestible fraction and complex organic fraction. The rapidly digestible fraction in FW is often carbohydrates, and the complex organic fractions are lipids and proteins. Hence, complete biomethane potential is not achieved without pretreatment. With the help of pretreatment, biodegradability of recalcitrant organic fraction of FW can be increased significantly. It is a well-understood fact that in the case of complex substrates such as lignocellulosic substrate, hydrolysis is a rate-limiting phase, whereas for rapidly digestible substrates such as FW, methanogenesis is the rate-limiting step (Li et al. 2018). The efficiency of hydrolysis can be decisively affected by the operating temperature and nature of the organic matter (Srisowmeya et al. 2020). Various methods have been used to speed up the hydrolysis rate.

Often during the physical pretreatments, the size of substrates is reduced and the morphological structure of the substrates is also changed and therefore increases its solubilization (Ma et al. 2018). An increment of 28% in methane has been noted while reducing the particle size (by mechanical grinding) of the FW by 53%. Nevertheless, excessive reduction of the particle size has resulted in the accumulation of VFA and later high methane content. Hence, during the mechanical pretreatment, the primary objective should be to optimize the particle size of the substrate.

The solubility and accessibility of the FW can also be increased by ultrasonication. Ultrasonication reduces the complexity of the substrate by reducing its particle size mechanically. By ultrasonication, methane yield can be increased by 1.21–1.58 times (Nasr et al. 2012). For rapidly digestible substrates such as kitchen waste, microwave pretreatments along with electromagnetic energy have neither been found much effective for hydrolysis nor these are found economically feasible (Shahriari et al. 2013).

During the thermal pretreatment, the surface area of the organic matter is increased and therefore increasing the contact between microorganisms and the organic matter, thereby leading to better methane yield. Longer retention time (>4 h) and higher temperature (>120 °C) during the pretreatment have shown adverse effects on proteins and carbohydrates and resulted in bioproducts such as melanoidins and amodori which are difficult to degrade under anaerobic conditions (Vavilin et al. 2008). Additionally, during longer retention time and thermal pretreatment, loss of volatile solids and sugar occurs (Eskicioglu et al. 2006). Therefore, thermal pretreatment for longer retention time and a higher temperature is not recommended (Ariunbaatar et al. 2014b). The FW with the higher complex fraction ozonation pretreatment is more appropriate (Ariunbaatar et al. 2014b).

Pretreatment with the help of hydrolytic enzymes is also a highly efficient technique (Ma et al. 2018). It can increase the hydrolytic efficiency by substrate-specific action of enzymes owing to high energy recovery. Although pretreatment with the help of hydrolytic enzymes ensures enhanced methane yield, it has economical limitations on a commercial scale (Ma et al. 2018).

Due to the wide diversity of FW, it is very difficult to choose the most suitable pretreatment method though the application of pretreatments is important to ensure the utmost efficiency and methane yield (Ariunbaatar et al. 2014a).

4.4.2 Pretreatment of Lignocellulosic Biomass

In case of lignocellulosic waste, the biodegradable fraction becomes unavailable for the microorganisms involved in the process, and hence, it results in lower methane yield. Sometimes, even this may become the reason for the accumulation of inhibitory compounds within the digester. Therefore, various process enhancement techniques are recommended which increases the hydrolysis rate and overall degradability of the substrate. It is recommended that pretreatment must fulfill certain criteria such as (a) during the process, there should not be any formation of inhibitory substances; (b) there should not be much loss of carbohydrate; and (c) it should be cost-effective. The fundamental of pretreatment of any lignocellulosic biomass includes two processes, i.e., separation of lignin from the overall structure and exposing the rest matrix to degrading enzymes and disruption of the lignocellulosic matrix into cellulose, hemicellulose, and lignin (Sun and Cheng 2002; Vivekanand et al. 2012). Alike FW, hydrolysis is a rate-limiting step in the case of lignocellulosic biomasses, especially in cases of recalcitrant agriculture residues.

Figure 4.4 represents the various pretreatment techniques for lignocellulosic biomass. Often physical, chemical, thermophysical, thermochemical, and biological pretreatments are done to treat the lignocellulosic biomasses. A very basic and preliminary technique to break the lignin structure is grinding (size reduction).

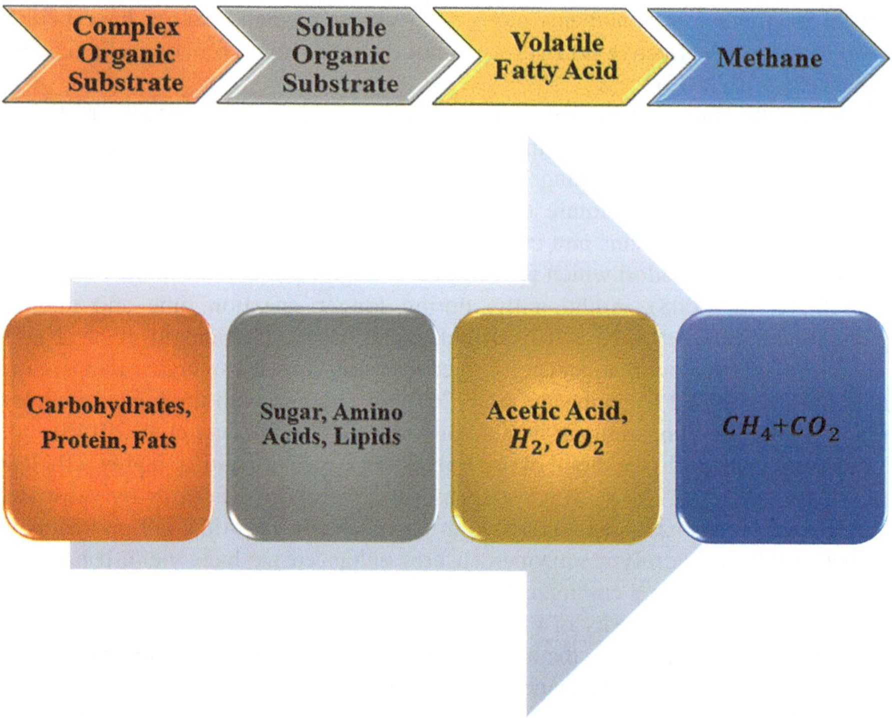

Fig. 4.4 Pretreatment techniques for lignocellulosic biomass

Grinding helps in increasing the surface area, porosity, altering the polymerization degree, and diminishing the crystallinity of the biomass (Hajji and Rhachi 2013; Zhang and Banks 2013; Lindmark et al. 2014).

Various chemical agents are also used as a catalyst for disrupting and delignification of the bond of the lignocellulosic matrix in various biomass substrates (Boonterm et al. 2016). Different acids used during the pretreatment process are HNO_3, H_2O_2, H_2SO_4, HCl, etc.

Lime, ammonia, NaOH, Na_2CO_3, etc., are used in alkaline pretreatments. Pretreatment by alkaline agents can increase the surface area, porosity, altering the polymerization degree, and disrupting the lignin of the biomass. Aqueous ethanol and acetone–butanol–ethanol are used for pretreatments as an organic solvent.

Although the usage of chemical agents is simple and effective, sometimes it is observed that these pretreatments produce inhibitory compounds that may further need treatment or they may inhibit the anaerobic digestion process.

Temperature is also used as a tool for pretreatment of different lignocellulosic substrates, and such methods are referred to as thermal pretreatment. Thermal pretreatment can increase the porosity of the surface and enhance the destruction of the lignin layer. Any liquid at higher temperatures hydrolyses the lignocellulosic fraction of the biomass. This is because at high temperature and pressure, water molecules break down into OH^- and H_3O^+ that further assists in catalytic conversion of lignocellulosic biomass. Delignification and improved porosity can be achieved at a higher temperature and short reaction time with dilute H_2SO_4, ammonia recycle percolation (APR), or steam explosion. Usually, after thermal pretreatments, an increased cellulosic content is achieved. Moreover, during this process, removal of some hemicellulosic content also takes place, which further helps in increase of the surface for enzymatic attack.

Under optimum microwave intensity and irradiance time, the hemicellulose and cellulose removal efficiency can be improved up to 30.6–43.3% (Ma et al. 2009). In fact, with the help of microwave pretreatment, a delignification of 6% can be achieved and as a result of that, hydrolysis is improved (Zhu et al. 2005). Although furan derivatives, phenolic substances, etc., are some inhibitory compounds that are produced during pretreatment with microwave irradiation, which later disturbs the anaerobic digestion. Therefore, such inhibitors are also a decisive step during bioconversion of lignocellulosic substrate to methane (Palmqvist and Hahn-Hägerdal 2000; Putro et al. 2016). The production of inhibitory compounds and higher operation cost and energy demand are the biggest drawbacks of physical, thermal, and chemical pretreatments.

On the other hand, biological pretreatments with microbes and enzymes provide the significantly environment friendly solution for the bioconversion of lignocellulosic substrates. Though, one of the main problems with pretreatments with microbes and enzymes is lesser surface area accessibility of microbes and microbial products on lignocellulose for efficient conversion to hydrolytic products (Kainthola et al. 2019b). The microbial route was found to be one of the economic and effective ways for the delignification and cellulolytic hydrolysis (Ghosh and Bhattacharyya 1999).

Microorganisms such as white, brown, and soft-rot fungi are involved in lignin and hemicellulose degradation. Temperature and pH during the anaerobic digestion process can hinder the biological pretreatment process. In such circumstances, extremophiles are also found to be a good alternative because they can sustain even in the harsh environment. During the last decade, various microorganisms have been developed that can sustain in an extreme environment and work efficiently. *Clostridium thermocellum*, *Caldicellulosiruptor saccharolyticus*, and *Caldicellulosiruptor bescii* DSM 6725 are some thermophilic bacteria that have gained huge attention in the past 10 years (Li et al. 2014).

4.5 Co-Digestion and its Advantages

When two substrates are digested simultaneously in an anaerobic digester for the production of biogas, the process is generally referred to as anaerobic co-digestion. In the last few decades, anaerobic co-digestion has gained ample popularity amongst the researchers and industries because anaerobic co-digestion has improved the process in various aspects when compared to mono-digestion. As in most of the cases, co-digestion provides missing nutrients, buffer, and sometimes moisture content required in the digester which has a positive synergic effect on the overall process (Mata-Alvarez et al. 2000).

We have already discussed the importance of the C/N ratio in the anaerobic digestion process. Mixing any substrate such as any lignocellulosic substrate whose C/N is comparatively high with a substrate whose C/N is low such as FW can optimize the overall C/N of the process.

In case of lignocellulosic substrates, high C/N ratio, lignin percentage, and contamination with pesticides can be resolved with the help of anaerobic co-digestion (Kainthola et al. 2019a). Some of the advantages of anaerobic co-digestion are (a) enhancement of the overall process stabilization, (b) weakening the inhibitory effects, (c) establishment of adequate moisture content within the digester, (d) higher OLR when compared to mono-digestion, (e) positive synergism during the digestion, (f) micronutrient and macronutrient balance, (f) enhance the economic feasibility of the process, (g) enhanced biomethane potential, and (h) improved digestibility of the individual substrate (cellulose and hemicellulose) and buffering capacity (Griffin et al. 1998; Zheng et al. 2014; Mata-Alvarez et al. 2014).

4.6 Recent Developments in Co-Digestion of Lignocellulosic Biomass and Food Wastes

In the last few years, researchers have performed extensive research on the co-digestion of lignocellulosic biomass and food wastes. In this regard, Kainthola et al. (2020) co-digested rice straw with food waste for the determination of methane yield for various C/N (i.e., 25, 30, and 35) ratios using a 1 L anaerobic digester. They have found almost similar methane yield for all the C/N ratios, i.e., 294.17 ± 3.78 L/KgVS. Besides, they have reported 71.09% more methane yield when compared to mono-digestion. However, in the same study, during the optimization of the process (i.e., pH = 7.32, C/N = 30, and F/M = 1.87), co-digestion resulted in 94.41% more methane yield when compared to mono-digestion (Kainthola et al. 2020).

Mu et al. (2020) used urban-derived food waste and co-digested it with yard waste. In this investigation, they found co-digestion a more promising alternative when compared to mono-digestion. They have found that due to co-digestion various parameters such as C/N ratio and buffering capacity have improved. The mono-digestion of yard waste resulted in a yield of 49.0 ± 5.0 mL methane/g VS, while co-digestion of yard waste and food waste resulted in 360.0 ± 30.2 mL methane/g VS (Mu et al. 2020).

David et al. (2018) co-digested three types of lignocellulosic substrates (corn stover (CS), Prairie cordgrass (PCG), and unbleached paper (UBP)) with food waste at thermophilic temperature. During this investigation, they reported that co-digestion can overcome the limitation of mono-digestion of individual substrates specifically, low buffering capacity, accumulation of VFA, and low C/N in case of

FW. All lignocellulosic wastes co-digested with food wastes have shown synergetic enhancement in methane yield. However, the highest methane yield was reported for the combination of FW-PCG-CS followed by FW-PCG. A better volatile solid reduction was found in those two mixtures when compared to mono-digestion. David et al. (2018) also stated that pretreatment of lignocellulosic substrates increased the readily available sugar for the anaerobic digestion; however, it increased the cost of the overall process. David et al. (2018) also conducted their investigation without any pretreatment of the lignocellulosic substrate and emphasized the fact that during the co-digestion, consortia can play a vital role if pretreatment is not performed. They also reported that although maintaining thermophilic temperature during the digestion process will increase the costs of the overall process, it provides extra advantages of digesting the substrate at higher loading rate and at lesser retention time (David et al. 2018).

Helenas Perin et al. (2020) studied the influence of garden waste on the anaerobic digestion of food waste. In this study, they noted 86 L/d biogas production, at OLR of 0.47 L/g VS in specific methane yield when compared to mono-digestion of food waste (17 L/d biogas production at OLR 0.006 L/g VS in specific methane yield), thus indicating the possibility of optimization of the overall process (Helenas Perin et al. 2020).

Panigrahi et al. (2020) studied the co-digestion of food waste & yard waste and stated that it is an efficient technique for sustainable bioenergy generation. They reported for maximum methane production, high C/N ratio, and recalcitrant nature of yard waste are a huge bottleneck. Therefore, they thermally pretreated the yard in this study, and further, it was co-digested with the food waste to enhance nutrient balance for the overall methane production. Besides, the optimization of F/M (food/microorganism) was also performed. They reported the highest methane potential of 431 mL/gVS when F/M ratio was 1.5 (Panigrahi et al. 2020).

Shi et al. (2018) investigated the co-digestion of wheat straw and FW using five different ratios at mesophilic and thermophilic temperatures. They reported that the synergic effects improved the overall stability and performance of the process at the same (OLR = 3.0 g VS/L/d). Both the reactors of mono-digestion showed system instability. However, reactors running at thermophilic temperature have shown 4.9–14.8% higher methane yield when compared to mesophilic reactors (Shi et al. 2018).

Tayyab et al. (2019) investigated the biomethane potential of pretreated *Parthenium* weed and also studied its co-digestion with catering food. They set up various lab-scale digesters with different mixing ratios (0:100, 20:80, 60:40, 40:60, 80:20, and 100:0 on total solid basis) for the determination of methane yield and to study the effect of co-digestion. They observed that the reactor with 60% catering food and 40% pretreated Parthenium weed yielded maximum accumulative biogas (5532 mL/L). On the basis of their experimental study, Tayyab et al. (2019) concluded that pretreated *Parthenium* weed as a potential substrate if co-digested with catering food waste.

Zou et al. (2020) aimed to accelerate the hydrolysis of corn cob during anaerobic digestion with the help of FW. The authors used FW as an acidic agent for the pretreatment of the corn cob. This is attributed to the fact that during the anaerobic digestion of FW, acidification occurs which can accelerate the hydrolysis of lignocellulose. In the beginning, the optimum mixing ratio of FW, corn cob, was reported as 1:3. The hydrolysis rate was increased by 28% when compared to mono-digestion of corn cob. A reduction of 6.7% in cellulose crystallinity and 13.2% in cellulose was also achieved at this mixing ratio. However, during the stage of methane generation, the mixing ratio of food waste and corn cob reported as 1:6 has shown maximum methane production as 401.6 mL/g·VS. During the kinetic study of cellulose/hemicellulose degradation, it was found that pretreatment of corn cob with food waste improved the degradation of cellulose (Zou et al. 2020).

4.7 Conclusion

This chapter focuses on the different types of types of lignocellulosic (LW) and food wastes (FW) that can be utilized as substrates for the production of biogas through anaerobic digestion under various temperature ranges. The process of mono-digestion of both the substrates (LW and FW) has several disadvantages. Therefore, the pretreatment techniques have been recognized as an important step before the digestion process of both substrates. There is clear scientific evidence present in which pretreatment has been found to be a recognized technique in the context of improved specific biomethane potential. Nevertheless, pretreatment has not proved to be cost-effective and environmental friendly for anaerobic digestion. On the other hand, the co-anaerobic digestion of LW with FW proved to be a more promising alternative when compared to mono-anaerobic digestion of an individual for synergistic enhancements in the context to trace elements, buffering capacity, high easily biodegradable components, and C/N ratio.

References

Achinas S, Achinas V, Euverink GJW (2017) A technological overview of biogas production from biowaste. Engineering 3:299–307. https://doi.org/10.1016/J.ENG.2017.03.002

Amigun B, Sigamoney R, von Blottnitz H (2008) Commercialisation of biofuel industry in Africa: a review. Renew Sust Energ Rev 12:690–711. https://doi.org/10.1016/j.rser.2006.10.019

Appels L, Baeyens J, Degrève J, Dewil R (2008) Principles and potential of the anaerobic digestion of waste-activated sludge. Prog Energy Combust Sci 34:755–781. https://doi.org/10.1016/j.pecs.2008.06.002

Ariunbaatar J, Panico A, Esposito G et al (2014a) Pretreatment methods to enhance anaerobic digestion of organic solid waste. Appl Energy 123:143–156. https://doi.org/10.1016/j.apenergy.2014.02.035

Ariunbaatar J, Panico A, Frunzo L et al (2014b) Enhanced anaerobic digestion of food waste by thermal and ozonation pretreatment methods. J Environ Manag 146:142–149. https://doi.org/10.1016/j.jenvman.2014.07.042

Atelge MR, Krisa D, Kumar G et al (2020) Biogas production from organic waste: recent Progress and perspectives. Waste Biomass Valorization 11:1019–1040. https://doi.org/10.1007/s12649-018-00546-0

Babel S, Fukushi K, Sitanrassamee B (2004) Effect of acid speciation on solid waste liquefaction in an anaerobic acid digester. Water Res 38:2417–2423. https://doi.org/10.1016/j.watres.2004.02.005

Banks CJ, Humphreys PN (1998) The anaerobic treatment of a ligno-cellulosic substrate offering little natural pH buffering capacity. Water Sci Technol 38:29–35. https://doi.org/10.1016/S0273-1223(98)00494-6

Bong CPC, Lim LY, Lee CT et al (2018) The characterisation and treatment of food waste for improvement of biogas production during anaerobic digestion – a review. J Clean Prod 172:1545–1558. https://doi.org/10.1016/j.jclepro.2017.10.199

Boonterm M, Sunyadeth S, Dedpakdee S et al (2016) Characterization and comparison of cellulose fiber extraction from rice straw by chemical treatment and thermal steam explosion. J Clean Prod 134:592–599. https://doi.org/10.1016/j.jclepro.2015.09.084

Börjesson P, Mattiasson B (2008) Biogas as a resource-efficient vehicle fuel. Trends Biotechnol 26:7–13. https://doi.org/10.1016/j.tibtech.2007.09.007

Bouallagui H, Torrijos M, Godon JJ et al (2004) Two-phases anaerobic digestion of fruit and vegetable wastes: bioreactors performance. Biochem Eng J 21:193–197. https://doi.org/10.1016/j.bej.2004.05.001

Bryant MP (1979) Microbial methane production—theoretical aspects. J Anim Sci 48:193–201. https://doi.org/10.2527/jas1979.481193x

Chandra R, Takeuchi H, Hasegawa T (2012) Methane production from lignocellulosic agricultural crop wastes: a review in context to second generation of biofuel production. Renew Sust Energ Rev 16:1462–1476. https://doi.org/10.1016/j.rser.2011.11.035

Chen D, Gao A, Cen K et al (2018) Investigation of biomass torrefaction based on three major components: hemicellulose, cellulose, and lignin. Energy Convers Manag 169:228–237. https://doi.org/10.1016/j.enconman.2018.05.063

Choudhary A, Kumar A, Govil T et al (2020a) Sustainable production of biogas in large bioreactor under psychrophilic and mesophilic conditions. J Environ Eng 146:04019117. https://doi.org/10.1061/(ASCE)EE.1943-7870.0001645

Choudhary A, Kumar A, Kumar S (2020b) Techno-economic analysis, kinetics, global warming potential comparison and optimization of a pilot-scale unheated semi-continuous anaerobic reactor in a hilly area: for north Indian hilly states. Renew Energy 155:1181–1190. https://doi.org/10.1016/j.renene.2020.04.034

Choudhary A, Kumar A, Kumar S (2020c) National municipal solid waste energy and global warming potential inventory: India. J Hazard Toxic Radioact Waste 24:06020002. https://doi.org/10.1061/(ASCE)HZ.2153-5515.0000521

Coelho NMG, Droste RL, Kennedy KJ (2011) Evaluation of continuous mesophilic, thermophilic and temperature phased anaerobic digestion of microwaved activated sludge. Water Res 45:2822–2834. https://doi.org/10.1016/j.watres.2011.02.032

David A, Govil T, Tripathi AK et al (2018) Thermophilic anaerobic digestion: enhanced and sustainable methane production from co-digestion of food and lignocellulosic wastes. Energies 11:2058. https://doi.org/10.3390/en11082058

Dell'Omo PP, Spena VA (2020) Mechanical pretreatment of lignocellulosic biomass to improve biogas production: comparison of results for giant reed and wheat straw. Energy 203:117798. https://doi.org/10.1016/j.energy.2020.117798

Demirer GN, Chen S (2005) Two-phase anaerobic digestion of unscreened dairy manure. Process Biochem 40:3542–3549. https://doi.org/10.1016/j.procbio.2005.03.062

Deublein D, Steinhauser A (2008) History and status to date in other countries. In: Deublein D, Steinhauser A (eds) Biogas from waste and renewable resources. Wiley-VCH Verlag GmbH & Co, Mörlenbach, pp 35–42

Dolan T, Cook MB, Angus AJ (2011) Financial appraisal of wet mesophilic AD technology as a renewable energy and waste management technology. Sci Total Environ 409:2460–2466. https://doi.org/10.1016/j.scitotenv.2011.03.011

Eskicioglu C, Kennedy KJ, Droste RL (2006) Characterization of soluble organic matter of waste activated sludge before and after thermal pretreatment. Water Res 40:3725–3736. https://doi.org/10.1016/j.watres.2006.08.017

Ferdeş M, Dincă MN, Moiceanu G et al (2020) Microorganisms and enzymes used in the biological pretreatment of the substrate to enhance biogas production: a review. Sustainability 12:7205. https://doi.org/10.3390/su12177205

Fernandes TV, Klaasse Bos GJ, Zeeman G et al (2009) Effects of thermo-chemical pre-treatment on anaerobic biodegradability and hydrolysis of lignocellulosic biomass. Bioresour Technol 100:2575–2579. https://doi.org/10.1016/j.biortech.2008.12.012

Fisgativa H, Tremier A, Dabert P (2016) Characterizing the variability of food waste quality: a need for efficient valorisation through anaerobic digestion. Waste Manag 50:264–274. https://doi.org/10.1016/j.wasman.2016.01.041

Forgács G (2012) Biogas production from Citrus wastes and chicken feather: pretreatment and co-digestion

Ghosh A, Bhattacharyya BC (1999) Biomethanation of white rotted and brown rotted rice straw. Bioprocess Eng 20:297–302. https://doi.org/10.1007/s004490050594

Griffin ME, McMahon KD, Mackie RI, Raskin L (1998) Methanogenic population dynamics during start-up of anaerobic digesters treating municipal solid waste and biosolids. Biotechnol Bioeng 57:342–355. https://doi.org/10.1002/(SICI)1097-0290(19980205)57:3<342::AID-BIT11>3.0.CO;2-I

Hajji A, Rhachi M (2013) The influence of particle size on the performance of anaerobic digestion of municipal solid waste. Energy Procedia 36:515–520. https://doi.org/10.1016/j.egypro.2013.07.059

Helenas Perin JK, Biesdorf Borth PL, Torrecilhas AR et al (2020) Optimization of methane production parameters during anaerobic co-digestion of food waste and garden waste. J Clean Prod 272:123130. https://doi.org/10.1016/j.jclepro.2020.123130

Hobsen PN, Bousfield S, Summers R (1981) Methane production from agricultural and domestic wastes. Springer

Jain SR, Mattiasson B (1998) Acclimatization of methanogenic consortia for low pH biomethanation process. Biotechnol Lett 20:771–775. https://doi.org/10.1023/B:BILE.0000015920.45724.29

Jiang X, Sommer SG, Christensen KV (2011) A review of the biogas industry in China. Energy Policy 39:6073–6081. https://doi.org/10.1016/j.enpol.2011.07.007

Kabir MM, del Pilar Castillo M, Taherzadeh MJ, Horváth IS (2013) Effect of the N-Methylmorpholine-N-oxide (NMMO) pretreatment on anaerobic digestion of Forest residues. Bioresources 8:5409–5423

Kabir MM, Niklasson C, Taherzadeh MJ, Horváth IS (2014) Biogas production from lignocelluloses by N-methylmorpholine-N-oxide (NMMO) pretreatment: effects of recovery and reuse of NMMO. Bioresour Technol 161:446–450. https://doi.org/10.1016/j.biortech.2014.03.107

Kabir MM, Forgács G, Sárvári Horváth I (2015) Biogas from lignocellulosic materials. In: Karimi K (ed) Lignocellulose-based bioproducts. Springer International Publishing, Cham, pp 207–251

Kainthola J, Kalamdhad AS, Goud VV (2019a) A review on enhanced biogas production from anaerobic digestion of lignocellulosic biomass by different enhancement techniques. Process Biochem 84:81–90. https://doi.org/10.1016/j.procbio.2019.05.023

Kainthola J, Kalamdhad AS, Goud VV, Goel R (2019b) Fungal pretreatment and associated kinetics of rice straw hydrolysis to accelerate methane yield from anaerobic digestion. Bioresour Technol 286:121368. https://doi.org/10.1016/j.biortech.2019.121368

Kainthola J, Kalamdhad AS, Goud VV (2020) Optimization of process parameters for accelerated methane yield from anaerobic co-digestion of rice straw and food waste. Renew Energy 149:1352–1359. https://doi.org/10.1016/j.renene.2019.10.124

Kanokwan B (2006) Online monitoring and control of the biogas process. Technical University of Denmark

Kashyap DR, Dadhich KS, Sharma SK (2003) Biomethanation under psychrophilic conditions: a review. Bioresour Technol 87:147–153. https://doi.org/10.1016/S0960-8524(02)00205-5

Kim IS, Hwang MH, Jang NJ et al (2004) Effect of low pH on the activity of hydrogen utilizing methanogen in bio-hydrogen process. Int J Hydrog Energy 29:1133–1140. https://doi.org/10.1016/j.ijhydene.2003.08.017

Koniuszewska I, Korzeniewska E, Harnisz M, Czatzkowska M (2020) Intensification of biogas production using various technologies: a review. Int J Energy Res 44:6240–6258. https://doi.org/10.1002/er.5338

Lehtomäki A, Huttunen S, Rintala JA (2007) Laboratory investigations on co-digestion of energy crops and crop residues with cow manure for methane production: effect of crop to manure ratio. Resour Conserv Recycl 51:591–609. https://doi.org/10.1016/j.resconrec.2006.11.004

Leung DYC, Wang J (2016) An overview on biogas generation from anaerobic digestion of food waste. Int J Green Energy 13:119–131. https://doi.org/10.1080/15435075.2014.909355

Li W, Zheng P, Guo J et al (2014) Characteristics of self-alkalization in high-rate denitrifying automatic circulation (DAC) reactor fed with methanol and sodium acetate. Bioresour Technol 154:44–50. https://doi.org/10.1016/j.biortech.2013.11.097

Li Y, Jin Y, Borrion A et al (2017) Effects of organic composition on mesophilic anaerobic digestion of food waste. Bioresour Technol 244:213–224. https://doi.org/10.1016/j.biortech.2017.07.006

Li W, Loh K-C, Zhang J et al (2018) Two-stage anaerobic digestion of food waste and horticultural waste in high-solid system. Appl Energy 209:400–408. https://doi.org/10.1016/j.apenergy.2017.05.042

Lindmark J, Lagerkvist A, Nilsson E et al (2014) Evaluating the effects of electroporation pre-treatment on the biogas yield from ley crop silage. Appl Biochem Biotechnol 174:2616–2625. https://doi.org/10.1007/s12010-014-1213-7

Liu Y, Fang J, Tong X et al (2019) Change to biogas production in solid-state anaerobic digestion using rice straw as substrates at different temperatures. Bioresour Technol 293:122066. https://doi.org/10.1016/j.biortech.2019.122066

Ma H, Liu W-W, Chen X et al (2009) Enhanced enzymatic saccharification of rice straw by microwave pretreatment. Bioresour Technol 100:1279–1284. https://doi.org/10.1016/j.biortech.2008.08.045

Ma C, Liu J, Ye M et al (2018) Towards utmost bioenergy conversion efficiency of food waste: pretreatment, co-digestion, and reactor type. Renew Sust Energ Rev 90:700–709. https://doi.org/10.1016/j.rser.2018.03.110

Mara D, Horan NJ (2003) Handbook of water and wastewater microbiology. Elsevier

Mata-Alvarez J, Macé S, Llabrés P (2000) Anaerobic digestion of organic solid wastes. An overview of research achievements and perspectives. Bioresour Technol 74:3–16. https://doi.org/10.1016/S0960-8524(00)00023-7

Mata-Alvarez J, Dosta J, Macé S, Astals S (2011) Codigestion of solid wastes: a review of its uses and perspectives including modeling. Crit Rev Biotechnol 31:99–111. https://doi.org/10.3109/07388551.2010.525496

Mata-Alvarez J, Dosta J, Romero-Güiza MS et al (2014) A critical review on anaerobic co-digestion achievements between 2010 and 2013. Renew Sust Energ Rev 36:412–427. https://doi.org/10.1016/j.rser.2014.04.039

Meng Y, Li S, Yuan H et al (2015) Evaluating biomethane production from anaerobic mono- and co-digestion of food waste and floatable oil (FO) skimmed from food waste. Bioresour Technol 185:7–13. https://doi.org/10.1016/j.biortech.2015.02.036

Mittal K (1996) Biogas systems: principles and applications. New Age International Limited Publishers, New Delhi

Mu L, Zhang L, Zhu K et al (2020) Anaerobic co-digestion of sewage sludge, food waste and yard waste: synergistic enhancement on process stability and biogas production. Sci Total Environ 704:135429. https://doi.org/10.1016/j.scitotenv.2019.135429

Naik SN, Goud VV, Rout PK, Dalai AK (2010) Production of first and second generation biofuels: a comprehensive review. Renew Sust Energ Rev 14:578–597. https://doi.org/10.1016/j.rser.2009.10.003

Nasr N, Elbeshbishy E, Hafez H et al (2012) Comparative assessment of single-stage and two-stage anaerobic digestion for the treatment of thin stillage. Bioresour Technol 111:122–126. https://doi.org/10.1016/j.biortech.2012.02.019

Nghiem LD, Koch K, Bolzonella D, Drewes JE (2017) Full scale co-digestion of wastewater sludge and food waste: bottlenecks and possibilities. Renew Sust Energ Rev 72:354–362. https://doi.org/10.1016/j.rser.2017.01.062

Palmqvist E, Hahn-Hägerdal B (2000) Fermentation of lignocellulosic hydrolysates. II: inhibitors and mechanisms of inhibition. Bioresour Technol 74:25–33. https://doi.org/10.1016/S0960-8524(99)00161-3

Panigrahi S, Sharma HB, Dubey BK (2020) Anaerobic co-digestion of food waste with pretreated yard waste: a comparative study of methane production, kinetic modeling and energy balance. J Clean Prod 243:118480. https://doi.org/10.1016/j.jclepro.2019.118480

Phuttaro C, Sawatdeenarunat C, Surendra KC et al (2019) Anaerobic digestion of hydrothermally-pretreated lignocellulosic biomass: influence of pretreatment temperatures, inhibitors and soluble organics on methane yield. Bioresour Technol 284:128–138. https://doi.org/10.1016/j.biortech.2019.03.114

Pramanik SK, Suja FB, Zain SM, Pramanik BK (2019) The anaerobic digestion process of biogas production from food waste: prospects and constraints. Bioresour Technol Rep 8:100310. https://doi.org/10.1016/j.biteb.2019.100310

Putro JN, Soetaredjo FE, Lin S-Y et al (2016) Pretreatment and conversion of lignocellulose biomass into valuable chemicals. RSC Adv 6:46834–46852. https://doi.org/10.1039/C6RA09851G

Rajendran K, Aslanzadeh S, Taherzadeh MJ (2012) Household biogas digesters—a review. Energies 5:2911–2942. https://doi.org/10.3390/en5082911

Resch C, Wörl A, Waltenberger R et al (2011) Enhancement options for the utilisation of nitrogen rich animal by-products in anaerobic digestion. Bioresour Technol 102:2503–2510. https://doi.org/10.1016/j.biortech.2010.11.044

Schink B (1997) Energetics of syntrophic cooperation in methanogenic degradation. Microbiol Mol Biol Rev 61:262–280

Shahriari H, Warith M, Hamoda M, Kennedy K (2013) Evaluation of single vs. staged mesophilic anaerobic digestion of kitchen waste with and without microwave pretreatment. J Environ Manag 125:74–84. https://doi.org/10.1016/j.jenvman.2013.03.042

Shi X, Guo X, Zuo J et al (2018) A comparative study of thermophilic and mesophilic anaerobic co-digestion of food waste and wheat straw: process stability and microbial community structure shifts. Waste Manag 75:261–269. https://doi.org/10.1016/j.wasman.2018.02.004

Sitorus B, Sukandar, Panjaitan SD (2013) Biogas recovery from anaerobic digestion process of mixed fruit -vegetable wastes. Energy Procedia 32:176–182. https://doi.org/10.1016/j.egypro.2013.05.023

Speece RE (2008) Anaerobic biotechnology for industrial wastewater treatment. https://pubs.acs.org/doi/pdf/10.1021/es00115a001. Accessed 23 Sept 2020

Srisowmeya G, Chakravarthy M, Nandhini Devi G (2020) Critical considerations in two-stage anaerobic digestion of food waste—a review. Renew Sust Energ Rev 119:109587. https://doi.org/10.1016/j.rser.2019.109587

Sun Y, Cheng J (2002) Hydrolysis of lignocellulosic materials for ethanol production: a review. Bioresour Technol 83:1–11. https://doi.org/10.1016/S0960-8524(01)00212-7

Taconi KA, Zappi ME, Todd French W, Brown LR (2008) Methanogenesis under acidic pH conditions in a semi-continuous reactor system. Bioresour Technol 99:8075–8081. https://doi.org/10.1016/j.biortech.2008.03.068

Taherzadeh MJ, Karimi K (2008) Pretreatment of lignocellulosic wastes to improve ethanol and biogas production: a review. Int J Mol Sci 9:1621–1651. https://doi.org/10.3390/ijms9091621

Tampio E, Marttinen S, Rintala J (2016) Liquid fertilizer products from anaerobic digestion of food waste: mass, nutrient and energy balance of four digestate liquid treatment systems. J Clean Prod 125:22–32. https://doi.org/10.1016/j.jclepro.2016.03.127

Tayyab A, Ahmad Z, Mahmood T et al (2019) Anaerobic co-digestion of catering food waste utilizing Parthenium hysterophorus as co-substrate for biogas production. Biomass Bioenergy 124:74–82. https://doi.org/10.1016/j.biombioe.2019.03.013

Vavilin VA, Fernandez B, Palatsi J, Flotats X (2008) Hydrolysis kinetics in anaerobic degradation of particulate organic material: an overview. Waste Manag 28:939–951. https://doi.org/10.1016/j.wasman.2007.03.028

Vivekanand V, Ryden P, Horn SJ et al (2012) Impact of steam explosion on biogas production from rape straw in relation to changes in chemical composition. Bioresour Technol 123:608–615. https://doi.org/10.1016/j.biortech.2012.06.088

Wang L-H, Wang Q, Cai W, Sun X (2012) Influence of mixing proportion on the solid-state anaerobic co-digestion of distiller's grains and food waste. Biosyst Eng 112:130–137. https://doi.org/10.1016/j.biosystemseng.2012.03.006

Wang L, Shen F, Yuan H et al (2014) Anaerobic co-digestion of kitchen waste and fruit/vegetable waste: lab-scale and pilot-scale studies. Waste Manag 34:2627–2633. https://doi.org/10.1016/j.wasman.2014.08.005

Ward AJ, Hobbs PJ, Holliman PJ, Jones DL (2008) Optimisation of the anaerobic digestion of agricultural resources. Bioresour Technol 99:7928–7940. https://doi.org/10.1016/j.biortech.2008.02.044

Wiegel J, Mesbah NM, Abou-El-Ela SH (2007) Life at extreme limits: 2005:275–275. https://doi.org/10.11490/isea.2005.0.275.0

Xu F, Li Y, Ge X et al (2018) Anaerobic digestion of food waste – challenges and opportunities. Bioresour Technol 247:1047–1058. https://doi.org/10.1016/j.biortech.2017.09.020

Yong Z, Dong Y, Zhang X, Tan T (2015) Anaerobic co-digestion of food waste and straw for biogas production. Renew Energy 78:527–530. https://doi.org/10.1016/j.renene.2015.01.033

You Z, Pan S-Y, Sun N et al (2019) Enhanced corn-stover fermentation for biogasproduction by NaOH pretreatment with CaO additive and ultrasound. J Clean Prod 238:117813. https://doi.org/10.1016/j.jclepro.2019.117813

Zhang Y, Banks CJ (2013) Impact of different particle size distributions on anaerobic digestion of the organic fraction of municipal solid waste. Waste Manag 33:297–307. https://doi.org/10.1016/j.wasman.2012.09.024

Zhang C, Su H, Baeyens J, Tan T (2014) Reviewing the anaerobic digestion of food waste for biogas production. Renew Sust Energ Rev 38:383–392. https://doi.org/10.1016/j.rser.2014.05.038

Zheng Y, Zhao J, Xu F, Li Y (2014) Pretreatment of lignocellulosic biomass for enhanced biogas production. Prog Energy Combust Sci 42:35–53. https://doi.org/10.1016/j.pecs.2014.01.001

Zhou J, Zhang R, Liu F et al (2016) Biogas production and microbial community shift through neutral pH control during the anaerobic digestion of pig manure. Bioresour Technol 217:44–49. https://doi.org/10.1016/j.biortech.2016.02.077

Zhu S, Wu Y, Yu Z et al (2005) Pretreatment by microwave/alkali of rice straw and its enzymic hydrolysis. Process Biochem 40:3082–3086. https://doi.org/10.1016/j.procbio.2005.03.016

Zou H, Jiang Q, Zhu R et al (2020) Enhanced hydrolysis of lignocellulose in corn cob by using food waste pretreatment to improve anaerobic digestion performance. J Environ Manag 254:109830. https://doi.org/10.1016/j.jenvman.2019.109830

Chapter 5
Current Status and Prospects of Biohydrogen Production Process

Chandan Mahata and Debabrata Das

Abstract Biohydrogen is considered a fuel for the future due to its unique attributes in clean energy generation, waste management, and high energy content. Recently, its economic production has gained considerable attention from numerous scientists and industrialists. This chapter addresses microbiological, biochemical, molecular biological, and other perspectives related to biological hydrogen production (BHP). Process parameters such as pH, substrate type, temperature, agitation speed, hydraulic retention time, and hydrogen partial pressure greatly influence the dark fermentation process. Therefore, several optimization approaches, including statistical and artificial intelligence, have been demonstrated. Additionally, different kinetic models associated with substrate degradation, cell mass growth, and product formation in dark fermentation have been discussed in detail. This chapter also discusses different types of reactors and their suitability for biological hydrogen production. The viability of any process relies on its ability to be applied to the industrial level. Therefore, the scale-up of the biohydrogen production process has been exemplified. In summary, this chapter presents a holistic overview of the biohydrogen production process and highlights recent scientific findings and achievements.

Keywords Biohydrogen · Dark fermentation · Kinetic analysis · Process optimization · Scale-up

C. Mahata
Advanced Technology Development Centre, Indian Institute of Technology, Kharagpur, West Bengal, India

D. Das (✉)
Department of Biotechnology, Indian Institute of Technology, Kharagpur, West Bengal, India

J. K. Saini, R. K. Sani (eds.), *Microbial Biotechnology for Renewable and Sustainable Energy*, Clean Energy Production Technologies,
https://doi.org/10.1007/978-981-16-3852-7_5

5.1 Introduction

In history, the evolution of lives on the Earth has taken place according to the laws of nature. Nevertheless, most of the subsequent changes have been caused by humans and their activities. The present generation needs to be more vigilant in its behavior for sustaining the world's future. Among all the major issues, climate change, fossil fuel depletion, pollution, and biodiversity loss are the major challenges in the twenty-first century. All these challenges are interlinked with the rapid increase in the human population. The scenario of energy demand and production plays a vital role in the development and civilization of humankind. Nowadays, most of the energy is derived from fossil-based fuels such as crude oil, petroleum, and natural gas, which are becoming depleted rapidly (Tapia-Venegas et al. 2015). Additionally, fossil fuels, on combustion, are mainly responsible for the excessive emission of greenhouse gases. These gaseous emissions have severely affected the atmosphere and are significantly attributed to the impacts of climate change. According to the report of the Intergovernmental Panel on Climate Change (IPCC 2019), CO_2 emission needs to be diminished from its current level by about 45% by the year 2030 to keep global warming to 1.5 °C. Therefore, researchers are focusing on carbon-neutral renewable fuels. Hydrogen, a carbon-free fuel, can be considered a promising energy source mainly due to its high energy density (142 kJ/g) (Zheng et al. 2014), sustainability (Kumar et al. 2017), and nonpolluting nature (Das 2009). Presently, about 95% of the commercially available hydrogen(H_2) is produced from conventional technologies using non-renewable resources such as natural gas, coal, heavy oil, and naphtha (Balachandar et al. 2019; Das and Veziroglu 2008). The conventional processes for H_2 production are methane-steam reforming, coal gasification, pyrolysis, thermal cracking, and water splitting (Das et al. 2008). These processes are either thermochemical or electrochemical, which are energy-consuming and not environmentally sustainable. In contrast, biological processes of H_2 production are mainly performed at ambient conditions; thus, they are less energy-intensive and eco-friendly (Das and Veziroğlu 2001). Additionally, these processes can utilize waste feedstock for hydrogen production, which facilitates resource recovery from waste materials (Das 2009).

The main goal of the biohydrogen production processes is to make the process commercially feasible. This chapter focuses on the current and future directions of biohydrogen production processes. The chapter also discusses the potential strategies for the enhancement of biohydrogen production.

5.2 Biological Processes and Their Potentiality in Hydrogen Production

Hydrogen production is essentially sustainable and environment-friendly via biological routes. A diverse range of feedstock such as domestic waste, industrial effluents, agricultural residue, municipal solid waste, and even water can be utilized for hydrogen production. The common biohydrogen production processes are direct biophotolysis, indirect biophotolysis, photo-fermentation, dark fermentation (DF), and microbial electrolysis (Fig. 5.1). The processes can be broadly classified as light-dependent and light-independent. Biophotolysis (direct and indirect) and photo-fermentation are light-dependent, whereas dark fermentation and electrohydrogenesis do not require a light source. Photolysis is driven by green algae or blue-green algae (cyanobacteria) while photo-fermentation is performed by sulfur and nonsulfur bacteria. Similarly, acidogenic bacteria and exoelectrogenic bacteria play important roles in dark fermentation and microbial electrolysis, respectively.

5.2.1 *Direct Biophotolysis*

This method adopts the same pathways as used in plants and algal photosynthesis but modifies them to produce hydrogen gas rather than carbon-based biomass. The photosynthesis process takes place using chlorophyll, which has magnesium in its

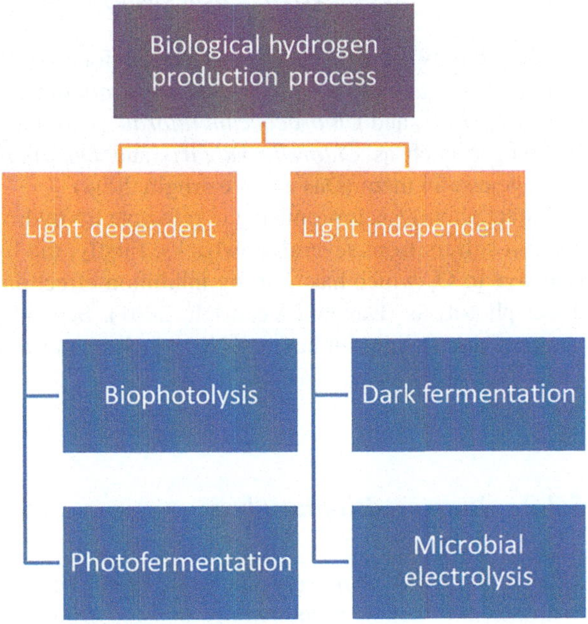

Fig. 5.1 Classification of biological hydrogen production processes

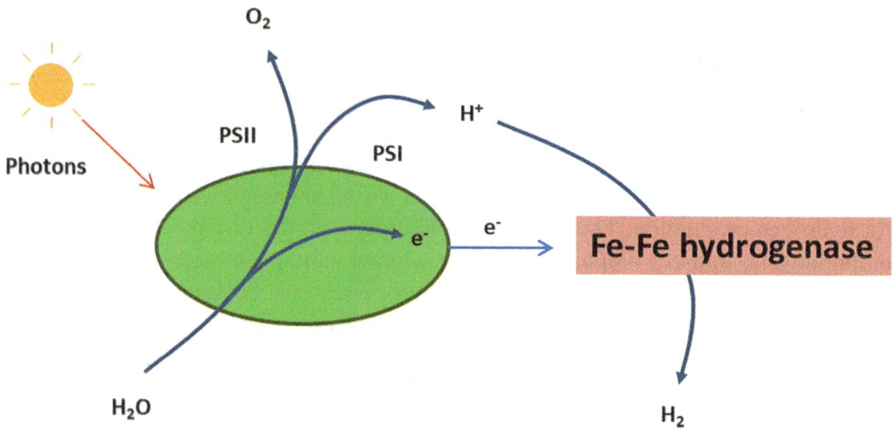

Fig. 5.2 Mechanism of direct biophotolysis

center. The degradation of water molecules to H_2 and O_2 occurs during photosynthesis in the presence of sunlight (photons). Hydrogen ions are generated by solar photons in the reducing site of photosystem I (PSI) under anaerobic conditions or when excessive energy is captured. It is further transformed into H_2 gas in a medium with electrons provided by the reduced enzyme of the algal cell (Fig. 5.2). Simultaneously, molecular oxygen is produced at the oxidizing side of photosystem II (PSII). Overall, the reaction can be illustrated as follows:

$$2H_2O + \text{light energy} \rightarrow 2H_2 + O_2 \tag{5.1}$$

The activity of hydrogenase has been found in several green algae such as *Chlamydomonas reinhardtii*, *Scenedesmus obliquus*, *Chlorella fusca*, *Platymonas subcordiformis*, and *Chlorococcum littorale* (Das and Veziroglu 2008). In contrast, microalgae such as *Chlorella vulgaris* and *Dunaliella salina* do not have Fe-Fe hydrogenase in them (Das and Veziroglu 2008).

Direct photolysis is promising in principle for hydrogen generation. However, the process suffers from several drawbacks. Firstly, the hydrogenase enzyme is highly sensitive to O_2 which has a strong inhibition effect on hydrogen production during direct photolysis (Das and Veziroğlu 2001). Secondly, a lower hydrogen yield is obtained due to light limitations. Nevertheless, the challenges need to be tackled to make the process more feasible.

5.2.2 *Indirect Biophotolysis*

Indirect photolysis also occurs under sunlight like direct photolysis. In this process, hydrogen production is temporally isolated from O_2-evolving photosynthesis by

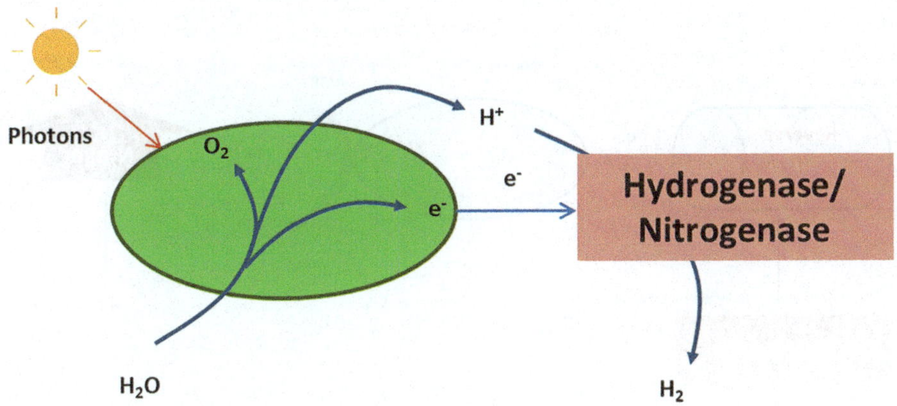

Fig. 5.3 Mechanism of indirect biophotolysis

sulfur depletion or repletion. It is a two-stage process. In stage 1, CO_2 is fixed by cyanobacteria for carbohydrates' biosynthesis (Eq. 5.2). This is followed by (in stage 2) the stored carbohydrates that are fermented to produce hydrogen with the help of H_2-producing enzymes (Eq. 5.3). In this process, unlike direct photolysis, the presence of nitrogenase enzymes can fix the atmospheric N_2 during hydrogen production (Fig. 5.3). It could be possible to separate these two stages by cultivating the microalgae in separate aerobic and anaerobic phases. In this process, hydrogen can be produced by hydrogenase or nitrogenase enzymes. Like hydrogenase, nitrogenase is also inhibited by oxygen evolution.

$$6H_2O + 6CO_2 + \text{light energy} \rightarrow C_6H_{12}O_6 + 6O_2 \quad (5.2)$$

$$C_6H_{12}O_6 + 6H_2O + \text{light energy} \rightarrow 12H_2 + 6CO_2 \quad (5.3)$$

This process is mainly driven by a diverse group of cyanobacteria species, which may be either N_2 fixing or non-N_2 fixing. The N_2 fixing cyanobacteria are *Calothrix* sp., non-marine *Anabaena* sp., and *Oscillatoria* sp., whereas the non-N_2 fixing cyanobacteria are *Gloebacter* sp., *Synechococcus* sp., and marine *Anabaena* sp. (Das and Veziroglu 2008). Similar to direct photolysis, it has several practical limitations, which challenge the scale-up and commercialization of the process.

5.2.3 Photofermentation

Photofermentation is a series of biochemical reactions in which organic substances like short-chain volatile fatty acids, such as acetic acid, are converted to hydrogen, manifested by a diverse group of photosynthetic bacteria under anaerobic conditions. Numerous strains of photosynthetic bacteria, including green sulfur bacteria, purple sulfur/non-sulfur bacteria, can produce hydrogen through photofermentation (Ding

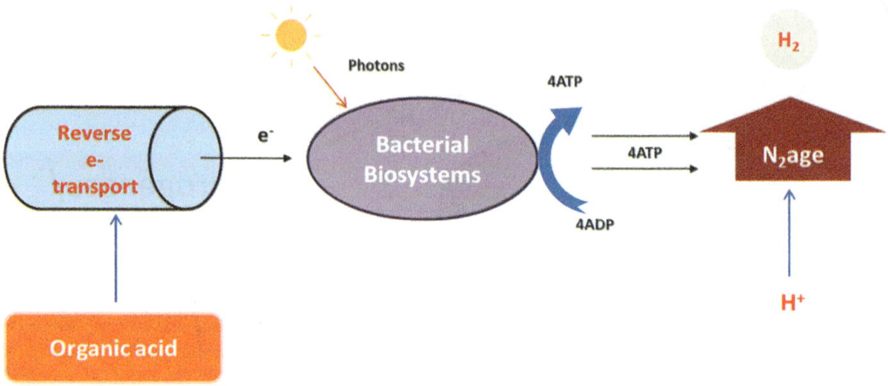

Fig. 5.4 Mechanism of photofermentation

et al. 2016). Nevertheless, researchers mainly focus on purple nonsulfur (PNS) bacteria due to a wide variety of feedstock consumption. PNS bacteria such as *Rhodobacter capsulatus*, *Rhodobacter sphaeroides*, *Rhodospirillum rubrum*, *Rhodopseudomonas palustris* are responsible for hydrogen generation. Unlike photosynthesis, photosystem I (PSI) is absent in photofermentative PNS bacteria. Therefore, it could not produce oxygen during hydrogen production. Generation of hydrogen in PNS bacteria is mainly mediated by nitrogenase under nitrogen limiting conditions (Eq. 5.4), whereas uptake hydrogenase consumes hydrogen. The photofermentative hydrogen production by nitrogenase could be illustrated in Eq. 5.4 shown below (Fig. 5.4):

$$2H^+ + 2e^- + 4ATP + light \overset{\text{nitrogenase}}{\rightarrow} H_2 + 4ADP + 4P_i \tag{5.4}$$

The stoichiometric equation for acetic acid as an organic acid can be written as follows:

$$CH_3COOH + 2H_2O \rightarrow 4H_2 + 2CO_2 \tag{5.5}$$

In this process, a massive amount of ATP (4 mol) is required for 1 mol of hydrogen production. As a result, strict control of the reaction environment is necessary (Koku et al. 2002). Therefore, despite high hydrogen yield, the process has several bottlenecks, such as high energy consumption, low photosynthetic conversion, and low volumetric production rate (Veeravalli et al. 2019).

5.2.4 *Microbial Electrolysis Cell*

Microbial electrolysis cell (MEC), a modification microbial fuel cell (MFC), is a bioelectrochemical system that can convert organic matter to molecular hydrogen with the help of exoelectrogenic bacteria by applying an external electric current (Logan and Regan 2006). The system comprises three main parts: anode, cathode, and proton exchange membrane (PEM). PEM permits only protons to flow through it by restricting electrons. In the anode, the organic substance is oxidized and produces electrons and protons by exoelectrogens. Oxidation of organic matter in the anode is not thermodynamically spontaneous ($\Delta G^0 > 0$). Therefore, the external voltage supply is recommended to force the reaction. The minimum theoretical voltage of 0.11 V is required to make a spontaneous reaction (Das and Veziroglu 2008). The protons move from anode to cathode through PEM, whereas electrons are transferred through an external circuit (Fig. 5.5). Hydrogen gas is generated through the reduction of hydrogen ions by electrons. The most common exoelectrogens are *Shewanella* sp., *Burkholderia* sp., *Geobacter* sp., *Pseudomonas* sp., *Rhodoferax ferrireducens*, *Escherichia coli*, and *Citrobacter* sp. (Feng et al. 2014). Carbon paper, carbon cloth, and graphite can be used as an anode, whereas graphite, titanium, and platinum can be employed as a cathode (Kadier et al. 2016; Kundu et al. 2013). The high cost of conventional cathode materials drives the research into biocathode as a substitute (Kundu et al. 2013).

The overall reaction can be represented as following Eq. 5.6:

Anode chamber:

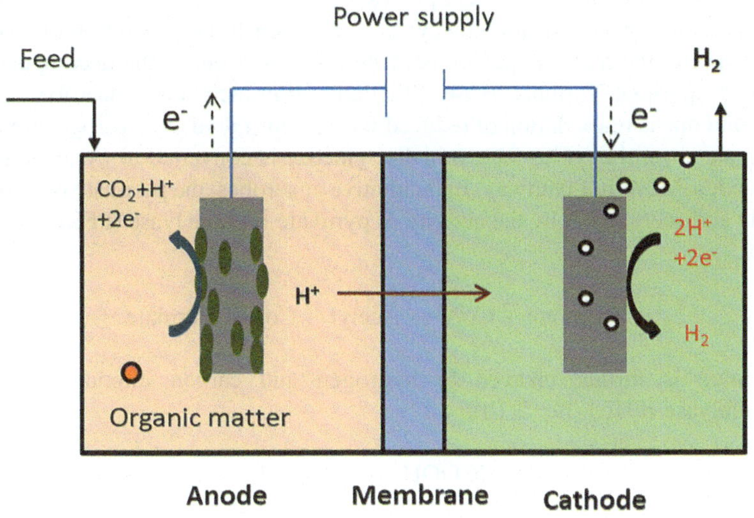

Fig. 5.5 Schematic of the microbial electrolysis cell

$$CH_3COOH + 2H_2O \rightarrow 2CO_2 + 8e^- + 8H^+ \quad (5.6)$$

Cathode chamber:

$$8e^- + 8H^+ \rightarrow 4H_2 \quad (5.7)$$

Overall,

$$CH_3COOH + 2H_2O \rightarrow 2CO_2 + 4H_2 \quad (5.8)$$

Although this process is suitable for wastewater treatment along with bioenergy generation, it suffers from several drawbacks, such as scale-up problems, low hydrogen production rate, high cost of the membrane, and external energy source.

5.2.5 *Dark Fermentation*

Dark fermentation is an anaerobic conversion of organic substances, mainly carbohydrates, to H_2 gas exhibited by various acidogenic bacteria (Das et al. 2008). Under anaerobic condition, microorganism generates energy for cells in the form of ATP by blocking the Tricarboxylic Acid (TCA) cycle. Consequently, the produced extra electron is used for the production of metabolic end products such as volatile fatty acids and ethanol. The process has several advantages over other biohydrogen production processes due to its high production rate and yield (Table 5.1). Additionally, it has no light limitations like photolysis and photofermentation, as dark fermentation is a light-independent process.

Two distinct biochemical pathways can accomplish the generation of molecular hydrogen with the help of specific enzymes. The first one is the decomposition of formate by pyruvate formate- lyase (PFL) present in facultative anaerobes, whereas the second one is re-oxidation of reduced ferredoxin (Fdred) by hydrogenase present in obligate anaerobes (Fig. 5.6). Initially, glucose is converted to pyruvate through the Embden-Meyerhof pathway. In facultative anaerobes, the pyruvate is oxidized to formate and acetyl-CoA by the activity of pyruvate formate lyase (PFL) as shown in Eq. 5.9.

$$\text{Pyruvate} + \text{CoA} \rightarrow \text{Acetyl} - \text{CoA} + \text{Formate} \quad (5.9)$$

Formate is further cleaved to hydrogen and carbon dioxide by formate hydrogenlyase (FHL) (Eq. 5.10).

$$HCOOH \rightarrow CO_2 + H_2 \quad (5.10)$$

Table 5.1 Performance of different biological hydrogen production processes

Methods	Microorganisms	Substrate	Hydrogen production rate (L L^{-1} day^{-1})	Hydrogen yield (mol H_2 mol^{-1} substrate)	References
Direct photolysis	*Chlamydomonas reinhardtii*	Tris-acetate-phosphate (TAP) medium	0.02	–	Ban et al. (2019)
	Co-culture of *Chlamydomonas reinhardtii* and *Bradyrhizobium japonicum*	TAP medium	0.01	–	Wu et al. (2012)
	Rhodobacter sphaeroides	Malic acid	0.26	2.71	Gilbert et al. (2011)
	Rhodobacter sphaeroides	2-hydroxybutanedioic acid	0.15	4.5	Basak and Das (2009)
Indirect photolysis	*Anabaena variabilis*	N-free BG-11 medium	0.48	–	Liu et al. (2006)
	Anabaena cylindrica	Nutrient medium	0.20	–	Neil et al. (1976)
	Wild-type *Synechocystis* sp. PCC 6803	BG11 medium	300[a]	–	
Photo-fermentation	*Rhodobacter capsulatus* DSM 1710	Acetate, Glutamate	0.39	0.60	Boran et al. (2010)
	Rhodopseudomonas palustris 42OL	Malate, Glutamate	0.70	–	Ren et al. (2009)
	Anabaena variabilis PK84	Carbon dioxide	0.48	–	Chen et al. (2006)
	Rhodopseudomonas faecalis RLD-53	Acetate	0.69	3.12	Xie et al. (2012)
	Rhodobacter capsulatus YO3	Malate, Glutamate	0.21	0.35	Sözen et al. (2005)
MEC	*Shewanella oneidensis* MR-1	Sodium acetate	0.69	2.6	Hu et al. (2008)
	Mixed consortia (dominated by *Proteobacteria* sp.)	Acetate	0.052	2.1	Chae et al. (2008)
	Exoelectrogens	Sodium acetate	0.33	-	Rozendal et al. (2007)

(continued)

Table 5.1 (continued)

Methods	Microorganisms	Substrate	Hydrogen production rate (L L^{-1} day^{-1})	Hydrogen yield (mol H_2 mol^{-1} substrate)	References
	Exoelectrogenic bacteria	Swine wastewater	1.0	1.2	Wagner et al. (2009)
	Geobacter sp.	Fermentation effluent (buffered)	1.52	–	Lu et al. (2009)
	Waste activated sludge	Sodium acetate	1.76	1.2^{b}	Liu et al. (2012a)
Dark fermentation	*Enterobacter aerogenes*	Glucose	9.35	1.09	Fabiano and Perego (2002)
	Enterobacter cloacae IIT-BT 08	Cellobiose	15.6	5.4	Kumar and Das (2000)
	Enterobacter cloacae IIT-BT 08	Sucrose	15.84	6.0	Kumar and Das (2000)
	Mixed culture	Starch	5.6	1.1	Lin et al. (2008)
	Thermoanaerobacterium-rich sludge	Palm oil mill effluent	6.1	2.24	O-Thong et al. (2007)
	Thermophilic mixed culture	Rice spent wash	4.03	464^{c}	Roy et al. (2012)

[a]nmol H_2 mg $chla^{-1}$ h^{-1}
[b]mL H_2/mg COD
[c]mL H_2/g carbohydrate

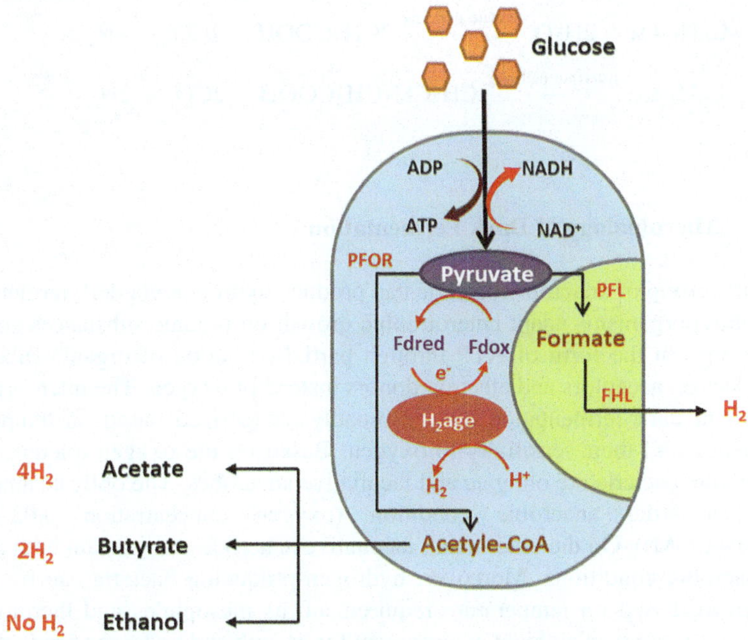

Fig. 5.6 Biochemical pathway of dark fermentation

The second type of biochemical reaction is observed in obligate anaerobes, where pyruvate is converted to acetyl-CoA by pyruvate-ferredoxin oxidoreductase (PFOR). Ferredoxin (Fd) is reduced during the oxidation of pyruvate to acetyl-CoA. When the organic acid is accumulated, reduced Fd (Fdred) is oxidized by Fe-Fe hydrogenase and subsequently, molecular hydrogen is formed (Fig. 5.6) (Das and Veziroglu 2008). The overall reaction can be represented as follows:

$$\text{Pyruvate} + \text{CoA} + 2\text{Fd}_{\text{ox}} \overset{\text{PFOR}}{\rightarrow} \text{Acetyl} - \text{CoA} + 2\text{Fd}_{\text{red}} + \text{CO}_2 \tag{5.11}$$

$$2\text{H}^+ + \text{Fd}_{\text{red}} \rightarrow \text{H}_2 + \text{Fd}_{\text{ox}} \tag{5.12}$$

The stoichiometry of the process shows that 4 mols of hydrogen are generated from 1 mol of glucose when pyruvate is oxidized to acetate as the only metabolic product, whereas it produces 2 mol of hydrogen when pyruvate is converted to butyrate (Eqs. 5.13 and 5.14). A few microorganisms follow mixed acid pathways. Hydrogen yield depends on the acetate-to-butyrate ratio. Nevertheless, if the end metabolites are ethanol, lactic acid, and propionic acid, no hydrogen formation occurs.

$$C_6H_{12}O_6 + 2H_2O \xrightarrow{\text{Acetate pathway}} 2CH_3COOH + 2CO_2 + 4H_2 \quad (5.13)$$

$$C_6H_{12}O_6 \xrightarrow{\text{Butyrate pathway}} CH_3CH_2CH_2COOH + 2CO_2 + 2H_2 \quad (5.14)$$

5.2.5.1 Microbiology of Dark Fermentation

A diverse group of anaerobic bacteria can produce hydrogen via dark fermentation. These microorganisms adapt heterotrophic growth on organic substances and generate energy in the form of ATP through partial oxidation of organic substances using electron acceptors and electron donors instead of oxygen. The microorganism involved in dark fermentation can be broadly categorized based on temperature dependency and their sensitivity to oxygen. Based on the oxygen tolerance, dark fermentative bacteria are obligate and facultative anaerobes. The obligate anaerobes require a strict anaerobic condition (oxygen concentration 0.02–0.04% (0.24–0.48 mM)). On the other hand, facultative anaerobes can sustain both aerobic and anaerobic conditions. Moreover, hydrogen-producing bacteria can be further categorized, based on temperature requirement, as mesophiles and thermophiles. Mesophiles require an ambient environment for growth and hydrogen production. In contrast, thermophiles adapt to high temperatures (>45 °C) for their growth. Naturally, a mixed microbial community serves a beneficial role in the generation of hydrogen from various complex wastes (Mishra et al. 2015). The selection of microorganisms depends on the substrate used.

Facultative Anaerobic bacteria

In an aerobic environment, facultative anaerobes can generate energy in the form of ATP in aerobic respiration, while in anaerobic conditions, ATP is produced by these anaerobes through anaerobic fermentation. The most common facultative anaerobes are *Enterobacter* sp. that can produce hydrogen under an anaerobic environment. The species could possess either formate hydrogen lyase (FHL) or Fe-Fe hydrogenase, which is mainly accountable for a high rate of H_2 formation. The most commonly used bacteria are *Enterobacter cloacae* IIT-BT 08 and *Enterobacter aerogenes* E.82005 (Kumar and Das 2000; Tanisho et al. 1987). Usually, facultative microorganisms are preferred primarily because of their ease of control and sustainability in the lower partial pressure of hydrogen (Nakashimada et al. 2002).

Obligate Anaerobic bacteria

Recently, obligate anaerobes have gained considerable attention from researchers because they can consume a variety of carbohydrates, including waste materials.

Furthermore, they can also produce a high rate of H_2 in comparison to facultative bacteria. The most commonly used obligate anaerobe is *Clostridium* sp. H_2 production usually takes place in the exponential growth phase of the microorganism. In the starvation phase, the metabolic pathway could alter from acidogenesis to solventogenesis (Han and Shin 2004). *Clostridium paraputrificum*, *C. tyrobutyricum*, *C. thermocellum*, *C. thermolacticum*, *C. acetobutylicum*, and *C. saccharoperbutylacetonicum* are promising examples of obligate anaerobic bacteria, which can form spores under harsh conditions. A diverse group of *Clostridium* species can generate H_2 in the range of 1.46–2.8 mol mol^{-1} glucose (Lin et al. 2007; Oh et al. 2009). Some of the obligate anaerobes are thermophiles, which are mainly available in the hot areas around the Earth, such as thermal baths and deep-sea vents. The composition of the growth medium for thermophile bacteria depends on the source of the bacteria isolated. Anaerobes isolated from the hot-springs area need high sulfur concentration, whereas anaerobes collected from deep-sea vents require high sodium chloride concentration in the medium (Schröder et al. 1994; Van Niel et al. 2002). Reducing agents such as L-cysteine hydrochloride could be added to remove trace amounts of O_2 from the hydrogen-producing medium (Singh et al. 2019; Roy et al. 2014). Hydrogen production using thermophiles is much more thermodynamically favorable than using mesophiles (Roy et al. 2014). Some typical examples of the thermophiles genus are *Thermoanaerobacter*, *Caldicellulosiuptor*, *Thermoanaerobacterium*, and *Thermotoga* (Roy et al. 2014; Slobodkin et al. 1999; Van Ooteghem et al. 2002).

Mixed Culture

Recently, the application of mixed consortium and co-culture has gained considerable attention for hydrogen production from complex substrates, such as industrial effluent, domestic waste, and agricultural residue (Mishra et al. 2015, 2017; Singh et al. 2013). Mixed consortia consist of a variety of bacteria that secrete various types of hydrolytic enzymes. Mixed consortia can therefore efficiently use various complex substrates present in wastewater (Mishra et al. 2015). Furthermore, dark fermentative hydrogen can be generated in a non-sterile and less regulated condition using mixed consortia, which could facilitate the scale and commercialization (Tomczak et al. 2018). Hydrogen-producing mixed inoculum can be isolated from anaerobic digester of various organic materials, such as cow dung, sewage sludge, industrial effluent (Mishra et al. 2015; Kumari and Das 2017; Tang et al. 2008). Apart from the H_2 producing bacteria, some H_2 consuming bacteria, such as homoacetogen and methanogen, are also present in the culture. Therefore, an effective pretreatment is required to inhibit the H_2 consuming bacterial activity, as well as enrich anaerobic spore-forming bacteria. Usually, pretreatment methods include heat (O-Thong et al. 2009), acid and base stock and base shock (O-Thong et al. 2009; Yang and Wang 2018), and electric field (Jeong et al. 2013). However, heat shock microbial culture has the best performance in a higher yield of H_2 production. Therefore, this technique is mostly used for the treatment of mixed cultures (Kumari and Das 2017). Moreover, heat shock treatment is simple and effective. It requires around 100 °C for 10–120 min in order to suppress nonspore-forming bacteria (Kumari and Das 2017; Barros and Silva 2012).

5.3 Theoretical Considerations

5.3.1 Kinetic Analysis

The Monod growth model can explain the relationship between limiting-substrate concentration and specific growth of microorganism rate according to Eq. 5.15:

$$\mu = \frac{\mu_{\max} S}{K_S + S} \tag{5.15}$$

where μ is the specific growth rate (h^{-1}), $\mu_{\max}$ is the maximum specific growth rate (h^{-1}), K_S is half-velocity constant (g VSS L^{-1}), S is the concentration of limiting substrate for cell growth (g COD L^{-1}).

The Monod model can be linearized in the form of a Lineweaver-Burk plot (Eq. 5.16) to evaluate kinetic constants.

$$\frac{1}{\mu} = \frac{K_S}{\mu_{\max}} \frac{1}{S} + \frac{1}{\mu_{\max}} \tag{5.16}$$

The Logistic model can be employed to evaluate the microbial growth kinetics (Eq. 5.17) (Gilbert et al. 2011).

$$\frac{dX}{dt} = k_c X \left(1 - \frac{X}{X_{\max}}\right) \tag{5.17}$$

where k_c represents the specific growth rate (h^{-1}), X is biomass concentration (g L^{-1}), and $X_{\max}$ indicates the maximum biomass concentration (g L^{-1}).

By integrating Eq. 5.17 and simplifying, biomass concentration can be expressed as shown in Eq. 5.18 given below.

$$X = \frac{X_0 \exp(k_c t)}{1 - \frac{X_0}{X_{\max}}(1 - \exp(k_c t))} \tag{5.18}$$

where X_0 represents the initial cell mass concentration (g VSS L^{-1}).

Substrate consumption can be analyzed by first-order reaction kinetics using Eq. 19 (Najafpour et al. 2004).

$$-\frac{dS}{dt} = kS \tag{5.19}$$

where S represents the concentration of substrate used (g L^{-1}) and k is rate constant (h^{-1}).

Further, substrate utilization for biomass formation and cell maintenance is determined by Pirt model as shown in Eq. 5.20 (Pirt 1965).

$$\frac{1}{Y_{X/S}} = \frac{1}{Y_{X/S(g)}} + \frac{m}{\mu} \tag{5.20}$$

where $Y_{X/S}$ and $Y_{X/S(g)}$ represent the apparent growth yield (g g^{-1}) and true growth yield (g g^{-1}), respectively. μ and m indicate specific growth rate (h^{-1}) and maintenance coefficient (g g^{-1} h^{-1}), respectively.

H_2 production kinetics is analyzed by the modified Gompertz equation (Eq. 5.21) (Jia et al. 2014).

$$H = Pexp\left\{-\exp\left[\frac{R_m \times e}{P}(\lambda - t) + 1\right]\right\} \tag{5.21}$$

where H represents the cumulative H_2 production (mL L^{-1}) at any time t (h); P and R_m indicate the H_2 production potential (mL L^{-1}) and the maximum H_2 production rate (mL L^{-1} h^{-1}), respectively; λ represents the lag time (h) for H_2 production.

Furthermore, the Luedeking Piret model can be used to determine the relationship between cell mass formation and H_2 production (Eq. 5.22) (Luedeking and Piret 2000).

$$\frac{1}{X}\frac{dP}{dt} = \alpha\left(\frac{1}{X}\frac{dX}{dt}\right) + \beta \tag{5.22}$$

where $(1/X)(dP/dt)$ (g H_2 g^{-1} h^{-1}) and $(1/X)(dX/dt)$ (h^{-1}) are specific product and biomass formation rate, respectively; α (g g^{-1} H_2) and β (h^{-1}) are growth and nongrowth associated coefficients, respectively.

The kinetic parameters of the aforementioned kinetic models can be determined by linear and nonlinear regression.

5.3.2 *Material and Energy Analysis*

For the assessment of the functionality and viability of any emerging technology, a techno-economic evaluation is required. It can be carried out by several means, such as material and energy analysis.

Material analysis is a crucial aspect of tracking different materials during the fermentation, considering all input, output, and accumulated materials involved in the process. Material analysis of any process provides a general idea about substrate utilization and product formation potential. The mechanism of hydrogen production can be confirmed by material analysis. In dark fermentation, for instance, the ratio of accumulated acetate and butyrate in the fermentation broth can reveal the dominant biochemical pathway during the fermentation. For pure substrate, having a known molecular formula, elemental balance is performed. On the other hand, total chemical oxygen demand (COD) is considered for the complex substrate, such as organic waste or agricultural residue. In the case of COD balance, the amount of all individual products accumulated need to be expressed in terms of COD. For example, the conversion factor for hydrogen is 8 g COD g^{-1} H_2.

Energy analysis can be conducted based on gaseous energy recovery. Furthermore, the gaseous energy recovery can be calculated in terms of substrate added or the total energy required for the process (Eqs. 5.23 and 5.24).

$$\text{Energy recovery} = \frac{\text{Energy content of hydrogen produced}}{\text{Energy content of substarte consumed}} \tag{5.23}$$

$$\text{Energy recovery} = \frac{\text{Energy content of hydrogen produced}}{\text{Total process energy requirement including substrate}} \tag{5.24}$$

Kumari and Das (2015) calculated the theoretical maximum energy recovery from dark fermentation as 34.1%.

5.4 Effect of Physicochemical Parameters on Dark Fermentative Hydrogen Production

The performance of dark fermentation depends on different physicochemical parameters, such as pH, temperature, medium composition, partial pressure of hydrogen, soluble metabolic products, and hydraulic retention time (HRT).

5.4.1 pH

The pH of the hydrogen-producing medium is one of the dominant factors influencing the functionality of the hydrogenase regulating the metabolic pathway of dark fermentation. All enzymes have their optimal range of pH, in which the activity of the enzyme is its maximum. If acid accumulation increases in the fermentation broth, it results in a decreased pH. Consequently, the metabolic pathway of hydrogen production shifts towards solventogenesis. Khanal et al. (2004) reported that the shifting of the metabolic pathway occurs below 4.5 pH (Khanal et al. 2004). Similarly, several studies have stated that the optimum pH for hydrogen production varies near 6 (Cao and Zhao 2009; Van Ginkel et al. 2001).

5.4.2 Temperature

An environmental condition such as temperature dramatically influences dark fermentative hydrogen production because the growth of microorganisms is affected by temperature. According to temperature tolerance, hydrogen-producing bacteria may be mesophiles (25–45 °C) or thermophiles (>45 °C). In general, most of the studies (nearly 73%) have been conducted using mesophiles (Li and Fang 2007). Previous studies have revealed that temperature has a significant influence on microbial-specific growth rate and substrate utilization rate during dark fermentation. However, deactivation of the hydrogen-producing enzyme is started above the optimum temperature. Activation and deactivation energy of hydrogenase can be determined using the Arrhenius equation (Singh et al. 2019). For instance, Singh et al. (2019) evaluated the activation energy for mesophilic bacteria as 58.8 kJ mol^{-1}. This study also showed that the deactivation of hydrogen-producing bacteria started above the threshold temperature of 37 °C. On the other hand, Lee et al. (2006) observed that dark fermentative hydrogen production was most efficient at 40 °C. Therefore, the optimization of process temperature is critical for enhanced hydrogen production.

5.4.3 Medium Composition

Hydrogen-producing medium mainly comprises C-source, N-source, minerals, and vitamins. Each element has its role in fermentation. C-source, the sole element in the medium, is required for cell mass growth, product formation, and energy generation in terms of ATP. N-source is essential for protein synthesis and hence growth, whereas minerals and vitamins act as co-factors in the metabolic pathways. Furthermore, the carbon-to-nitrogen (C/N) ratio performs a vital part in the synthesis of H_2 (Kumari and Das 2017). Therefore, an appropriate combination of C- and N-sources is required for hydrogen production. Similarly, a suitable concentration of trace metals such as Fe, Ca, Na, Cu, Ni, Mg, K, and vitamins in a hydrogen-producing medium stimulates the generation of molecular hydrogen (Sekoai and Daramola 2018; Lin and Lay 2005).

5.4.4 Feedstock

Several studies have considered simple carbohydrates such as xylose, fructose, glucose, sucrose, and arabinose because of their ease of utilization by microorganisms (Pan et al. 2008; Abreu et al. 2012; Jayasinghearachchi et al. 2012). These pure substrates however lead to high process costs. In contrast, organic waste has significant COD, which is detrimental to the ecosystem. This could be considered as a promising feedstock for dilution factor (DF) (Mishra et al. 2015). The use of waste for the production of hydrogen, therefore, has double benefit of bioremediation and energy generation. Earlier, various organic wastes such as distillery effluent (Balachandar et al. 2019), rice winery wastewater (Yu et al. 2002), household wastewater (Van Ginkel et al. 2005), food waste (Elbeshbishy et al. 2011), and paper mill wastewater (Lin et al. 2006) were used as the sole substrate for the generation of H_2. Nevertheless, these feedstocks may not comprise the nutrients required for the growth of the microorganisms. Therefore, several studies have been conducted on co-substrates such as agricultural residue, water hyacinth (Mishra et al. 2017; Varanasi et al. 2018) for H_2 production processes. The selection of the co-substrate is primarily based on the suitable C/N ratio. Mishra et al. (2017) investigated the application of de-oiled cake, as a supplement, for dark fermentative H_2 production and observed the maximum results of 3.38 L H_2 L^{-1} using groundnut de-oiled cake (GDOC) as a supplement with distillery effluent (Mishra et al. 2017).

5.4.5 *Hydrogen Partial Pressure*

The H_2 partial pressure in the fermenter is a crucial parameter that influences the rate of hydrogen production because the metabolic pathway is highly influenced by the hydrogen partial pressure. Accumulation of H_2 gas in the headspace of the reactor can increase partial pressure. According to Le Chatelier's principle, the generation of hydrogen will be suppressed at the high partial pressure of hydrogen, and consequently, the metabolic pathway will be shifted toward alcohol production (Das 2017). Continuous removal of hydrogen could reduce the partial pressure, resulting in negating inhibition effect. Mandal et al. (2006) examined the effect of partial pressure on dark fermentation by developing a vacuum system inside the bioreactor. Their study revealed that the maximum rate of hydrogen production can be obtained at 380 mmHg pressure. On the other hand, some researchers reported that nitrogen sparging during fermentation could be an effective approach to negate the effect of hydrogen accumulation (Mizuno et al. 2000; Tanisho et al. 1998). However, the main bottleneck of sparging nitrogen is the dilution of hydrogen gas, resulting in high separation costs.

5.4.6 *Soluble Metabolic Products*

In dark fermentation, soluble end-metabolites, produced along with hydrogen, greatly influence hydrogen production. The major metabolic products are volatile fatty acids such as acetic acid, butyric acid, propionic acid, etc., and ethanol. Toward the starvation phase, the ionic strength of fermentation broth escalates attributed to the accumulation of the metabolites, resulting in cellular lysis. Due to cell disruption, high maintenance energy is required to restore its physiological balance. Lee et al. (2002) evaluated the inhibition effect of the end-metabolites on dark fermentative hydrogen production by externally adding acetic acid, butyric acid, propionic acid, and ethanol to the medium (Lee et al. n.d.). The study concluded that the addition of these volatile fatty acids and alcohol has an adverse effect on H_2 generation.

5.4.7 *Hydraulic Retention Time (HRT)*

During the continuous operation of the hydrogen-producing reactor, HRT is a crucial factor influencing the rate of hydrogen production. Mathematically, it is inversely proportional to dilution rate and hence the specific growth rate of microorganisms. The physical significance of HRT is that it is the measure of substrate residence time in the reactor. Several studies have shown that lowing HRT could increase the rate of hydrogen production (Tomczak et al. 2018; Zhang et al. 2006; Baima Ferreira Freitas et al. 2020). Additionally, the strategy of lowing HRT could separate the slow-growing methanogens from hydrogen-producing bacteria in the mixed consortium. Nevertheless, hydrogen production could be ceased below optimum HRT because of cell mass washout.

5.4.8 *Agitation Speed*

Agitation speed plays a vital role in any fermentation process. Agitation in suspended culture provides adequate mixing, heat, and mass transfer. Furthermore, the agitation could reduce the partial pressure of hydrogen by removing it from the liquid phase. Agitation is one of the most crucial design parameters that influences the scaling-up of the process. An optimum agitation speed ensures a homogeneous suspension of nutrients in the medium. At lower agitation speed, microorganisms may settle down, resulting in reduced hydrogen production. However, higher agitation, above the optimum point, can cause cell damage due to unreasonable shear stress. Ghosh et al. (2018) observed the highest hydrogen production of 3.42 L L^{-1} at the agitation speed of 200 rpm. Recently, Mahata et al. (2020) found the agitation speed of 180 rpm to be optimum for dark fermentative hydrogen production.

5.4.9 *Inoculum Age and Size*

Apart from the source of inoculum, pre-culture age and size have a significant effect on dark fermentation (Pandey et al. 2019). Inoculum age is the time required to grow the culture before its use for hydrogen production. An optimum inoculum age indicates the most active phase of the microorganism. Likewise, hydrogen production also depends on inoculum size. Hydrogen production could be increased by increasing inoculum size. Above the optimal point, however, more carbon is devoted to cell mass growth gather than to product formation. Kotay and Das (2007) investigated the effect of inoculum age and size and identified the optimum inoculum age and size at 14 h and 10% v/v, respectively. Nevertheless, the study also revealed that these parameters have less impact on dark fermentative hydrogen production than other parameters.

5.5 Optimization of the Process Parameters for the Dark Fermentation

Over the last few decades, several studies have been conducted to improve dark fermentative hydrogen production using various optimization strategies. In order to maximize H_2 yield or production rate, several parameters such as pH, temperature, substrate concentration, C/N ratio, HRT, and hydrogen partial pressure have been considered as independent variables. The optimization, based on the experimental design, can be broadly classified into two categories: "one-variable-at-a-time" (single parameter optimization) and "multi-variable-at-a-time" approach (multi-parameter optimization). Additionally, there are several experimental designs such as Plackett–Burman and Taguchi orthogonal design, which are employed to select the most influential parameters.

Single parameter optimization, a traditional optimization approach, involves the variation of a single process parameter at a time while maintaining the other parameters constant. This is a widely used method because of its simplicity in design. However, interactive effects among the selected independent variables cannot be elucidated clearly and would be imprecise for optimal points (Jo et al. 2008; Karthic et al. 2013). Additionally, this classical method requires enormous experimental trials, resulting in a long time for optimization.

Design of experiments (DOE) for multiparameter optimization can be performed by several fractional designs such as central composite designs (CCD) and Box-Behnken designs (BBD). These designs can simultaneously handle a maximum of up to ten factors. The experimental data are further analyzed by response surface methodology (RSM) to obtain the optimum points of process parameters and the cumulative effect of their mutual interaction. RSM is a set of statistical and mathematical approaches that examine the relationship between many independent variables and assesses the optimum experimental condition. The RSM develops an empirical model in the form of a second-order polynomial equation (Eq. 5.25) to explain the behavior of responses with independent variables.

$$Y = C_0 + \sum_{i=1}^{n} C_i X_i + \sum_{i=1}^{n} C_{ii} X_i^2 + \sum\nolimits_{i=1;j=1;i\neq j}^{n} C_{ij} X_i X_j \tag{5.25}$$

where Y represents the response modeled by RSM, n is the number of the independent variables, C_0 is the constant, C_i is the coefficient for linear relation, C_{ii} is the coefficient for quadratic relation, C_{ij} is the coefficient of interactive part, and X is the uncoded level of the input variable.

The significance of each term in the equation is estimated using analysis of variance (ANOVA). Several studies have successfully employed the RSM technique for the improvement of hydrogen production (Guo et al. 2009; Vi et al. 2017; Xing et al. 2011). Nevertheless, one major drawback of RSM is its inability to model highly non-linear responses accurately (Nath and Das 2011). As biological processes such as dark fermentation are extremely non-linear, RSM, sometimes, may fail to model the system because of its restriction in a second-order polynomial.

Recently, artificial intelligence (AI)-based optimization techniques have been studied to overcome the limitation of statistical techniques. AI-based optimization has several advantages over RSM: (1) AI does not require any prior knowledge about the system, and (2) it has universal approximation capability, whereas RSM is restricted in a quadratic function. Many studies reported that AI is far more suitable for response optimization than statistical approaches (Karthic et al. 2013; Ardabili et al. 2018). Fundamentally, it provides two tools: (1) modeling tool which establishes the relationship among the process variables and provides adequate objective functions, (2) optimization tools that search for an optimal solution using the objective function. Artificial neural networks (ANN) and support vector machines (SVM) are the most popular modeling approaches available in AI. Previously, several studies have employed the ANN model in dark fermentation (Nath and Das 2011; Ardabili et al. 2018; Nasr et al. 2013; Sewsynker and Gueguim Kana 2016). More recently, Mahata et al. (2020) revealed the suitability of the SVM model in dark fermentative hydrogen production. The study suggested that the SVM model could possess better prediction accuracy than by ANN and RSM. Once the model with desire accuracy is developed, it is further used as an objective function in optimization tools to obtain the optimal point. Several optimization tools such as genetic algorithm (GA), particle swarm optimization (PSO), artificial ant colony (AAC), and simulated annealing (SA) in AI can be applied. Many researchers have coupled the ANN with GA for the maximization of hydrogen production (Nath and Das 2011; Wang and Wan 2009a, b). Recently, PSO has been employed in dark fermentation (Mahata et al. 2020). The study revealed that PSO could exhibit the optimal solution faster as compared to GA. However, AAC and SA have not been explored yet for H_2 production.

5.6 Effect of Bioreactor Configurations on the Biohydrogen Production

Several experiments on hydrogen production have been conducted in batch, semi-continuous, and continuous modes of operations. Preliminary studies such as characterization of inoculum and optimization of culture conditions are usually conducted with the batch reactor. However, its performance is inefficient because of a lower rate of hydrogen production. On the contrary, continuous operation shows higher H_2 production in comparison with batch mode. Additionally, the continuous operation could hold a particular phase of the microorganism for an infinite period. On the other hand, the semi-continuous operation is employed when the substrate inhibition effect is observed. The most commonly used reactor configuration for hydrogen production is a continuous stirred tank reactor (CSTR). Apart from CSTR, other reactor configurations such as packed bed reactor (PBR), fluidized bed reactor (FBR), anaerobic sequencing batch reactor (ASBR), and up-flow anaerobic sludge blanket (UASB) reactor are employed for hydrogen production. Several studies have mentioned that higher H_2 yield could be obtained using these reactors attributed to higher physical retention of hydrogen-producing bacteria.

5.6.1 Continuous Stirred Tank Reactor (CSTR)

CSTR is widely used because of its simple design, mixing efficiency, and ease of operation. Under constant mixing hydrodynamics, an appropriate substrate-microbes contact can be achieved inside the reactor. Nonetheless, cell mass washout could be observed at short HRTs, resulting in a stoppage of hydrogen production. In general, the concentration of biomass in CSTR varies in the range of 1–4 g VSS L^{-1} (Show et al. 2010). On the other hand, the use of granular sludge as an inoculum could increase the biomass retention capability (Show et al. 2007). Previous studies reported that CSTR using granular sludge could be operated up to the lowest HRT of 0.5 h without failure (Show et al. 2007; Zhang et al. 2007). Another way of improving cell mass retention is the employment of a settling tank in the effluent and followed by, recycling the settled biomass by passing through an activation chamber (Khanal et al. 2006).

5.6.2 Packed Bed Reactor (PBR)

PBR could retain a high concentration of biomass inside the reactor; hence, it is one of the possible solutions to the problem associated with CSTR. The reactor is supported by packing materials within the reactor, which plays a pivotal role in cell mass retention and hydrogen production. However, the hydrodynamics of mixing is less turbulent, resulting in a higher pH gradient along the reactor length and higher hydrogen gas holdup. Consequently, the H_2 production rate and substrate conversion efficiency decrease. On the other hand, recirculation of liquid effluent can be recommended to maintain higher hydrogen production and substrate conversion (Tomczak et al. 2018). Kumar and Das (2001) investigated packed bed reactors with various geometric configurations such as tubular, tapered, and rhomboid. The study revealed that the rhomboid with convergent-divergent shape had superior performance in hydrogen production (1.60 L L^{-1} h^{-1}) as compared to tubular

($1.40\ L\ L^{-1}\ h^{-1}$) and tapered ($1.46\ L\ L^{-1}\ h^{-1}$) reactor. This result could be due to the better mixing phenomenon owing to lower gas-holdup and higher substrate-microbes contact. Additionally, the study also showed that coconut coir, as a supporting material, could exhibit better hydrogen production than other lignocellulosic carrier materials.

5.6.3 *Fluidized Bed Reactor (FBR)*

FBR, a three-phase system, is the combination of CSTR and PBR, which provides excellent mixing characteristics. Previously, this reactor configuration has been extensively employed in biological wastewater treatment due to its potentiality in high organic loading rate. In FBR, microorganisms are immobilized on the solid supports to form a granular or biofilm. Lin et al. (2009) highlighted that attached sewage sludge in FBR could efficiently produce hydrogen at the HRT of 2–6 h with the maximum H_2 yield of 4.26 mol H_2 mol^{-1} sucrose. Zhang et al. (2007) achieved the maximum H_2 production rate of 2.36 L H_2 $L^{-1}\ h^{-1}$ at 1 h HRT using biofilm culture propagated on activated carbon in FBR. However, the main drawback of this system is the high energy demand required to maintain its fluidization.

5.6.4 *Anaerobic Sequencing Batch Reactor (ASBR)*

This system has a unique feature to retain high cell mass by segregating the operation into four cyclic stages, such as feed, reaction, settling, and decant. Previously, the reactor was used for wastewater treatment. Recently, it has gained significant attention for biohydrogen production along with waste treatment (Vijaya Bhaskar et al. 2008; Maaroff et al. 2019). In order to sustain the reactor performance, pH plays the most important role in the system (Kim et al. 2010). Chen et al. (2009) achieved the highest H_2 yield of 1.86 mol H_2 mol^{-1} sucrose at the operational condition of 4 h cyclic time, 16 HRT, and pH 4.9.

5.6.5 *Up-Flow Anaerobic Sludge Blanket (UASB) Reactor*

The up-flow anaerobic sludge blanket (UASB) reactor is an extensively and widely used economically viable technology developed by Gatze Lettinga for wastewater treatment due to its high conversion efficiency and supreme operational stability. Over the last five decades, the UASB process has been successfully employed for the anaerobic treatment of different types of wastewater and simultaneous methane production by promoting the development of granular sludge with an excellent settling ability (Parawira et al. 2006; Bourque et al. 2008). In recent years, it has been demonstrated that the UASB system is also a promising module for H_2 production (Sivagurunathan et al. 2016; Jung et al. 2010). Successful and efficient operation of UASB reactor depends on the formation of high-strength granular sludge. Extra-cellular polymeric substance (EPS) secreted by bacteria acts as a bio-glue, which could facilitate microbial aggregation, resulting in sludge bed development (Jung et al. 2011). EPS in the sludge mainly comprises carbohydrates and protein; it plays a crucial role in the immobilization of hydrogen-producing bacteria and stability for the long-term operation of the UASB reactor (Lu et al. 2015). Recently, researchers have suggested that the UASB reactor could be promising for a high rate of hydrogen production even at low HRT without manual immobilization (Lu et al. 2015; Chang and Lin 2004; Mahmod et al. 2019). More recently, Sivagurunathan et al. (2016) observed the maximum H_2 production rate of 56.8 L H_2 L^{-1} day^{-1} from galactose in the UASB reactor at 2 h HRT (Sivagurunathan et al. 2016). However, the major bottleneck of the UASB reactor is the long start-up period for microbial granulation (Liu et al. 2012b). To overcome this drawback, some studies recommended the addition of microbial carriers such as activated carbon, carbon nanotubes, and filter sponge in the blanket zone to fasten the film formation (Liu et al. 2012b; Lee et al. 2004).

5.7 Scaling up of the Biohydrogen Production Processes

Dark fermentative H_2 production from organic waste has tremendous potential to replace conventional energy sources in the future. Presently, this process is not technologically viable on a large scale. Therefore, there is an enormous scope to study the scale-up of bioreactors for dark fermentation using cheap feedstock such as organic waste and residue. The purpose of scaling up is to acquire a condition similar to that of a smaller reactor. During the scale-up of dark fermentation, there are several approaches such as geometric similarity, constant power number, constant agitation speed, and constant mixing time to magnify the reactor volume. To date, Vatsala et al. (2008) reported the performance of the largest reactor (100 m^3) for H_2 production from distillery effluent using co-cultures of *Citrobacter freundii* 01, *Rhodopseudomonas palustris* P2, and *Enterobacter aerogenes* E10 (Vatsala et al. 2008). The researchers estimated the rate of hydrogen production as 0.53 kg H_2 h^{-1}. Recently, researchers at the Indian Institute of Technology, Kharagpur explored the feasibility of a 10 m^3 bioreactor for H_2 production via dark fermentative from cane molasses and groundnut de-oiled cake as a co-substrate using *Enterobacter cloacae* IIT-BT 08 (Balachandar et al. 2019). The pilot-scale study reported the maximum hydrogen production of 76.2 m^3 with the COD conversion efficiency of 37.9%. Furthermore, several studies have attempted to scale up this process as listed in Table 5.2. However, detailed "techno-economic analysis (TEA) and life cycle assessment (LCA)" are still needed to be explored.

Table 5.2 Scale-up studies for the biohydrogen production process

Reactor types	Reactor volume (m^3)	Feedstock	Microorganism	H_2 production rate ($L\ L^{-1}\ day^{-1}$)	H_2 yield	References
CSTR	0.025	Fruits and vegetables wastes	Digested sludge	0.72	72.6 mL H_2/g VSS	Morra et al. (2014)
CSTR	0.03	Waste sugar	*Clostridium* sp.	–	2.93 mol H_2/mol hexose	Krupp and Widmann (2009)
CSTR	0.2	Food waste	Sewage sludge	1.1	66.7 l/kgTVS	Cavinato et al. (2012)
Sequencing batch biofilm	0.02	Food waste	Acidogenic mixed consortia	11.6	–	Pasupuleti et al. (2014)
Agitated granular sludge bed reactor	0.4	Molasses	Mixed consortia	15.59	1.04 mol H_2/mol sucrose	Lin et al. (2011)
CSTR	1.48	Molasses	Anaerobic sludge	5.57	26.13 mol/kg COD removed	Ren et al. (2006)
Batch	10	Cane molasses + groundnut de-oiled cake	*Enterobacter cloacae* IIT-BT 08	7.31	18.74 mol/kg COD	Balachandar et al. (2019)

5.8 Major Challenges and Perspectives in Biohydrogen Production

There are numerous studies in the literature for the enhancement of H_2 production, including the genetic modification of hydrogen-producing microorganisms, development of bioreactors, and selection of feedstock and process modification. However, the process is not commercially viable on a large scale due to some technological challenges. Major challenges in the improvement of biohydrogen production can be summarized as follows (Das et al. 2008):

- There is a lack of knowledge on industrially applicable robust microorganisms that could be engineered to produce more than 4 mols hydrogen from 1 mol glucose.
- Feedstock sterilization involved in the biohydrogen production process is an energy-consuming step. Therefore, an abundance study is required using non-sterile feedstock.
- The process efficiency and the hydrogen yield depend on the sensitivity of hydrogenase to H_2 and O_2 partial pressure.
- Usually, a major portion of the substrate is devoted to soluble metabolites production rather than hydrogen. To overcome this, research should focus on the metabolic engineering of the biochemical pathway.
- There is no significant literature on the economic understanding of the integrated H_2 generation system, such as dark fermentation-photo fermentation and dark fermentation-MEC (microbial electrolysis cells).
- Various engineering issues such as novel bioreactor for long-term hydrogen production, scale-up for commercial application, separation of CO_2, process optimization, need to be addressed.

In the future, hydrogen can be utilized in the internal combustion engine and fuel cell in the automobile sectors. Biohydrogen has a great potential to replace conventional energy sources such as fossil fuels. Nevertheless, its production process must overcome the aforementioned limitations in order to compete with conventional energy sources in the fuel market. Future biohydrogen production technology should also consider social acceptance, economic feasibility, and government policy. At the same time, the government should also provide research subsidies on this technology.

5.9 Biohythane Process

As per the stoichiometry of dark fermentation, a maximum of 34.1% of energy as hydrogen can be recovered from the substrate used (Kumari and Das 2015). Hence, the process efficiency of hydrogen production is significantly low. Several volatile fatty acids remain in the fermentation broth after the biohydrogen fermentation process which is a good feedstock for the biomethanation process. Biohydrogen production followed by biomethanation process is known as "Biohythane process". So, this process can increase the overall energy recovery to a great extent. After hydrogen, methane has the second highest energy content (55 kJ g^{-1}). Hythane is a mixture of hydrogen (5–30%) and methane (80–95%). Hythane® is a trademark first introduced by Hydrogen Component Inc. (HCI) (Bolzonella et al. 2018). Production of hythane through biological route is comprehensively called "Biohythane" (Liu et al. 2013). Biohythane has several advantages over methane as a fuel for IC (Internal combustion) engines such as higher combustion rate, improved lean flammability, and enhanced fuel flaming speed. Nowadays, this two-stage biohythane process is being widely accepted energy-producing process because of its viability on a commercial scale.

5.10 Conclusion

Biohydrogen can be considered as a promising alternative energy, which can offer clean and sustainable fuel currency. Among all the biological processes, dark fermentation has gained considerable attention from researchers. Numerous studies have been conducted to improve hydrogen production, considering process optimization, inoculum development, reactor design, substrate selection. Still, however, the process suffers from several technological limitations due to its lower hydrogen yield. To overcome this, researchers should focus on the genetic and metabolic engineering of the microbial strain. In addition, an integrated system, such as DF-photofermentation, simultaneous dark fermentation (DF) and MEC (DF-MEC), and "Biohythane", is recommended for achieving enhanced energy recovery. Besides, it is also essential to scale-up the study, including appropriate techno-economic and life cycle analysis, to access its potentiality in commercial hydrogen production.

References

Abreu AA, Karakashev D, Angelidaki I, Sousa DZ, Alves M (2012) Biohydrogen production from arabinose and glucose using extreme thermophilic anaerobic mixed cultures. Biotechnol Biofuels 5:6. https://doi.org/10.1186/1754-6834-5-6

Ardabili SF, Najafi B, Shamshirband S, Bidgoli M, Chand R, Kwok-Wing Chau D et al (2018) Engineering applications of computational fluid mechanics computational intelligence approach for modeling hydrogen production: a review computational intelligence approach for modeling hydrogen production: a review. Eng Appl Comput Fluid Mech 12:438–458. https://doi.org/10.1080/19942060.2018.1452296

Baima Ferreira Freitas I, Aparecida de Menezes C, Luiz Silva E (2020) An alternative for value aggregation to the sugarcane chain: biohydrogen and volatile fatty acids production from sugarcane molasses in mesophilic expanded granular sludge bed reactors. Fuel 260:116419. https://doi.org/10.1016/j.fuel.2019.116419

Balachandar G, Varanasi JL, Singh V, Singh H, Das D (2019) Biological hydrogen production via dark fermentation: a holistic approach from lab-scale to pilot-scale. Int J Hydrog Energy 45:5202–5215. https://doi.org/10.1016/j.ijhydene.2019.09.006

Ban S, Lin W, Luo J (2019) Ca $^{2+}$ enhances algal photolysis hydrogen production by improving the direct and indirect pathways. Int J Hydrog Energy 44:1466–1473. https://doi.org/10.1016/j.ijhydene.2018.11.075

Barros AR, Silva EL (2012) Hydrogen and ethanol production in anaerobic fluidized bed reactors: performance evaluation for three support materials under different operating conditions. Biochem Eng J 61:59–65. https://doi.org/10.1016/J.BEJ.2011.12.002

Basak N, Das D (2009) Photofermentative hydrogen production using purple non-sulfur bacteria Rhodobacter sphaeroides O.U.001 in an annular photobioreactor: a case study. Biomass Bioenergy 33:911–919. https://doi.org/10.1016/j.biombioe.2009.02.007

Bolzonella D, Battista F, Cavinato C, Gottardo M, Micolucci F, Lyberatos G et al (2018) Recent developments in biohythane production from household food wastes: a review. Bioresour Technol 257:311–319. https://doi.org/10.1016/j.biortech.2018.02.092

Boran E, Özgür E, Van Der Burg J, Yücel M, Gündüz U, Eroglu I (2010) Biological hydrogen production by Rhodobacter capsulatus in solar tubular photo bioreactor. J Clean Prod 18:S29–S35. https://doi.org/10.1016/j.jclepro.2010.03.018

Bourque JS, Guiot SR, Tartakovsky B (2008) Methane production in an UASB reactor operated under periodic mesophilic-thermophilic conditions. Biotechnol Bioeng 100:1115–1121. https://doi.org/10.1002/bit.21850

Cao X, Zhao Y (2009) The influence of sodium on biohydrogen production from food waste by anaerobic fermentation. J Mater Cycles Waste Manag 11:244–250. https://doi.org/10.1007/s10163-009-0237-5

Cavinato C, Giuliano A, Bolzonella D, Pavan P, Cecchi F (2012) Bio-hythane production from food waste by dark fermentation coupled with anaerobic digestion process: a long-term pilot scale experience. Int J Hydrog Energy 37:11549–11555. https://doi.org/10.1016/j.ijhydene.2012.03.065

Chae KJ, Choi MJ, Lee J, Ajayi FF, Kim IS (2008) Biohydrogen production via biocatalyzed electrolysis in acetate-fed bioelectrochemical cells and microbial community analysis. Int J Hydrog Energy 33:5184–5192. https://doi.org/10.1016/j.ijhydene.2008.05.013

Chang F-Y, Lin C-Y (2004) Biohydrogen production using an up-flow anaerobic sludge blanket reactor. Int J Hydrog Energy 29:33–39. https://doi.org/10.1016/S0360-3199(03)00082-X

Chen CY, Lee CM, Chang JS (2006) Feasibility study on bioreactor strategies for enhanced photohydrogen production from *Rhodopseudomonas palustris* WP3-5 using optical-fiber-assisted illumination systems. Int J Hydrog Energy 31:2345–2355. https://doi.org/10.1016/j.ijhydene.2006.03.007

Chen WH, Sung S, Chen SY (2009) Biological hydrogen production in an anaerobic sequencing batch reactor: pH and cyclic duration effects. Int J Hydrog Energy 34:227–234. https://doi.org/10.1016/j.ijhydene.2008.09.061

Das D (2009) Advances in biohydrogen production processes: an approach towards commercialization. Int J Hydrog Energy 34:7349–7357. https://doi.org/10.1016/J.IJHYDENE.2008.12.013

Das D (2017) A road map on biohydrogen production from organic wastes. INAE Letters 2:153–160. https://doi.org/10.1007/s41403-017-0031-y

Das D, Veziroğlu TN (2001) Hydrogen production by biological processes: a survey of the literature. Int J Hydrog Energy 26:13–28. https://doi.org/10.1016/S0360-3199(00)00058-6

Das D, Veziroglu TN (2008) Advances in biological hydrogen production processes. Int J Hydrog Energy 33:6046–6057. https://doi.org/10.1016/j.ijhydene.2008.07.098

Das D, Khanna N, Nejat Veziroğlu T (2008) Recent developments in biological hydrogen production processes. Chem Ind Chem Eng Q 14:57–67. https://doi.org/10.2298/CICEQ0802057D

Ding C, Yang KL, He J (2016) Biological and fermentative production of hydrogen. In: Handbook of biofuels production: processes and technologies, 2nd edn. Elsevier Inc., pp 303–333. https://doi.org/10.1016/B978-0-08-100455-5.00011-4

Elbeshbishy E, Hafez H, Dhar BR, Nakhla G (2011) Single and combined effect of various pretreatment methods for biohydrogen production from food waste. Int J Hydrog Energy 36:11379–11387. https://doi.org/10.1016/J.IJHYDENE.2011.02.067

Fabiano B, Perego P (2002) Thermodynamic study and optimization of hydrogen production by Enterobacter aerogenes. Int J Hydrog Energy 27:149–156. https://doi.org/10.1016/S0360-3199(01)00102-1

Feng C, Li J, Qin D, Chen L, Zhao F, Chen S et al (2014) Characterization of exoelectrogenic bacteria enterobacter strains isolated from a microbial fuel cell exposed to copper shock load. PLoS One 9:e113379. https://doi.org/10.1371/journal.pone.0113379

Ghosh S, Dutta B, Banerjee A, Roy S (2018) Bio-H2 production using de-oiled cake as cheap nitrogen source and subsequent electricity generation by a hybrid system. Bioresour Technol Rep 4:1–8. https://doi.org/10.1016/J.BITEB.2018.09.005

Gilbert JJ, Ray S, Das D (2011) Hydrogen production using *Rhodobacter sphaeroides* (O.U. 001) in a flat panel rocking photobioreactor. Int J Hydrog Energy 36:3434–3441. https://doi.org/10.1016/j.ijhydene.2010.12.012

Guo W-Q, Ren N-Q, Wang X-J, Xiang W-S, Ding J, You Y et al (2009) Optimization of culture conditions for hydrogen production by *Ethanoligenens harbinense* B49 using response surface methodology. Bioresour Technol 100:1192–1196. https://doi.org/10.1016/J.BIORTECH.2008.07.070

Han SK, Shin HS (2004) Biohydrogen production by anaerobic fermentation of food waste. Int J Hydrog Energy 29:569–577. https://doi.org/10.1016/j.ijhydene.2003.09.001

Hu H, Fan Y, Liu H (2008) Hydrogen production using single-chamber membrane-free microbial electrolysis cells. Water Res 42:4172–4178. https://doi.org/10.1016/j.watres.2008.06.015

Jayasinghearachchi HS, Sarma PM, Lal B (2012) Biological hydrogen production by extremely thermophilic novel bacterium *Thermoanaerobacter mathranii* A3N isolated from an oil-producing well. Int J Hydrog Energy 37:5569–5578. https://doi.org/10.1016/J.IJHYDENE.2011.12.145

Jeong D-Y, Cho S-K, Shin H-S, Jung K-W (2013) Application of an electric field for pretreatment of a seeding source for dark fermentative hydrogen production. Bioresour Technol 139:393–396. https://doi.org/10.1016/J.BIORTECH.2013.04.039

Jia X, Li M, Xi B, Zhu C, Yang Y, Xia T et al (2014) Integration of fermentative biohydrogen with methanogenesis from fruit-vegetable waste using different pre-treatments. Energy Convers Manag 88:1219–1227. https://doi.org/10.1016/j.enconman.2014.02.015

Jo JH, Lee DS, Park D, Choe W-S, Park JM (2008) Optimization of key process variables for enhanced hydrogen production by *Enterobacter aerogenes* using statistical methods. Bioresour Technol 99:2061–2066. https://doi.org/10.1016/J.BIORTECH.2007.04.027

Jung KW, Kim DH, Shin HS (2010) Continuous fermentative hydrogen production from coffee drink manufacturing wastewater by applying UASB reactor. Int J Hydrog Energy 35:13370–13378. https://doi.org/10.1016/j.ijhydene.2009.11.120

Jung KW, Kim DH, Shin HS (2011) A simple method to reduce the start-up period in a H2-producing UASB reactor. Int J Hydrog Energy 36:1466–1473. https://doi.org/10.1016/j.ijhydene.2010.09.095

Kadier A, Simayi Y, Abdeshahian P, Azman NF, Chandrasekhar K, Kalil MS (2016) A comprehensive review of microbial electrolysis cells (MEC) reactor designs and configurations for sustainable hydrogen gas production. Alex Eng J 55:427–443. https://doi.org/10.1016/j.aej.2015.10.008

Karthic P, Joseph S, Arun N, Kumaravel S (2013) Optimization of biohydrogen production by *Enterobacter* species using artificial neural network and response surface methodology. J Renew Sustain Energy 5:033104. https://doi.org/10.1063/1.4803746

Khanal SK, Chen WH, Li L, Sung S (2004) Biological hydrogen production: effects of pH and intermediate products. Int J Hydrog Energy 29:1123–1131. https://doi.org/10.1016/j.ijhydene.2003.11.002

Khanal SK, Chen W-H, Li L, Sung S (2006) Biohydrogen production in continuous-flow reactor using mixed microbial culture. Water Environ Res 78:110–117. https://doi.org/10.2175/106143005x89562

Kim DH, Kim SH, Kim KY, Shin HS (2010) Experience of a pilot-scale hydrogen-producing anaerobic sequencing batch reactor (ASBR) treating food waste. Int J Hydrog Energy 35:1590–1594. https://doi.org/10.1016/j.ijhydene.2009.12.041

Koku H, Erolu I, Gündüz U, Yücel M, Türker L (2002) Aspects of the metabolism of hydrogen production by *Rhodobacter sphaeroides*. Int J Hydrog Energy 27:1315–1329. https://doi.org/10.1016/S0360-3199(02)00127-1

Kotay SM, Das D (2007) Microbial hydrogen production with Bacillus coagulans IIT-BT S1 isolated from anaerobic sewage sludge. Bioresour Technol 98:1183–1190. https://doi.org/10.1016/J.BIORTECH.2006.05.009

Krupp M, Widmann R (2009) Biohydrogen production by dark fermentation: experiences of continuous operation in large lab scale. Int J Hydrog Energy 34:4509–4516. https://doi.org/10.1016/j.ijhydene.2008.10.043

Kumar N, Das D (2000) Enhancement of hydrogen production by Enterobacter cloacae IIT-BT 08. Process Biochem 35:589–593. https://doi.org/10.1016/S0032-9592(99)00109-0

Kumar N, Das D (2001) Continuous hydrogen production by immobilized Enterobacter cloacae IIT-BT 08 using lignocellulosic materials as solid matrices. Enzym Microb Technol 29:280–287. https://doi.org/10.1016/S0141-0229(01)00394-5

Kumar G, Sivagurunathan P, Pugazhendhi A, Thi NBD, Zhen G, Chandrasekhar K et al (2017) A comprehensive overview on light-independent fermentative hydrogen production from wastewater feedstock and possible integrative options. Energy Convers Manag 141:390–402. https://doi.org/10.1016/j.enconman.2016.09.087

Kumari S, Das D (2015) Improvement of gaseous energy recovery from sugarcane bagasse by dark fermentation followed by biomethanation process. Bioresour Technol 194:354–363. https://doi.org/10.1016/J.BIORTECH.2015.07.038

Kumari S, Das D (2017) Improvement of biohydrogen production using acidogenic culture. Int J Hydrog Energy 42:4083–4094. https://doi.org/10.1016/J.IJHYDENE.2016.09.021

Kundu A, Sahu JN, Redzwan G, Hashim MA (2013) An overview of cathode material and catalysts suitable for generating hydrogen in microbial electrolysis cell. Int J Hydrog Energy 38:1745–1757. https://doi.org/10.1016/j.ijhydene.2012.11.031

Lee KS, Wu JF, Lo YS, Lo YC, Lin PJ, Chang JS (2004) Anaerobic hydrogen production with an efficient carrier-induced granular sludge bed bioreactor. Biotechnol Bioeng 87:648–657. https://doi.org/10.1002/bit.20174

Lee KS, Lin PJ, Chang JS (2006) Temperature effects on biohydrogen production in a granular sludge bed induced by activated carbon carriers. Int J Hydrog Energy 31:465–472. https://doi.org/10.1016/j.ijhydene.2005.04.024

Lee YJ, Miyahara T, Noike T (n.d.) Effect of pH on microbial hydrogen fermentation. https://doi.org/10.1002/jctb.623

Li C, Fang HHP (2007) Fermentative hydrogen production from wastewater and solid wastes by mixed cultures. Crit Rev Environ Sci Technol 37:1–39. https://doi.org/10.1080/10643380600729071

Lin CY, Lay CH (2005) A nutrient formulation for fermentative hydrogen production using anaerobic sewage sludge microflora. Int J Hydrog Energy 30:285–292. https://doi.org/10.1016/j.ijhydene.2004.03.002

Lin CY, Hung CH, Chen CH, Chung WT, Cheng LH (2006) Effects of initial cultivation pH on fermentative hydrogen production from xylose using natural mixed cultures. Process Biochem 41:1383–1390. https://doi.org/10.1016/j.procbio.2006.01.021

Lin PY, Whang LM, Wu YR, Ren WJ, Hsiao CJ, Li SL et al (2007) Biological hydrogen production of the genus *Clostridium*: metabolic study and mathematical model simulation. Int J Hydrog Energy 32:1728–1735. https://doi.org/10.1016/j.ijhydene.2006.12.009

Lin CY, Chang CC, Hung CH (2008) Fermentative hydrogen production from starch using natural mixed cultures. Int J Hydrog Energy 33:2445–2453. https://doi.org/10.1016/j.ijhydene.2008.02.069

Lin CN, Wu SY, Chang JS, Chang JS (2009) Biohydrogen production in a three-phase fluidized bed bioreactor using sewage sludge immobilized by ethylene-vinyl acetate copolymer. Bioresour Technol 100:3298–3301. https://doi.org/10.1016/j.biortech.2009.02.027

Lin CY, Wu SY, Lin PJ, Chang JS, Hung CH, Lee KS et al (2011) A pilot-scale high-rate biohydrogen production system with mixed microflora. Int J Hydrog Energy 36:8758–8764. https://doi.org/10.1016/j.ijhydene.2010.07.115

Liu J, Bukatin VE, Tsygankov AA (2006) Light energy conversion into H_2 by *Anabaena variabilis* mutant PK84 dense cultures exposed to nitrogen limitations. Int J Hydrog Energy 31:1591–1596. https://doi.org/10.1016/j.ijhydene.2006.06.025

Liu W, Huang S, Zhou A, Zhou G, Ren N, Wang A et al (2012a) Hydrogen generation in microbial electrolysis cell feeding with fermentation liquid of waste activated sludge. Int J Hydrog Energy 37:13859–13864. https://doi.org/10.1016/j.ijhydene.2012.04.090

Liu Z, Lv F, Zheng H, Zhang C, Wei F, Xing XH (2012b) Enhanced hydrogen production in a UASB reactor by retaining microbial consortium onto carbon nanotubes (CNTs). Int J Hydrog Energy 37:10619–10626. https://doi.org/10.1016/j.ijhydene.2012.04.057

Liu Z, Zhang C, Lu Y, Wu X, Wang L, Wang L et al (2013) States and challenges for high-value biohythane production from waste biomass by dark fermentation technology. Bioresour Technol 135:292–303. https://doi.org/10.1016/j.biortech.2012.10.027

Logan BE, Regan JM (2006) Electricity-producing bacterial communities in microbial fuel cells. Trends Microbiol 14:512–518. https://doi.org/10.1016/j.tim.2006.10.003

Lu L, Ren N, Xing D, Logan BE (2009) Hydrogen production with effluent from an ethanol-H_2-coproducing fermentation reactor using a single-chamber microbial electrolysis cell. Biosens Bioelectron 24:3055–3060. https://doi.org/10.1016/j.bios.2009.03.024

Lu X, Zhen G, Estrada AL, Chen M, Ni J, Hojo T et al (2015) Operation performance and granule characterization of upflow anaerobic sludge blanket (UASB) reactor treating wastewater with starch as the sole carbon source. Bioresour Technol 180:264–273. https://doi.org/10.1016/j.biortech.2015.01.010

Luedeking R, Piret EL (2000) Kinetic study of the lactic acid fermentation. Batch process at controlled pH. Biotechnol Bioeng 67:636–644. https://doi.org/10.1002/(SICI)1097-0290(20000320)67:6<636::AID-BIT3>3.0.CO;2-U

Maaroff RM, Md Jahim J, Azahar AM, Abdul PM, Masdar MS, Nordin D et al (2019) Biohydrogen production from palm oil mill effluent (POME) by two stage anaerobic sequencing batch reactor (ASBR) system for better utilization of carbon sources in POME. Int J Hydrog Energy 44:3395–3406. https://doi.org/10.1016/j.ijhydene.2018.06.013

Mahata C, Ray S, Das D (2020) Optimization of dark fermentative hydrogen production from organic wastes using acidogenic mixed consortia. Energy Convers Manag 219:113047. https://doi.org/10.1016/j.enconman.2020.113047

Mahmod SS, Azahar AM, Tan JP, Jahim JM, Abdul PM, Mastar MS et al (2019) Operation performance of up-flow anaerobic sludge blanket (UASB) bioreactor for biohydrogen production by self-granulated sludge using pre-treated palm oil mill effluent (POME) as carbon source. Renew Energy 134:1262–1272. https://doi.org/10.1016/j.renene.2018.09.062

Mandal B, Nath K, Das D (2006) Improvement of biohydrogen production under decreased partial pressure of H_2 by *Enterobacter cloacae*. Biotechnol Lett 28:831–835. https://doi.org/10.1007/s10529-006-9008-8

Mishra P, Roy S, Das D (2015) Comparative evaluation of the hydrogen production by mixed consortium, synthetic co-culture and pure culture using distillery effluent. Bioresour Technol 198:593–602. https://doi.org/10.1016/j.biortech.2015.09.074

Mishra P, Balachandar G, Das D (2017) Improvement in biohythane production using organic solid waste and distillery effluent. Waste Manag 66:70–78. https://doi.org/10.1016/J.WASMAN.2017.04.040

Mizuno O, Dinsdale R, Hawkes FR, Hawkes DL, Noike T (2000) Enhancement of hydrogen production from glucose by nitrogen gas sparging. Bioresour Technol 73:59–65. https://doi.org/10.1016/S0960-8524(99)00130-3

Morra S, Arizzi M, Allegra P, La Licata B, Sagnelli F, Zitella P et al (2014) Expression of different types of [FeFe]-hydrogenase genes in bacteria isolated from a population of a bio-hydrogen pilot-scale plant. Int J Hydrog Energy 39:9018–9027. https://doi.org/10.1016/j.ijhydene.2014.04.009

Najafpour G, Younesi H, Mohamed AR (2004) Effect of organic substrate on hydrogen production from synthesis gas using *Rhodospirillum rubrum*, in batch culture. Biochem Eng J 21:123–130. https://doi.org/10.1016/j.bej.2004.06.001

Nakashimada Y, Rachman MA, Kakizono T, Nishio N (2002) Hydrogen production of *Enterobacter aerogenes* altered by extracellular and intracellular redox states. Int J Hydrog Energy 27:1399–1405. https://doi.org/10.1016/S0360-3199(02)00128-3

Nasr N, Hafez H, El Naggar MH, Nakhla G (2013) Application of artificial neural networks for modeling of biohydrogen production. Int J Hydrog Energy 38:3189–3195. https://doi.org/10.1016/J.IJHYDENE.2012.12.109

Nath K, Das D (2011) Modeling and optimization of fermentative hydrogen production. Bioresour Technol 102:8569–8581. https://doi.org/10.1016/J.BIORTECH.2011.03.108

Neil G, Nicholas DJD, O'Bockris JM, McCann JF (1976) The photosynthetic production of hydrogen. Int J Hydrog Energy 1:45–48. https://doi.org/10.1016/0360-3199(76)90008-2

Oh SE, Zuo Y, Zhang H, Guiltinan MJ, Logan BE, Regan JM (2009) Hydrogen production by *Clostridium acetobutylicum* ATCC 824 and megaplasmid-deficient mutant M5 evaluated using a large headspace volume technique. Int J Hydrog Energy 34:9347–9353. https://doi.org/10.1016/j.ijhydene.2009.09.084

O-Thong S, Prasertsan P, Intrasungkha N, Dhamwichukorn S, Birkeland NK (2007) Improvement of biohydrogen production and treatment efficiency on palm oil mill effluent with nutrient supplementation at thermophilic condition using an anaerobic sequencing batch reactor. Enzym Microb Technol 41:583–590. https://doi.org/10.1016/j.enzmictec.2007.05.002

O-Thong S, Prasertsan P, Birkeland N-K (2009) Evaluation of methods for preparing hydrogen-producing seed inocula under thermophilic condition by process performance and microbial community analysis. Bioresour Technol 100:909–918. https://doi.org/10.1016/J.BIORTECH.2008.07.036

Pan CM, Fan YT, Xing Y, Hou HW, Zhang ML (2008) Statistical optimization of process parameters on biohydrogen production from glucose by *Clostridium* sp. Fanp2. Bioresour Technol 99:3146–3154. https://doi.org/10.1016/J.BIORTECH.2007.05.055

Pandey A, Srivastava S, Rai P, Duke M (2019) Cheese whey to biohydrogen and useful organic acids: a non-pathogenic microbial treatment by *L. acidophilus*. Sci Rep 9:1–9. https://doi.org/10.1038/s41598-019-42752-3

Parawira W, Murto M, Zvauya R, Mattiasson B (2006) Comparative performance of a UASB reactor and an anaerobic packed-bed reactor when treating potato waste leachate. Renew Energy 31:893–903. https://doi.org/10.1016/j.renene.2005.05.013

Pasupuleti SB, Sarkar O, Venkata Mohan S (2014) Upscaling of biohydrogen production process in semi-pilot scale biofilm reactor: evaluation with food waste at variable organic loads. Int J Hydrog Energy 39:7587–7596. https://doi.org/10.1016/j.ijhydene.2014.02.034

Pirt SJ (1965) The maintenance energy of bacteria in growing cultures. Proc R Soc Lond Series B 163:224–231. https://doi.org/10.1098/rspb.1965.0069

Ren N, Li J, Li B, Wang Y, Liu S (2006) Biohydrogen production from molasses by anaerobic fermentation with a pilot-scale bioreactor system. Int J Hydrog Energy 31:2147–2157. https://doi.org/10.1016/j.ijhydene.2006.02.011

Ren NQ, Liu BF, Zheng GX, Xing DF, Zhao X, Guo WQ et al (2009) Strategy for enhancing photo-hydrogen production yield by repeated fed-batch cultures. Int J Hydrog Energy 34:7579–7584. https://doi.org/10.1016/j.ijhydene.2009.07.030

Roy S, Ghosh S, Das D (2012) Improvement of hydrogen production with thermophilic mixed culture from rice spent wash of distillery industry. Int J Hydrog Energy 37:15867–15874. https://doi.org/10.1016/j.ijhydene.2012.08.016

Roy S, Vishnuvardhan M, Das D (2014) Improvement of hydrogen production by newly isolated *Thermoanaerobacterium thermosaccharolyticum* IIT BT-ST1. Int J Hydrog Energy 39:7541–7552. https://doi.org/10.1016/j.ijhydene.2013.06.128

Rozendal RA, Hamelers HVM, Molenkamp RJ, Buisman CJN (2007) Performance of single chamber biocatalyzed electrolysis with different types of ion-exchange membranes. Water Res 41:1984–1994. https://doi.org/10.1016/j.watres.2007.01.019

Schröder C, Selig M, Schönheit P (1994) Glucose fermentation to acetate, CO2 and H2 in the anaerobic hyperthermophilic eubacterium *Thermotoga maritima*: involvement of the Embden-Meyerhof pathway. Arch Microbiol 161:460–470. https://doi.org/10.1007/BF00307766

Sekoai PT, Daramola MO (2018) Effect of metal ions on dark fermentative biohydrogen production using suspended and immobilized cells of mixed bacteria. Chem Eng Commun 205:1011–1022. https://doi.org/10.1080/00986445.2018.1428958

Sewsynker Y, Gueguim Kana EB (2016) Intelligent models to predict hydrogen yield in dark microbial fermentations using existing knowledge. Int J Hydrog Energy 41:12929–12940. https://doi.org/10.1016/J.IJHYDENE.2016.05.250

Show KY, Zhang ZP, Tay JH, Tee Liang D, Lee DJ, Jiang WJ (2007) Production of hydrogen in a granular sludge-based anaerobic continuous stirred tank reactor. Int J Hydrog Energy 32:4744–4753. https://doi.org/10.1016/j.ijhydene.2007.07.005

Show KY, Zhang ZP, Tay JH, Liang DT, Lee DJ, Ren N et al (2010) Critical assessment of anaerobic processes for continuous biohydrogen production from organic wastewater. Int J Hydrog Energy 35:13350–13355. https://doi.org/10.1016/j.ijhydene.2009.11.110

Singh L, Siddiqui MF, Ahmad A, Rahim MHA, Sakinah M, Wahid ZA (2013) Biohydrogen production from palm oil mill effluent using immobilized mixed culture. J Ind Eng Chem 19:659–664. https://doi.org/10.1016/j.jiec.2012.10.001

Singh V, Singh H, Das D (2019) Optimization of the medium composition for the improvement of hydrogen and butanol production using *Clostridium saccharoperbutylacetonicum* DSM 14923. Int J Hydrog Energy 44:26905–26919. https://doi.org/10.1016/j.ijhydene.2019.08.125

Sivagurunathan P, Anburajan P, Kumar G, Kim SH (2016) Effect of hydraulic retention time (HRT) on biohydrogen production from galactose in an up-flow anaerobic sludge blanket reactor. Int J Hydrog Energy 41:21670–21677. https://doi.org/10.1016/j.ijhydene.2016.06.047

Slobodkin AI, Tourova TP, Kuznetsov BB, Kostrikina NA, Chernyh NA, Bonch-Osmolovskaya EA (1999) *Thermoanaerobacter siderophilus* sp. nov., a novel dissimilatory Fe(III)- reducing, anaerobic, thermophilic bacterium. Int J Syst Bacteriol 49:1471–8. https://doi.org/10.1099/00207713-49-4-1471

Sözen A, Arcaklioğlu E, Özalp M, Kanit EG (2005) Solar-energy potential in Turkey. Appl Energy 80:367–381. https://doi.org/10.1016/j.apenergy.2004.06.001

Tang G-LL, Huang J, Sun Z-JJ, Tang Q-QQ, Yan C-HH, Liu G-QQ (2008) Biohydrogen production from cattle wastewater by enriched anaerobic mixed consortia: influence of fermentation temperature and pH. J Biosci Bioeng 106:80–87

Tanisho S, Suzuki Y, Wakao N. Fermentative hydrogen evolution by Enterobacter aerogenes strain E.82005. Int J Hydrog Energy 1987;12:623–627. doi:https://doi.org/10.1016/0360-3199(87)90003-6

Tanisho S, Kuromoto M, Kadokura N (1998) Effect of CO2 removal on hydrogen production by fermentation. Int J Hydrog Energy 23:559–563. https://doi.org/10.1016/s0360-3199(97)00117-1

Tapia-Venegas E, Ramirez-Morales JE, Silva-Illanes F, Toledo-Alarcón J, Paillet F, Escudie R et al (2015) Biohydrogen production by dark fermentation: scaling-up and technologies integration for a sustainable system. Rev Environ Sci Biotechnol 14:761–785. https://doi.org/10.1007/s11157-015-9383-5

Tomczak W, Ferrasse J-H, Giudici-Orticoni M-T, Soric A (2018) Effect of hydraulic retention time on a continuous biohydrogen production in a packed bed biofilm reactor with recirculation flow of the liquid phase. Int J Hydrog Energy 43:18883–18895. https://doi.org/10.1016/J.IJHYDENE.2018.08.094

Van Ginkel S, Sung S, Lay JJ (2001) Biohydrogen production as a function of pH and substrate concentration. Environ Sci Technol 35:4726–4730. https://doi.org/10.1021/es001979r

Van Ginkel SW, Oh SE, Logan BE (2005) Biohydrogen gas production from food processing and domestic wastewaters. Int J Hydrog Energy 30:1535–1542. https://doi.org/10.1016/j.ijhydene.2004.09.017

Van Niel EWJ, Budde MAW, De Haas G, Van der Wal FJ, Claassen PAM, Stams AJM (2002) Distinctive properties of high hydrogen producing extreme thermophiles, *Caldicellulosiruptor saccharolyticus* and *Thermotoga elfii*. Int J Hydrog Energy 27:1391–1398. https://doi.org/10.1016/S0360-3199(02)00115-5

Van Ooteghem SA, Beer SK, Yue PC (2002) Hydrogen production by the thermophilic bacterium *Thermotoga neapolitana*. Appl Biochem Biotechnol 98–100:177–189. https://doi.org/10.1385/ABAB:98-100:1-9:177

Varanasi JL, Kumari S, Das D (2018) Improvement of energy recovery from water hyacinth by using an integrated system. Int J Hydrog Energy 43:1303–1318. https://doi.org/10.1016/j.ijhydene.2017.11.110

Vatsala TM, Raj SM, Manimaran A (2008) A pilot-scale study of biohydrogen production from distillery effluent using defined bacterial co-culture. Int J Hydrog Energy 33:5404–5415. https://doi.org/10.1016/j.ijhydene.2008.07.015

Veeravalli SS, Shanmugam SR, Ray S, Lalman JA, Biswas N (2019) Biohydrogen production from renewable resources. In: Advanced bioprocessing for alternative fuels, biobased chemicals, and bioproducts: technologies and approaches for scale-up and commercialization. Elsevier, pp 289–312. https://doi.org/10.1016/B978-0-12-817941-3.00015-2

Vi LVT, Salakkam A, Reungsang A (2017) Optimization of key factors affecting bio-hydrogen production from sweet potato starch. Energy Procedia 138:973–978. https://doi.org/10.1016/J.EGYPRO.2017.10.092

Vijaya Bhaskar Y, Venkata Mohan S, Sarma PN (2008) Effect of substrate loading rate of chemical wastewater on fermentative biohydrogen production in biofilm configured sequencing batch reactor. Bioresour Technol 99:6941–6948. https://doi.org/10.1016/j.biortech.2008.01.029

Wagner RC, Regan JM, Oh SE, Zuo Y, Logan BE (2009) Hydrogen and methane production from swine wastewater using microbial electrolysis cells. Water Res 43:1480–1488. https://doi.org/10.1016/j.watres.2008.12.037

Wang J, Wan W (2009a) Optimization of fermentative hydrogen production process using genetic algorithm based on neural network and response surface methodology. Int J Hydrog Energy 34:255–261. https://doi.org/10.1016/J.IJHYDENE.2008.10.010

Wang J, Wan W (2009b) Application of desirability function based on neural network for optimizing biohydrogen production process. Int J Hydrog Energy 34:1253–1259. https://doi.org/10.1016/J.IJHYDENE.2008.11.055

Wu S, Li X, Yu J, Wang Q (2012) Increased hydrogen production in co-culture of *Chlamydomonas reinhardtii* and *Bradyrhizobium japonicum*. Bioresour Technol 123:184–188. https://doi.org/10.1016/j.biortech.2012.07.055

Xie G-J, Liu B-F, Guo W-Q, Ding J, Xing D-F, Nan J et al (2012) Feasibility studies on continuous hydrogen production using photo-fermentative sequencing batch reactor. Int J Hydrog Energy 37:13689–13695. https://doi.org/10.1016/J.IJHYDENE.2012.02.107

Xing Y, Fan SQ, Zhang JN, Fan YT, Hou HW (2011) Enhanced bio-hydrogen production from corn stalk by anaerobic fermentation using response surface methodology. Int J Hydrog Energy 36:12770–12779. https://doi.org/10.1016/j.ijhydene.2011.07.065

Yang G, Wang J (2018) Kinetics and microbial community analysis for hydrogen production using raw grass inoculated with different pretreated mixed culture. Bioresour Technol 247:954–962. https://doi.org/10.1016/J.BIORTECH.2017.09.041

Yu H, Zhu Z, Hu W, Zhang H (2002) Hydrogen production from rice winery wastewater in an up-flow anaerobic reactor by using mixed anaerobic cultures. Int J Hydrog Energy 27:1359–1365. https://doi.org/10.1016/S0360-3199(02)00073-3

Zhang H, Bruns MA, Logan BE (2006) Biological hydrogen production by Clostridium acetobutylicum in an unsaturated flow reactor. Water Res 40:728–734. https://doi.org/10.1016/j.watres.2005.11.041

Zhang ZP, Tay JH, Show KY, Yan R, Tee Liang D, Lee DJ et al (2007) Biohydrogen production in a granular activated carbon anaerobic fluidized bed reactor. Int J Hydrog Energy 32:185–191. https://doi.org/10.1016/j.ijhydene.2006.08.017

Zheng H-S, Guo W-Q, Yang S-S, Feng X-C, Du J-S, Zhou X-J et al (2014) Thermophilic hydrogen production from sludge pretreated by thermophilic bacteria: analysis of the advantages of microbial community and metabolism. Bioresour Technol 172:433–437. https://doi.org/10.1016/J.BIORTECH.2014.09.020

Chapter 6
Biorefining of Lignin Wastes: Modularized Production of Value-Added Compounds

Tanvi Govil, Magan Vaughn, David R. Salem, and Rajesh K Sani

Abstract Lignin, an aromatic polymer present in lignocellulosic biomasses, is conventionally viewed as a waste by-product of the pulp, paper, and other industries that use plant biomass as feedstocks. More recently, lignin has been reported as a renewable feedstock whose valorization generates several renewable aromatic intermediates, which can be used as carbon sources to synthesize other value-added

T. Govil
Department of Chemical and Biological Engineering, South Dakota Mines, Rapid City, SD, USA

Composite and Nanocomposite Advanced Manufacturing—Biomaterials Center, Rapid City, SD, USA

M. Vaughn
Department of Chemical and Biological Engineering, South Dakota Mines, Rapid City, SD, USA

D. R. Salem (✉)
Department of Chemical and Biological Engineering, South Dakota Mines, Rapid City, SD, USA

Composite and Nanocomposite Advanced Manufacturing—Biomaterials Center, Rapid City, SD, USA

Department of Materials and Metallurgical Engineering, South Dakota Mines, Rapid City, SD, USA
e-mail: David.Salem@sdsmt.edu

R. K. Sani (✉)
Department of Chemical and Biological Engineering, South Dakota Mines, Rapid City, SD, USA

Composite and Nanocomposite Advanced Manufacturing—Biomaterials Center, Rapid City, SD, USA

Department of Chemistry, Biology, and Health Sciences, South Dakota Mines, Rapid City, SD, USA

BuG ReMeDEE Consortium, Rapid City, SD, USA
e-mail: Rajesh.Sani@sdsmt.edu

J. K. Saini, R. K. Sani (eds.), *Microbial Biotechnology for Renewable and Sustainable Energy*, Clean Energy Production Technologies,
https://doi.org/10.1007/978-981-16-3852-7_6

chemicals, polymers, fuels, and other oxidized products. In this endeavor, the lignin's recalcitrance, complex structure, and heterogeneity are major impediments that not only make microbial depolymerization exceptionally challenging but also generate phenolic compounds which inhibit microbial fermentation. Furthermore, during lignin's breakdown, a heterogeneous complex mixture of low-molecular-weight aromatic monomers is released, representing additional hurdles when striving to recover the desired compound selectively. Nevertheless, in nature, some micro-organisms are competent to funnel the heterogeneous stream of central aromatic intermediates into a single, pure compound. Centered on lignin, this chapter starts with a generalized description of the structure and types of commercially available lignin (e.g., lignosulphonates or sulfonated lignin, kraft lignin, soda lignin, organosolv lignin, and biorefinery lignin). Next, the slate of aromatic intermediatory compounds formed down the lignin-degrading β-ketoadipate pathway is briefly presented. Finally, this chapter summarizes few case studies related to the production of a high value-added chemical, vanillin, and a biopolymer, polyhydroxyalkanoates (PHAs), using lignin or its derivatives.

Keywords Biorefining · Lignin · Polyhydroxyalkanoates · Valorization · Vanillin

6.1 Introduction

Lignocellulosic biomass (LCB) from agricultural and forest wastes is an abundant renewable feedstock, well-suited for biofuel, biopolymers, and other biomaterials production. LCB is largely composed of cellulose, hemicellulose, and lignin. Normally, cellulose and hemicellulose cover approximately two-thirds of LCB. The remaining material is lignin and its derivatives. Chemical or physiochemical pre-treatment makes the cellulose and hemicellulose fraction readily available for enzymatic saccharification and subsequent production of liquid and gaseous biofuels (e.g., alcohols, hydrogen). Lignin and its derivatives are released as by-products (referred to as "crude lignin"). Annually, the total amount of lignin generated commercially as a bioproduct in biorefineries and other industries such as paper, pulp, and the ethanol industry is estimated to be 10^8 tons (Bajwa et al. 2019; Xu et al. 2019). Much of this lignin is consumed on-site, being formed into pellets for steam and electricity generation. However, when burned as a fuel, crude lignin does not reach its full commercial potential and releases greenhouse gases into the environment. Hence, alternative usage of lignin and its derivatives to synthesize biobased products is desirable and is associated with the lowest ecological impacts. Lignin's valorization for new product development such as aromatic macromolecules, biopolymers, biofuels, and bio-oils is essential to the aging pulp and paper industries, which must expand their commodities portfolio to retain their vitality.

Today, the isolation of lignin from the lignocellulosic biomasses is no longer a barrier, as specific industrial pulping processes—mainly the sulfite, soda, kraft, and organosolv processes—are well established as a technology (see Sect. 6.2 for further details). However, further depolymerization and fragmentation of lignin into its constituent aromatic building blocks or its derivatives are still a challenge. Lignin is highly heterogeneous and non-uniform in its composition, and its thermochemical valorization leads to the synthesis of multiple product species, necessitating rigorous separation and purification measures to obtain a single target product (Xu et al. 2019). For controlled degradation of lignin, microbial-based methods are becoming increasingly attractive because microbes possess (a) the ability to make and break bonds selectively and (b) diverse metabolic pathways, which can be used to channel lignin's assorted aromatic building blocks, derived intermediates, and other fermentation residuals into a particular target compound. Nevertheless, lignin's depolymerization and subsequent fragmentation release certain phenolics into the media that often inhibit the fermentation and product formation. Furthermore, many of the targeted chemicals are toxic to microbial metabolism. Hence, the use of cell-free systems, such as enzyme extracts or microbial strains having tolerance to the phenolics or aromatic compounds, is being investigated by scientists for attaining the desired conversion rate and the associated molar yields. Efforts to discover alternative microbial pathways that can lead to easy bioconversions while avoiding the generation of side products are also an area of active research.

Today, the production and separation of high added-value compounds from lignin due to their chemistry and properties is an evolving domain of valuable research to both scientific and industrial communities. For sustainable industrial biorefineries, it is important that all the chief components of the lignocellulose, including lignin, be transformed to higher value compounds. This paradigm shift is essential for biorefineries and forestry-based industries to stay competitive. In this chapter, a comprehensive summary of the biological solutions to unlock the potential of lignin for the production of a wide range of bio compounds is presented.

6.2 Structure and Types of Lignin

Lignin is a high-molecular-weight macromolecular found in the cell wall of woods and plants. It is the third largest naturally occurring polymer on the earth after cellulose and chitin (Abdelaziz et al. 2019). Lignin is extremely branched in its structure, composed of three primary phenylpropanoid alcohols (monolignol phenolic units): coniferyl alcohol (G), sinapyl alcohol (S), and p-coumaryl alcohol (H), that are modified with a variety of functional groups such as methoxy ($O–CH_3$), carboxyl (R-COOH), hydroxyl (R-OH) and carbonyl (C=O) (Chatterjee and Saito 2015). There are some complex linkages between the three phenylpropanoid units that provide lignin a dense, hydrophobic structure resistant to depolymerization by enzymes (Abdelaziz et al. 2016; Chatterjee and Saito 2015; Horwath 2015). These include the carbon-carbon bond (C-C), ether bond (C-O-C), C-O bonds of α- and β-arylalkyl ethers, and bonds that OH groups can make with other polysaccharides (cellulose and hemicellulose) and proteins (extensins) in the plant's cell wall. The β-O-4 linkage dominates, represent approximately 45–50% of the linkages, followed by 5–5 (18–25%), β-5 (9–12%), and the β-1 (7–10%) linkage (Strassberger et al. 2014). While these tight linkages necessitate harsh pretreatments to depolymerize lignin, they make lignin one the most sturdy macromolecule on the planet and are estimated to hold about 95 billion tons of carbon (Chatterjee and Saito 2015). Due to its high carbon content, lignin is considered a significant replacement for fossil fuels. It has also been explored as a renewable bioresource for the synthesis of carbon-based chemicals, and has promising potential as a constituent in polymer blends and composites. However, the ultimate end usage of lignin depends on its properties, which vary according to source and type. Commercially available lignin includes lignosulphonates, kraft lignin, soda lignin, organosolv lignin, and biorefinery lignin.

6.2.1 Lignosulphates

In the industries which separate lignin from cellulose, different oxidative pulping methods are employed. The first and the most widely adopted separation methodology is sulfite pulping, where sulfur dioxide or an acidic bisulfite/sulfite solution is used to soften the plant material and remove lignin as lignosulphonates (Hintz 2001). Lignosulphates are sulfur-containing lignin that behaves as a polar ionic molecule, soluble in water but insoluble in organic solvents. Their single most extensive usage is in the concrete industry, where lignosulphonates are used as plasticizers and allow concrete to be made with 15% less water, which also creates more rigid concrete while retaining its capacity to flow (Gargulak et al. 2015; Vazquez and Pique 2016). Lignosulphonates have the power to decrease the viscosity of the solutions and provide hydration. They also find usage in the production of linoleum flooring and plaster (Vazquez and Pique 2016). Lignosulphonates possess good binding, dispersing, and emulsifying abilities that make them useful additives in tanning, leather, pesticides, fertilizers, *wax emulsions, dyes, pigments, oil drilling mud,* coal briquettes, *and food industry (*WebArchive 2003*).*

Chemical oxidation of sulfonated lignin at elevated temperature (up to 160 °C) and pressure (10 bar) has a value for the production of low-molecular-weight artificial flavoring agents such as vanillin and vanillic acid (4-hydroxy-3-methoxy benzoic *acid*) (Pacek et al. 2013; Richter et al. 1945). The global market potential of vanillic acid (in 2020) was USD 1189.9 million (Fox40 2020). Dimethyl sulphoxide (DMSO) is another high-utilization chemical derived from lignosulphonates (Macfarlane et al. 2014). Hence, lignosulphonates have immense market potential. However, the sulfite coking process by which they are produced is environmentally costly, and lignosulphonates contain sulfur and hemicellulose (Abdelaziz et al. 2016), making them less pure. It is their property of being soluble in water that distinguishes them from other lignin streams. Due to their anionic sulfate groups, lignosulphonates can scavenge metals and keep them dissolved in solutions. This property has an advantage in agriculture where lignosulphonates can keep metals available to plants (preventing them from precipitating out as insoluble compounds), and also in drinking water systems, where scaly metals deposition on the walls of the water systems can be avoided (WebArchive 2003). Moreover, with their ability to immobilize metals, the utilization of lignosulphonates for bioremediation of wastewater cannot be ruled out.

6.2.2 *Kraft Lignin*

Compared to sulfite pulping, the more modern pulping process is the Kraft process, where the plant material is treated with a mixture of water, sodium hydroxide (NaOH), and sodium sulfide (Na_2S) at 140–180 °C to break the linkages binding lignin with cellulose and hemicellulose. The lignin extracted during this process is known as Kraft lignin which accounts for about 85% of the total lignin production in the world (Chen 2015). Approximately 630,000 tons of kraft lignin is produced annually, and most of its utilization is for combustion. Its high value utilization as an additive and binder for improvement of the properties of resins, foams, printing inks, fertilizers, and adhesives, only stands at 2% of the total kraft lignin produced (Macfarlane et al. 2014). Like lignosulphonate, kraft lignin has reactive sulfate groups attached to its phenyl rings. Also, it is associated with hemicellulose, making it impure. Nevertheless, there is an ongoing push to use kraft lignin for producing activated carbon and as a low-cost raw material for carbon fiber synthesis. It has an expected compound annual growth rate (CAGR) of 7% (John 2020).

6.2.3 *Soda or Alkaline Lignin*

Produced by one of the simplest pulping processes, soda or alkaline lignin is produced as a by-product in the pulp industry using sodium hydroxide (NaOH) or a mixture of NaOH and anthraquinone at 150–170 °C (Macfarlane et al. 2014). On average, soda pulping yields approximately 80% of lignin from the wood samples (USDA 2020). Performing a steam explosion before soda pulping can enhance the lignin extraction percentage to 90% (USDA 2020). While the lignin from this process is sulfur-free, it still contains hemicellulose. It has comparatively high aliphatic and phenolic content than lignosulphonates and kraft lignin (Abdelaziz et al. 2016). Hence, aromatic resins are the key chemicals expected to be derived from the use of soda lignin as the substrate. Also, because they are free from sulfur, soda lignin can be utilized as binders in animal feed (Macfarlane et al. 2014).

6.2.4 *Organosolv Lignin*

Organosolv is the most eco-friendly pulping process that uses organic solvents (e.g., ethanol, acetone, methanol, acetylene glycol, etc.) to efficiently depolymerize wood into lignin, hemicellulose, and cellulose fractions. Organosolv lignin is of high quality and purity, with <1 wt.% residual carbohydrate content (Strassberger et al. 2014). Organosolv lignin is free from sulfur and other impurities. Hence, it can directly be used to produce specialty value-added products via a more environmentally friendly method. Organosolv lignin has a highly homogenous nature, with its composition being much closer to native lignin (Tribot et al. 2019). It is very rich in phenolic content and is highly hydrophobic, which allows it to be spun into fibers directly, without blending with other polymers (Macfarlane et al. 2014). Also, because of its purity, organosolv lignin can be exploited for its antioxidant, antibiotic, and antitumor properties in the cosmetics, medicinal, and pharmaceutical sectors (Macfarlane et al. 2014). To date, organosolv lignin is the most attractive lignin in terms of its quality. However, the commercial realization of the organosolv processes is presently marginal, probably as a result of high process costs, and is not readily available at volume (Abdelaziz et al. 2016; Thoresen et al. 2020).

6.2.5 Biorefinery Lignin

The lignin produced as a by-product in biorefineries using second-generation lignocellulosic feedstocks is another source of lignin, with an annual production of approximately 100 kilotons (Bajwa et al. 2019). The biorefinery lignin, produced by a hydrolytic pretreatment (acidic, thermal, or enzymatic) of underutilized lignocellulose is essentially sulfur-free. However, some hemicellulose is still linked with it through ester, glycosidic, ether, or carbon-carbon bonds (Hansen et al. 2013). These bonds are responsible for the hydrophilic surface properties of lignin, which limits its applications in some industries, including the range of polymer resins it can be effectively combined within biocomposite applications. Nevertheless, biorefinery lignin holds the potential to be precipitated from the rest of the biomass in the form of an insoluble, amorphous, solid residue that has a significant amount of protein attached to it and that can be used as animal feed (Hansen et al. 2013). In the future, removing carbohydrate impurities and polishing lignin by enzymes could yield low-molecular-weight lignin with higher purity and hydrophobicity suitable for various industrial applications.

Over the years, lignin separated from lignocellulosic biomass by different biomass conversion technologies has been given distinct names, as in cellulolytic enzyme lignin, produced by cellulolytic enzyme treatment of pretreated agricultural residue (Tian et al. 2017; Zhang et al. 2010); Bjorkman lignin, produced by treatment of lignocellulose with neutral organic solvents (Bjorkman 1954; Obst and Kirk 1988); Klason lignin, produced by treatment of lignocellulose with sulphuric acid followed by removal of ash (Chen 2015; Obst and Kirk 1988), etc. (Retsina et al. 2013).

From this section, it is evident that varieties of industrial lignins exist in the market whose properties and structures depend on the pulping or the coking process (Table 6.1). These lignins are directly suitable for a range of applications, from

Table 6.1 Comparison of lignins generated from lignocellulosic via various cooking processes

Industrial lignin	Molecular Weight (kDa) (USDA 2020)	Impurity	Sulfur content (%) (Abdelaziz et al. 2016)	Hemicellulose content	Phenolic content	Solubility in organic solvents	Worldwide Production (kilotons/year) (Miller 2016)	Price (USD/ton) (2019) (Tribot et al. 2019)
Lignosulfonates	20,000–50,000	Sulphur, and hemicellulose	5–6	High	Low	Insoluble. Soluble in water	1000	300–2700
Kraft lignin	100–3000	Sulphur, and hemicellulose	1–2	High	Low	Insoluble. Soluble in water above pH 11	75	250–500
Soda or alkaline lignin	800–3000	Hemicellulose, ash	Nil	Average	High	Partly soluble	5–10	200–300
Organosolv lignin	500–4000	None	Nil	Very low (<1%)	Highest	Highly soluble	3	280–520
Cellulolytic lignin	Variable	Hemicellulose, ash	Nil	High	Variable	Insoluble	100	Not determined

low-density combustion fuel to binder and blender additives to make high-strength and durable concrete, cardboard, and papers. Lately, the use of lignin in its purest form as an antioxidant has been recognized. Furthermore, in 2007, scientists from the Pacific Northwest National Laboratory (PNNL) released a report evaluating the opportunity for using lignin for the derivation of certain macromolecules, higher aromatic monomers, and oxidized compounds by breaking the lignin's polymeric structure (Holladay et al. 2007). The report presented a case that lignin produced in biorefineries has a high economic opportunity to generate large revenues (USD 12–35 billion) by producing a variety of co-products (Holladay et al. 2007). This strategy would entail the choice of technology for selective fragmentation of lignin into low molecular weight monomers and oligomers, which can be further biocatalyzed by microorganisms, via cellular assimilation, into value-added compounds.

Frequently, the microorganisms that depolymerize lignin into its constituent monolignols proceed to polymerize these intermediates into other renewable chemicals using their versatile metabolic pathways. Section 6.3 touches on some of the high-value lignin monomers and oligomers synthesized by microbial/enzymatic-based lignin depolymerization without going into the technical details of these processes. Section 6.4 discusses some final value-added chemicals produced by microorganisms via cellular assimilation of such lignin intermediates via the β-ketoadipate pathway.

6.3 Mono-and Oligomers as Intermediates from Lignin Depolymerization

In nature, microorganisms existing in symbiotic association with plants have evolved mechanisms to degrade and utilize lignocellulosic biomass. These plant degrading microbes release a repertoire of extracellular enzymes collectively termed ligninolytic enzymes (comprising laccase, superoxide dismutase, oxidoreductases, and peroxidases) that oxidize phenolic units in lignin, but not the non-phenolic compounds (Datta et al. 2017; Govil et al. 2020a; Janusz et al. 2017). Some mediators such as p-coumaric acid, 2,2′-azino-di(3-ethylbenzthiazoline-6-sulfonic acid, vanillin, and syringaldehyde, can enhance the oxidative capacity of laccase itself to oxidize the non-phenolic units in lignin (Abdelaziz et al. 2016). In addition, some low-molecular-weight secondary metabolites produced by microbes during lignin degradation, such as benzoic acid, veratryl alcohol (MnO_2), oxalate, and 2-chloro-1,4-dimethoxybenzene, can aid the further breakdown of the phenolic and non-phenolic groups in lignin (Datta et al. 2017; Janusz et al. 2017; Shimada et al. 1981). These metabolites are slowly metabolized by the microorganisms and accumulate in the reaction solutions. Other metabolites such as flavonoids, tannins, and lignans are parts of plants themselves. They are also known to initiate bond scissions in lignin via an oxidative and reductive cascade of reactions (Janusz et al. 2017).

Over the years, many studies have reported the biodegradation of lignin by microorganisms and their enzymes, and significant progress has been made in understanding these processes (Abdelaziz et al. 2016; Lee et al. 2019). Literature is also available that details the pathways for lignin valorization. The pathways suggest that during the breakdown of lignin, various low-molecular-weight aromatic compounds can be formed depending on the composition of the lignin and the depolymerization method used. The most common of these monolignols are p-coumaric acid, caffeic acid, ferulic acid, guaiacol, syringic acid, syringaldehyde, phenol, benzoic acid, and vanillic acid (Abdelaziz et al. 2016). Tang et al. (2015) reported production of 8.04 mg, 0.88 mg, 0.63 mg, 0.34 mg, and 0.29 mg of hydroxybenzoic acid, syringaldehyde, vanillic acid, *p*-coumaric acid, and ferulic acid, respectively, from each gram of oil palm empty fruit bunch lignin. Chen et al. (1982) reported the generation of vanillic acid, veratric acid, and various benzoic acid derivatives during degradation of spruce wood lignin by white-rot fungus *Phanerochaete chrysosporium* (Chen et al. 1982). The production of phenolic oligomers containing about seven phenylpropane units from kraft lignin when subjected to degradation by the fungus *Trametes versicolor* has also been reported (Reid 1998). Guaiacol, benzoic acid, and vanillic acid were identified as significant intermediates when a member of the genus *Acetoanaerobium* degraded kraft lignin. In the same study, ferulic acid, syringic acid, and benzenepropanoic acid were also detected, which were considered the final products after intermediate stage degradation (Duan et al. 2016). Consistent with these observations, similar degradation products have also been found in other studies, some of which are summarized in Table 6.2.

Commercially, many of these degradation products of lignin are valued chemicals, and they can find industrial applications. For instance, syringaldehyde is an aromatic aldehyde with valued antioxidant, bioactive (antimicrobial), and antioncogenic activity and is used in cosmetics, pharmaceuticals, food, paper, and pulp industries. Moreover, syringaldehyde is also a promising laccase and peroxidase mediator that can enhance these enzymes' activity by almost six-fold (Mohamad Ibrahim et al. 2012). p-coumaric acid possesses excellent anti-infection, anti-inflammatory, and antioxidant activities that can help it protect against conditions of oxidative stress (Shen et al. 2019). Ferulic acid also possesses antithrombic, antimicrobial, anticancer, antidiabetic, and immunostimulant properties, and finds applications in cosmetics, pharmaceuticals, food, and health industries (Kumar and Pruthi 2014). Guaiacol and its derivatives are valuable as additives in mucoactive, antiseptic, and anesthetic agents. Indeed, lignin and the phenolic compounds derived from it have a total market value of approximately USD 732 million, projected to reach USD 913 million by 2025 (Bajwa et al. 2019). However, despite the potential to synthesize such promising aromatics, vanillin stays at present the only marketable aromatic product of lignin produced using microbial sources.

Table 6.2 Aromatic intermediary metabolites produced from biological lignin depolymerization

Source of lignin	Microbe/enzyme	Conditions	Product	Reference
Oil palm empty fruit bunch	Cutinase and manganese peroxidase	55 °C, pH 8.0	Hydroxybenzoic acid, syringaldehyde, vanillin, *p*-coumaric acid, and ferulic acid	Tang et al. (2015)
Spruce wood	*Phanerochaete chrysosporium*	30 °C, pH 7.0	Vanillic acid, veratric acid, and benzoic acid	Chen et al. (1982)
Kraft lignin	*Trametes versicolor*	40 °C, pH 4.8	Seven phenylpropane oligomer	Reid (1998)
Kraft lignin	*Acetoanaerobium* sp. WJDL-Y2	40 °C	Ferulic acid, syringic acid, and benzenepropanoic acid	Duan et al. (2016)
Biorefinery lignin	*Pseudomonas putida* and *Rhodococcus* RHA1	30 °C	Propiophenone and benzoic acid derivatives	Ahmad et al. (2010)
Kraft lignin	*Bacillus* sp. and *Aneurinibacillus aneurinilyticus*	55 °C, pH 7.0	Trans-4- hydroxycinnamic acid, 3,4,5-trimethoxy benzaldehyde, gallic acid and ferulic acid	Raj et al. (2007)
Kraft lignin	*Comamonas* sp. B-9	30 °C, pH 7.0	Ethanediol, 3, 5-dimethyl-benzaldehyde and phenethyl alcohol	Chen et al. (2012)
Kraft lignin	*Dysgonomonas* sp. WJDL-Y1	33 °C, pH 6.8	Vanillic acid, syringic acid, ferulic acid, and benzoic acid	Jing et al. (2016)
Wastepaper	*Aeromonas formicans*	30 °C, pH 7.2	Benzoic acid, Vanillic acid, Protocatechuic acid, Syringic acid, cinnamic acid and ferulic acid	Gupta et al. (2001)
Milled wood lignins	β-O-4-cleaving enzymes from *Novosphingobium sp.*	15 °C, pH 8.5	Guaiacylhydroxylpropanone and Syringylhydroxylpropanone	Ohta et al. (2017)
Softwood lignin	*E. coli* recombinantly expressing β-O-4-cleaving enzymes	30 °C, pH 8.0	Vanillin	Reiter et al. (2013)
Wheat straw lignocellulose	Mutant *Rhodococcus jostii* RHA1	30 °C, pH 8.0	Vanillin	Sainsbury et al. (2013)

6.4 Final Value-Added Chemicals Production from Lignin Intermediates

Ideally, the production of these lignin intermediates is linked to each other via the β-ketoadipate pathway. One intermediate gets converted to the other rapidly via a cascade of reactions. Hence, most of the time, lignin breakdown is a complex mixture of low-molecular-weight aromatic monomers. It is, in fact, difficult to synthesize a particular aromatic compound from lignin at sufficient concentration. This has a negative prospect for the valorization of lignin to generate a definable aromatic compound. However, Ohta et al. (2017) showed that highly specific aromatic monomers can be synthesized from lignin by controlling the reactions using selective enzymes. In their study, the authors achieved the exclusive synthesis of monomers with a phenylpropane moiety (e.g., guaiacylhydroxylpropanone (GHP), syringylhydroxylpropanone, SHP)) using four β-O-4-cleaving enzymes (two short-chain dehydrogenase/reductase, two glutathione *S*-transferases with β-etherase activity) isolated from a *Novosphingobium* strain in one pot (Ohta et al. 2017). Prim et al. (2003) showed that the production of 4-ethyl phenol, a phenolic compound responsible for aroma in wine, is possible from hydroxycinnamic acids (e.g., ferulic, p-coumaric, sinapyl acid, and caffeic acids) using phenolic acid decarboxylase from *Bacillus* sp. BP-7, with no accumulation of any side products (Prim et al. 2003). A similar conversion was achieved using a decarboxylase gene from *Bacillus licheniformis* (Hu et al. 2015). Furthermore, phenolic acid decarboxylase activity from *Bacillus amyloliquefaciens* has been demonstrated for the selective synthesis of p-hydroxystyrene from p-coumaric acid (Jung et al. 2013).

6.4.1 Specific Case of Vanillin Production

As mentioned earlier, because of lignin's heterogeneity, many side products are formed along with the desired compound during lignin processing and intermediate production. To avoid this issue, studies have been conducted to examine microbial enzymes which can bioconvert a particular metabolite into other monomeric aromatic compounds. In this regard, processes that lead to the production of vanillic acid are one of the most intensively studied enzymes and enzymatic processes. Vanillic acid is a food preservative and a natural precursor for vanillin production - an aromatic compound responsible for the characteristic vanilla flavor, which has enormous consumer demand. Organic vanillin obtained from plants is high-priced (approximately USD 3000/kg) and has a net worth of more than USD1 billion annually (Li and Rosazza 2000). Vanillin synthesized chemically is comparatively cheaper (USD 11/kg) though it is not considered natural under US legislation (Ashengroph et al. 2011). Hence, cheaper production of natural vanillin using microbial transformations has increasingly been attempted and has immense economic potential, owing to the spur in demand of bio-vanillin as a flavoring base in foods, pharmaceuticals, and cosmetic industries (Luziatelli et al. 2019) (Fig. 6.1).

For years, the most exploited feedstock for vanillic acid's biological production is ferulic acid (a hydroxycinnamic acid found attached to hemicellulose via several ester linkages). Some of the microbial enzymes that have been studied for this purpose include enzymes belonging to the superfamily feruloyl enoyl-SCoA hydratase/lyase (EC 4.2.1.101) and feruloyl-CoA synthetase (EC 6.2.1.34)) expressed by *Pseudomonas fluorescens* (Gasson et al. 1998; Leonard et al. 2006), *Pseudomonas* sp. Strain HR199 (Overhage et al. 1999), *Pseudomonas putida*

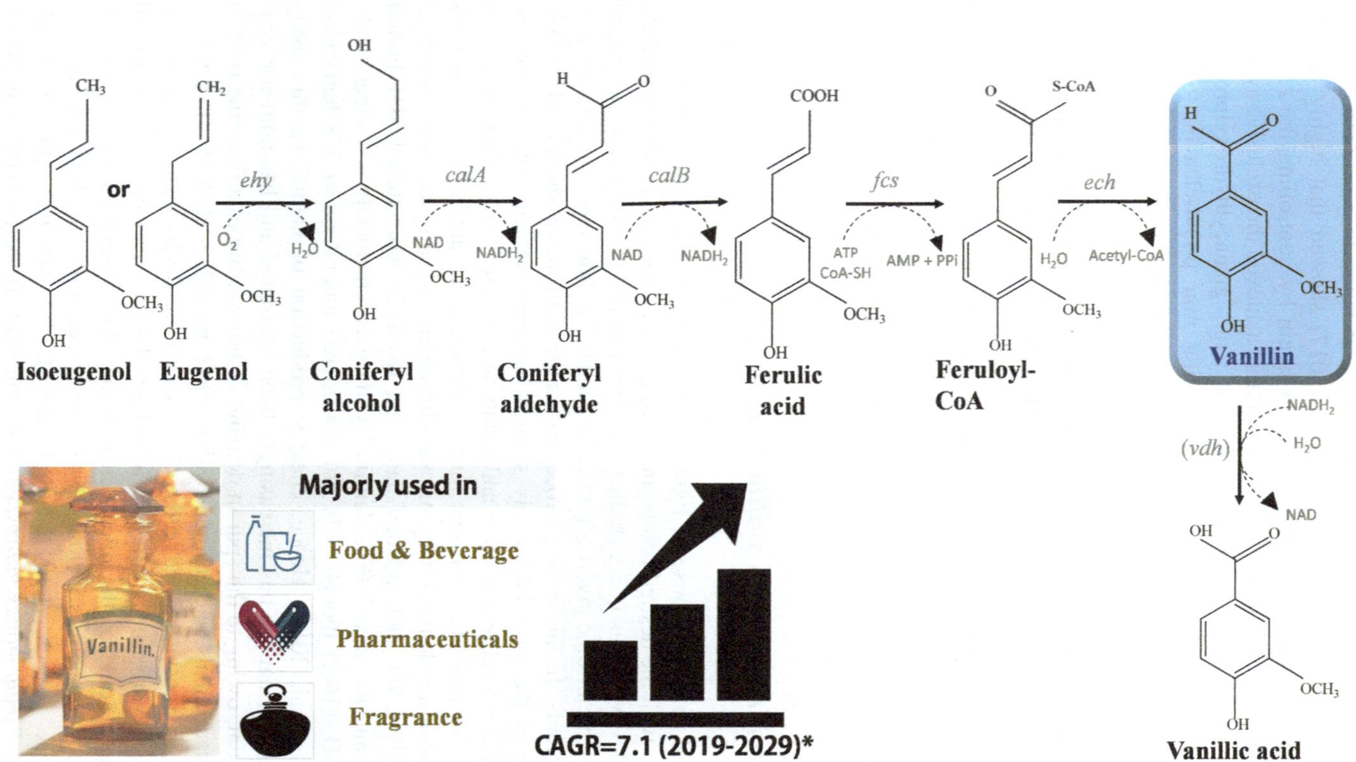

Fig. 6.1 Production of vanillin from Ferulic acid and Isoeugenol. *ehy* Eugenol hydroxylase, *calA* Coniferyl alcohol dehydrogenase, *calB* Coniferyl aldehyde dehydrogenase, *fcs* Feruloyl-CoA synthetase, *ech* Enoyl-SCoA hydratase/lyase, *vdh* Vanillin dehydrogenase. (Source of data depicting compound annual growth rate of Vanillin: Persistence Market Research Report, released Match 2020 (PMR 2020))

KT2440 (Plaggenborg et al. 2003), plant species *Glechoma hederacea* and *Vanilla planifolia* (Gallage et al. 2014; Negishi et al. 2009), *Amycolatopsis* sp. strain HR167 (Achterholt et al. 2000), *Streptomyces* sp. NL15-2K (Nishimura et al. 2018; Yang et al. 2013) and strain V1 (Hua et al. 2007b), *Streptomyces setonii* (Muheim and Lerch 1999), *Delftia acidovorans* (Plaggenborg et al. 2001), *Rhodotorula rubra* (Huang et al. 1993), *Halomonas elongate* (Abdelkafi et al. 2006), *Bacillus subtilis* (Gurujeyalakshmi and Mahadevan 1987), and *Aspergillus niger* CGMCC0774 and *Pycnoporus cinnabarinus* CGMCC1115 (Zheng et al. 2007). In the two-step conversion process, feruloyl-CoA synthetase (*fcs*) converts ferulic acid into feruloyl CoA and subsequently enoyl-SCoA hydratase/lyase (*ech*) mediates the conversion of feruloyl CoA to vanillin and acetyl CoA (Fig. 6.1). Table 6.3 summarizes the studies where the natural host strain or the recombinant strains expressed the enzymes that could transform ferulic acid to vanillin and yields up to 28.3 g/L were achieved. In the future, cloning an efficient ferulic acid esterase gene that can release ferulic acid from hemicellulose (Bugg et al. 2011) in such clones can create a path for producing vanillin directly from hemicellulose.

As such, ferulic acid is an excellent nontoxic precursor for vanillin synthesis. A high concentration of ferulic acid can be fed to the microorganisms without inhibiting microbial growth (Muheim and Lerch 1999). However, ferulic acid is costly, and this has prompted research where eugenol (a plant-derived phenylpropanoid) is being considered as a cheaper (USD 5/kg) and a more abundant vanillin precursor. Several microbial strains and their enzymes have been shown to aid in the biotransformation of eugenol into vanillin, including *Rhodococcus rhodochrous* (Chatterjee et al. 1999), *Serratia marcescens* DSM30126 (Rabenhorst and Hopp 1991), strains of the genus *Bacillus* (Hua et al. 2007a; Shimoni et al. 2000; Zhang et al. 2006), *strains of the genus Pseudomonas* (Kasana et al. 2007; Unno et al. 2007; Yamada et al. 2007), and *Candida galli* strain PGO6 (Ashengroph et al. 2011). Lately, the use of cell-free extracts rich in characteristic enzymes for biocatalysis of eugenol into vanillin has also been attempted. For example, vanillyl alcohol oxidase enzyme (vaoA) from *Penicillium simplicissimum* CBS 170.90 has successfully been used for the biocatalytic conversion of 1 g/L of eugenol into vanillin (0.24 g/L, the molar yield of 10%) and vanillic acid (1.1 g/L, the molar yield of 44%) (Ashengroph et al. 2011). *vaoA* (Vanillyl-alcohol oxidase) gene when cloned into a recombinant *Escherichia coli*, that had already been transformed with the genes encoding coniferyl alcohol dehydrogenase and coniferyl aldehyde dehydrogenase from *Pseudomonas* sp. strain HR199, produced 0.3 g of vanillin per liter of the fermentation medium from eugenol (Overhage et al. 2003).

Several attempts have been made to produce vanillin with a minimum of co-products from lignin as the starting substrate. For example, Reiter et al. (2013) recombinantly expressed three β-O-4-cleaving enzymes, a Cα-dehydrogenase, a β-etherase, and a glutathione lyase from the proteobacterium *Sphingobium* sp. *SYK6* in *E. coli* BL21, and obtained the production of 58.2 mg/L of vanillin from softwood lignin with a small amount of ferulic acid as the co-product (Reiter et al. 2013). Exclusive production of vanillin from wheat straw lignocellulose was also reported in a study by Sainsbury et al. (2013), where a mutant strain of

Table 6.3 Case studies detailing vanillin production by biological means from lignin or its intermediates

Host/Enzyme	Substrate	System	Conditions	Vanillin concentration (g/L)	Substrate concentration (g/L)	Time	Molar yield (%)	References
Aspergillus niger CGMCC0774 and *Pycnoporus cinnabarinus* CGMCC1115	Ferulic acid	Growing cells	30 ° C, pH 7.2, 150 rpm	2.2	4	72	57.7	Zheng et al. (2007)
Streptomyces sp. strain V1	Ferulic acid	Growing cells	Fed-batch 30 ° C, pH 5.8, 200 rpm	19.2	9	18	58	Hua et al. (2007b)
Streptomyces setonii	Ferulic acid	Growing cells	Fed-batch 30 ° C, pH 7.2, 150 rpm	6.41	12	54	68	Muheim and Lerch (1999)
Bacillus subtilis	Ferulic acid	Biofilm	35 °C, pH 9.0, 200 rpm	1.84	2	20	60.43	Yan et al. (2016)
Halomonas elongate	Ferulic acid	Resting cells	37 ° C, pH 7.2, 150 rpm	0.65	0.97	14	86	Abdelkafi et al. (2006)
Recombinant *Escherichia coli* XL1-Blue	Ferulic acid	Resting cells	30 ° C, 150 rpm	0.3	5	15	ND	Overhage et al. (2003)
Recombinant *E. coli* strain JM109	Ferulic acid	Resting cells	30 ° C, 150 rpm	2.52	0.21	6	50	Barghini et al. (2007)
Recombinant *E. coli* DH5alpha	Ferulic acid	Growing cells	37 ° C, 180 rpm	5.14	3.0	24	86.6	Lee et al. (2009)
Recombinant *E. coli* NTG-VR1	Ferulic acid	Growing cells	37 ° C, 180 rpm	2.9	10	48	62%	Yoon et al. (2007)
Recombinant *E. coli* FR13	Ferulic acid	Resting cells	30 ° C, pH 9.0	4.2	4.5	24	68%	Luziatelli et al. (2019)
Engineered Amycolatopsis sp. ATCC 39116	Ferulic acid	Growing cells	Fed-batch 45 ° C, 600 rpm	22.3	10	20	94.9	Fleige et al. (2016)

Recombinant *Pseudomonas putida* KT2440	Ferulic acid	Resting cells	30 ° C	1.30	2.0	3	86%	Graf and Altenbuchner (2014)
Recombinant *P. fluorescens* BF13 3	Ferulic acid	Growing cells	3 L stirred tank reactor 30 ° C, 200 rpm	1.28	2.0	24	81%	Di Gioia et al. (2011)
Recombinant *Pediococcus acidilactici* BD16	Ferulic acid	Growing cells	37 ° C, 150 rpm	0.48	0.2	0.33	ND	Kaur et al. (2014)
Pseudomonas putida IE27	Isoeugenol	Resting cells	20 ° C, 200 rpm with 10% DMSO	16.1	24	24	71	Yamada et al. (2007)
Candida galli strain PGO6	Isoeugenol	Resting cells	30 ° C, pH 7.0, 200 rpm	1.12	5	60	25.7	*Candida galli* strain PGO6
Pseudomonas resinovorans SPR1	Isoeugenol	Resting cells	28 ° C, 200 rpm	0.24	1	30	10	Ashengroph et al. (2011)
Psychrobacter sp. strain CSW4	Isoeugenol	Resting cells	28 ° C, 200 rpm	13.8	10	48	10	Ashengroph et al. (2012)
Pseudomonas chlororaphis CDAE5	Isoeugenol	Growing cells	25 ° C, 180 rpm	1.2	10	24	12.6	Kasana et al. (2007)
Trichosporon asahii	Isoeugenol	Growing cells	28 ° C, pH 5.8, 200 rpm	2.4	5	24	52.5	Ashengroph and Amini (2017)
Soybean lipoxygenase	Isoeugenol	Enzyme extract	28 ° C, pH 9.0, 180 rpm	2.68	30	120	ND	Liu et al. (2020)
Halomonas sp. B15	Isoeugenol	Growing cells	33 ° C, pH 7.0, 150 rpm	0.365	1.5	112	36.5	Vyrides et al. (2015)
Recombinant *E. coli* (BL21)	Isoeugenol	Growing cells	20 ° C, 180 rpm	28.3	37	6	81	Yamada et al. (2008)
Recombinant *E. coli* BL21	Softwood lignin	Growing cells	30 °C, pH 8.0	0.089	25	50	ND	Reiter et al. (2013)

(continued)

Table 6.3 (continued)

Host/Enzyme	Substrate	System	Conditions	Vanillin concentration (g/L)	Substrate concentration (g/L)	Time	Molar yield (%)	References
Engineered Rhodococcus jostii RHA1	Wheat straw lignin	Growing cells	30 ° C, pH 8.0, 180 rpm	0.096	25	144	ND	Sainsbury et al. (2013)

DMSO Dimethyl Sulfoxide, *ND* Not determined

Rhodococcus jostii RHA1 deficient in gene vanillin dehydrogenase was found to accumulate up to 96 mg/L of vanillin, together with minor quantities of ferulic acid and 4-hydroxybenzaldehyde (Sainsbury et al. 2013). These studies show that the exclusive production of specific phenolic compounds from lignin is possible, provided the application of targeted pathway engineering can control the biocatalytic routes for lignin breakdown. However, the reported yield and concentration of products produced via this route are below 1 g/L and do not qualify as economically efficient biotransformation. In these studies, the reported yield of vanillin production of less than 1 g/L and associated efficiency of less than 10%, was due to the toxic effects of vanillin on the cellular systems at concentrations above 1 g/L, leading to its quick metabolization by the microorganism into vanillic acid and vanillyl alcohol using vanillin dehydrogenase (EC 1.2.1.67) (Ashengroph and Amini 2017). Hence, in microbial hosts, the respective alcohol or the respective acid typically gets accumulated rather than vanillin.

To enhance the bioconversion yield of vanillin from eugenol or ferulic acid, the use of static growth conditions is the first strategy adopted where resting cells have been shown to delay degradation of vanillin to vanillic acid. Here, with the resting cells of *Psychrobacter* sp. strain CSW4, a vanillin concentration of 1.28 g/L (molar yield of 13.8%) was achieved by Ashengroph et al. (2012). Secondly, researchers have also, from time to time, reported the isolation of a few strains from nature that are resistant to the toxic effects of vanillin, and hence can accumulate more vanillin in a reduced reaction time. *Pseudomonas chlororaphis* CDAE5 is one such strain that has demonstrated the potential to be a suitable candidate for biotechnological production of vanillin from isoeugenol. Another study has shown *Pseudomonas chlororaphis* CDAE5 to produce 1.2 g/L of vanillin, with a molar yield of 13% (Kasana et al. 2007). Much higher productivity of vanillin has been reported by Ashengroph and Amini (2017), where the yeast *Trichosporon asahii* transformed 5 g/L of isoeugenol into 2.4 g/L of vanillin with a 52.5% molar yield (Ashengroph and Amini 2017). Even cell-free extracts rich in a designated enzyme have been investigated for the process. Enzyme lipoxygenase is useful for this biocatalysis approach, where Liu et al. (2020) tested a method for the synthesis of vanillin from isoeugenol and eugenol using soybean lipoxygenase (lipoxidase). The reported production of vanillin in this study was 2.68 g/L. A European patent dating back to 1991 claimed a process for the preparation of vanillin (10–15 g/L) from eugenol or isoeugenol using lipoxidase (Markus et al. 1992), an enzyme from *Glycine max* (soybean) that is now commercially available from Sigma Aldrich (L6632 and L7395).

To augment these processes and improve productivity, the metabolic engineering of microbial strains already known to have the tolerance to vanillin toxicity has been documented in the literature. Specifically, *Amycolatopsis* sp. ATCC 39116, a gram-positive Actinobacteria, was engineered with the deletion of its vanillin dehydrogenase-encoding (vdh) gene that codes for vanillin catabolism enzyme, vanillin dehydrogenase. This mutation decreased the catabolism of vanillin to vanillic acid by 90%, and resulted in an increase of the total vanillin production to 2.2 g/L from ferulic acid, with a molar yield of 80.9% (Fleige et al. 2016). The same group achieved a vanillin concentration of 19.3 g/L (molar yield of 94.9%) by constitutively expressing two of the vanillin anabolism genes *fcs* (coding for feruloyl-coenzyme A (CoA) synthetase) and *ech* (enoyl-CoA hydratase/aldolase) in the same stain of *Amycolatopsis* sp. 39116 (Fleige et al. 2016). The transcription of *ech* and *fcs* eliminated the adaptation phase in the host. Moreover, by using an improved fed-batch feeding strategy, the group could attain an even higher concentration of vanillin, 22.3 g/L, which is the highest vanillin concentration reported to date from any of the native wild host strains (Fleige et al. 2016). This study shows that improvements in vanillin yield using whole cells are possible through the right combination of strategies involving optimization of the fermentation parameters with resting cells and metabolic engineering. The identical strategy of inactivating vanillin dehydrogenase and overexpressing feruloyl-coenzyme A (CoA) synthetase and enoyl-CoA hydratase/aldolase in *Pseudomonas* sp. have been documented by Graf and Altenbuchner (2014). However, here the authors observed vanillin metabolism to vanillic acid, despite knockout of the vanillin dehydrogenase gene (*vdh*). Hence, additional inactivation of a molybdate transporter gene was done in their study, which led to the complete prevention of vanillin degradation. However, the concentration of vanillin achieved in their study was only 1.2 g/L (Graf and Altenbuchner 2014). This indicates that each microbial strain has a characteristic tolerance to the amount of vanillin it can produce in the system. The highest concentration of vanillin production of vanillin has been achieved at 1.2 g/L with any *Pseudomonas* strain to date.

Genetic engineering strategies have also been tried with recombinant *E. coli* (transformed with vanillin synthesizing genes) as the preferred candidate for cost-effective vanillin synthesis because it has a well-studied and understood fermentation process and has no vanillin degradation pathway (Lee et al. 2009). Lee et al. induced an *E. coli* host transformed with feruloyl-CoA synthetase (*fcs*) and enoyl-CoA hydratase/aldolase (*ech*) genes to produce more vanillin by amplifying a *glt* gene encoding citrate synthase in it. During the vanillin synthase from ferulic acid, acetyl-CoA is a concomitant by-product whose accumulation impedes feruloyl-CoA's forward reaction to vanillin. The enzyme citrate synthase bio transforms acetyl-CoA into CoA and helps the vanillin synthesis reaction to be pulled forward by eliminating product inhibition. Therefore, in their study, by overexpressing the *gltA* gene, 1.98 g/L of vanillin was produced, which was almost twofold more than the vanillin production of 0.91 g/L obtained by the *E. coli* without *gltA* amplification (Lee et al. 2009). In another study, Yoon et al. (2007) followed a two-step strategy to enhance vanillin production in an *E. coli* harboring *fcs* and *ech* genes transformed from *Amycolatopsis* sp. strain HR104. First, they generated mutants of *E. coli* that were vanillin resistant, and second, they used XAD-2 resin for the adsorption and removal of released toxic vanillin from the medium. This combined engineering strategy increased the vanillin production from the recombinant host to 2.9 g/L, which was three-fold higher than that for its wild-type strain without the use of the resin (Yoon et al. 2007). Indeed, the utility of adsorbent resins with microporous structures can be observed from a study by Hua et al. (2007a, b) where vanillin produced by ferulic acid biotransformation by *Streptomyces* sp. strain V-1 was adsorbed on the resins, leading to high production of 19.2 g/L along with ease of its downstream processing (Hua et al. 2007b). With a similar concept of obtaining the product adsorbed onto a resin (HD8), Zhao et al. (2006) also obtained decent production of vanillin (8.1 g/L) from isoeugenol. Recently, Luziatelli et al. (2019) reported production of approximately 4.9 g/L concentration of vanillin from ferulic acid in recombinant *E.coli* transformed with *fcs* and *ech* genes from a *Pseudomonas* strain, using the concept of resting cells, optimization of the bioprocess variables after using response surface methodology (RSM), and a unique solid-liquid separation system which had ferulic acid entrapped into 1.75% w/v agarose gel cylinders (Luziatelli et al. 2019). Here, in contrast to the product's adsorption, the substrate was immobilized for its steady release in the media. Overall, these studies largely demonstrate that product inhibition could be well sidestepped by the addition of adsorbent resins in the fermentation systems.

Finally, in terms of using genetically engineered strains as the hosts, the study by Yamada et al. (2007) is worth mentioning in which the authors cloned a rare isoeugenol monooxygenase gene from a *Pseudomonas putida* strain IE27 into *E. coli* BL21. With the expression of just a single gene in *E. coli*, the concentration of 28.3 g/L of vanillin was realized from 230 mM isoeugenol in 6 h (Yamada et al. 2008). The achieved concentration of 28.3 g/L was the highest concentrations of vanillin ever reported in the literature from the use of either recombinant or native cells or cell-free extracts. Although it was nearly close to the production attained using wild *Amycolatopsis* sp. 39116 (22.3 g/L) by Fleige et al. (2016). The use of *E. coli* as the host eliminates the complications associated with the use of *Amycolatopsis* like microorganisms that are spore formers and do not have the requisite Generally recognized As Safe (GRAS) status. Infact, Kaur et al. (2014) have reported heterologous expression of *fcs* and *ech* genes in a lactic acid bacterium, *Pediococcus acidilactici* BD16, with GRAS status. In their study, the authors could recover 3.14 mM of vanillin within 20 min from 1.08 mM ferulic acid (Kaur et al. 2014).

Nevertheless, the recombinant strains are associated with certain disadvantages such as their genetic instability, inappropriate genetic tools, and high cost associated with the cloning, transformation, or recombination. Hence, immobilization of vanillin-producing microbial cells as biofilms has gained attention as yet another strategy due to their exceptional operational stabilities when persistent bioconversion times are required, high cell concentrations, and tolerance against harsh environments. Yan et al. (2016) attempted biocatalysis of ferulic acid to vanillin in a packed bed bioreactor, which had *Bacillus subtilis* cells immobilized as biofilms on the carbon fiber textiles (CFT); their vanillin's reported production was 1.84 g/L, with the hydraulic retention time of just 20 h (Yan et al. 2016). Therefore, their process represents a faster production of vanillin in stable biotransformation where recycling or recovery of the immobilized biomass presents a potential economic advantage.

6.4.2 *Specific Case of Polyhydroxyalkanoates (PHA) Production*

Within the biopolymer group, polyhydroxyalkanoates or PHAs that are synthesized directly by microorganisms, are the plastic materials of the 21st century. PHAs are deposited intracellularly within the bacteria during the stress conditions, as energy storage or carbon reserves, (Getachew and Woldesenbet 2016). Their monomer building blocks, formed mainly from saturated or unsaturated hydroxy alkanoic acids, can vary in length from C3 to C14 carbon atoms with a variety of straight or branched chain aliphatic or aromatic side groups. Typically, the structure of PHAs depends on the feedstock monomers available together with the substrate specificity of the PHA synthase (PhaC).

There are reports available where microorganisms have been isolated in nature that grow on lignin as the sole carbon sources and transform derivatives of lignin to (*R*)-3-hydroxyacyl-CoA (3HA-CoA) via fatty acid de novo biosynthesis pathways for the biosynthesis of PHA. *Pandoraea* sp. ISTKB is one such strain tested for its ability to degrade lignin and use the released derivatives for PHA (Kumar et al. 2017). For this testing, the authors grew ISTKB on kraft lignin and its lignin derivatives, particularly syringol, vanillic acid, 4-hydroxybenzoic acid (4-HBA), *p*-coumaric acid, and 2,6-dimethoxyphenol (DMP), as the only carbon sources in the media, aerobically under nutrient-limited conditions for 6 days at 30 °C, pH 8, and 185 rpm. The concentration of PHA that *Pandoraea* sp. ISTKB accumulated was 246 mg/L with 4-HBA, followed by 170 mg/L with *p*-coumaric acid, 72 mg/L with *p*-coumaric acid, 69 mg/L with DMP, and 18 mg/L with kraft lignin (Kumar et al. 2017). Their results indicate that the bacterium's PHA accumulation decreased with an increase in the substrate's structural complexity. 4-HBA is an intermediate produced down the *p*-coumaric acid degradation pathway and seemingly had the simplest structure. Hence, ISTKB accumulated maximum PHA of the type poly (3-hydroxybutyric-co-hydroxyvaleric) acid *P*(*HB-co-HV*) with 4-HBA, and least with lignin (Kumar et al. 2017). Previously, with *Ralstonia eutropha* H16 as the model strain, Satoshi et al. (2014) reported similar results. The authors tested the capacity of H16 to synthesize *P* (*HB-co-HV*) from a variety of lignin derivatives. However, it was with 4-HBA and 3-HBA as the substrates that maximum PHA at 63 wt.% and 65 wt.% was accumulated by *R. eutropha* H16 (Tomizawa et al. 2014). With other intermediates, such as vanillic acid, ferulic acid, and *p*-coumaric acid, cell growth inhibition and PHA accumulation were observed (Tomizawa et al. 2014). Inhibition of cell growth was also observed in a γ-proteobacterium marine isolate *Oceanimonas doudoroffii* in the presence of p-coumaric acid, vanillic acid, ferulic acid, caffeic acid, and gallic acid. With *O. doudoroffii*, the authors reported reasonable PHA production (short-chain length polyhydroxyvalerate (PHV), 1.9 wt.% with sinapinic acid, and 2.7 wt.% with syringic acid (Numata and Morisaki 2015), which are again the simpler derivatives produced during lignin degradation.

In contrast to the aforementioned observations, where microorganisms tend to be inhibited by lignin, strains of the genus *Pseudomonas* have been reported for PHA production from lignin itself. For instance, *P. putida* KT2440 has been demonstrated to produce 150 mg/L medium chain-length (C6-C14) polyhydroxyalkanoates (*mcl*-PHAs) from alkaline pretreated corn liquor rich in lignin (32% wt./wt.) and extractives of lignin (23% wt./wt. of *p*-coumaric acid, vanillic acid, and ferulic acid) by Linger et al. (2014). The *mcl*-PHA biopolymer produced by *P. putida* KT2440 has a molecular weight of 124 kDa and has side chains comprising 3-hydroxydecanoic acid (55%), 3-hydroxyoctanoic acid (22%), 3-hydroxydodecanoic acid (16%), 3-hydroxytetradecancoic acid (4%), and 3-hydroxyhexanoic acid (3%) (Linger et al. 2014). Alkaline pretreated lignin (APL) from corn stover was also used to support *P. putida* KT2440, *P. putida* mt-2, and *Cupriavidus necato* growth and PHA accumulation (Salvachúa et al. 2015). Therein, the *P. putida* KT2440 and *P. putida* mt-2 synthesized 52 mg/L and 60 mg/L of mcl-PHA, respectively. *C. necator* was found to accumulate 162 mg/L of short-chain length polyhydroxy butyrate (PHB) (Salvachúa et al. 2015). More recently, *Pandoraea* sp. B-6 has also been shown to carry the potential to bio convert kraft lignin into PHA. Strain B-6 was shown to degrade 40% of kraft lignin in barely 4 days, with a resultant 24.7% accumulation of scl-PHB (Liu et al. 2019a). Similar findings were reported by Shi et al. (2017) with *Cupriavidus basilensis* B-8, which could accumulate 128 mg/L of PHB from kraft lignin (without any pretreatment) as the sole carbon source in 7 days (Shi et al. 2017). The authors in the study also highlighted the utility of fed-batch in enhancing PHB production during the bioconversion, as 319.4 mg/L of PHB was reported with 5 g/L of lignin using the fed-batch mode of fermentation (Shi et al. 2017).

Overall, these studies lay a solid foundation for pursuing bioconversion of lignin-rich streams to value-added biopolymers (Fig. 6.2). Generally, lignin and its aromatic intermediates are recalcitrant and toxic, and have been found to impede the fermentation and bioconversions in the host. In the future, optimization of culture conditions, the use of innovative fermentation modes, isolating new microbial strains, and metabolically engineering existing strains may prove useful for enhancing the yield and percent accumulation of PHA inside the host when lignin or its intermediates are used as the carbon source. Recently, CRISPR/Cas9n-based tool was used to engineer *Pseudomonas putida* KT2440 to produce a higher amount of mcl-PHA (270 mg/L) using ferulic acid as the feedstock. In yet another bioengineering study, Lin et al. (2016) improved the tolerance and productivity of *Pseudomonas putida* strain A514 toward lignin and its derivative vanillin by overexpressing a gene that codes for the VanAB enzyme of the β-ketoadipate pathway that is explicitly induced in the presence of vanillic acid (Lin et al. 2016). The group further channelized the vanillin bioconversion towards mcl-PHA synthesis by overexpressing phaB and phaC genes in A514. The results indicated that the modified A514 could accumulate 65 mg/L PHA, with a yield of 73.5% per CDW compared with 54% in the wild strain of A514. After this two-step metabolic engineering, A514 was also able to accumulate 75 mg/L mcl PHA (C8-C14) with kraft lignin, which is significantly more recalcitrant than the processed APL lignin (Lin et al. 2016). In addition, the authors reported that enhanced PHA production occurred through the complete growth cycle for both nitrogen-limiting and nitrogen-excess conditions. Moreover, the composition of the produced mcl-PHA

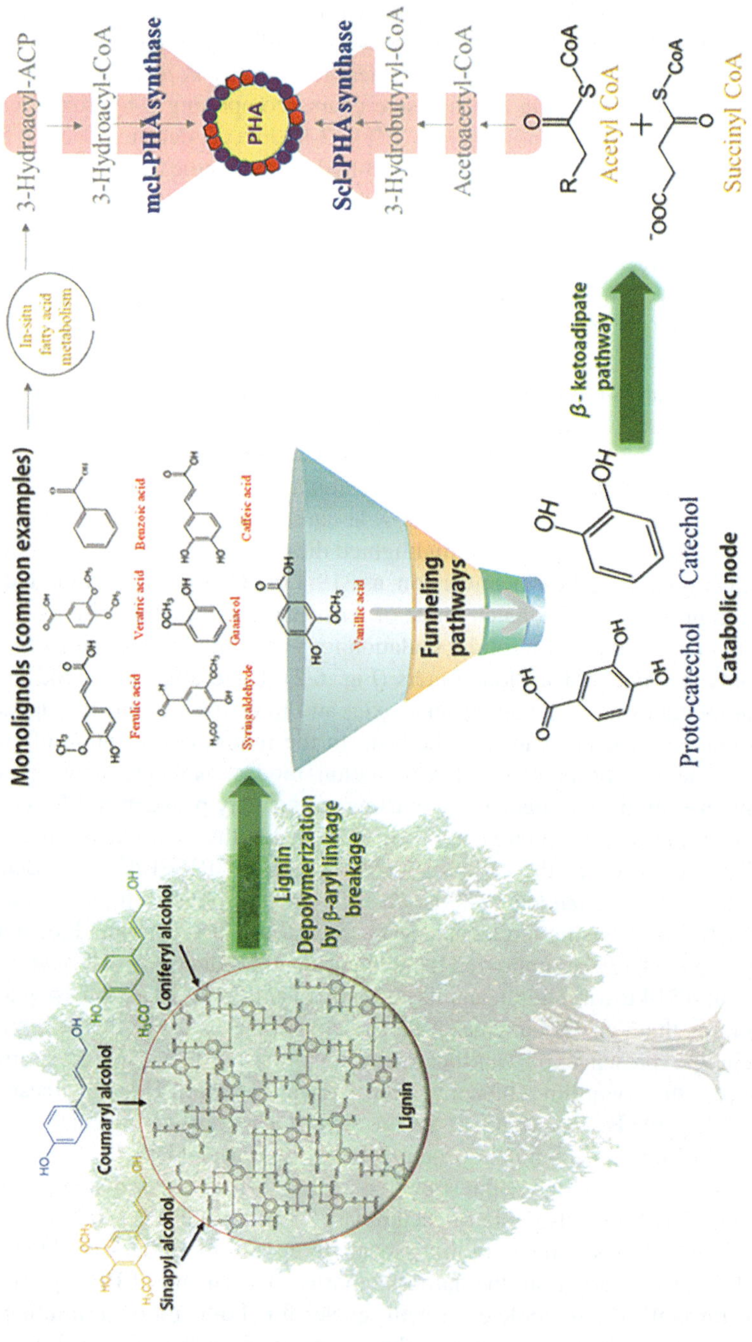

Fig. 6.2 General scheme for bioconversion of lignin-rich streams into polyhydroxyalkanoates

had side chains ranging from C8-C14, which in itself is valued highly for its ability to be a precursor for the synthesis of jet fuels (C8-C16) (Lin et al. 2016). Wang et al. (2018) enhanced the PHA production of *Pseudomonas putida* strain A514 to 246 mg/L from vanillin by further overexpression of long-chain fatty acid-CoA ligase (*alkK*), and 3-hydroxyacyl-acyl carrier protein (ACP) thioesterase (*phaG*). In the same study, these genes were identified after an extensive study of genomics, transcriptomics, and proteomics data of A514 grown on vanillin under nitrogen-limited conditions (Wang et al. 2018). Thus, molecular insights based on omics study can help identify schemes to enhance PHA production from lignin or its derivatives.

Generally, for synthesizing high-molecular-weight PHA copolymers, expensive odd fatty acids (propionic acid, valeric acid) are supplemented in the medium as co-carbon substrates (Govil et al. 2020b). This co-addition of additional feedstocks not only increases the cost of production of PHA copolymer but also has a low yield and accumulation of copolymers compared to the addition of homopolymers. However, the studies discussed in this section suggest that the lignin or its aromatic derivatives can serve as low-cost platform precursors for synthesizing not only scl-copolymers like P(HV) and *P(HB-co-HV)* but also mcl-PHAs with side chains ranging from 3-hydroxyhexanoic acid to hydroxy-tetra decanoate.

Currently, the valorization process of lignin to PHA has an important drawback. The concentration and the yield of the produced PHA are low, being in the tens or low hundreds of milligrams per liter. The principal reason for this low yield is related to the low reactivity and assimilation of lignin by the microbial hosts. However, a study published in 2019 showed that co-utilization of lignin with a limited amount of glucose could facilitate lignin biocatalysis to PHA. The authors used this concept to produce the record titer of 1.5 g/L of PHA by the synergistic bioconversion of lignin and residual sugar released during corn stover pretreatment and hydrolysis by the *Pseudomonas putida* strain KT2440 (Liu et al. 2019b). This shows that lignin-based biorefinery sustainability is conceivable by applying innovative yet straightforward concepts.

6.5 Conclusion and Future Directions

Lignin, a polyaromatic macromolecule, is one of the most underutilized fractions of the lignocellulosic biomasses, whose valorization to fine chemicals has a much higher economic and environmental benefit than burning it for heat and electricity. The abundance of aromatic monomers in its skeletal structure makes lignin a promising substrate for biocatalysis into an array of value-added products such as aromatic biomolecules, biopolymers, bio-oil, and biofuels. Today, lignin is also considered a commendable, environmental-friendly component or additive in the preparation of epoxy resins, fire-retardants, antioxidants, adhesives, and concrete admixtures. Hence, the production of lignin-derived co-products can support, and enhance the profitability of second-generation biorefineries, and related industries.

With the advancement of multi-omics knowledge, the sophisticated metabolic pathways essential for lignin degradation are being elucidated in detail. In the future, engineering of these pathways in microbes for the overproduction of useful intermediates such as ferulic acid, vanillin, and guaiacol is foreseen. In addition, the central target for the near future should be the improvement of the technology for separating lignin efficiently and cost-effectively from lignocellulosic biomass.

Acknowledgments We acknowledge support from the CNAM-Bio-Center and the South Dakota Governor's Office of Economic Development. The authors also gratefully acknowledge support from the National Science Foundation (Award #1736255, #1849206, and #1920954) and the Department of Chemical and Biological Engineering at the South Dakota Mines.

Conflicts of Interest The authors declare no conflict of interest.

References

Abdelaziz OY, Brink DP, Prothmann J, Ravi K, Sun M, García-Hidalgo J, Sandahl M, Hulteberg CP, Turner C, Lidén G, Gorwa-Grauslund MF (2016) Biological valorization of low molecular weight lignin. Biotechnol Adv 34(8):1318–1346

Abdelaziz OY, Meier S, Prothmann J, Turner C, Riisager A, Hulteberg CP (2019) Oxidative depolymerisation of lignosulphonate lignin into low-molecular-weight products with Cu–Mn/δ-Al2O3. Top Catal 62(7):639–648

Abdelkafi S, Sayadi S, Ben Ali Gam Z, Casalot L, Labat M (2006) Bioconversion of ferulic acid to vanillic acid by Halomonas elongata isolated from table-olive fermentation. FEMS Microbiol Lett 262(1):115–120

Achterholt S, Priefert H, Steinbüchel A (2000) Identification of Amycolatopsis sp. strain HR167 genes, involved in the bioconversion of ferulic acid to vanillin. Appl Microbiol Biotechnol 54 (6):799–807

Ahmad M, Taylor CR, Pink D, Burton K, Eastwood D, Bending GD, Bugg TD (2010) Development of novel assays for lignin degradation: comparative analysis of bacterial and fungal lignin degraders. Mol Biosyst 6(5):815–821

Ashengroph M, Amini J (2017) Bioconversion of isoeugenol to vanillin and vanillic acid using the resting cells of *Trichosporon asahii*. 3 Biotech 7(6):358–358

Ashengroph M, Nahvi I, Zarkesh-Esfahani H, Momenbeik F (2011) *Pseudomonas resinovorans* SPR1, a newly isolated strain with potential of transforming eugenol to vanillin and vanillic acid. New Biotechnol 28(6), 656–64

Ashengroph M, Nahvi I, Zarkesh-Esfahani H, Momenbeik F (2012) Conversion of isoeugenol to vanillin by Psychrobacter sp. strain CSW4. Appl Biochem Biotechnol 166(1):1–12

Bajwa DS, Pourhashem G, Ullah AH, Bajwa SG (2019) A concise review of current lignin production, applications, products and their environmental impact. Ind Crop Prod 139:111526

Barghini P, Di Gioia D, Fava F, Ruzzi M (2007) Vanillin production using metabolically engineered Escherichia coli under non-growing conditions. Microb Cell Factories 6(1):13

Bjorkman A (1954) Isolation of lignin from finely divided wood with neutral solvents. Nature 174 (4440):1057–1058

Bugg TD, Ahmad M, Hardiman EM, Rahmanpour R (2011) Pathways for degradation of lignin in bacteria and fungi. Nat Prod Rep 28(12):1883–1896

Chatterjee S, Saito T (2015) Lignin-derived advanced carbon materials. ChemSusChem 8 (23):3941–3958

Chatterjee, T., De, B.K., and , Bhattacharyya, D.K. 1999. Microbial conversion of isoeugenol to vanillin by Rhodococcus rhodochrous. Indian J Chem Sect B, 38B(5), 538–541

Chen H (2015) Chapter 3—Lignocellulose biorefinery feedstock engineering. In: Chen H (ed) Lignocellulose biorefinery engineering. Woodhead Publishing, Sawston, pp 37–86

Chen CL, Chang HM, Kirk TK (1982) Aromatic acids produced during degradation of lignin in spruce wood by Phanerochaete chrysosporium. Department of Wood and Paper Science, North Carolina State University, Raleigh, NC; Forest Products Lab., USDA Forest Service, Madison, WI

Chen YH, Chai LY, Zhu YH, Yang ZH, Zheng Y, Zhang H (2012) Biodegradation of Kraft lignin by a bacterial strain Comamonas sp. B-9 isolated from eroded bamboo slips. J Appl Microbiol 112(5):900–906

Datta R, Kelkar A, Baraniya D, Molaei A, Moulick A, Meena RS, Formanek P (2017) Enzymatic degradation of lignin in soil: a review. Sustainability 9(7):1163

Di Gioia D, Luziatelli F, Negroni A, Ficca AG, Fava F, Ruzzi M (2011) Metabolic engineering of Pseudomonas fluorescens for the production of vanillin from ferulic acid. J Biotechnol 156 (4):309–316

Duan J, Huo X, Du WJ, Liang JD, Wang DQ, Yang SC (2016) Biodegradation of Kraft lignin by a newly isolated anaerobic bacterial strain, Acetoanaerobium sp. WJDL-Y2. Lett Appl Microbiol 62(1):55–62

Fleige C, Meyer F, Steinbüchel A (2016) Metabolic engineering of the Actinomycete Amycolatopsis sp. strain ATCC 39116 towards enhanced production of natural vanillin. Appl Environ Microbiol 82(11):3410–3419

Fox40 (2020) Vanillic Acid (CAS 121-34-6) Market Size Latest Report 2020 Segment by manufacturers, Type, Applications and Dynamics

Gallage NJ, Hansen EH, Kannangara R, Olsen CE, Motawia MS, Jørgensen K, Holme I, Hebelstrup K, Grisoni M, Møller BL (2014) Vanillin formation from ferulic acid in Vanilla planifolia is catalysed by a single enzyme. Nat Commun 5:4037

Gargulak JD, Lebo SE, McNally TJ (2015) Lignin. in: Kirk-Othmer encyclopedia of chemical technology, pp 1–26

Gasson MJ, Kitamura Y, McLauchlan WR, Narbad A, Parr AJ, Parsons EL, Payne J, Rhodes MJ, Walton NJ (1998) Metabolism of ferulic acid to vanillin. A bacterial gene of the enoyl-SCoA hydratase/isomerase superfamily encodes an enzyme for the hydration and cleavage of a hydroxycinnamic acid SCoA thioester. J Biol Chem 273(7):4163–4170

Getachew A, Woldesenbet F (2016) Production of biodegradable plastic by polyhydroxybutyrate (PHB) accumulating bacteria using low cost agricultural waste material. BMC Res Notes 9:509

Govil T, Saxena P, Samanta D, Singh SS, Kumar S, Salem DR, Sani RK (2020a) Adaptive enrichment of a thermophilic bacterial isolate for enhanced enzymatic activity. Microorganisms 8(6):871

Govil T, Wang J, Samanta D, David A, Tripathi A, Rauniyar S, Salem DR, Sani RK (2020b) Lignocellulosic feedstock: a review of a sustainable platform for cleaner production of nature's plastics. J Clean Prod 270:122521

Graf N, Altenbuchner J (2014) Genetic engineering of Pseudomonas putida KT2440 for rapid and high-yield production of vanillin from ferulic acid. Appl Microbiol Biotechnol 98(1):137–149

Gupta VK, Minocha AK, Jain N (2001) Batch and continuous studies on treatment of pulp mill wastewater by *Aeromonas formicans*. J Chem Technol Biotechnol 76(6):547–552

Gurujeyalakshmi G, Mahadevan A (1987) Dissimilation of ferulic acid byBacillus subtilis. Curr Microbiol 16(2):69–73

Hansen MAT, Jørgensen H, Laursen KH, Schjoerring JK, Felby C (2013) Structural and chemical analysis of process residue from biochemical conversion of wheat straw (*Triticum aestivum* L.) to ethanol. Biomass Bioenergy 56:572–581

Hintz HL (2001) Paper: pulping and bleaching. In: Buschow KHJ, Cahn RW, Flemings MC, Ilschner B, Kramer EJ, Mahajan S, Veyssière P (eds) Encyclopedia of materials: science and technology. Elsevier, Oxford, pp 6707–6711

Holladay JE, Bozell JJ, White JF, Johnson D (2007) Top value-added chemicals from biomass. Volume II—results of screening for potential candidates from biorefinery lignin. Pacific Northwest National Laboratory

Horwath W (2015) Chapter 12—Carbon cycling: the dynamics and formation of organic matter. In: Paul EA (ed) Soil microbiology, ecology and biochemistry, 4th edn. Academic, Boston, pp 339–382

Hu H, Li L, Ding S (2015) An organic solvent-tolerant phenolic acid decarboxylase from Bacillus licheniformis for the efficient bioconversion of hydroxycinnamic acids to vinyl phenol derivatives. Appl Microbiol Biotechnol 99(12):5071–5081

Hua D, Ma C, Lin S, Song L, Deng Z, Maomy Z, Zhang Z, Yu B, Xu P (2007a) Biotransformation of isoeugenol to vanillin by a newly isolated *Bacillus pumilus* strain: identification of major metabolites. J Biotechnol 130(4):463–470

Hua D, Ma C, Song L, Lin S, Zhang Z, Deng Z, Xu P (2007b) Enhanced vanillin production from ferulic acid using adsorbent resin. Appl Microbiol Biotechnol 74(4):783–790

Huang Z, Dostal L, Rosazza JP (1993) Mechanisms of ferulic acid conversions to vanillic acid and guaiacol by Rhodotorula rubra. J Biol Chem 268(32):23954–23958

Janusz G, Pawlik A, Sulej J, Swiderska-Burek U, Jarosz-Wilkolazka A, Paszczynski A (2017) Lignin degradation: microorganisms, enzymes involved, genomes analysis and evolution. FEMS Microbiol Rev 41(6):941–962

Jing D, Jidong L, Yiping W, Wenjing D (2016) Kraft lignin biodegradation by Dysgonomonas sp. WJDL-Y1, a new anaerobic bacterial strain isolated from sludge of a pulp and paper mill. J Microbiol Biotechnol 26(10):1765–1773

John B (2020) Kraft lignin market analysis share size and growth demand by 2020–2025

Jung DH, Choi W, Choi KY, Jung E, Yun H, Kazlauskas RJ, Kim BG (2013) Bioconversion of p-coumaric acid to p-hydroxystyrene using phenolic acid decarboxylase from B. amyloliquefaciens in biphasic reaction system. Appl Microbiol Biotechnol 97(4):1501–1511

Kasana RC, Sharma UK, Sharma N, Sinha AK (2007) Isolation and identification of a novel strain of Pseudomonas chlororaphis capable of transforming isoeugenol to vanillin. Curr Microbiol 54 (6):457–461

Kaur B, Chakraborty D, Kumar B (2014) Metabolic engineering of Pediococcus acidilactici BD16 for production of vanillin through ferulic acid catabolic pathway and process optimization using response surface methodology. Appl Microbiol Biotechnol 98(20):8539–8551

Kumar N, Pruthi V (2014) Potential applications of ferulic acid from natural sources. Biotechnol Rep (Amst) 4:86–93

Kumar M, Singhal A, Verma PK, Thakur IS (2017) Production and characterization of polyhydroxyalkanoate from lignin derivatives by Pandoraea sp. ISTKB. ACS Omega 2 (12):9156–9163

Lee EG, Yoon SH, Das A, Lee SH, Li C, Kim JY, Choi MS, Oh DK, Kim SW (2009) Directing vanillin production from ferulic acid by increased acetyl-CoA consumption in recombinant *Escherichia coli*. Biotechnol Bioeng 102(1):200–208

Lee S, Kang M, Bae J-H, Sohn J-H, Sung BH (2019) Bacterial valorization of lignin: strains, enzymes, conversion pathways, biosensors, and perspectives. Front Bioeng Biotechnol 7:209–209

Leonard PM, Brzozowski AM, Lebedev A, Marshall CM, Smith DJ, Verma CS, Walton NJ, Grogan G (2006) The 1.8 A resolution structure of hydroxycinnamoyl-coenzyme A hydratase-lyase (HCHL) from Pseudomonas fluorescens, an enzyme that catalyses the transformation of feruloyl-coenzyme A to vanillin. Acta Crystallogr D Biol Crystallogr 62 (Pt 12):1494–1501

Li T, Rosazza JP (2000) Biocatalytic synthesis of vanillin. Appl Environ Microbiol 66(2):684–687

Lin L, Cheng Y, Pu Y, Sun S, Li X, Jin M, Pierson EA, Gross DC, Dale BE, Dai SY, Ragauskas AJ, Yuan JS (2016) Systems biology-guided biodesign of consolidated lignin conversion. Green Chem 18(20):5536–5547

Linger JG, Vardon DR, Guarnieri MT, Karp EM, Hunsinger GB, Franden MA, Johnson CW, Chupka G, Strathmann TJ, Pienkos PT, Beckham GT (2014) Lignin valorization through integrated biological funneling and chemical catalysis. Proc Natl Acad Sci U S A 111 (33):12013–12018

Liu D, Yan X, Si M, Deng X, Min X, Shi Y, Chai L (2019a) Bioconversion of lignin into bioplastics by Pandoraea sp. B-6: molecular mechanism. Environ Sci Pollut Res Int 26(3):2761–2770

Liu Z-H, Shinde S, Xie S, Hao N, Lin F, Li M, Yoo CG, Ragauskas AJ, Yuan JS (2019b) Cooperative valorization of lignin and residual sugar to polyhydroxyalkanoate (PHA) for enhanced yield and carbon utilization in biorefineries. Sustain Energy Fuels 3(8):2024–2037

Liu H-M, Zou Y, Yao C-Y, Yang Z (2020) Enzymatic synthesis of vanillin and related catalytic mechanism. Flavour Fragr J 35(1):51–58

Luziatelli F, Brunetti L, Ficca AG, Ruzzi M (2019) Maximizing the efficiency of vanillin production by biocatalyst enhancement and process optimization. Front Bioeng Biotechnol 7(279)

Macfarlane AL, Mai M, Kadla JF (2014) Chapter 20—Bio-based chemicals from biorefining: lignin conversion and utilisation. In: Waldron K (ed) Advances in biorefineries. Woodhead Publishing, Sawston, pp 659–692

Markus HP, Peters MLJ, Roos R (1992) Process for the preparation of phenylaldehydes, (Ed.) Q.I. BV, Vol. EP0542348A2. Europe

Miller J (2016) Lignin: technology, applications, and markets. RISI Latin American Pulp & Paper Outlook Conference

Mohamad Ibrahim, M.N., Sriprasanthi, R.B., Shamsudeen, S., Adam, F., and , Bhawani, S.A. 2012. A concise review of the natural existance, synthesis, properties, and applications of syringaldehyde. Bioresour Technol, 7(3), 4377–4399

Muheim A, Lerch K (1999) Towards a high-yield bioconversion of ferulic acid to vanillin. Appl Microbiol Biotechnol 51(4):456–461

Negishi O, Sugiura K, Negishi Y (2009) Biosynthesis of vanillin via ferulic acid in Vanilla planifolia. J Agric Food Chem 57(21):9956–9961

Nishimura M, Kawakami S, Otsuka H (2018) Molecular cloning and characterization of vanillin dehydrogenase from Streptomyces sp. NL15-2K. BMC Microbiol 18(1):154–154

Numata K, Morisaki K (2015) Screening of marine Bacteria to synthesize Polyhydroxyalkanoate from lignin: contribution of lignin derivatives to biosynthesis by *Oceanimonas doudoroffii*. ACS Sustain Chem Eng 3(4):569–573

Obst JR, Kirk TK (1988) Isolation of lignin. In: Methods in enzymology, vol. 161. Academic, pp 3-12

Ohta Y, Hasegawa R, Kurosawa K, Maeda AH, Koizumi T, Nishimura H, Okada H, Qu C, Saito K, Watanabe T, Hatada Y (2017) Enzymatic specific production and chemical functionalization of Phenylpropanone platform monomers from lignin. ChemSusChem 10(2):425–433

Overhage J, Priefert H, Steinbüchel A (1999) Biochemical and genetic analyses of ferulic acid catabolism in Pseudomonas sp. strain HR199. Appl Environ Microbiol 65(11):4837–4847

Overhage J, Steinbüchel A, Priefert H (2003) Highly efficient biotransformation of eugenol to ferulic acid and further conversion to vanillin in recombinant strains of Escherichia coli. Appl Environ Microbiol 69(11):6569–6576

Pacek AW, Ding P, Garrett M, Sheldrake G, Nienow AW (2013) Catalytic conversion of sodium lignosulfonate to vanillin: engineering aspects. Part 1. Effects of processing conditions on vanillin yield and selectivity. Ind Eng Chem Res 52(25):8361–8372

Plaggenborg R, Steinbüchel A, Priefert H (2001) The coenzyme A-dependent, non-beta-oxidation pathway and not direct deacetylation is the major route for ferulic acid degradation in Delftia acidovorans. FEMS Microbiol Lett 205(1):9–16

Plaggenborg R, Overhage J, Steinbüchel A, Priefert H (2003) Functional analyses of genes involved in the metabolism of ferulic acid in Pseudomonas putida KT2440. Appl Microbiol Biotechnol 61(5):528–535

PMR (2020) Bio vanillin market, vol. 2021. Persistence Market Research

Prim N, Pastor FI, Diaz P (2003) Biochemical studies on cloned Bacillus sp. BP-7 phenolic acid decarboxylase PadA. Appl Microbiol Biotechnol 63(1):51–56

Rabenhorst J, Hopp R (1991) Process for the preparation of vanillin Patent application, EP0405197

Raj A, Krishna Reddy MM, Chandra R (2007) Identification of low molecular weight aromatic compounds by gas chromatography–mass spectrometry (GC–MS) from Kraft lignin degradation by three Bacillus sp. Int Biodeterior Biodegradation 59(4):292–296

Reid ID (1998) Fate of residual lignin during delignification of Kraft pulp by trametes versicolor. Appl Environ Microbiol 64(6):2117–2125

Reiter J, Strittmatter H, Wiemann LO, Schieder D, Sieber V (2013) Enzymatic cleavage of lignin β-O-4 aryl ether bonds via net internal hydrogen transfer. Green Chem 15(5):1373–1381

Retsina T, Pylkkanen V, Nelson K, Szczepanik M (2013) Processes and apparatus for lignin separation in biorefineries, vol. US20140034047A1. Granbio Intellectual Property Holdings LLC. USA

Richter JS, Brink DL, Diddams DG, Peter O (1945) Process for making vanillin, vol. US2434626A. Salvo Chemical Corp. USA

Sainsbury PD, Hardiman EM, Ahmad M, Otani H, Seghezzi N, Eltis LD, Bugg TD (2013) Breaking down lignin to high-value chemicals: the conversion of lignocellulose to vanillin in a gene deletion mutant of Rhodococcus jostii RHA1. ACS Chem Biol 8(10):2151–2156

Salvachúa D, Karp EM, Nimlos CT, Vardon DR, Beckham GT (2015) Towards lignin consolidated bioprocessing: simultaneous lignin depolymerization and product generation by bacteria. Green Chem 17(11):4951–4967

Shen Y, Song X, Li L, Sun J, Jaiswal Y, Huang J, Liu C, Yang W, Williams L, Zhang H, Guan Y (2019) Protective effects of p-coumaric acid against oxidant and hyperlipidemia-an in vitro and in vivo evaluation. Biomed Pharmacother 111:579–587

Shi Y, Yan X, Li Q, Wang X, Liu M, Xie S, Chai L, Yuan J (2017) Directed bioconversion of Kraft lignin to polyhydroxyalkanoate by Cupriavidus basilensis B-8 without any pretreatment. Process Biochem 52:238–242

Shimada M, Nakatsubo F, Kirk TK, Higuchi T (1981) Biosynthesis of the secondary metabolite veratryl alcohol in relation to lignin degradation in Phanerochaete chrysosporium. Arch Microbiol 129(4):321–324

Shimoni E, Ravid U, Shoham Y (2000) Isolation of a Bacillus sp. capable of transforming isoeugenol to vanillin. J Biotechnol 78(1):1–9

Strassberger Z, Tanase S, Rothenberg G (2014) The pros and cons of lignin valorisation in an integrated biorefinery. RSC Adv 4(48):25310–25318

Tang P-L, Hassan O, Maskat MY, Badri K (2015) Production of monomeric aromatic compounds from oil palm empty fruit bunch fiber lignin by chemical and enzymatic methods. Biomed Res Int 2015:–891539

Thoresen PP, Matsakas L, Rova U, Christakopoulos P (2020) Recent advances in organosolv fractionation: towards biomass fractionation technology of the future. Bioresour Technol 306:123189

Tian D, Hu J, Chandra RP, Saddler JN, Lu C (2017) Valorizing recalcitrant cellulolytic enzyme lignin via lignin nanoparticles fabrication in an integrated biorefinery. ACS Sustain Chem Eng 5 (3):2702–2710

Tomizawa S, Chuah J-A, Matsumoto K, Doi Y, Numata K (2014) Understanding the limitations in the biosynthesis of polyhydroxyalkanoate (PHA) from lignin derivatives. ACS Sustain Chem Eng 2(5):1106–1113

Tribot A, Amer G, Abdou Alio M, de Baynast H, Delattre C, Pons A, Mathias J-D, Callois J-M, Vial C, Michaud P, Dussap C-G (2019) Wood-lignin: supply, extraction processes and use as bio-based material. Eur Polym J 112:228–240

Unno T, Kim SJ, Kanaly RA, Ahn JH, Kang SI, Hur HG (2007) Metabolic characterization of newly isolated Pseudomonas nitroreducens Jin1 growing on eugenol and isoeugenol. J Agric Food Chem 55(21):8556–8561

USDA (2020) United States Department of Agriculture Agricultural Marketing Service|National Organic Program Document Cover Sheet

Vazquez A, Pique TM (2016) Chapter 5—Biotech admixtures for enhancing Portland cement hydration. In: Pacheco-Torgal F, Ivanov V, Karak N, Jonkers H (eds) Biopolymers and biotech admixtures for eco-efficient construction materials. Woodhead Publishing, Sawston, pp 81–98

Vyrides I, Agathangelou M, Dimitriou R, Souroullas K, Salamex A, Ioannou A, Koutinas M (2015) Novel Halomonas sp. B15 isolated from Larnaca Salt Lake in cyprus that generates vanillin and vanillic acid from ferulic acid. World J Microbiol Biotechnol 31(8):1291–1296

Wang X, Lin L, Dong J, Ling J, Wang W, Wang H, Zhang Z, Yu X (2018) Simultaneous improvements of Pseudomonas cell growth and polyhydroxyalkanoate production from a lignin derivative for lignin-consolidated bioprocessing. Appl Environ Microbiol 84(18):e01469–e01418

WebArchive (2003) Lignins—products with many uses, vol. 2020. Way Back Machine. Lignin Institute

Xu Z, Lei P, Zhai R, Wen Z, Jin M (2019) Recent advances in lignin valorization with bacterial cultures: microorganisms, metabolic pathways, and bio-products. Biotechnol Biofuels 12(1):32

Yamada M, Okada Y, Yoshida T, Nagasawa T (2007) Biotransformation of isoeugenol to vanillin by Pseudomonas putida IE27 cells. Appl Microbiol Biotechnol 73(5):1025–1030

Yamada M, Okada Y, Yoshida T, Nagasawa T (2008) Vanillin production using Escherichia coli cells over-expressing isoeugenol monooxygenase of Pseudomonas putida. Biotechnol Lett 30 (4):665–670

Yan L, Chen P, Zhang S, Li S, Yan X, Wang N, Liang N, Li H (2016) Biotransformation of ferulic acid to vanillin in the packed bed-stirred fermentors. Sci Rep 6(1):34644

Yang W, Tang H, Ni J, Wu Q, Hua D, Tao F, Xu P (2013) Characterization of two Streptomyces enzymes that convert Ferulic acid to vanillin. PLoS One 8(6):e67339

Yoon SH, Lee EG, Das A, Lee SH, Li C, Ryu HK, Choi MS, Seo WT, Kim SW (2007) Enhanced vanillin production from recombinant E. coli using NTG mutagenesis and adsorbent resin. Biotechnol Prog 23(5):1143–1148

Zhang Y, Xu P, Han S, Yan H, Ma C (2006) Metabolism of isoeugenol via isoeugenol-diol by a newly isolated strain of Bacillus subtilis HS8. Appl Microbiol Biotechnol 73(4):771–779

Zhang A, Lu F, Sun R-C, Ralph J (2010) Isolation of cellulolytic enzyme lignin from wood Preswollen/dissolved in dimethyl sulfoxide/N-Methylimidazole. J Agric Food Chem 58 (6):3446–3450

Zhao L-Q, Sun Z-H, Zheng P, He J-Y (2006) Biotransformation of isoeugenol to vanillin by Bacillus fusiformis CGMCC1347 with the addition of resin HD-8. Process Biochem 41 (7):1673–1676

Zheng L, Zheng P, Sun Z, Bai Y, Wang J, Guo X (2007) Production of vanillin from waste residue of rice bran oil by Aspergillus niger and Pycnoporus cinnabarinus. Bioresour Technol 98 (5):1115–1119

Chapter 7
Understanding the Potential Applications of Biofilms as Industrial "Cell Factories"

Tanvi Govil, Saveena Solanki, Zachary Hogan, Sudhir Kumar, David R. Salem, and Rajesh K Sani

Abstract There are quite a few noteworthy factors to be weighed in selecting a microbial strain for the production of a bioproduct in a laboratory or an industry. These factors range from the choice of the carbon source to be fed (feedstock) and

---

T. Govil
Department of Chemical and Biological Engineering, South Dakota Mines, Rapid City, SD, USA

Composite and Nanocomposite Advanced Manufacturing—Biomaterials Center, Rapid City, SD, USA

S. Solanki · S. Kumar
Department of Biotechnology and Bioinformatics, Jaypee University of Information Technology, Solan, Himachal Pradesh, India

Z. Hogan
Department of Mechanical Engineering, South Dakota Mines, Rapid City, SD, USA

D. R. Salem (✉)
Department of Chemical and Biological Engineering, South Dakota Mines, Rapid City, SD, USA

Composite and Nanocomposite Advanced Manufacturing—Biomaterials Center, Rapid City, SD, USA

Department of Materials and Metallurgical Engineering, South Dakota Mines, Rapid City, SD, USA
e-mail: david.salem@sdsmt.edu

R. K. Sani (✉)
Department of Chemical and Biological Engineering, South Dakota Mines, Rapid City, SD, USA

Composite and Nanocomposite Advanced Manufacturing—Biomaterials Center, Rapid City, SD, USA

Department of Chemistry, Biology, and Health Sciences, South Dakota Mines, Rapid City, SD, USA

BuG ReMeDEE Consortium, Rapid City, SD, USA
e-mail: rajesh.sani@sdsmt.edu

J. K. Saini, R. K. Sani (eds.), *Microbial Biotechnology for Renewable and Sustainable Energy*, Clean Energy Production Technologies,
https://doi.org/10.1007/978-981-16-3852-7_7

fermentation conditions to be operated to the final product to be synthesized. For an economical process that has a high yield with a minimum of operational costs, bacteria enclosed as sessile cells inside biofilms have gained considerable interest. Cells immobilized as biofilms in a reactor exhibit a distinct ability to carry on a fermentative process continuously for a long time with a reduced lag phase and operating time. Biofilm cells also exhibit higher resistance to metabolic stress as opposed to their free cell planktonic suspensions, while being tolerant to toxic levels of solvents and inhibitors. These features of the biofilm mode of growth have positioned them as "cell factories" for sustainable production of certain commercial value biofuels, organic acids, vitamins, amino acids, bioplastics, and also enzymes. Further, biofilms have proven themselves useful for the treatment of soil and water against recalcitrant organic pollutants and toxic metals. This book chapter summarizes some such positive applications associated with biofilms as industrial cell factories for bioremediation of the environment and production of value-added products. The chapter also reviews the benefits of biofilms in agriculture as biofertilizers to improve crop ecosystems.

Keywords Biofilms · Bioremediation · Industrial workhouse · Sustainable · Value-added products

7.1 Introduction

In nature, there are two different physiological lifestyles that a bacterium can adopt: (1) a unicellular planktonic mode of growth, where the bacteria thrive as free-floating dispersed cells in the medium; (2) as biofilms, where the cells grow in as a sessile syntrophic consortium that is reversibly attached or to a stationary substratum through a self-produced extracellular polymeric substance (EPS) (Govil et al. 2019). These lifeforms have their associated benefits and hardships, and a bacterium living in a biofilm typically has substantially distinct characteristics from its free-living form. Lately, the importance of the biofilm mode of growth in biotechnological applications has attracted increased attention.

A biofilm is a small habitat for microorganisms that can attach to a surface by excreting a sticky EPS substance that encompasses the bacteria in a matrix. A biofilm can be composed of a single microbial species or a conglomerate of species. In most cases, biofilms primarily comprise bacteria, but they can also include other organisms such as archaea, protozoa, fungi, or algae (Hall and Mah 2017). Together, the participating strains create a complex micro-environment comparable to the system existing within the multicellular organisms. In the biofilms, the microbes interact with each other through an intricate communication system called quorum sensing—a scheme based on chemical signals to coordinate different gene expressions (e.g., *N*-Acyl homoserine lactone-mediated quorum sensing in Gram-negative bacteria) (Berlanga et al. 2012). The surface association of a biofilm via EPS provides adequate strength and resilience to the cells to maintain their assembly against dispersion under shear forces; this allows the host cells to last for much longer time, in a relatively synchronized state, on a surface that supports their proliferation (Berlanga et al. 2012).

The EPS biofilm covering comprises a variety of carbohydrates, proteins, glycoproteins, and glycolipids, which hold the biofilm cells in sync and preserve water. This EPS biofilm habitat (a) keeps the extracellular enzymes close to the biofilm's cells through its sorption properties, thereby facilitating the digestion of dissolved colloidal and solid biopolymers; (b) sequesters dissolved and particulate nutrients from the surroundings; (c) protects the biofilm cells from environments which are hostile to the microbial population by reducing the diffusion of certain toxic antimicrobial compounds, like antibiotics from the surroundings; (d) expels waste and toxic compounds formed inside the biofilms by the residing microbes (Berlanga et al. 2012; Flemming et al. 2007; Jachlewski et al. 2015; Jefferson 2004). All these factors allow microorganisms inside the biofilm to live at high cell densities while maintaining their stationary, dormant, or slow-growing subpopulation. In addition, different microbes in biofilms establish a balanced consumption rate of electron donors/acceptors from a substrate to gain energy (Bruce and Perry 2001).

Altogether, enclosed within the polymer EPS matrix, a biofilm reflects a successful symbiotic microbial consortium, wherein the cells synchronize distinct physiological processes and cooperative activities. The formed biofilms can endure antimicrobial agents and disinfectants at 10–1000 times concentrations that are considered necessary to eradicate genetically parallel planktonic cells. Consequently, it has been well documented that biofilms endanger human health and cause billions of dollars of loss to industrial productivity, with particularly grave and persistent consequences in healthcare (Jefferson 2004), food processing industries (Galié et al. 2018), and most industrial water-based process, including drinking water treatment distribution systems and pulp and paper manufacturing (Muhammad et al. 2020).

However, it has recently been recognized that biofilms can have a highly constructive impact as well. The density of microbial cells in the biofilms is high enough to have the power to immobilize toxic organic hydrocarbons and inorganic metals. Biofilms also provide controlled growth conditions to the cells, needed for optimized production of the desired compound (Berlanga et al. 2012). Thus, biofilm systems are remarkably suitable for bioremediation of recalcitrant compounds from hazardous waste sites (Singh et al. 2006), biofiltering of industrial and municipal water and wastewater (Asri et al. 2019), and forming bio-barriers to efficiently *prevent* the propagation of *contaminants* towards the *groundwater* (EPA 2013). Also, the potential of biofilms for the industrial production of various fermentation products (Berlanga et al. 2012) is gaining interest. This book chapter summarizes some of these positive applications of biofilms as environmental and industrial materials to address emerging biotechnological problems.

7.2 Use of Biofilms in Bioremediation of Environment

7.2.1 Remediation of Organic Hydrocarbons

The pollution of the environment by noxious and persistent chemical contaminants like polycyclic aromatic hydrocarbons (PAHs) has increased during decades of industrialization. Amongst the numerous methods known for remediating contaminants, in situ bioremediation processes are efficient, ecologically benign, and economically viable. Moreover, the potential of biofilm cultures to be grown under controlled laboratory conditions as a bioremediation tool has recently been recognized. The biofilm cultures have a rapid pollutant degradation activity, which may perhaps be attributed to the high microbial density, robustness, superior adaptation, and survival (notably, stress endurance), and functional capabilities of the constituent cells encapsulated within the EPS matrix, which behave as a physical protective barrier for the cells (Singh et al. 2006). The microbial cells, naturally immobilized within the biofilms, also eliminate the requirement of artificial cell immobilization. Using biofilm for bioremediation is, therefore, an advanced and cost-effective method, which can be commercially deployed for the immobilization (by biosorption) and the biomineralization of recalcitrant organic compounds, xenobiotics, and toxic metals (Fida et al. 2012; Mitra and Mukhopadhyay 2016; Shukla et al. 2014; Singh et al. 2006).

A consortium of cultures isolated from an activated sludge sample and *Pseudomonas putida* grown in a rotating perforated tube biofilm reactor (RTBR) was used to remove a chlorinated aromatic hydrocarbon (2,4-dichlorophenol (DCP)) from synthetic wastewater, and nearly 100% DCP was removed (Kargi and Eker 2005). Similarly, a hydrogenotrophic biofilm of microbes isolated from sources like the anaerobic sludge from a swine wastewater treatment plant (Chang et al. 2004) and rhizosphere of *Phragmites australis* (Caldeira et al. 1999) was used to remove 2-chlorophenol (2-CP) and 4-chlorophenol (4-CP) with efficiencies of >90% in laboratory-scale bioreactors. On-site, pilot-scale degradation of trichlorophenolic compounds from contaminated groundwaters using biofilms of *Pseudomonas* sp. and *Rhodococcus* sp. has also been demonstrated (Puhakka et al. 1995). Other halogenated phenols whose mineralization has been shown using biofilms of *Pseudomonas* sp., *Sphingomonas* sp., *Rhodococcus* sp., *Alcaligenes* sp., *Caulobacter* sp., *Providentia* sp., and *Variovorax* sp. include benzopyrene, phenanthracene, 2-hydroxytoluene, naphthene, hemellitol, n-alkanes, toluene, and tetrachloromethane (Singh et al. 2006). A study conducted by Luke and Burton on phenol bioremediation showed that a fungal biofilm of *Neurospora crassa* was able to maintain its catalytic activity much longer (eight-fold) than the time noted for its planktonic counterpart, without degradation or decontamination (Luke and Burton 2001). These studies establish that microbial biomasses immobilized as biofilms can sustain their cell density and productivity for continuous bioremediation of organic pollutants from natural contaminated environments over considerable time.

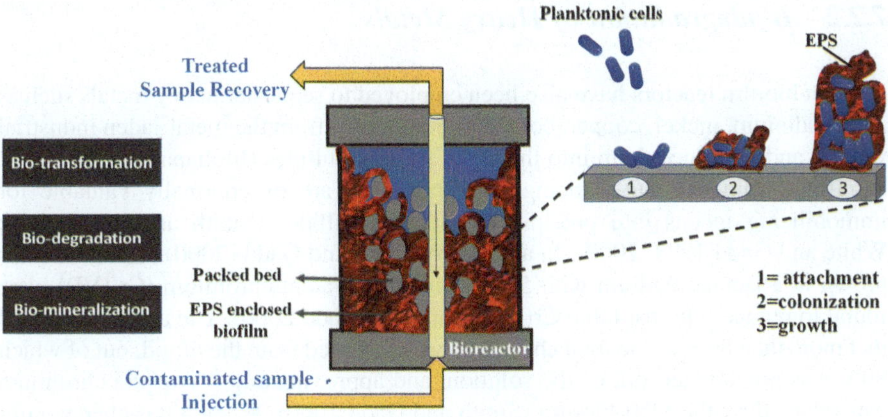

Fig. 7.1 Bioremediation of contaminated samples (soil/wastewater) by microbial biofilms in a bioreactor

Other than the halogenated and phenolic compounds, the bioremediation of phenoxy herbicides like 1-(3-chlorophenyl)piperazine and 2-(2,4-dichlorophenoxy) acetic acid was reported in a packed bioreactor with granular activated carbon (GAC) as a support matrix using a consortium of herbicide-degrading bacteria (Kye-Heon and Tuovinen 1994). Even the degradation of an azo dye, acid orange, from municipal wastewaters was reported by Zhang et al. (1995) using *Methylosinus trichosporium* in an RDBRs setup. Degradation efficiencies higher than 60% were achieved at all loading rates (Zhang et al. 1995). Synchronized biodegradation and adsorption of textile dye Everzol Turquoise Blue were examined using the white-rot fungus *Trametes versicolor* biofilm in an activated sludge set up and showed the highest color removal efficiency (82%) (Kapdan and Kargi 2002).

Biofilms are efficient in removing toxic pollutants from not only liquid contaminants but also from waste gases. This versatility of biofilm biomasses can be witnessed from the study by Hekmat et al. (2004) where a multispecies biofilm was assessed for the treatment of a mixture of aromatic hydrocarbons called Solvesso100 with an unsterilized bench-scale Trickle Bed reactor (Hekmat et al. 2004). In this study, the bioreactor was filled with 10% cells and 90% EPS secreted by these cells (Hekmat et al. 2004). The rhamnolipid surfactants contained in the EPS were said to provide the hydrophobic surfaces for further attachment and growth of biofilm cells. Further, the polyhydroxyalkanoates (PHAs) stored within the biofilm-EPS matrix acted as an energy reserve, providing buffer against fluctuating organic load in the pollutant gas feed. This maintained the performance of the reactor during the entire time course of pollutant's degradation.

Figure 7.1 illustrates that biofilm growth and metabolism in bioreactors attached to packing beads of some sort are essential considerations in the remediation of polycyclic aromatic hydrocarbons from soil and wastewaters.

7.2.2 Biodegradation of Heavy Metals

Distinct biofilm reactors have also been employed to sequester heavy metals such as zinc, cadmium, nickel, copper, cobalt, uranium, etc., from the metal-laden industrial wastes, and to reduce them into the metal form of sulfides (Mohapatra et al. 2020). Biofilm forming sulfate-reducing bacteria (SRB) are exceptionally valuable for immobilizing metals into precipitates of metal sulfides (Smith and Gadd 2000; White and Gadd 1998, 2000). In a study by Smith and Gadd (2000), SRBs biofilms grown in a lactate medium with 500 μmol/L hexavalent chromium (Cr (VI)) were found to reduce it to insoluble Cr(III) (Smith and Gadd 2000). The authors reported that more than 80% of the total chromium was removed from the liquid, out of which 80% was precipitated out of the solution, and approximately 8% of the chromium was retained by the SRB biofilm (Smith and Gadd 2000). When a flat-plate continuous-flow reactor containing SRB biofilms was fed with 126 μM U(VI), 88–96% of U was removed from the solution and immobilized in the biofilms (Beyenal et al. 2004). In another study, continuous cultures of SRB biofilms were shown to accumulate up to 200 μmol of cadmium (Cd) in the form of cadmium sulfide (CdS). Electron microscopy of the biofilm sections imaged the accumulation of CdS in the superficial coat of the biofilm, implying that the possible mechanism of Cd uptake was its entrapment (after precipitation) at the biofilm surface (White and Gadd 1998). The same research group later studied the bioprecipitation of copper ions as copper sulfide by SRB biofilms (White and Gadd 2000). In this study, a concurrent increase in the carbohydrates and protein content in the EPS layer of the biofilm was detected, which signaled the role of EPS in the entrapment of precipitated copper sulfide at the surface of the biofilms (White and Gadd 2000). Other bacterial strains that have been studied for the sequestering of heavy metals by its EPS includes *Pseudomonas aeruginosa* (P8) biofilms remediating zinc, lead (Meliani and Bensoltane 2016), and lanthanum (Langley and Beveridge 1999); *Pseudomonas putida* biofilms for elemental mercury (Wagner-Döbler et al. 2000); *Burkholderia cepacia* biofilms accumulating lead (Templeton et al. 2001), and *Rhodotorula mucilaginosa* biofilm for remediating mercury, copper, and lead, with an efficiency of up to 95% (Grujić et al. 2017).

Microbial outer coverings, as encapsulating cell walls or plasma membranes, generally adsorb metal ions on the surface via ionizable groups in their structure (carboxyl, hydroxyl, amino, and phosphate groups) (Singh et al. 2006). For instance, adsorption of polycrystalline uranium salts to monophosphate groups of the lipopolysaccharide (LPS) in the cell wall of *Citrobacter* sp. N14 was observed by Macaskie et al. (2000). The role of acid phosphatase, secreted and localized on the biofilm cells' outer membrane, was shown to further consolidate the uranium deposition by forming ammonium uranylphosphate ($NH_4UO_2PO_4$) film (Jeong et al. 1997; Macaskie et al. 2000). Labrenz et al. (2000) showed the bioremediation of zinc as sphalerite (ZnS) aggregates by aerotolerant SRB, *Desulfobacteriaceae* in a natural biofilm (Labrenz et al. 2000).

Recent studies suggest that some bacterial strains use their electrochemically active biofilms (EAB) as an electron exchanger (Erable et al. 2010). This electrocatalytic property of biofilms is becoming increasingly utilized in bioelectrochemical systems to drive electron transfer across electrodes for bioremediation, biosynthesis, and biohydrogen production (Kiran and Patil 2019). Li et al. (2008) used microbial fuel cell (MFC) to reduce 99% Cr^{6+} in real electroplating wastewater and generate 1600 mW/m^2 of electricity simultaneously (Li et al. 2008). The conductive biofilms of autotrophic and heterotrophic denitrifying bacteria as collaborative partners in an MFC have also been used for denitrification of the domestic sewage with >80% (Cong et al. 2013) and >97% efficiency (Zhao et al. 2011). In both these studies, the authors used an air-lift internal loop biofilm reactor with aeration set in the center of the reactor, which supported the growth of aerobic and anoxic biofilms simultaneously. This led to synchronized nitrification and denitrification in the same reactor with the total nitrogen removal percentage being <25% and >80% with axonic and anoxic biofilms, respectively (Zhang et al. 2013). Hence, the studies reviewed in this section illustrate that biofilms are essential catalysts for cleaning up contaminated environments. The use of electrochemically active biofilms can result in synergetic increases in heavy metals' remediation by manyfold.

7.3 Application of Biofilms for the Biosynthesis of Value-Added Products

Enclosed within the EPS matrix, biofilms maintain a balanced and controlled environment for their cell growth. Their extracellular matrix protects the bacteria against the extreme conditions of osmolarity, pH, temperature, and the presence of the toxic substance. This property of the biofilms has been exploited to maximize the fermentative production of various biological compounds by its member microbes.

7.3.1 Application of Biofilms for Production of Biofuels

A number of laboratory-scale studies illustrating the utility of biofilms of *Zymomonas mobilis* for ethanol production are available. Ethanol productivity of 105 g/L/h. was achieved in the attached film expanded bed (AFEB) fermenter, which had vermiculite as the support material for attachment of the *Zymomonas mobilis* biofilms. In this study, ethanol's productivity achieved with the *Z. mobilis* planktonic cells was <4 g/L/h (Bland et al. 1982). Krug and Daugulis (1983) achieved 377 g/L/h of ethanol productivity with 80% glucose conversion efficiency from a *Z. mobilis* biofilm on ion-exchange resin after 200 h of operation (Krug and Daugulis 1983).In another study, experiments were conducted with *Z. mobilis* ATCC 331821 in a packed-bed biofilm reactor with polypropylene and soybean hull-zein composites. The highest ethanol productivity achieved in the biofilm reactor was 536 g/L/h, outperforming the ethanol productivity of 5 g/L/h in suspension-culture reactors by 100-fold (Kunduru and Pometto 1996). In the same study, with an *S. cerevisiae* ATCC 24859 as the fermenting strain, the total ethanol production was <76 g/L/h, on the plastic composite-support with a 45% yield (Kunduru and Pometto 1996). Furthermore, with a similar concept of immobilizing cells on plastic composite supports (PCS), biofilms of *Actinobacillus succinogenes*, was used for succinic acid production. Here also, while the highest concentration of succinic acid was 10.4 g/L with the biofilm cells, the concentration was lower at 7 g/L with suspended-cell batch fermentations (Urbance et al. 2004). These results demonstrate that biofilm processes have the capability for greater productivity of biofuels than free-cell suspension processes. In yet another study, acetone and butanol's continuous production was shown from *Clostridium acetobutylicum* ATCC 824 biofilms attached to beechwood shavings in a glass reactor. In this study, the author showed that maintaining nutritional conditions in the reactor is essential for retaining a stable biofilm in the reactor (Förberg and Häggström 1985).

In principle, the use of lignocellulosic biomasses (LCB) presents the prospective for reduced biofuel production costs, as these substrates are cheap and available abundantly. Their use removes the competition for use of food crops like sugarcane, maize, and corn for the generation of fermentative products (Govil et al. 2020). However, LCB requires a pretreatment to achieve the desired yield of monomeric sugars for the microbial host to produce the biofuels, which generate numerous lignin derivatives such as furfural, hydroxymethylfurfural (HMF), levulinic acid, and certain phenolics. These co-compounds are inhibitory for microbial growth and often lower the biomass conversion efficiency (Govil et al. 2020). Recently, numerous detoxification methods have been tested to remove the inhibitory compounds from the LCB hydrolysates prior to their application in the fermentation process. Nevertheless, all these supplementary steps affect the final production cost. Hence, more recently, the concept of growing different viable cells as biofilms, within which the microbes encapsulated by EPS may tolerate the LCB inhibitors better than their planktonic counterparts, is gaining attention. Using this concept, strides were made when *Z. mobilis* grown as a biofilm was used to improve ethanol production from rice bran hydrolysate (RBH). The achieved ethanol concentration was higher than that obtained with its planktonic cells (Todhanakasem et al. 2014) and had a reported ethanol production of 13.40 ± 2.43 g/L, representing $72.47 \pm 6.13\%$ theoretical yield with the *Z. mobilis* ZM4, an obligatory fermentative alpha-proteobacterium biofilm, inoculated on RBH. Planktonic cells of ZM4 produced almost no ethanol (0.432 ± 0.29 g/L). In their later 2016 study, the same research group developed a composite of polystyrene and corn silk for improved growth of *Z. mobilis* TISTR 551 biofilms and obtained up to 0.51 g ethanol/g glucose consumed under batch fermentation (Todhanakasem et al. 2016). This study, therefore, suggests that fermentative microorganism biofilms could work as an efficient biocatalyst in the production of biofuels from lignocellulosic substrates, since biofilms can tolerate the toxic LCB inhibitors released in the hydrolysates or the fermentative broths. Biofilms also create an anaerobic environment, which aids in ethanol production.

In another study, a 2.2 L fluidized bed reactor was used for the two-phase production of ethanol by *Zymomonas mobilis* from unsterile hydrolyzed starch (Weuster-Botz et al. 1993). The results of this study indicated that the biofilm set up was able to operate stably for more than 4 months in an un-sterile environment of glucose from the hydrolyzed starch, yielding 50 g/L ethanol at a rate of 13 g/L/h (Weuster-Botz et al. 1993), demonstrating the reactors' resilience for long-term experimental studies, even with an unsterilized fermenting medium.

Furthermore, microbial biofilms are acknowledged to be more resistant to toxic amounts of ethanol in the fermentation broth, for which most of the ethanologenic strains are intolerant. Zhou et al. (2008) developed a recombinant ethanologenic *Escherichia coli* B strain KO11, whose biofilms were shown to have 2.3- to 15-fold higher survival rates than those of free suspended cells. With the biofilms of KO11 immobilized on porous glass microspheres in a fluidized bed, a stable ethanol yield of >85% was maintained for the duration of the fermentation (10 days) while, with the planktonic cells of KO11, the ethanol yield declined to <60% (Zhou et al. 2008). A similar approach was demonstrated by Qureshi et al. (2004) using recombinant *Escherichia coli* FBR-5 strain biofilms for ethanol production from pentose sugars (xylose) and corn fiber hydrolysates. The FBR-5 biofilms clay brick particles operated continuously for more than 80 days without hindrance, causing improved productivity of 2.21 g/L/h or a concentration of 27.7 g/L, whereas the ethanol concentration achieved in batch fermentation with free cells was only 0.28 g/L/h (Qureshi et al. 2004).

Structured packing of biofilms has also been recognized to aid in transformations that require a gas-liquid exchange. Hickey et al. synthesized ethanol, butanol, hexanol, acetic acid, and butyric acid from synthesis gas (a mixture of carbon monoxide and hydrogen) with >90% efficiency (Hickey et al. 2011). Here, the biofilm membrane reactor had anaerobic microorganisms supported as a concentrated layer on a microporous membrane, which facilitated diffusion of CO and H_2 from the gas side across from the membrane to the biofilm side, where the attached microbes transformed them into liquid products. More information about this invention is available in the related patent (US20090029434A1). The method is being commercialized by Synata Bio, Inc. (USA) as part of their Gas to Liquid (GLT) platform for converting syngas to liquids (Hickey et al. 2011).

The utility of biofilms for biofuel production has also been demonstrated by Zhang et al. (2019), who conducted continuous production of up to 57.6 mL/L/h hydrogens (yield of 1.8 ± 0.1 mol H_2/mol glucose) in a photobioreactor with a photosynthetic bacterium, *Rhodopseudomonas palustris* grown as a biofilm. In their study, the photobioreactor had a high surface-to-volume ratio that enriched the biofilm density besides enhancing the carbon and other mass transfer ratios (Zhang et al. 2019). Also, immobilization of cells within a biofilm prevented the cells from being washed off with the suspended broth, improved tolerance to the metabolic stress (e.g., pH, organic loading rates), and improved H_2 production stability during long-term operation (Zhang et al. 2019). Similar results were reported by Guo et al.

(2015) who used a biofilm photobioreactor with additional rough surfaces and equipped with an optical fiber as the light source to enhance photo-hydrogen production from *Rhodopseudomonas palustris* CQK 01. In this study, Guo reported enhanced H_2 production at a rate of 1.75 mmol/L/h and a yield of 1.3 mol H_2/mol glucose with reasonably steady long-term performance (Guo et al. 2011, 2015). These studies suggest that biofilm-based photobioreactors are promising for the scale-up of H_2 gas production due to the long-term stability of the production systems and yields reaching a theoretical maximum. Moreover, van Groenestijn et al. (2009) reestablished that biofilm reactors are useful for non-sterile laboratory and commercial applications. Here, the thermophilic bacterium *Caldicellulosiruptor saccharolyticus* produced 22 mmol of hydrogen/L of filter bed volume in 1 h (100 mol H_2/day) from sucrose in a non-axenic (microbiologically open) 400 L trickle bed biofilm reactor. The yield obtained in this study (2.8 mol H_2/mol hexose converted) is claimed to be much higher than that achieved in similar studies with a similar reactor (van Groenestijn et al. 2009). It thus established the feasibility of operating large volume biofilm reactors without sterilization at elevated temperatures, which is of great industrial significance.

Another novel application from this field comes from the successful design of an autotrophic biocathode where H_2 was produced as a sole product from an electroactive biofilm grown on carbon dioxide as the only carbon source (Jourdin et al. 2015). The electroactive bacterial biofilm grown on the cathode in this study was found to avoid the development of H_2-consuming microbes such as acetogenins and methanogens. This sustained the hydrogen-producing consortium's bio-electrocatalytic activity for >1 year on the biocathode with an estimated per day hydrogen yield of 9.2 L/m^2 (Jourdin et al. 2015). The microbial electrolysis (MEB) concept with an enriched microbial consortium sustained as a biofilm on the anode has also been shown to result in high hydrogen productivity of 9.35 L/day per L of switchgrass-derived bio-oil (Lewis et al. 2018). Studies such as these serve as a foundation for harnessing the capability of electroactive biofilms for hydrogen production from renewable carbon sources. Table 7.1 summarizes few studies that are discussed in Sect. 3.1, highlighting the reactor configurations along with yields or productivity obtained with biofilm vs. planktonic setups wherever applicable.

7.3.2 Application of Biofilms for Production of Biochemicals

Since biofilms are recognized for their superior resistance to antagonistic conditions, in contrast to free-living planktonic cells, Li et al. (2006) studied the ability of *Z. mobilis* biofilms to tolerate the toxic substrate benzaldehyde and bioprocess it to benzyl alcohol. The *Z. mobilis* as biofilm was found to tolerate up to 50 mM concentration of benzaldehyde for 1 h while retaining about 45% of their metabolic potential when planktonic cells were fully inactivated. When provided glucose in the medium, *Z. mobilis* biofilms in continuous fermentation converted 10 mM

Table 7.1 Summary of studies employing biofilms for the synthesis of biofuels

Product	Organism	Reactor	Bed material	Substrate	Productivity of the biofilm	Productivity of the planktonic cells	Reference
Ethanol	*Z. mobilis*	Attached film expanded bed (AFEB) fermenter	Vermiculite	Glucose	105 g/L/h	<4 g/L/h	Bland et al. (1982)
Ethanol	*Z. mobilis*	Column bioreactor	Ion-exchange resin XE 352 and IRA-938	Glucose	377 g/L/h	Not evaluated	Krug and Daugulis (1983)
Ethanol	*Z. mobilis* ATCC 331821	Trickling bed biofilm reactor	Polypropylene and soybean hull-zein composites	Glucose	536 g/L/h	5 g/L/h	Kunduru and Pometto (1996)
Ethanol	*S. cerevisiae* ATCC 24859	Packed biofilm reactor	Soybean hull and soybean flour composite	Glucose	76 g/L/h	5 g/L/h	Kunduru and Pometto (1996)
Ethanol	*Z. mobilis ZM4*	Batch reactor	Plastic surface	Rice bran hydrolysate	13.40 ± 2.43 g/L	0.432 ± 0.29 g/L	Todhanakasem et al. (2014, 2016)
Ethanol	*Z. mobilis* TISTR 551	Batch reactor	Composite of polystyrene and corn silk	Rice bran hydrolysate	0.40 ± 0.15 g ethanol/g glucose	0.02 ± 0.02 g ethanol/g glucose	Todhanakasem et al. (2016)
Ethanol	*Z. mobilis*	Fluidized bed reactor	Macroporous glass carriers	Unsterile hydrolyzed starch	13 g/L/h	Not evaluated	Weuster-Botz et al. (1993)
Ethanol	*Escherichia coli* B strain KO11	Vertical tubular fermenter operated as a liquid-fluidized bed	Porous glass microspheres	Xylose	85 g/L	60 g/L	Zhou et al. (2008)

(continued)

Table 7.1 (continued)

Product	Organism	Reactor	Bed material	Substrate	Productivity of the biofilm	Productivity of the planktonic cells	Reference
Ethanol	Recombinant *Escherichia coli* FBR-5	Continuous biofilm reactor	Clay brick	Corn fiber hydrolysate	2.21 g/L/h	0.28–0.90 g/L/h	Qureshi et al. (2004)
Butanol, Acetone and Ethanol	*Clostridium acetobutylicum* ATCC 824	Glass reactor	Beechwood shavings	Glucose	36.6 g/L/day with a product ratio of 6:3:1	Not evaluated	Förberg and Häggström (1985)
Ethanol, butanol, and other liquids	Anaerobic microorganisms	Membrane reactor	Membrane	Synthesis gas	Not available	Not available	Hickey et al. (2011)
Hydrogen	*Rhodopseudomonas palustris*	Alveolar panel photobioreactor	Polymethine methacrylate	Glucose	57.6 mL/L/h	24.36 mL/L/h	Zhang et al. (2019)
Hydrogen	*Rhodopseudomonas palustris* CQK 01	Photobioreactor	Glass beads	Glucose	1.75 mmoles/L/h or 63.6 mL/L/h or	0.74mmoles/L/h or 26.9 mL/L/h	Guo et al. (2015)
Hydrogen	Electroactive bacterial biofilm consortium	Graphite plates as a cathode electrode	Conductive carbon paint	Carbon-dioxide	9.2 L/m^2	Not evaluated	Jourdin et al. (2015)
Hydrogen	Electroactive bacterial biofilm consortium	Microbial electrolysis chamber	Not applicable	Switchgrass-derived bio-oil	9.35 L/day	Not evaluated	Lewis et al. (2018)
Hydrogen	*Caldicellulosiruptor saccharolyticus*	Non-axenic trickle bed reactor	Polyurethane foam	Sucrose	22 mmoles/L/h	Not evaluated	van Groenestijn et al. (2009)

benzaldehyde into 9 mM benzyl alcohol over a 45-h period, representing a yield of 90% (w/w) and volumetric productivity of 10 mM/h or 25.9 g/L/day (Li et al. 2006).

The robustness and utility of biofilms as catalysts for long-term yet stable bioconversion of biologically challenging toxic reactants was also displayed by Gross et al. (2007), where solvent-tolerant *Pseudomonas* sp. strain, VLB120DeltaC, was used for the biotransformation of styrene to (S)-styrene oxide in a tubular reactor (Gross et al. 2007). In this reaction, both the substrate (styrene) and product ((S)-styrene oxide) were volatile, which would not be as much toxic as the water-soluble compounds. Still, the styrene oxide production process in the continuous biofilm reactor with the *Pseudomonas* strain was stable for 60 days, yielding a maximal volumetric styrene oxide productivity of 16 g/L/day. This operational period of the reactor represented significant enhancement over a standard 10–50 h operation with a batch reactor for suspension cultures (Gross et al. 2007).

Glyoxylic acid is another high-value industrial chemical with disparate usage in food, pharmaceutical, cosmetics, and agricultural industries. Zhong Li (2012) studied the microbial production of glyoxylic acid from ethylene glycol by *Pseudomonas putida* JM37 in a trickle bed reactor. The operation was stable for over 2 months leading to steady-state productivity of up to 1.6 g/L/h (Zhong Li 2012). Overall, these studies establish the potential of biofilms as "bio-factories" for the continuous production of biologically challenging fine chemicals without biocatalyst degeneration or contamination (Zhong Li 2012).

Likewise, some organic acids whose continuous production has been documented using cultures grown as biofilms include succinic acid (Brink and Nicol 2014; Ferone et al. 2018; Longanesi et al. 2018; Salvachúa et al. 2016; Urbance et al. 2004), lactic acid (Cuny et al. 2019; Dagher et al. 2010; Demirci et al. 1993, 2003; Ho et al. 1997; Rangaswamy and Ramakrishna 2008; Schlegel et al. 2017), fumaric acid (Cao et al. 1996, 1997), and propionic acid (Ozadali 1997). In these studies, the authors achieved higher volumetric productivity of the respective acids, compared to that using a stirred vessel with cells in the suspended state. Since biofilms support high cell density fermentation by entrapping the cells disconnected from the liquid broth and concentrated in the fermenter (Brink and Nicol 2014), the improved stability of the cells in the biofilm reactors support continuous fermentation with a shortened lag phase (Ho et al. 1997), months of stability with no apparent fluctuations in both acids and biomass levels (Qureshi et al. 2005), as well presenting survival advantages to the cells in a toxic environment (Dagher et al. 2010). These factors make biofilms a useful biocatalyst for enhancing the industrial production economics of biobased platform chemicals.

7.3.3 Application of Biofilms for Production of Food Products

An example of the utility of biofilms in the food industry for the production of value-added products is the production of vinegar (or acetic acid) in a 60 m^3 trickle-bed reactor (TBR) using biofilms of acetic acid-producing bacteria grown on beechwood shavings (Tayyab 1990). In this work, *Acetobacter* was employed to oxidize an ethanolic solution to acetic acid in a TBR and an eventual vinegar concentration of 120 g/L has been reported (Tayyab 1990). According to the authors, vinegar produced by acetic acid bacteria in TBR is superior in its aroma relative to vinegar generated by the cells suspended in a submerged process (Tayyab 1990). Also, because TBRs do not need mechanical agitation, the total energy consumption in TBRs is lower than that required to run a continuous or batch fermenter. This has a supplementary gain of reducing the cost of vinegar produced, which has prompted the emergence of over 300 wood shavings-based trickle-bed plants in Eastern Europe, producing high-value, superior-flavored vinegar (Rosche et al. 2009).

The applicability of biofilms for the biosynthesis of vitamin (Menaquinone-7 (MK7) has been demonstrated by Mahdinia and his group (Mahdinia et al. 2019a). MK7, in its transform, is a high-value vitamin required in the diet for maintaining healthy bones and the cardiovascular system (Ravishankar et al. 2015). Individual food sources such as cheese, meat, and soybean have MK5, but at low concentrations (for example, 1.5 μg of MK-7 per 100 g of the cheese) (Mahdinia et al. 2019a). External supplementation of this vitamin in food products is quite expensive at approximately $1200/kg in a 0.1% formulation (Berenjian et al. 2015). Recently, cheaper production of trans MK-7 using bacterial hosts, especially members of genus *Bacillus subtilis,* via static fermentation operations in biofilm reactors has been acknowledged (Mahdinia et al. 2019a). Mahdinia's group produced 35.5 mg/L of MK-7 from glycerol using the *B. subtilis natto* biofilms, which was 2.3-fold higher than the concentration achieved in suspended-cell bioreactors (Mahdinia et al. 2017, 2018, 2019b). This research established the potential of biofilm reactors for static fermentative biosynthesis of vitamins.

Another example that exemplifies the usefulness of biofilms in the food sector comes from the production of "*kombucha,*" a traditional fermented tea with an acidic and effervescent taste. This tea is produced during the synergistic growth of a community of (acetic acid-producing) bacteria and (ethanol fermenting) yeast as a biofilm on a starter solution (sucrose-rich black or green tea). Initially, the yeast starts the reaction by producing an invertase enzyme that reduces sucrose to its monomeric components (glucose and fructose). Subsequently, while yeast utilizes these reduced sugars to synthesize ethanol, the acetic acid bacteria convert the produced ethanol into acetic acid. The acetic acid bacteria also produce other acids during their glucose and fructose consumption, namely, gluconic acid, with lesser proportions of tartaric, malic, and citric acid (May et al. 2019; Tran et al. 2020). This makes "*Kombucha*" a slightly carbonated, sweet, and sour drink, which also contains several amino acids, vitamins, antioxidants, and hydrolytic enzymes.

As with the role of biofilms in the making of *kombucha*, it has been stated that the participating bacterium in the consortium also generates a cellulose pellicle during its growth, which supports the development of a thick surface biofilm. The biofilm floats at the liquid-air interface and offers the following advantages: (a) gives shelter to the community from invasion by microbial competitors from the environment (Villarreal-Soto et al. 2018), (b) establishes an anaerobic surface barrier, assisting yeast in its anaerobic ethanol fermentation, while simultaneously providing greater access to oxygen for the bacterial community embedded within it (May et al. 2019), (c) acts as storage for resources, and (d) functions like a mother scaffold that promotes cross-kingdom interactions and ease of further inoculations (Tran et al. 2020). Thus, *kombucha* can serve as a model system highlighting the critical role biofilms can play in the food sector to produce products with manifold health benefits.

7.3.4 Application of Biofilms for Production of PHAs

Polyhydroxyalkanoates (PHAs) are intracellular bio-polyesters accumulated by some microorganisms as energy storage reserves and are generally produced during environmental stress conditions involving limited nutrients, especially nitrogen, but with excess carbon (Govil et al. 2020). Being an intracellular biopolymer, the content of PHA produced by a microbe is highly dependent on the dry cell mass. Biofilms can serve as an effective means of increasing the percentage of biomass attainable in a reaction under nutrient-limited conditions. Biofilms provide a natural surface for accumulating cells by immobilizing them, and its EPS barrier helps in maintaining the required nutrient-limited state. Khiyami et al. (2011) evaluated the efficiency of a *Bacillus* SA biofilm grown with date syrup as a low-cost carbon source for accumulating PHA (Khiyami et al. 2011). The production of PHB reached the maximum amount at around 30 h with 15% (v/v) date syrup. Cell density was 7.3 g/L with a PHB content of 70.5% (w/w) (Khiyami et al. 2011), compared with a relative cell density and PHA content of 6.4 g/L and 63.6% (w/w), respectively, when *Bacillus* SA was grown as suspended cells. The somewhat higher cell dry mass and PHA accumulation from date-syrup grown *Bacillus* SA may arise from the biofilm's resistance to the effects of toxic compounds such as acetic acid, propionic acid, butyric acid, and formic acid released during the hydrolysis of the date syrup (Khiyami et al. 2011). This study implies that PHA-producing biofilms may resist inhibitors released during lignocellulosic biomass biodegradation, providing a possible strategy to enhance the cell growth and PHA accumulation of microbes utilizing cheap and abundantly available agricultural raw materials as carbon sources.

Another example that illustrates the relevance of biofilms for the production of PHAs comes from studies conducted with low-temperature psychrophiles. It has earlier been shown that PHA accumulation increases the fitness for survival during growth at low temperatures (Koller 2017; Obruca et al. 2016). Tribelli and López

(2011) carried out experiments with *Pseudomonas extremaustralis* sp. grown on sodium octanoate and reported that PHA accumulation and EPS formation grow in parallel under cold conditions. The *Pseudomonas sp. strain* cells protected by biofilms accumulated more PHAs while producing enhanced biosurfactants (Tribelli et al. 2012; Tribelli and López 2011). The authors did not report PHA yield or productivity in their studies, but such studies demonstrate that the biofilm lifestyle of benthic communities of psychrophiles might provide a vital source for isolating high PHA-producing candidates. It is expected that similar studies in the future, with other extremophilic biofilm cultures, will prove fruitful.

7.3.5 Application of Biofilms for Production of Enzymes

The enzymes used for the delignification of lignocellulosic substrates include laccase, peroxidases (manganese peroxidase, lignin peroxidase, dye-decolorizing peroxidase, and versatile peroxidase), H_2O_2-generating oxidases (glucose oxidase, methanol oxidase, aryl alcohol oxidase, and glyoxal oxidase), etc. and have wide biotechnological applications. These enzymes can also degrade diverse dyes and xenobiotic compounds, making them also highly suited to application in bioremediation of the environment (Mendonça Maciel et al. 2010). Considering the demand for these enzymes in many industries, and their environmental benefits, an efficient system for scaling up their production is of great research interest. Khiyami et al. (2006) evaluated one such configuration where a stirred tank biofilm reactor was used to increase the production of ligninolytic enzymes in *Phanerochaete chrysosporium,* a fungal host (Khiyami et al. 2006). In their study, the group reported the highest lignin peroxidase (LiP) and manganese peroxidase (MnP) activity to be 47.3 U/L and 63.2 U/L, respectively. These activities were achieved in the early growth phase of the fungus; the sixth day for LiP, and the third day for MnP. By contrast, during batch culture operations, the group achieved LiP activity of 30.0 U/L and MnP activity of 49.3 U/L. Hence, this study established the utility of biofilm configurations for higher enzyme production levels, with the additional possibility of decreasing the fermentation time, which could lead to substantial cost savings for the industry. Table 7.2 lists some of the studies employing biofilms for the synthesis of other value-added products.

7.4 Application of Biofilms in Agriculture

Interaction between plants and associated microbes (fungi/bacteria) is very much dependent on the physical interaction between the two systems, and rhizobium-forming bacteria (rhizobacteria) have been known to stimulate plant growth as well protect them against soilborne pathogens (Ghiasian 2020). Rhizobacteria establish a symbiotic relationship with their plant host by forming biofilms that enable root

Table 7.2 Summary of studies employing biofilms for the synthesis of other value-added products

Product	Organism	Reactor	Bed Material	Substrate	Productivity of the biofilm	Productivity of the planktonic cells	Reference
Benzyl alcohol	*Z. mobilis*	Packed bed bioreactor	Glass beads	Benzaldehyde	10 mM/h or 25.9 g/L/day	4.9 mM/h or 12.7 g/L/day	Li et al. (2006)
Styrene oxide	*Pseudomonas* sp. strain VLB120DeltaC	Tubular glass reactor	Silicon tubing	Styrene	16 g/L/day	Not evaluated	Gross et al. (2007)
Glyoxylic acid	*Pseudomonas putida* JM37	Trickle bed reactor	Plastic composite support	Ethylene glycol	1.6 g/L/h	Not evaluated	Zhong Li (2012)
Succinic acid	*Actinobacillus succinogenes*	Continuous bioreactor	Composite of polypropylene (PP), soybean hulls, bovine albumin, soybean flour, yeast extract, dried red blood cells, and peptone	Glucose	10.4 g/L	7 g/L	Urbance et al. (2004)
Pyruvic acid	Recombinant *Acetobacter xylinum*	Stirred vessel (SC)	Self-produced cellulose nanofibers	D-Alanine (amino acid -substrate?)	0.14 mM/mM	0.39 mM/mM	Setyawati et al. (2009)
Dihydroxy -acetone	*Gluconobacter oxydans*	Packed bed bubble col-umn bioreactor	Silicone-coated Ralurings	Glycerol	0.85 kg/kg	0.87 kg/kg	Hekmat et al. (2003)
Vinegar (or acetic acid)	Mixed consortia	Trickle-bed reactor	Beechwood shavings	Beechwood shavings	120 g/L	Not evaluated	Tayyab (1990)
Vitamin (Menaquinone-7)	*Bacillus subtilis* natto	Biofilm reactor	Plastic composite support	Glucose	35.5 mg/L	15.43 mg/L	Mahdinia et al. (2017, 2018, 2019a, b)

(continued)

Table 7.2 (continued)

Product	Organism	Reactor	Bed Material	Substrate	Productivity of the biofilm	Productivity of the planktonic cells	Reference
Nisin	*Lactococcus lactis* subsp. *lactis* ATCC 11454	Packed bed bioreactor	Cotton fabric	Lactose Whey permeate	2.6×10^4 AU/mL 5.1×10^4 AU/mL	Not evaluated	Liu et al. (2005)
PHA	*Bacillus* SA	Bioflow 10 benchtop fermenter	Plastic composite support tubes	Date syrup	70.5% (w/w)	63.6%	Khiyami et al. (2011)
PHA	*Pseudomonas extremaustralis* sp.	Microtiter plate	Polystyrene	Sodium octanoate	Not evaluated	Not evaluated	Tribelli et al. (2012), Tribelli and López (2011)
Ligninolytic enzymes	*Phanerochaete chrysosporium*	Stirred tank biofilm reactor	Plastic composite support	Veratryl alcohol	47.3–63.2 U/mL	30–49.3 U/mL	Khiyami et al. (2006)

attachment, nodule formation, root hair development, and curling. Rhizobacteria as biofilms increase cell aggregation and boost nutrient cycling. The EPS matrix also enhances the endurance of the bacterial cells to drought situations and lengthens their survivability and residence time in soils (Ghiasian 2020; Pandit et al. 2020). Such naturally active biofilms, attached to plant roots, can improve crop ecosystems, crop quality, and protect the host plant against various environmental stresses and pathogenic attacks (Pandit et al. 2020).

Biofilms with biofertilizer capability can fix atmospheric nitrogen, mobilize minerals such as phosphorus, increase soil carbon content, and enhance the production of growth hormones that regulate and promote plant growth (Velmourougane et al. 2017). Studies have shown that the use of biobased fertilizers could reduce the use of chemical fertilizers by 50% in many crops under field conditions, together with several additional beneficial roles considered necessary for the agroecosystem (Seneviratne and Kulasooriya 2013). More details on the present status and prospects associated with biofilms in agriculture can be found in some recently published review articles on the topic (Ghiasian 2020; Pandit et al. 2020; Velmourougane et al. 2017). It is suggested that in the future, microbial biofilm-based agents should be incorporated as next-generation products for sustainable agricultural practices in crop nutrition, crop protection, soil quality improvement, and bioremediation programs.

7.5 Conclusion

Interest in biofilms in biotechnology and bio-based industries has grown significantly over recent years. Several studies report the benefits associated with biofilms as bioremediation agents for recalcitrant organic pollutants, xenobiotics, toxic metals, and wastewater treatment. Biofilms can also provide important biocatalysts for enhancing the industrial economics of fine-chemical production. This high-value application arises from the distinct capability of biofilms to enable (a) greater productivity of fermentative products than free-cell suspension processes; (b) long-term, stable, and continuous fermentation studies and production, even using unsterilized fermenting media; and (c) higher tolerance of toxic amounts of inhibitors and solvents than their planktonic counterparts. Biofilm reactors have the additional benefit of aiding downstream processing. Therefore, microbial biofilms in a bioreactor may serve as "bio-factories" for continuous production of biologically challenging biobased platform biofuels, biochemicals, bioplastics (e.g. PHA), as well as enzymes, without biocatalyst degeneration or contamination. Various factors play an important role in realizing "biofilm factories," such as selecting a suitable biofilm-forming microbial strain, developing new support materials, designing suitable reactors, and establishing operating conditions for ideal mass transfer and productivity. Overall, biofilm-based cell factories offer excellent systems for scientific inquiry due to their direct applicability to environmental science, biotechnology, health, and industry. Increasing awareness among biotechnologists of the countless applications of beneficial biofilms for the execution of highly efficient processes, promises to ensure their transition to widespread utilization in the sustainable production of value-added products.

Acknowledgments We acknowledge support from the CNAM-Bio-Center and the South Dakota Governor's Office of Economic Development. The authors also gratefully acknowledge support from the National Science Foundation (Award #1736255, #1849206, and #1920954) and the Department of Chemical and Biological Engineering at the South Dakota Mines, USA.

Conflicts of Interest The authors declare no conflict of interest.

References

Asri M, Elabed S, Ibnsouda Koraichi S, El Ghachtouli N (2019) Biofilm-based systems for industrial wastewater treatment. In: Hussain CM (ed) Handbook of environmental materials management. Springer International Publishing, Cham, pp 1767–1787

Berenjian A, Mahanama R, Kavanagh J, Dehghani F (2015) Vitamin K series: current status and future prospects. Crit Rev Biotechnol 35(2):199–208

Berlanga M, Minana-Galbis D, Domenech O, Guerrero R (2012) Enhanced polyhydroxyalkanoates accumulation by Halomonas spp. in artificial biofilms of alginate beads. Int Microbiol 15 (4):191–199

Beyenal H, Sani RK, Peyton BM, Dohnalkova AC, Amonette JE, Lewandowski Z (2004) Uranium immobilization by sulfate-reducing biofilms. Environ Sci Technol 38(7):2067–2074

Bland RR, Chen HC, Jewell WJ, Bellamy WD, Zall RR (1982) Continuous high rate production of ethanol by Zymomonas mobilis in an attached film expanded bed fermentor. Biotechnol Lett 4 (5):323–328

Brink HG, Nicol W (2014) Succinic acid production with Actinobacillus succinogenes: rate and yield analysis of chemostat and biofilm cultures. Microb Cell Factories 13(1):111

Bruce ER, Perry LM (2001) Environmental biotechnology: principles and applications. McGraw-Hill Education, New York

Caldeira M, Heald SC, Carvalho MF, Vasconcelos I, Bull AT, Castro PML (1999) 4-Chlorophenol degradation by a bacterial consortium: development of a granular activated carbon biofilm reactor. Appl Microbiol Biotechnol 52(5):722–729

Cao N, Du J, Gong CS, Tsao GT (1996) Simultaneous production and recovery of fumaric acid from immobilized Rhizopus oryzae with a rotary biofilm contactor and an adsorption column. Appl Environm Microbiol 62(8):2926–2931

Cao N, Du J, Chen C, Gong CS, Tsao GT (1997) Production of fumaric acid by immobilized rhizopus using rotary biofilm contactor. Appl Biochem Biotechnol 63–65:387–394

Chang C-C, Tseng S-K, Chang C-C, Ho C-M (2004) Degradation of 2-chlorophenol via a hydrogenotrophic biofilm under different reductive conditions. Chemosphere 56(10):989–997

Cong Y, Xu Q, Feng H, Shen D (2013) Efficient electrochemically active biofilm denitrification and bacteria consortium analysis. Bioresour Technol 132:24–27

Cuny L, Pfaff D, Luther J, Ranzinger F, Ödman P, Gescher J, Guthausen G, Horn H, Hille-Reichel A (2019) Evaluation of productive biofilms for continuous lactic acid production. Biotechnol Bioeng 116(10):2687–2697

Dagher SF, Ragout AL, Siñeriz F, Bruno-Bárcena JM (2010) Cell immobilization for production of lactic acid biofilms do it naturally. Adv Appl Microbiol 71:113–148

Demirci A, Pometto AL, Johnson KE (1993) Lactic acid production in a mixed-culture biofilm reactor. Appl Environ Microbiol 59(1):203–207

Demirci A, Cotton JC, Pometto AL 3rd, Harkins KR, Hinz PN (2003) Resistance of Lactobacillus casei in plastic-composite-support biofilm reactors during liquid membrane extraction and optimization of the lactic acid extraction system. Biotechnol Bioeng 83(7):749–759

EPA (2013) Introduction to in situ bioremediation of groundwater. Office of Solid Waste and Emergency Response. EPA 542-R-13-018, United States Environmental Protection Agency

Erable B, Duţeanu NM, Ghangrekar MM, Dumas C, Scott K (2010) Application of electro-active biofilms. Biofouling 26(1):57–71

Ferone M, Raganati F, Ercole A, Olivieri G, Salatino P, Marzocchella A (2018) Continuous succinic acid fermentation by Actinobacillus succinogenes in a packed-bed biofilm reactor. Biotechnol Biofuels 11(1):138

Fida TT, Breugelmans P, Lavigne R, Coronado E, Johnson DR, van der Meer JR, Mayer AP, Heipieper HJ, Hofkens J, Springael D (2012) Exposure to solute stress affects genome-wide expression but not the polycyclic aromatic hydrocarbon-degrading activity of Sphingomonassp. Strain LH128 in biofilms. Appl Environ Microbiol 78(23):8311–8320

Flemming H-C, Neu TR, Wozniak DJ (2007) The EPS matrix: the "house of biofilm cells". J Bacteriol 189(22):7945–7947

Förberg C, Häggström L (1985) Control of cell adhesion and activity during continuous production of acetone and butanol with adsorbed cells. Enzym Microb Technol 7(5):230–234

Galié S, García-Gutiérrez C, Miguélez EM, Villar CJ, Lombó F (2018) Biofilms in the food industry: health aspects and control methods. Front Microbiol 9:898

Ghiasian M (2020) Chapter 9—Microbial biofilms: beneficial applications for sustainable agriculture. In: Rastegari AA, Yadav AN, Yadav N, editors. New and future developments in microbial biotechnology and bioengineering. Elsevier, pp 145–155

Govil T, Sharma W, Chauhan NK, Kumar S, Salem DR, Sani RK (2019) "MINES" method for genomic DNA extraction from deep biosphere biofilms. J Microbiol Methods 167:105730

Govil T, Wang J, Samanta D, David A, Tripathi A, Rauniyar S, Salem DR, Sani RK (2020) Lignocellulosic feedstock: a review of a sustainable platform for cleaner production of nature's plastics. J Clean Prod 270:122521

Gross R, Hauer B, Otto K, Schmid A (2007) Microbial biofilms: new catalysts for maximizing productivity of long-term biotransformations. Biotechnol Bioeng 98(6):1123–1134

Grujić S, Vasić S, Radojević I, Čomić L, Ostojić A (2017) Comparison of the Rhodotorula mucilaginosa biofilm and planktonic culture on heavy metal susceptibility and removal potential. Water Air Soil Pollution 228(2):73

Guo CL, Zhu X, Liao Q, Wang YZ, Chen R, Lee DJ (2011) Enhancement of photo-hydrogen production in a biofilm photobioreactor using optical fiber with additional rough surface. Bioresour Technol 102(18):8507–8513

Guo CL, Cao HX, Guo FQ, Huang CL, Wang HG, Rao ZH (2015) Enhanced photo-H2 production by unsaturated flow condition in continuous culture. Biotechnol Lett 37(2):359–366

Hall CW, Mah TF (2017) Molecular mechanisms of biofilm-based antibiotic resistance and tolerance in pathogenic bacteria. FEMS Microbiol Rev 41(3):276–301

Hekmat D, Bauer R, Fricke J (2003) Optimization of the microbial synthesis of dihydroxyacetone from glycerol with Gluconobacter oxydans. Bioprocess Biosyst Eng 26(2):109–116

Hekmat D, Feuchtinger A, Stephan M, Vortmeyer D (2004) Biofilm population dynamics in a trickle-bed bioreactor used for the biodegradation of aromatic hydrocarbons from waste gas under transient conditions. Biodegradation 15(2):133–144

Hickey R, Datta R, Tsai SP, Basu R (2011) Membrane supported bioreactor for conversion of syngas components to liquid products

Ho KL, Pometto AL 3rd, Hinz PN (1997) Optimization of L-(+)-lactic acid production by ring and disc plastic composite supports through repeated-batch biofilm fermentation. Appl Environ Microbiol 63(7):2533–2542

Jachlewski S, Jachlewski WD, Linne U, Bräsen C, Wingender J, Siebers B (2015) Isolation of extracellular polymeric substances from biofilms of the Thermoacidophilic archaeon Sulfolobus acidocaldarius. Front Bioeng Biotechnol 3:123

Jefferson KK (2004) What drives bacteria to produce a biofilm? FEMS Microbiol Lett 236 (2):163–173

Jeong BC, Hawes C, Bonthrone KM, Macaskie LE (1997) Localization of enzymically enhanced heavy metal accumulation by Citrobacter sp. and metal accumulation in vitro by liposomes containing entrapped enzyme. Microbiology 143(7):2497–2507

Jourdin L, Freguia S, Donose BC, Keller J (2015) Autotrophic hydrogen-producing biofilm growth sustained by a cathode as the sole electron and energy source. Bioelectrochemistry 102:56–63

Kapdan IK, Kargi F (2002) Simultaneous biodegradation and adsorption of textile dyestuff in an activated sludge unit. Process Biochem 37(9):973–981

Kargi F, Eker S (2005) Removal of 2,4-dichlorophenol and toxicity from synthetic wastewater in a rotating perforated tube biofilm reactor. Process Biochem 40(6):2105–2111

Khiyami MA, Pometto AL, Kennedy WJ (2006) Ligninolytic enzyme production by Phanerochaete chrysosporium in plastic composite support biofilm stirred tank bioreactors. J Agric Food Chem 54(5):1693–1698

Khiyami MA, Al-Fadual SM, Bahklia AH (2011) Polyhydroxyalkanoates production via Bacillus PCS biofilm and date palm syrup. J Med Plant Res 5(14):3312–3320

Kiran R, Patil SA (2019) Microbial electroactive biofilms. In: Introduction to biofilm engineering, vol. 1323. American Chemical Society, pp 159–186

Koller M (2017) Production of poly hydroxyalkanoate (PHA) biopolyesters by extremophiles. MOJ Poly Sci 1(2):69–85

Krug TA, Daugulis AJ (1983) Ethanol production using Zymomonas mobilis immobilized on an ion exchange resin. Biotechnol Lett 5(3):159–164

Kunduru MR, Pometto AL (1996) Continuous ethanol production by Zymomonas mobilis and Saccharomyces cerevisiae in biofilm reactors. J Indus Microbiol 16(4):249–256

Kye-Heon O, Tuovinen OH (1994) Biodegradation of the phenoxy herbicides MCPP and 2,4-D in fixed-film column reactors. Int Biodeterior Biodegradation 33(1):93–99

Labrenz M, Druschel GK, Thomsen-Ebert T, Gilbert B, Welch SA, Kemner KM, Logan GA, Summons RE, Stasio GD, Bond PL, Lai B, Kelly SD, Banfield JF (2000) Formation of sphalerite (ZnS) deposits in natural biofilms of sulfate-reducing bacteria. Science 290 (5497):1744–1747

Langley S, Beveridge TJ (1999) Metal binding by Pseudomonas aeruginosa PAO1 is influenced by growth of the cells as a biofilm. Can J Microbiol 45(7):616–622

Lewis AJ, Campa MF, Hazen TC, Borole AP (2018) Unravelling biocomplexity of electroactive biofilms for producing hydrogen from biomass. Microb Biotechnol 11(1):84–97

Li XZ, Webb JS, Kjelleberg S, Rosche B (2006) Enhanced benzaldehyde tolerance in Zymomonas mobilis biofilms and the potential of biofilm applications in fine-chemical production. Appl Environ Microbiol 72(2):1639–1644

Li Z, Zhang X, Lei L (2008) Electricity production during the treatment of real electroplating wastewater containing Cr6+ using microbial fuel cell. Process Biochem 43(12):1352–1358

Liu X, Chung Y-K, Yang S-T, Yousef AE (2005) Continuous nisin production in laboratory media and whey permeate by immobilized Lactococcus lactis. Process Biochem 40(1):13–24

Longanesi L, Frascari D, Spagni C, DeWever H, Pinelli D (2018) Succinic acid production from cheese whey by biofilms of Actinobacillus succinogenes: packed bed bioreactor tests. J Chem Technol Biotechnol 93(1):246–256

Luke AK, Burton SG (2001) A novel application for Neurospora crassa: Progress from batch culture to a membrane bioreactor for the bioremediation of phenols. Enzym Microb Technol 29 (6):348–356

Macaskie LE, Bonthrone KM, Yong P, Goddard DT (2000) Enzymically mediated bioprecipitation of uranium by a Citrobacter sp.: a concerted role for exocellular lipopolysaccharide and associated phosphatase in biomineral formation. Microbiology 146(8):1855–1867

Mahdinia E, Demirci A, Berenjian A (2017) Strain and plastic composite support (PCS) selection for vitamin K (Menaquinone-7) production in biofilm reactors. Bioprocess Biosyst Eng 40 (10):1507–1517

Mahdinia E, Demirci A, Berenjian A (2018) Implementation of fed-batch strategies for vitamin K (menaquinone-7) production by Bacillus subtilis natto in biofilm reactors. Appl Microbiol Biotechnol 102(21):9147–9157

Mahdinia E, Demirci A, Berenjian A (2019a) Biofilm reactors as a promising method for vitamin K (menaquinone-7) production. Appl Microbiol Biotechnol 103(14):5583–5592

Mahdinia E, Demirci A, Berenjian A (2019b) Effects of medium components in a glycerol-based medium on vitamin K (menaquinone-7) production by Bacillus subtilis natto in biofilm reactors. Bioprocess Biosyst Eng 42(2):223–232

May A, Narayanan S, Alcock J, Varsani A, Maley C, Aktipis A (2019) Kombucha: a novel model system for cooperation and conflict in a complex multi-species microbial ecosystem. PeerJ 7: e7565

Meliani A, Bensoltane A (2016) Biofilm-mediated heavy metals bioremediation in PGPR Pseudomonas. J Bioremed Biodegr 7:370

Mendonça Maciel M, Castro e Silva A, Camarão Telles Ribeiro H (2010) Industrial and biotechnological applications of ligninolytic enzymes of the basidiomycota: a review. Electron J Biotechnol, 13(6)

Mitra A, Mukhopadhyay S (2016) Biofilm mediated decontamination of pollutants from the environment. AIMS Bioengineering 3(1):44–59

Mohapatra RK, Behera SS, Patra JK, Thatoi H, Parhi PK (2020) Chapter 17—Potential application of bacterial biofilm for bioremediation of toxic heavy metals and dye-contaminated environments. In: Yadav MK, Singh BP, editors. New and future developments in microbial biotechnology and bioengineering: microbial biofilms. Elsevier, pp 267–281

Muhammad MH, Idris AL, Fan X, Guo Y, Yu Y, Jin X, Qiu J, Guan X, Huang T (2020) Beyond risk: bacterial biofilms and their regulating approaches. Front Microbiol 11:928

Obruca S, Sedlacek P, Krzyzanek V, Mravec F, Hrubanova K, Samek O, Kucera D, Benesova P, Marova I (2016) Accumulation of poly(3-hydroxybutyrate) helps bacterial cells to survive freezing. PLoS One 11(6):e0157778

Ozadali F (1997) Development of a biofilm bioreactor for enhanced propionic and acetic acid production. vol. Doctor of Philosophy, Iowa State University

Pandit A, Adholeya A, Cahill D, Brau L, Kochar M (2020) Microbial biofilms in nature: unlocking their potential for agricultural applications. J Appl Microbiol 129(2):199–211

Puhakka JA, Melin ES, Järvinen KT, Koro PM, Rintala JA, Hartikainen P, Shieh WK, Ferguson JF (1995) Fluidized-bed biofilms for chlorophenol mineralization. Water Sci Technol 31 (1):227–235

Qureshi N, Brining HR, Iten LB, Dien BS, Nichols NN, Saha BC, Cotta MA (2004) Adsorbed cell dynamic biofilm reactor for ethanol production from xylose and corn fiber hydrolysate [abstract]. In: Great Lakes Regional American Chemical Society Symposium, p 179

Qureshi N, Annous BA, Ezeji TC, Karcher P, Maddox IS (2005) Biofilm reactors for industrial bioconversion processes: employing potential of enhanced reaction rates. Microb Cell Factories 4:24–24

Rangaswamy V, Ramakrishna SV (2008) Lactic acid production by Lactobacillus delbrueckii in a dual reactor system using packed bed biofilm reactor. Lett Appl Microbiol 46(6):661–666

Ravishankar B, Dound YA, Mehta DS, Ashok BK, de Souza A, Pan M-H, Ho C-T, Badmaev V, Vaidya ADB (2015) Safety assessment of menaquinone-7 for use in human nutrition. J Food Drug Anal 23(1):99–108

Rosche B, Li XZ, Hauer B, Schmid A, Buehler K (2009) Microbial biofilms: a concept for industrial catalysis? Trends Biotechnol 27(11):636–643

Salvachúa D, Smith H, St. John PC, Mohagheghi A, Peterson DJ, Black BA, Dowe N, Beckham GT (2016) Succinic acid production from lignocellulosic hydrolysate by Basfia succiniciproducens. Bioresour Technol 214:558–566

Schlegel C, Chodorski J, Huster M, Davoudi N, Huttenlochner K, Bohley M, Reichenbach I, Buhl S, Breuninger P, Müller-Renno C, Ziegler C, Aurich J, Antonyuk S, Ulber R (2017)

Analyzing the influence of microstructured surfaces on the lactic acid production of Lactobacillus delbrueckii lactis in a flow-through cell system. Eng Life Sci 17(8):865–873
Seneviratne G, Kulasooriya SA (2013) Reinstating soil microbial diversity in agroecosystems: the need of the hour for sustainability and health. Agric Ecosyst Environ 164:181–182
Setyawati MI, Chien L-J, Lee C-K (2009) Self-immobilized recombinant Acetobacter xylinum for biotransformation. Biochem Eng J 43(1):78–84
Shukla SK, Mangwani N, Rao TS, Das S (2014) Chapter 8—Biofilm-mediated bioremediation of polycyclic aromatic hydrocarbons. In: Das S (ed) Microbial biodegradation and bioremediation. Elsevier, Oxford, pp 203–232
Singh R, Paul D, Jain RK (2006) Biofilms: implications in bioremediation. Trends Microbiol 14 (9):389–397
Smith WL, Gadd GM (2000) Reduction and precipitation of chromate by mixed culture sulphate-reducing bacterial biofilms. J Appl Microbiol 88(6):983–991
Tayyab S (1990) Biotechnology: a textbook of industrial microbiology: by Wulf Crueger and Anneliese Crueger. Editor of the english edition: Thomas D Brock. pp 357. Sinauer associates, Sunderland, MA. 1990 ISBN 0-87893-131-7. Biochem Educ 18(3):151–152
Templeton AS, Trainor TP, Traina SJ, Spormann AM, Brown GE (2001) Pb(II) distributions at biofilm–metal oxide interfaces. Proc Natl Acad Sci U S A 98(21):11897–11902
Todhanakasem T, Sangsutthiseree A, Areerat K, Young GM, Thanonkeo P (2014) Biofilm production by Zymomonas mobilis enhances ethanol production and tolerance to toxic inhibitors from rice bran hydrolysate. New Biotechnol 31(5):451–459
Todhanakasem T, Tiwari R, Thanonkeo P (2016) Development of corn silk as a biocarrier for Zymomonas mobilis biofilms in ethanol production from rice straw. J Gen Appl Microbiol 62 (2):68–74
Tran T, Grandvalet C, Verdier F, Martin A, Alexandre H, Tourdot-Maréchal R (2020) Microbial dynamics between yeasts and acetic acid bacteria in Kombucha: impacts on the chemical composition of the beverage. Foods 9(7)
Tribelli PM, López NI (2011) Poly(3-hydroxybutyrate) influences biofilm formation and motility in the novel Antarctic species *Pseudomonas extremaustralis* under cold conditions. Extremophiles 15(5):541–547
Tribelli PM, Di Martino C, López NI, Raiger Iustman LJ (2012) Biofilm lifestyle enhances diesel bioremediation and biosurfactant production in the Antarctic polyhydroxyalkanoate producer *Pseudomonas extremaustralis*. Biodegradation 23(5):645–651
Urbance SE, Pometto AL 3rd, Dispirito AA, Denli Y (2004) Evaluation of succinic acid continuous and repeat-batch biofilm fermentation by Actinobacillus succinogenes using plastic composite support bioreactors. Appl Microbiol Biotechnol 65(6):664–670
van Groenestijn JW, Geelhoed JS, Goorissen HP, Meesters KPM, Stams AJM, Claassen PAM (2009) Performance and population analysis of a non-sterile trickle bed reactor inoculated with Caldicellulosiruptor saccharolyticus, a thermophilic hydrogen producer. Biotechnol Bioeng 102 (5):1361–1367
Velmourougane K, Prasanna R, Saxena AK (2017) Agriculturally important microbial biofilms: present status and future prospects. J Basic Microbiol 57(7):548–573
Villarreal-Soto SA, Beaufort S, Bouajila J, Souchard JP, Taillandier P (2018) Understanding Kombucha tea fermentation: a review. J Food Sci 83(3):580–588
Wagner-Döbler I, Lünsdorf H, Lübbehüsen T, von Canstein HF, Li Y (2000) Structure and species composition of mercury-reducing biofilms. Appl Environ Microbiol 66(10):4559–4563
Weuster-Botz D, Aivasidis A, Wandrey C (1993) Continuous ethanol production by Zymomonas mobilis in a fluidized bed reactor. Part II: process development for the fermentation of hydrolysed B-starch without sterilization. Appl Microbiol Biotechnol 39(6):685–690
White C, Gadd GM (1998) Accumulation and effects of cadmium on sulphate-reducing bacterial biofilms. Microbiology 144(5):1407–1415
White C, Gadd GM (2000) Copper accumulation by sulfate-reducing bacterial biofilms. FEMS Microbiol Lett 183(2):313–318

Zhang TC, Fu YC, Bishop PL, Kupferle M, FitzGerald S, Jiang HH, Harmer C (1995) Transport and biodegradation of toxic organics in biofilms. J Hazard Mater 41(2):267–285
Zhang C, Wang L, Yan N, Zhang Y, Liu R (2013) Air-lift internal loop biofilm reactor for realized simultaneous nitrification and denitrification. Bioprocess Biosyst Eng 36(5):597–602
Zhang C, Ma S, Wang G, Guo Y (2019) Enhancing continuous hydrogen production by photosynthetic bacterial biofilm formation within an alveolar panel photobioreactor. Int J Hydrog Energy 44(50):27248–27258
Zhao Y, Feng C, Wang Q, Yang Y, Zhang Z, Sugiura N (2011) Nitrate removal from groundwater by cooperating heterotrophic with autotrophic denitrification in a biofilm–electrode reactor. J Hazard Mater 192(3):1033–1039
Zhong Li X (2012) Application of microbial biofilms for the production of chemicals. In: School of biotechnology and biomolecular sciences. Doctor of Philosophy, The University of New South Wales. Sydney, Australia
Zhou B, Martin GJO, Pamment NB (2008) Increased phenotypic staffigbility and ethanol tolerance of recombinant Escherichia coli KO11 when immobilized in continuous fluidized bed culture. Biotechnol Bioeng 100(4):627–633

Chapter 8
Nanotechnological Interventions for Sustainable Production of Microbial Biofuel and Bioenergy

A. P. Ingle, F. A. F. Antunes, A. V. Paula, D. L. Flumignan, R. Terán-Hilares, R. R. Philippini, S. E. Martiniano, P. Abdeshahian, A. Hernandez-Perez, G. M. M. Silva, S. Sánchez-Muñoz, T. M. Rocha, D. R. Ribeaux, E. M. D. Oliveira, J. C. Santos, and S. S. da Silva

Abstract Energy plays a pivotal role in the socio-economic development of every country and serves as the backbone of any nation. However, a continuous increase in energy demand due to the ever-growing population and industrial globalization leads to a rapid depletion in sources of fossil fuels. In addition, the burning of fossil fuels has led to the emission of greenhouse gases which raised many environmental challenges such as climate change and global warming. All these concerns have pressed toward exploring sustainable and renewable energy sources in the form of bioenergies. Bioenergies mainly include the biofuels (bioethanol, biodiesel, bio-oils, bio hydrogens, methane, butanol, etc.) obtained from a variety of biological materials like biomass, algae, etc. Different conventional methods have been developed and routinely used for the production of second-generation biofuels. However, all such methods have certain limitations such as high energy demand and specialized processing equipment which ultimately escalate the associated cost. In this context, considering the widespread applications of nanotechnology in various fields including biofuel production, it is believed that the utilization of nanotechnology-based solutions would be promising alternatives. Application of different nanomaterials, particularly magnetic nanomaterials, in the development of nanocatalysts for biofuel production facilitates the easy recovery and reuse of the same nanocatalyst for multiple cycles which help to reduce the cost and make the process ecofriendly and economically viable. The present chapter mainly focuses on an overview of

A. P. Ingle (✉) · F. A. F. Antunes · R. Terán-Hilares · R. R. Philippini · S. E. Martiniano · P. Abdeshahian · A. Hernandez-Perez · G. M. M. Silva · S. Sánchez-Muñoz · T. M. Rocha · D. R. Ribeaux · J. C. Santos · S. S. da Silva
Department of Biotechnology, Engineering School of Lorena, University of Sao Paulo, Lorena, Brazil

A. V. Paula · D. L. Flumignan · E. M. D. Oliveira
Department of Biochemistry and Organic Chemistry, Institute of Chemistry, São Paulo State University (UNESP), Araraquara, SP, Brazil

J. K. Saini, R. K. Sani (eds.), *Microbial Biotechnology for Renewable and Sustainable Energy*, Clean Energy Production Technologies,
https://doi.org/10.1007/978-981-16-3852-7_8

biofuels and different conventional methods available for the production of nanomaterials. Apart from these, a special focus has been given on interventions of nanotechnology in the sustainable production of biofuels. Moreover, other aspects such as challenges in the application of nanotechnology in biofuels production are also discussed briefly.

Keywords Bioenergy · Biomass · Nanotechnology · Nanomaterials · Sustainable · Renewable · Global warming

8.1 Introduction

Environmental pollution is one of the most serious global challenges that humanity faces, attempting to preserve biodiversity, ecosystems, and human health worldwide (Xu et al. 2018). This problem has intensified over the last few years, with an increase of industrial and transport activities,that uses fossil fuels (Covert et al. 2016). Burning fossil fuels (e.g., diesel, gasoline, or coal) emits air pollutants, such as nitrogen dioxide (NO_2), sulfur dioxide (SO_2), and carbon dioxide (CO_2), which are released into the atmosphere (Mitchell et al. 2018). According to the NASA-Global Climate Change website (2020), the emission of these pollutant gases was strongly decreased, and the air quality was improved due to the recent lock-downs as a result of the spread of COVID-19. However, in a normal situation (Fig. 8.1), the accumulation of pollutant gases is worrying. One alternative to reduce the consumption of fossil fuels and, at the same time, mitigate the greenhouse effects is the use of alternative green fuels, such as hydrogen, biofuels (ethanol, biodiesel),

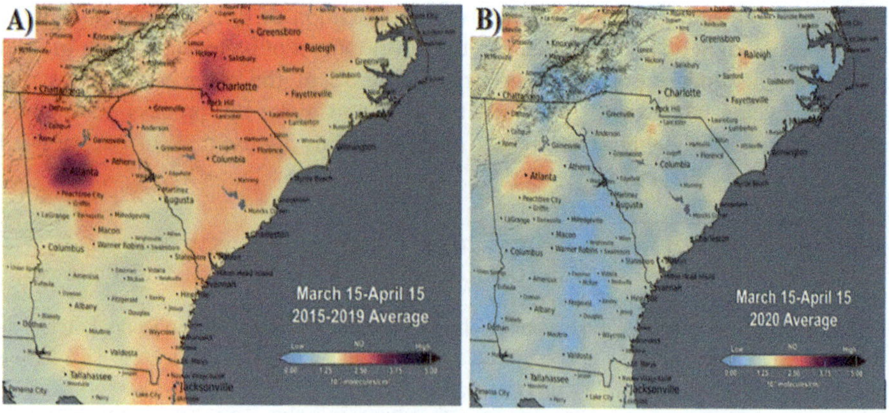

Fig. 8.1 Tropospheric NO_2 Column. (**a**) March 15–April 15, 2015–2019 Average and (**b**) March 15–April 15, 2020 Average, Southeast USA, With Cities. Pictures were obtained from the NASA-Global Climate Change website (2020)

fuel cells, etc. which have been extensively studied aiming to optimize their production in the pilot- or large-scale and their techno-economic viability.

Biofuels are classified as first- (ethanol, biodiesel, biogas, etc.), second- (bio-oil, lignocellulosic ethanol, butanol, etc.), third- (ethanol and biodiesel obtained from microorganism), and fourth-generation biofuels (biohydrogen, biomethane, and synthetic biofuels) (Itskos et al. 2016). Second-generation ethanol produced from lignocellulosic biomass has been extensively studied over the last few decades. According to the SCOPUS database, more than 500 articles on this topic have been published only in 2019 (SCOPUS 2020). Most of these studies have focused on the development of suitable and more efficient technologies for the deconstruction of recalcitrant biomass, the optimization of cellulose hydrolysis, and the optimization of the fermentation process. Recently, innovative technologies have attracted the interest of researchers. One of those is the promising use of nanoparticles in biofuel industries mainly due to their high surface area, reactivity, and functional properties, which promote the better performance of the process (Khan et al. 2019). In this context, to date various kinds of nanomaterials have successfully been used in different processes involved in biofuel production. For example, Ingle et al. (2020a, b) demonstrated the use of acid-functionalized magnetic nanoparticles (MNPs) for the pretreatment of lignocellulosic biomass. In another study, MNPs were used as support for immobilizing cellulase enzymes aiming at enzymatic hydrolysis of biomass (Gaikwad et al. 2018).

Nanotechnology has been applied in biodiesel and biohydrogen production processes, improving the recyclability of the catalyst and the performance of the process by increasing the activity and stability of immobilized enzymes such as lipases (Sarno and Iuliano 2019; Teo et al. 2019) or improving the stabilization of oil-in-methanol Pickering emulsions which can be used as interfacial catalysts in the transesterification reaction for biodiesel production (Peng et al. 2020). As already discussed in this section, the development of innovative technologies for biofuels production is a current challenge. Considering these facts, in the present chapter, we have discussed the concepts and applications of nanotechnology in biofuels production.

8.2 Biofuels: Green Alternative Fuels

A fuel produced using renewable biomass-based resources (plant biomass, microorganisms, or animal by-products) is referred to as a biofuel. Global biofuel production is mainly directed to the transportation sector and it is believed that an increase in the supply of these fuels is essential to assure both energy security and the reduction of greenhouse gas emissions (OECD/FAO 2019). According to Kamani et al. (2019), biofuels have the following benefits over fossil fuels:

- Biodegradability, renewability, and contribution towards a sustainable economy;
- Availability limited only by the amount of biomasses resources;

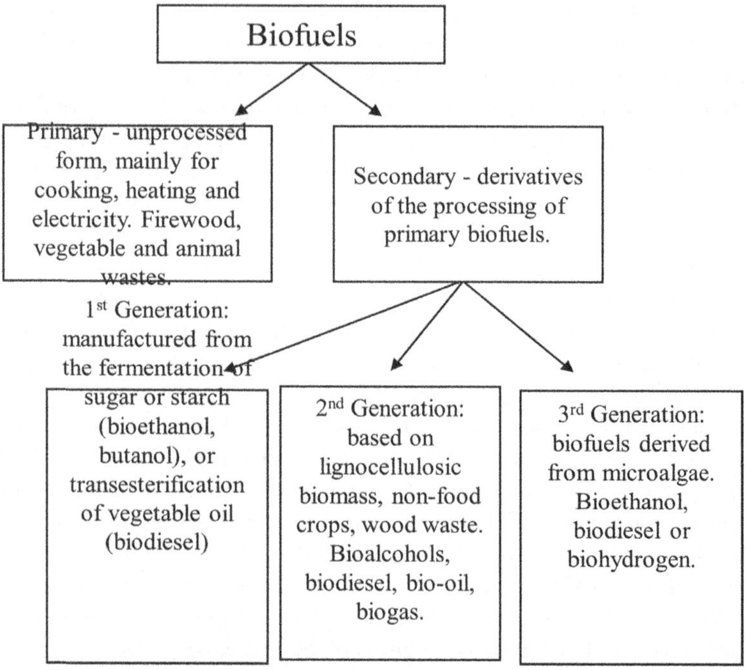

Fig. 8.2 General classification of biofuels (Source: Fatma et al. (2018), Kamani et al. (2018), Paul et al. (2019))

- Reduction of the environmental impacts related to agriculture wastes disposal;
- Lower impact on the environment as compared to fossil fuels.
- Achievement of energy security;
- Fortification of the economy by creating more opportunities related to agriculture and raising of agricultural incomes;
- Intensification of industrial investments;

Biomass is the important feedstock used for the production of the majority of biofuels (or biomass-based fuels) and is usually obtained through thermal, physical, or biological processes (Kamani et al. 2019). Despite a variety of definitions of biofuels found in literature, biofuels are generally classified by their chemical nature or are based on the feedstock source. Regarding their chemical nature, biofuels can be derived from alcoholic fermentation, from the esterification of vegetable oils or animal fat, or even from anaerobic digestion (Kamani et al. 2019; Roberts and Patterson 2014). Fig. 8.2 shows three generations of evolution of the feedstocks utilized for biofuel production.

8.3 Global Production of the Major Biofuels

Nowadays, approximately 10% of the world's total primary energy supply is represented by bioenergy, with a global production of 154 billion liters in 2018. Biofuels production is led by United States, Brazil, European Union, ASEAN, China, and India; it is mainly represented by bioethanol, biodiesel, and biogas, although other fuels exist in the state of solid (biochar), liquid (biobutanol, biomethanol, bio-oil, 2,5-dimethylfuran) or gas (biohydrogen) (Sindhu et al. 2019).

Bioethanol production relies on the alcoholic fermentation of plant biomass performed by yeasts. Globally, the most used crops for bioethanol production are corn, sugarcane, cassava, sugar beets, wheat, and other grains. Since bioethanol is mainly used for transportation, this biofuel offers an excellent opportunity to reduce the utilization of crude oil and to scale down CO_2 atmospheric accumulation, an imperative maneuver to mitigate the negative effects of the climatic crisis upon the environment and our society and economy (Kamani et al. 2019).

Biodiesel has originated from the transesterification of natural lipids present in plants such as soybean, rapeseed, canola, palm and corn, waste oils, or animal fat (Carvalheiro et al. 2008). Algae are especially suitable for biodiesel production due to their ability to consume atmospheric CO_2 to produce large amounts of oil: on a dry basis, the lipid content of microalgae biomass is between 20 and 50%, but under certain conditions, it can reach nearly 80% (Kamani et al. 2019; Nobre et al. 2013).

In the case of biodiesel, emissions of non-combusted hydrocarbons or CO are lower than conventional diesel as well as there is no sulfur or aromatic compounds in its composition. Furthermore, this biofuel outstands regarding its potential for industrial scale-up and has been broadly marketed in numerous countries such as the United States, European countries, Brazil, and Australia (Beschkov 2012; Kamani et al. 2019).

Biogas, on the other hand, is produced by the anaerobic digestion of biological wastes using microbes. Its main component is methane (50–80%) and minor constituents are CO_2 (30–50%), CO, H_2S, nitrogen, oxygen, hydrogen and ammonia (Chen et al. 2015). Agricultural waste treatment generates expressive volumes of biogas, which has a great heating power and can be used for heat or electricity generation and, in specific cases, for internal combusting engines (Beschkov 2012; Kamani et al. 2019). Besides biogas, biohydrogen is another important biofuel generated from gasification of biomass. Several studies have been done toward the sustainability of biohydrogen production. It is considered that the generation of a coproduct simultaneously with biohydrogen from biomass is a path to ensure the economic viability of the process (Sindhu et al. 2019). The production of important biofuels using conventional approaches has been discussed in the following section.

8.3.1 Bioethanol

Through the expansion of modern biorefineries concept and the exploitation of renewable bio-based fuels, the world's demand for more environmentally friendly, less hazardous, and sustainable sources of energy has become one of the major targets for a prosperous and ecological future (Boboescu et al. 2019). In accordance with this fact, bioethanol has been a long-studied biofuel worldwide and a variety of carbon sources have been utilized for its production. For instance, several countries such as India, Brazil, the USA, and many others have been applying crops for ethanol generation although from different raw materials comprising mainly sugarcane molasses, sugarcane stalk juice, and corn, respectively (Soam et al. 2018; Costa et al. 2015; Cheng and Timilsina 2011). On the other hand, using food crops as a source for biofuel production is considered first generation (1G) and competition may intensify between food and energy supply, thereby increasing the prices in the food market which can become a global issue (Lazar et al. 2018).

A solution to this problem is the substitution of the direct use of crops for agricultural wastes and food wastes such as lignocellulosic materials (e.g., sugarcane bagasse, wheat straw, corncob, rice straw, etc.) (Banerjee et al. 2010). For several years, these materials were considered as wastes but due to extensive efforts of scientists and researchers, now these materials can be utilized for the production of high-value products. Therefore, lignocellulosic biomasses represent one of the possible substrates for second-generation (2G) ethanol and biofuels in general. To summarize the key role of agro-wastes implementation, Table 8.1 briefly displays a variety of industrial bioproducts which have an overwhelmingly positive impact on realistic environmental problems.

Keeping this in mind, it is of great importance to comprehend how the substrate may influence the overall process of bioethanol synthesis. In the case of lignocellulosic materials, it is well known that its compact structure is a rigid and complex mixture of polysaccharides and a macromolecule is composed of cellulose (30–50%), hemicellulose (25–30%), and lignin (10–35%), respectively (Spyridon and Willem Euverink 2016). In brief, cellulose is a linear glucose polymer-bonded within ß-1,4-glycosidic linkages that provides a high degree of crystallinity due to the extensive hydrogen bonds among the hydroxyl groups, whereas hemicellulose is a heteropolymer of a short and highly branched chain of pentoses, hexoses sugars, with some traits of organic acids (Limayem and Ricke 2012). Furthermore, the macromolecule lignin is composed of 4-hydroxyphenylpropanoid units which are considered its precursors. These units are linked throughout the chain via ether (C-O-C) and carbon (C-C) bonds. Furthermore, its arrangement acts as a protection structure and ensures entrapment of the restrained molecules in accordance with the degree of entanglement among the polysaccharides and lignin (de Gonzalo et al. 2016).

These components are linked tightly together to form a recalcitrant structure to hydrolytic attack and non-readily bio-digestible biomass (Bugg et al. 2011). To

Table 8.1 Bioproducts synthesized using different lignocellulosic carbon sources

Biomass	Microorganism	Process technique	Product	Reference
Corncob hydrolysate	*S. bombicola* NBRC 10243	Submerged fermentation	Biosurfactant	Konishi et al. (2015)
Opuntia ficus-indica cladode	*Kluyveromyces marxianus*	Submerged fermentation	Ethanol	López-Domínguez et al. (2019)
Digestate (bio-waste)	*Bacillus thuringiensis*	Solid-stated fermentation	Biopesticide	Cerda et al. (2019)
Agave bagasse hydrolysate	*Yarrowia Lipolytica*	Submerged fermentation	Lipids	Niehus et al. (2018)
Sugarcane bagasse hydrolysate	*C. guilliermondii* FTI 20037	Submerged fermentation	Xylitol	Sarrouh and da Silva (2010)
Elephant grass	*S. cerevisiae* CAT-1	Submerged fermentation	Ethanol	Scholl et al. (2015)
Apple pomace	*A. niger* NRRL-567	Solid-state fermentation	Cellulase	Dhillon et al. (2012)
Pulp and paper solid waste	*Rhizopus oryzae* 1526	Solid-state fermentation	Fumaric acid	Das et al. (2016)
Olive pomace	*Xantophylomyces dendrorhous*/*Sporidiobolus salmonicolor*	Solid-state fermentation	Pigment (astaxanthin)	Eryılmaz et al. (2016)
Wheat straw	*Bacillus* sp. BBXS-2	Solid-state fermentation	Amylase	Qureshi et al. (2016)

enhance microbial digestibility, a wise step pretreatment is required to depolymerize, reduce the degree of crystallinity of cellulose and hemicellulose as well as remove the lignin fraction. Moreover, the disruption of the fibers also reduces its compactness which, in turn, facilitates microbial accessibility to the fermentable sugars (Rastogi and Shrivastava 2017; Hendriks and Zeeman 2009). The bottleneck of 2G bioethanol relies significantly on the pretreatment features and progress. Thus, to analyze whether the overall process is having a negative impact and to quantify energy requirements and greenhouse gases emission, currently, there are practical tools such as life cycle assessment (LCA) that evaluates environmental issues in any step of biofuel production, including measurement of downstream processing and waste materials generation (Cherubini et al. 2009; Dadak et al. 2016).

In alignment with the strategy of minimizing the deleterious effects of rendering the pretreatments of lignocellulosic biomass, several methods have been developed from the necessity to mitigate the excessive use of chemicals and energy. In this respect, pretreatment assays may be carried out by a variety of approaches, including chemical, physical, physicochemical, and biological. Each technique aims to exert distinct effects on the biomass having inherent advantages and disadvantages. The most common ones are mechanical comminution, irradiation, acid (sulphuric or hydrochloric acid), alkali (such as calcium hydroxide), steam explosion, or combined processes that demand large energy input and high-cost equipment utilization (Kumari and Singh 2018; Ruane et al. 2010). Likewise, biological pretreatment is based on the natural ability of microorganisms to degrade lignin via enzymatic performance in a step termed delignification. The cultivation and growth of the targeted cells may be performed under submerged or solid-state fermentation (Zabed et al. 2017; Yahmed et al. 2017; Mishra et al. 2017).

The aforementioned techniques are prerequisites to increase the availability of cellulose and hemicellulose for enzymatic hydrolysis necessary to the conversion of those into their respective fermentable sugars (Lamb et al. 2018). Specific enzymes can hydrolyze cellulose and hemicellulose to selectively release their monomeric sugars in relatively low temperatures ranging from 45 to 50 °C by the active sites of cellulases and hemicellulases (xylanases), respectively (Duff and Murray 1996). In summary, cellulase is a cocktail of enzymes that exert desirable effects onto cellulose molecules and typically involves the synergistic action of endoglucanase, exoglucanase, and ß-glucosidase (Sun and Cheng 2002). Endoglucanase is responsible to hydrolyze internal (ß-1,4) glycosidic bonds throughout the D-glucan polymer chain, producing cellodextrins out of the amorphous regions of cellulose, thereby releasing free chain ends, whereas exoglucanase cleaves cellobiose and cellotriose units from the non-reducing terminal. The response to this system generates dimers termed cellobiose as an output which is a disaccharide of glucose that is consecutively converted into glucose by the selective action of ß-glucosidase (Dotaniya et al. 2019; Zabed et al. 2017). To give a more illustrative representation of the cellulase mechanism, Fig. 8.3 displays the summarized dynamics of cellulose degradation according to the selectivity of each enzyme required.

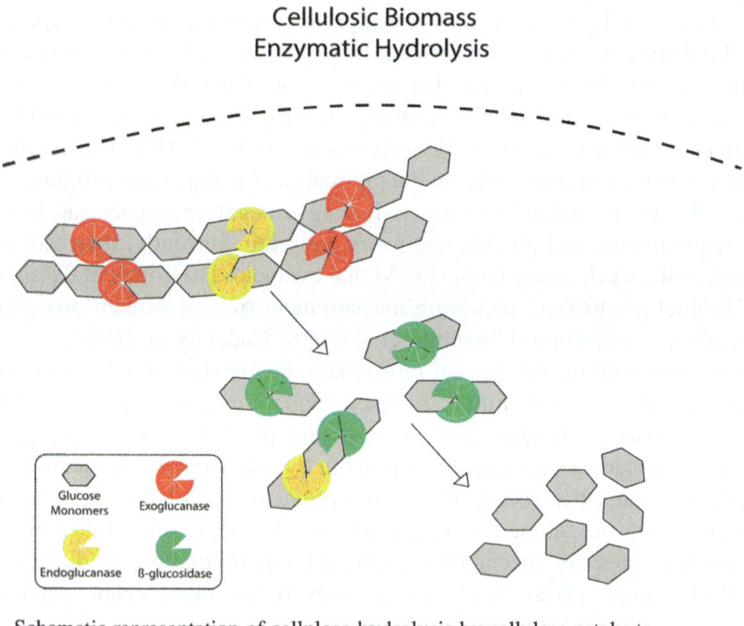

Fig. 8.3 Schematic representation of cellulose hydrolysis by cellulase catalysts

The enzymatic machinery to break down the heteropolymer hemicellulose is quite more complex due to its branched-chain and the specificity of the internal bonds. Therefore, the xylanase (hemicellulase) system contains usually endoxylanase, exoxylanase, ß-xylosidase, α-arabinofuranosidase, α-glucoronisidase, etc. Similarly, endo- and exo-xylanases catalyze selectively the breakdown of the main chain of xylans resulting in reduced size chains. Furtherly, ß-xylosidase cleaves xylo-oligosaccharides into xylose. The other enzymes rather act on the backbone of the xylan polymer and are responsible for the release of arabinose and 4-o-methyl glucuronic acid (Saha 2003).

The resulting concentration of pentoses and hexoses may vary according to the preceding pretreatment and the type of enzymes implied along with the hydrolysis. Therefore, the fermenting microorganism must be suitably selected in order to obtain maximum yield and productivity as well as avoid unwanted catabolic repression by the substrates and inhibitory compounds (Banerjee et al. 2010). The ability to co-assimilate C5 and C6 sugars is crucial for any bioethanol facility plant. For instance, the utilization of *Saccharomyces cerevisiae* and *Zymomonas mobilis* is frequently common to produce ethanol from hexoses; however, their inability to concomitantly consume pentoses delays the development of more robust processes. On the other hand, organisms that can ferment pentoses (e.g., *Pichiastipitis*, *Pachysolenthannopilus*, *Candida shehatae*) offer very low efficiency in the conversion factor (Hahn-Hägerdal et al. 2007). Yet, within the advances in metabolic engineering tools, pertinent efforts toward genetically modified microorganisms attempt to address this issue and to enhance co-assimilation of C5 and C6 sugars (Wackett 2011).

Contemporarily, fermentation processes may be carried out by several approaches including Separate Hydrolysis and Fermentation (SHF), Simultaneous Saccharification and Fermentation (SSF), Simultaneous Saccharification and Co-Fermentation (SSCF), and finally, a Consolidated Bioprocess (CBP) (Rastogi and Shrivastava 2017). SHF consists of rendering a two-stage process, wherein enzymatic hydrolysis is operated separately from fermentation. Albeit sugar accumulation throughout hydrolysis inhibits enzyme activity, positive aspects are encountered in this strategy, involving the implementation of optimal operation conditions of each stage (Vohra et al. 2014). SSF offers advantageous features such as reduction of inhibitors, less energy demand, and is economically attractive. It is performed simultaneously with the hydrolysis step at the same unit which in turn prevents undesired effects of sugar accumulation, thereby obtaining a higher ethanol yield conversion if compared to SHF (Foust et al. 2009; Brethauer and Wyman 2010).

Moreover, SSCF integrates C5 and C6 sugar assimilation into only one stage. By that, different methods may be reliable to concretize this operation which involves the use of a consortium of organisms having distinct metabolic pathways consuming synergistically both carbon sources. However, hexoses consumers grow faster, and it may lead to growth inhibition of pentose-utilizing microorganisms. Furthermore, one single bacteria or yeast may be genetically modified to efficiently incorporate C5 and C6 substrates rather than the use of capable natural-born wild strains that frequently lead to lower ethanol productivity (Sanchez and Cardona 2008).

Nevertheless, CBP is a robust attempt to integrate cellulolytic enzymes excretion, saccharification, and fermentation at the same operation step mediated uniquely by a microorganism community. The advantages rely strongly upon the fact that expenditures associated exclusively with enzyme production are avoided by combining those steps mentioned above. Aside from that, saccharification and fermentation are entirely compatible regarding operational parameters (Vohra et al. 2014). To gain insight, López-Domínguez and collaborators (2019), investigated the capability of *Acinetobacter pittii* and *Kluyveromyces marxianus* isolated from *Opuntia ficus-indica* toward decay of cladode to produce cellulase and simultaneously saccharify the targeted biomass and synthesize ethanol. The novelty of this study was the utilization of wild strains which possess naturally metabolic machinery that can achieve significant and promising yields of bioethanol in the near future.

To summarize, there is a broad scientific avenue favorable to the development and implementation of diverse techniques in the enzymatic and bioprocessing fields. The substitution of regular fossil fuels for biofuels still to some extent lacks optimization and cost-effectiveness. Therefore, further discussion in this chapter attempts to introduce the role of nanotechnology in enzymatic hydrolysis enhancement and bioconversion of ethanol.

8.3.2 Biodiesel and Biohydrogen

Nowadays, alternative energy resources such as wind, solar, and biofuel have grabbed the attention of scientists, researchers, and governments due to the rapid consumption of fossil resources, global climatic change, and the interest in more secure fuel supplies (Semwal et al. 2011; Chozhavendhan et al. 2020). Among renewable sources of energy, biodiesel has been considered a notable candidate to reduce environmental pollution and achieve sustainable development (Mahlia et al. 2020).

Biodiesel is typically produced through the transesterification process, in which triglycerides react with an alcohol in the presence of a catalyst to obtain mono-alkyl esters. These triglycerides may be obtained from micro-and macro-algae, fungi, animal fat, and vegetable oil, lignocellulose material, etc. (Sharma et al. 2008; Mahmudul et al. 2017). Since methanol is the most frequently used alcohol due to its low cost, other common names for biodiesel are fatty acid methyl esters (FAME) or B100, which means 100% FAME (Singh et al. 2020).

Biodiesel has many advantages such as it is eco-friendly, non-toxic, biodegradable; has a low emission profile, and is a renewable energy resource (Avhad and Marchetti 2015). In this sense, biodiesel is usually classified as first-, second-, and third-generation based on the raw materials used in its production. First-generation biodiesel is derived from edible feedstocks such as soybean oil, coconut oil, rapeseed oil, palm oil, sunflower oil, etc. (Mahdavi et al. 2015), while second-generation biodiesel is obtained from agricultural wastes and non-edible feedstocks such as neem oil, jatropha oil, nagchampa oil, karanja oil, etc. (Atabani et al. 2013). However, these categories generate conflict between land use and food supply (Mahlia et al. 2020). The case of third-generation biodiesel involves the use of high oil-content microalgae further alternate sources for biodiesel production (Leong et al. 2018). Moreover, a fourth classification has emerged from the metabolic engineering of photosynthetic organisms, which has been transformed through synthetic biology tools as another sustainable alternative (Chua et al. 2020).

On the other side, the biological production of hydrogen (biohydrogen) is another alternative that fits well with the renewable energy concept. Among known fuels, hydrogen has the highest gravimetric energy density and is compatible with electrochemical processes (Mudhoo et al. 2011). The conventional method of hydrogen generation is based on steam reforming or oxidation of natural gas and coal gasification. However, these primary sources for the production of hydrogen are nonrenewable and release carbon dioxide as a byproduct, which creates an environment negative effect (Hibino et al. 2018).

Thus, the sustainable production of hydrogen through biological routes such as photobiological and fermentative processes has been reported as a different approach (Rupprecht et al. 2006; Srivastava et al. 2020). Moreover, the generation of biohydrogen has also been reported through the combination of different methods. The advantages of these alternative processes include the production of hydrogen from renewable sources and the generation of emissions free of pollution (Singh et al. 2015; Sampath et al. 2020). The microorganisms involved in biohydrogen production are classified into two groups: photosynthetic and non-photosynthetic or fermentative hydrogen producers (Das and Veziroğlu 2001). Also, metabolic engineering has been an exceptional tool for improving the hydrogen productivity of available microbial sources rather than discover new strains (Chandrasekhar et al. 2015).

In the case of photobiological hydrogen production which includes bio photolysis, indirect bio photolysis, and photo fermentation, solar radiation is the driving force for the process. Among the microorganisms that are best suited for this light-dependent hydrogen production are some species of bacteria (purple-sulfur, and purple non-sulfur), algae, and cyanobacteria (Barbosa et al. 2001; Kovács et al. 2006). On the other hand, in dark fermentation or fermentative hydrogen production, the obligate anaerobes and the facultative anaerobes have been explored as producers for this purpose. The absence of energy light is the striking feature of this process. Since agricultural waste and organic waste generated from domestic and industrial activity can be decomposed through dark fermentation to produce hydrogen, this process is a particularly advantageous alternative (Guo et al. 2010; Łukajtis et al. 2018).

Thus, microbial electrolysis cells (MEC) represent a versatile technology for waste treatment processes. They were adapted from microbial fuel cells (MFCs) and the conversion of a wide range of organic substrates into hydrogen occurs under applied external potential (Cheng and Logan 2007; Chandrasekhar et al. 2015).

However, the microbial physiology, electrode materials, physicochemical transport processes, type of membrane used, and composition and concentration of the substrate are important factors that affect the performance of MEC and limit its commercial distribution (Hallenbeck 2011).

8.4 Limitations of Existing Conventional Methods

Though biofuels comprise a wide variety of energy sources derived from biomasses, such as bioethanol, biodiesel, biogas, biomethanol, bioethers, biohydrogen, and vegetable oils, the market seemed to be mainly focused on the first 3 i.e. bioethanol, biodiesel, biogas (Callegari et al. 2020). Currently, marketable biofuels are mostly produced from first-generation crops, which have similar drawbacks, related to limited availability and food competition, and, therefore, make room for second and third-generation feedstocks (Callegari et al. 2020). Among the second generation, biofuels derived from lignocellulosic byproducts and residues, driven by economic, environmental, and even social-political purposes have been widely explored in the last decades. Feedstocks have been selected based on their sustainability, energy content, local availability and distribution, and environmental and economic values (Karagiannidis and Perkoulidis 2009). Challenges related specifically to the feedstock have been addressed since their cost is an important issue in biofuels production technologies, such as new varieties with desirable characteristics, growing requirements, cultivation yields planting and harvesting techniques, and logistics, among others (Callegari et al. 2020; Shanmugam et al. 2020).

Extensive research has enabled important advancements in the processes for biofuels production from biomasses; however, there are still important technological barriers to overcome and to make them mature for commercial scale and competitive with fossil fuels (Khoo et al. 2020a, b). In this sense, the cost-effective release of fermentable carbohydrates from biomasses is one of the biggest challenges on biofuels production, with a high impact on the total process cost (Ingle et al. 2019a, b; Khoo et al. 2020a, b). The upstream steps include mainly biomass pretreatment and further hydrolysis of polymeric carbohydrates to release fermentable sugars, for which several methods, involving chemical, physical, biological methods and mixtures of them have been extensively studied. Despite the promising results obtained at laboratory and pilot scales with the conventional methods, the high cost jeopardizes their potential utilization at larger scales (Ingle et al. 2019a, b; Shanmugam et al. 2020). Most of the conventional methods are performed in intensive operation conditions, with high consumption of materials that are not recycled or are difficult to be reused, and generation of contaminating by-products and wastes, resulting in processes that are not economic and environmentally sustainable (Ingle et al. 2019a, b).

Particularly, in the polysaccharide (cellulose) hydrolysis after pretreatment, enzymatic technologies have been extensively studied, in order to increase hydrolysis efficiency and reduce enzyme-associated costs. In the technologies that have been mostly studied, enzymes cannot be reused or recycled, which increases the cost of this step and consequently of the process. Therefore, several studies have been focused on enzyme immobilization, in order to facilitate the separation of the enzymes and/or their reutilization in various sequential reactions, which, in turn, can reduce the overall process cost (Shanmugam et al. 2020).

In the particular case of biodiesel production, enzymatic transesterification is a remarkable alternative, since it is a less energy-intensive strategy, with higher selectivity, easier separation, less residual contamination when compared to chemically catalyzed processes. However, it has a main drawback also about the high cost associated with enzymes, which reduces its attractiveness to industrial applications (Callegari et al. 2020). Regarding biohydrogen production, which has been considered as the most efficient and cleanest form of energy, it still has important drawbacks to be addressed to achieve higher levels of readiness, such as low yield and high production cost (Shanmugam et al. 2020). According to these authors, several strategies for process intensification have been studied, including parameter optimization to improve the production rate, utilization of synthetic biology, and metabolic engineering.

Nanotechnology has the potential to increase the overall efficiency, feasibility, and sustainability of the biofuels production technologies, not only limited to the upstream steps but also the conversion processes and downstream (Ingle et al., 2019; Xu et al. 2019; Khoo et al. 2020a, b). Research and development on nanotechnology have grown expressively in the last years in different areas and with the participation of interdisciplinary and integrated science (Khoo et al. 2020a, b). For biofuels technology and regarding first the upstream steps, nanomaterials can be used for enzyme immobilization, named nano supports, which have advantages like large surface area, biocompatibility, non-toxic effects, a variety of physical and chemical properties that can enhance the activity of the enzyme, and the possibility of improving the recuperation and reuse of the enzymes (Rai et al. 2019; Khoo et al. 2020a, b; Shanmugam et al. 2020).

Nanomaterials can contribute not only as immobilization or encapsulation matrix for enzymes, promoting their reuse (Ingle et al. 2019a, b; Shanmugam et al. 2020) but also as nanocatalysts, which have been highlighted not only based on environmental and ecological issues compared to synthetic catalysts but also because, small particle size (related to their cell wall penetrating advantages), biodegradability, reusability and easy recuperation based on magnetic properties, functionalization possibilities, low price, and high availability (Ingle et al. 2019a, b; Xu et al. 2019; Shanmugam et al. 2020). However, some issues should be addressed regarding the safety and toxicity of various nanomaterials, nanoparticles aggregation problems, and synthesis costs (Ingle et al. 2019a, b; Khoo et al. 2020a, b).

Furthermore, in the case of biohydrogen production, nanotechnology strategies have also been studied as potentially cost-effective alternatives to improve the bioconversion step, since they can have a positive impact on the growth of the microorganism, the intracellular electron transfer, and the efficiency and protection of enzymes (oxygen-sensitive) involved in biohydrogen production (Yang and Wang 2018; Shanmugam et al. 2020). Moreover, nanotechnology strategies can improve the control of the operation conditions, such as illumination, temperature, and heat transfer, and even influence the bioreactor design (Shanmugam et al. 2020).

8.4.1 Socioeconomic and Environmental Considerations

It is an undeniable fact that an economy based on fossil fuels is no longer viable and a substantial amount of data, research studies, and public policies and future projected scenarios indicate that the shift to bioeconomy is a promising way to ensure welfare, economic, and food security to the human population. Regarding this conjecture, Johnson (2017) states that "A thriving bioeconomy that includes increasing reliance on biological processes and biobased products is a key element of the overall global sustainability transition."

"Implement green chemistry and sustainability principles" is not only enough to assure the success of a bioeconomy, but it is also necessary to establish coordinates and steps to make the transition from our present models to a sustainable economy. It is not only essential to develop a circular economy system, where waste generation is reduced to its minimum and all the possible uses of biomass are considered, but also to articulate social and economic sustainability in accordance with environmental health. Moreover, an integration between national and global policies is vital, along with the cooperation and comprehensive view between sectors that deal with different biomass uses (e.g., energy, transportation, agriculture, forestry) (Johnson 2017).

The production of first-generation biofuels is based on crops that are likewise used for human and animal feeding. Therefore, a concern has arisen that an increase in the production of these fuels can compromise food security (food versus fuel debate). According to Sindhu et al. (2019) life-cycle assessment (LCA) of first-generation biofuels indicates that, in most circumstances, there is a negative energy gain; however, second-generation fuel models suggest an increase in energy gain, while third-generation biofuels excel the previous categories in many aspects, such as CO_2 sequestration, expressive accumulation of neutral lipids, high biomass, and soil productivity (Sindhu et al. 2019).

Land use by biofuel crops is still a field of uncertainties, forasmuch as it is connected to a huge number of variables, for instance, demand for other applications, agriculture productivity, future demand for animal products, and the pressure upon natural environments that can be seen as idle lands (such as grasslands), which can result in biodiversity loss (OECD/FAO 2019; Sindhu et al. 2019). In this sense, it is crucial to develop public policies to regulate land use and assure the sustainability of biofuels; moreover, studies that aim at the production of biofuels with nonfood crops or lignocellulosic biomass must be supported and promoted.

8.5 Nanotechnology in Biofuels Production

Nanotechnology has emerged as a promising technology as far as biofuel industries are concerned. It is reported to have applications in the production of different biofuels like bioethanol, biodiesel, biohydrogen, etc.

8.5.1 Nanotechnology in Bioethanol Production

The use of nanotechnology in bioethanol production can improve the plant biomass pretreatment and its conversion into fermentable sugars as well as the fermentative process (Kushwaha et al. 2018). The recalcitrance properties in most agro-industrial wastes, especially in the lignocellulosic biomass, is still a bottleneck for its conversion into second-generation biofuels (Zuccaro et al. 2020) and the pretreatment plays an important role in the manufacturing process and product value. Nanomaterials can improve pretreatment efficiency and assists in bioethanol fermentation and recovery. The major applications of nanoparticles in bioethanol production are given in Fig. 8.4. Moreover, the reusability of nano compounds is an important advantage for the biofuels' economic viability (Beniwal et al. 2018).

Several types of nanoparticles have been studied for bioethanol production and are applied in biomass pretreatment for the recovery of the sugars in different lignocellulosic materials as feedstock. Pena et al. (2012 & 2014) studied the effects of different acid-functionalized nanoparticles for the pretreatment of wheat straw and corncob. Ingle et al. (2019a, b, 2020a, b) evaluated the pretreatment of sugarcane bagasse and sugarcane straw using two different acid-functionalized magnetic nanoparticles (alkyl sulfonic acid—Fe_3O_4_MNPs@Si@AS, and butylcarboxylic acid—Fe_3O_4_MNPs@Si@BCOOH), that presented maximum xylose recovery for

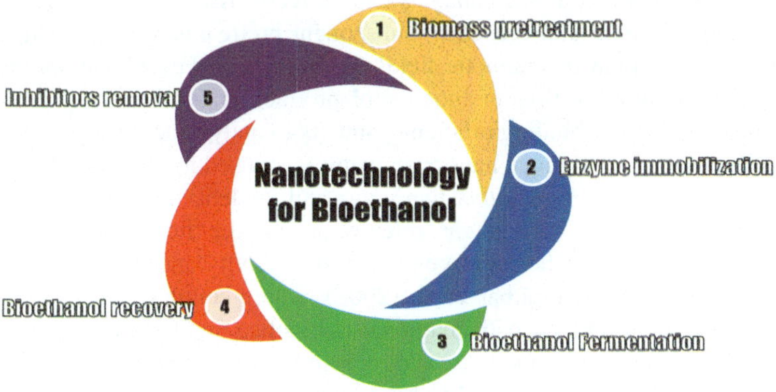

Fig. 8.4 Major applications of nanoparticles in bioethanol production

sugarcane bagasse (18.83 g/L and 18.67 g/L), and sugarcane straw (17.06 and 15.40) using the 500 mg/g of biomass.

Another utilization for nanoparticles in bioethanol production is for the immobilization of the enzymes. Enzymes are biological catalysts produced by bacteria and fungi and are a key factor for environment-friendly production biofuels because enzyme such as cellulases and hemicellulases play importanat role in the breakdown of cellulose and hemicellulose present in the lignocellulosic biomass (Mood et al. 2013). However, the utilization in the industrial scenario presents some obstacles to become economically viable, such as costly production and reuse of enzymes as they can contribute up to 30% of total processing cost in 2G sugars production (Sánchez-Ramírez et al. 2016; Chandel et al. 2018).

The immobilization of enzymes is an alternative for reducing costs with enzymes in an industrial scenario. Several supports can be used, such as inorganic materials, hybrid materials, polymers, and metal-organic frameworks (Suo et al. 2020). Immobilization methods vary in categories where the enzymes can be (1) bonded to support, which acts as a carrier or matrix, (2) entrapped in an encapsulation structure, or (3) cross-linked (Vaghari et al. 2015). The utilization of nanoparticles as an immobilizing agent presents several benefits to the enzymatic process. The immobilization of enzymes not only promotes increased yields and multiple cycles but is also presented as an environment-friendly alternative for enzyme application, also protecting them from inhibitory effects of alcohol and organic acids formed during fermentation (Sekoai et al. 2019). Cherian et al. (2015) studied the immobilization of cellulases using manganese dioxide (MnO_2) nanoparticles for the hydrolysis of sugarcane leaves to bioethanol (21.96 g/L), presenting 75% binding efficiency and 60% of catalytic activity, after five cycles. The biocompatibility, high specific surface area, stability and low toxicity, and resistance to mass transfer are highlighted, although the most prominent advantage is that immobilized enzymes can be recovered for repetitive applications in catalytic reactions, which can contribute to the overall reduction of costs in a biorefinery (Chandel et al. 2018; Suo et al. 2020).

The utilization of magnetic nanoparticles (MNPs) can be advantageous after the pretreatment of biomass as the catalysts can be recovered by the application of an external magnetic field and reused in subsequent pretreatment cycles (Ingle et al. 2020a). The utilization of magnetic fields in iron oxide (Fe_3O_4) nanoparticles for β-glucosidase immobilization in bioethanol production, studied by Verma et al. (2013), resulted in 93% binding efficiency and 50% catalytic activity after 16 cycles. Fe_3O_4 NPs and Fe_3O_4/Alginate nanocomposites were used for the immobilization of cellulases produced by *Aspergillus fumigatus* and evidenced an increased enzyme activity, resulting in a high sugar release during the rice straw pretreatment (Srivastava et al. 2015). The improvement in the activity and the thermal stability was also observed by Poorakbar et al. (2008), where cellulases from *Penicillium funiculosum* were employed with magnetic gold silica and showed a binding efficiency of 76% to the support matrix, and recycled for five cycles. Still, nickel oxide (NiO) nanoparticles were also used as bio-nanocatalysts in simultaneous saccharification and fermentation of potato peel waste was studied by Sanusi et al. (2020), and showed an increased bioethanol yield (19%). Even though nanoparticle utilization may be advantageous to bioethanol production, its use must be limited to its optimum values as it can inhibit the growth of microorganisms in higher concentrations (Sekoai et al. 2019).

Cells are microbial factories capable to synthesize enzymes for several industrial purposes. Though the nanomaterials use in enzyme immobilization, these compounds also act as supports to immobilize microorganisms (Rai et al. 2016a, b). Calcium alginate is commonly used as a matrix for cell immobilization, but the combination method with nano-structure materials has been demonstrated as promising alternatives for enhancing bioethanol production. Beniwal et al. (2018) achieved up to 0.42 g/g ethanol yield in 36 h with *Saccharomyces cerevisiae* and *Kluyveromyces marxianus* yeasts co-immobilized in calcium alginate using cheese whey as substrate. The authors immobilized β-galactosidase in a silicon dioxide nanoparticles matrix in a bioreactor for the same vessel hydrolysis and fermentation, demonstrating the nanoparticle reusability of 5 cycles. Besides increasing bioethanol yield in fermentation, nanoparticles could enhance the production of bioethanol in the syngas platform, as demonstrated by Kim et al. (2014) by using methyl-functionalized silica nanoparticles (0.3 wt %) during *Clostridium ljungdahlii* fermentation.

Another important use of nanomaterials is for bioethanol recovery from the broth. The presence of the bioethanol produced during the fermentation presents a negative effect on cell growth and viability, consequently decreasing the product yield (Xue et al. 2016). Pervaporation is considered a promising method for bioethanol recovery since it allows the integration of fermentation and biofuel recovery in situ (Fan et al. 2019). However, yeast cells can contaminate these membranes, fouling during the pervaporation, but the use of carbon nanotubes coupled in membrane filters assists the bioethanol recovery and enhances the antifouling performance (Xue et al. 2016). Besides, a nanofiltration membrane combined with a forward osmosis system was demonstrated to be effective for the removal of fermentation inhibitors and the concentration of fermentable sugars in rice straw hydrolysate (Shibuya et al. 2017). Nanotechnology enhances bioethanol production, especially assisting in enzyme immobilization (Rai et al. 2016a, b), helping to overcome bottlenecks and reducing costs in the manufacturing process.

Several factors such as the synthesis approach (co-precipitation method, thermal decomposition, microemulsion, hydrothermal synthesis, synthesis using biological organisms (fungi and algae), synthesis using plant materials, temperature range (100–700 °C), pressure, pH, and size may influence the performance of nanoparticles in fuels. These factors affect the morphology, size, and stability of nanoparticles as they have their advantages and disadvantages (Sekoai et al. 2019).

8.5.2 Nanotechnology in Biodiesel Production

The use of biofuels has been increasing over the last century; the ever-growing energetic demand, alongside environmental issues, has stimulated the search for alternative renewable fuel sources (Gardy et al. 2019). Biodiesel is a biodegradable, non-toxic, and environment-friendly alternative to petrol diesel. It consists of a mixture of monoalkyl esters derived from the esterification or transesterification of vegetable oils and animal fats with an excess of acyl acceptors, mostly short-chain alcohols, such as methanol or ethanol, with alkaline or acid catalysts. The fatty acid methyl or ethyl esters have properties similar to those of petrol diesel.

The biodiesel quality depends on several physicochemical properties, such as viscosity, specific mass, cetane number, cold flow plugging point, flash point, etc. The physicochemical properties and specifications limits are regulated by the National Agency of Petroleum, Natural Gas and Biofuels (ANP) in Brazil, European Standards (ES) in Europe, and the American Society for Testing and Materials (ASTM) in the USA. Biodiesel can be used directly in diesel engines or a mixture with petrol diesel. Several countries across the world have legally included biodiesel in the energetic matrix. In Brazil, biodiesel is obligatory mixed with diesel oil since 2008 and its use has increased currently to 12% v/v (B12), with a prediction of 20% (B20) in 2022 (Flumignan et al. 2012; ANP 2020).

The most common, though not exclusive, path for biodiesel synthesis is the reaction of feedstocks (in special, vegetable oils) with methanol and homogenous alkaline catalysts. Recent research shows emerging alternative methods to obtain biodiesel from sources like animal fats, residual oils, and other non-food feedstocks. The use of other synthesis routes, such as interesterification (with methyl acetate and dimethyl carbonate) and hydro-esterification (chemistry, enzymatic or supercritical) is also reported (Flumignan et al. 2012).

8.5.2.1 Biodiesel Feedstocks

Oils and fats are composed of triacylglycerides, which consist of three fatty acid chains esterified to a glycerol backbone. Generally, oils consist mostly of the unsaturated fatty acid chains and are in the liquid state, while fats have a majority of saturated fatty acid components and are solid at room temperature.

The use of crude vegetable oils in diesel engines is possible, but their high viscosity and cold flow behavior cause overall damage to the engines. Thus, it is more interesting to apply vegetable oils as a source to obtain biodiesel. Nowadays, biodiesel derives majorly from refined vegetable oils (soy, corn, rapeseed, sunflower, etc.), but the use of other feedstocks, such as residual oils and fats (waste cooking oil, fish oil, beef tallow, chicken fat, etc.) and non-food crude oils (jatropha, macaw, crambe, etc.) has been growing. Residual and non-food feedstocks are an appealing alternative for environmental and economic reasons. Nevertheless, there are limitations to the use of such feedstocks in the transesterification process employing the usual conditions. In the presence of homogeneous alkaline catalysts, high free fatty acid and water contents can shift the reactants towards the saponification side reaction (Gardy et al. 2019).

Moreover, heterotrophic microalgae can be considered a neutral source of bioenergy; hence, the fact that they consume CO_2 from the environment around them. In comparison to vegetable sources, microalgae growth is faster and cheaper, and its use as a source of bioenergy does not compete well with other industrial sectors (Zhang et al. 2013). Microalgae can accumulate up to 60% (w/w) of lipids, which can be extracted and converted into biodiesel. Also, recent studies show that microalgae biomass can be used in direct transesterification without the need for lipid extraction (Pandit and Fulekar 2017, 2019). In this context, microalgae are presented as an economic and environmentally interesting source for biodiesel production.

8.5.2.2 Catalysts for Biodiesel Production

Catalysts are applied in chemical reactions to conduct the synthesis of the products through a path that requires lower activation energy when compared to catalyst-free reaction, without being consumed. The occurrence of esterification and transesterification of oils and fats to obtain biodiesel requires the use of catalysts. More commonly, alkaline catalysts provide highly efficient ester conversion in relatively short reaction times, when compared to acid catalysts (Gardy et al. 2019).

Homogenous catalysts are in the same phase as reactants in the reaction medium, whereas heterogeneous catalysts are in different phases. The use of homogeneous catalysts is widely known, but can also cause corrosion of systems, soap formation and require tedious purification steps to achieve recovery of products, which increases both process cost and waste production. In this context, heterogeneous catalysts can also provide efficient conversions and are easily removed from the reaction medium with simple purification steps such as decantation, filtration, and centrifugation. Furthermore, recyclable heterogeneous catalysts may be presented as a more efficient, alternative industrial application (Gardy et al. 2019; Jain et al. 2014; De and Boxi 2020; Zhong et al. 2020).

8.5.2.3 Nanocatalysts for Biodiesel Production

The use of nanosized particles as catalysts instead of other heterogeneous catalysts is advantageous considering the high surface/volume ratio of nano-compounds as well as high selectivity, easier recovery, and overall stability of catalytic activity when applied in successive reactions. The nanocatalyst quality depends on the physical properties of the materials used, such as size, shape, active sites distribution, thermal stability, chemical stability, and spatial and electronic properties (Gardy et al. 2019; Jain et al. 2014).

Table 8.2 Metal oxide nanocatalysts for biodiesel production via transesterification process

Reference	Feedstock	Catalyst	Transesterification conditions	Biodiesel Production
Pandit and Fulekar (2017)	*A. obliquus* biomass	CaO eggshell waste (1.7% w/w)	Algae:MeOH 1:10 (w/v)/70 °C/3.6 h	91.86% conversion; 86.41% yield
Pandit and Fulekar (2019)	*S. armatus* biomass	CaO eggshell waste (1.61% w/w)	Algae:MeOH 1:10 (w/v)/70 °C/3.6 h	90.44% yield
De and Boxi (2020)	Palm oil	Cu impregnated TiO_2 (3% w/w)	Oil:MeOH 1:20/ 45 °C/45 min	90.93% yield
Tan et al. (2017)	WCO	CaO ostrich shell waste (1.50% w/v)	Oil:MeOH 1:10/ 65 °C/2 h	98.97% yield
Abdelhady et al. (2020)	Sunflower oil	CaO eggshell waste (1.50% w/v)	Oil:MeOH 1:4.5/ 75 °C/1 h	94.70% yield
		CaO beet sugar waste (1%)		93% conversion
Borah et al. (2018)	*M. ferrea* oil	Co doped ZnO (2.5% w/w)	Oil:MeOH 1:9/ 60 °C/3 h	98.03% conversion
Borah et al. (2019)	WCO	Zn doped CaO from waste eggshell (5% w/w)	Oil:MeOH 1:20/ 65 °C/4 h	96.74% conversion
Baskar et al. (2018)	Castor oil	Ni doped ZnO (11% w/w)	Oil:MeOH 1:8/ 55 °C/1 h	95.20% yield
Feyzi and Shahbazi (2015)	Refined vegetable oil blend	Cs-Ca/TiO_2-SiO_2	Oil:MeOH 1:12/ 60 °C/2 h	98% yield
Raj et al. (2019)	*N. oculata* lipid extract	PEG capped Mn-ZnO (3.5% w/w)	Oil:MeOH 1:15/ 60 °C/4 h	87.5% yield
Justine et al. (2020)	WCO	ZnO	Oil:MeOH 1:6/2 h	81.6%
		ZnO-SiO_2		54.6%
Botti et al. (2020)	Soybean oil	Na-geopolymer (3% w/w)	150% MeOH/70–75 °C	85.1–89.9% yield

The reaction will occur in the active sites distributed throughout the surface of the material. Thus, the smaller the size of the particle, the greater the surface area and the greater the catalytic activity achievable. Also, nanosized particles can be dissolved, precipitated, and crystallized successively, depending on the conditions of the medium, which makes recyclability easier. Nanocatalysts can be obtained through chemical, physical, and biological processes (Jain et al. 2014). Different types of nanotechnology-based heterogeneous catalysts for biodiesel synthesis through transesterification are explored hereafter. Table 8.2 summarizes results from transesterification of vegetable and waste cooking oil (WCO) as well as algal biomass and crude oil by applying different nanosized metal oxide particles and geopolymers.

Calcium oxide (CaO) based catalysts are derived from waste produced in agricultural and industrial activities, such as animal bones, egg and animal shells, paper

Table 8.3 Magnetic nanocatalysts for biodiesel production via transesterification process

Reference	Feedstock	Catalyst	Transesterification conditions	Biodiesel production
Liu et al. (2016)	Soybean oil	$MgFe_2O_4$@CaO (1% w/w)	Oil:MeOH 1:12/ 70 °C/3 h	98.3% yield
Mapossa et al. (2020)	Soybean oil	$Ni_{0.3}Zn_{0.7}Fe_3O_4$ (2% w/w)	Oil:MeOH 1:12/ 180 °C/1 h	94% yield
Feyzi and Norouzi (2016)	Sunflower oil	Ca/Fe_3O_4@SiO_2	Oil:MeOH 1:15/ 65 °C/5 h	97% yield
Baskar and Soumiya (2016)	Castor oil	Fe (II) doped ZnO (14% w/w)	Oil:MeOH 1:12/ 50 °C/55 min	91% yield
Alaei et al. (2018)	Sunflower oil	MgO/$MgFe_2O_4$ (4% w/w)	Oil:MeOH 1:12/ 110 °C/4 h	91.2% conversion
Amani et al. (2019)	Sunflower oil	MgO/$MgFe_2O_4$ (3% w/w)	Oil:MeOH 1:12/ 110 °C/3 h	92.5% conversion
Banerjee et al. (2019)	*N. oleoabundans* lipid extract	Fe_2O_3 (1% w/w)	Biomass:MeOH 1:5 (w/v)/65 °C/6 h	86% yield

industry, etc. Such catalysts are highly alkaline, relatively economical, and require mild reaction conditions to obtain efficient ester conversions. They are obtained through the calcination of materials, which convert $CaCO_3$ into CaO (Pandit and Fulekar 2017, 2019; Tan et al. 2017; Abdelhady et al. 2020).

Zinc oxide (ZnO) can be obtained through precipitation in an aqueous solution and annealing in a heated oven. Also, doping of CaO and ZnO with metals such as cobalt and nickel shows interesting results in biodiesel conversion from vegetable oils (Borah et al. 2018, 2019; Baskar et al. 2018). Titanium dioxide (TiO_2) nanoparticles are also widely used for catalysis in different industrial sectors, including biodiesel production (De and Boxi 2020; Feyzi and Shahbazi 2015). Nanocomposites and geopolymers (alkaline aluminosilicate powders) can also be applied to oil and fat conversion into methyl esters (Raj et al. 2019; Justine et al. 2020; Botti et al. 2020; Bai and Colombo 2018). MNPs are composed of elements with magnetic properties, most commonly of iron, nickel, and cobalt. They can be obtained through combustion, co-precipitation, and thermal decomposition, amongst others methods (Liu et al. 2016; Mapossa et al. 2020; Feyzi and Norouzi 2016; Baskar and Soumiya 2016; Alaei et al. 2018; Amani et al. 2019; Banerjee et al. 2019). The magnetic properties are interesting to reduce the cost and labor of purification processes; MNPs can be easily removed from the reaction medium by using a magnet to apply an external magnetic field. Table 8.3 summarizes the results of the transesterification catalyzed by MNPs.

The use of MNPs as catalysts for biodiesel production and also the use of biocatalysts is interesting considering chemical catalysis. The use of enzymes (lipases) as catalysts for transesterification of oils and fats, when compared to chemical alkaline or acid catalysts, provides higher product selectivity and is advantageous for avoiding soap formation and other contaminations. However, enzyme cost still limits the application in industrial scales. In this context, enzyme

Table 8.4 Enzymatic magnetic nanocatalysts for biodiesel production via transesterification process

Reference	Feedstock	Catalyst	Transesterification conditions	Biodiesel production
Nematian et al. (2020)	*C. vulgaris* lipid extract	*R. oryzae* lipase immobilized in Fe_3O_4 nanoparticles	Three-step addition MeOH/45 °C/24 h	69.8% conversion
Xie and Huang (2018)	Soybean oil	*C. rugosa* lipase immobilized in grapheme oxide/Fe_3O_4 nanocomposite	Three-step addition of MeOH/40 °C	92.8% yield
Xie and Huang (2020)	Soybean oil	*C. rugosa* lipase immobilized in poly(glycidyl methacrylate-*co*—methacrylic acid)/Fe_3O_4 nanocomposite	Three-step addition of MeOH/40 °C	92.8% yield
Badoei-dalfard et al. (2019)	WCO	Cross-linked lipase aggregates with Fe_3O_4 (0.3% w/w)	Oil:MeOH 1:3/35 °C/36 h	71% conversion
Ashjari et al. (2020)	WCO	*R. miehei*lipase immobilized in Fe_3O_4@SiO_2 nanoparticles (15.2% w/w)	Three-step addition of MeOH/40 °C/48 h	55.3% yield
		*T. languginosus*lipase immobilized in Fe_3O_4@SiO_2 nanoparticles (18.6% w/w)		81% yield

immobilization is an alternative to reduce overall cost, for it makes it possible to recycle and reuse the biocatalysts. Immobilization consists of attaching the enzymes to the pores and/or surface of a chosen support material and can also enhance enzyme stability and improve kinetics (Zhong et al. 2020; Nematian et al. 2020).

Table 8.4 summarizes the results of the transesterification catalyzed by enzymatic MNPs. It is worth mentioning that there are specific (i.e. *C. rugosa* and *T. languginosus*) and 1,3-specific (*R. miehei* and *R. oryzae*) lipases; specific lipases can achieve a full ester conversion, whereas 1,3-specific lipases can only convert 2/3 of the fatty acids from the triacylglyceride. Also, lipases are inactivated by high concentrations of methanol. Thus, the three-step addition of the solvent to the medium is important to achieve high yields. Also, the immobilization of lipases in MNPs makes it possible to recycle the biocatalysts for an average of 3–5 cycles without significant activity loss (Xie and Huang 2018, 2020; Nematian et al. 2020; Badoei-dalfard et al. 2019; Ashjari et al. 2020).

The immobilization of lipases for biodiesel production is a promising field. Other recent researchers are focusing on the development of nanoparticles as support for lipase immobilization, though still without application in the transesterification reaction for biodiesel production (Atiroglu 2020; Asmat and Husain 2019).

8.5.2.4 Nanotechnology in Biohydrogen Production

Fossil fuels lead to serious environmental problems, which are responsible to worsen the greenhouse effect; however, the continuous growth of the world population and industrialized economy made them indispensables (Gaurav et al. 2017; Moreira et al. 2017). Thus, fossil fuels such as oil, coal, and natural gas have been known as the main source of energy over the last century so that they have contributed to 80% of the total energy produced, and dependence on them is expected to decrease to 78% by 2040 (Höök and Tang 2013). Therefore, the establishment of alternative energies (biofuels) is a top priority in developments sectors and is a target of big research efforts directed through process intensification to enhance the efficiency of biomass conversion in biorefineries (Gaurav et al. 2017).

Biohydrogen is the most efficient and cleanest carbon-free energy, and it is considered a valuable and alternative fuels carrier to fossil ones (Kumar et al. 2019b; Sindhu et al. 2019). It also has the potential to reduce greenhouse gases emissions, especially from the energy and transportation sectors. Biohydrogen production has been attracting global attention due to its social, economic, and environmental merits, and due to its high content of energy with an approximate value of 122–141 kJ/g, which is higher than that of other fuels, such as methane (55.65 kJ/g) and ethanol (29.7 kJ/g).

Hydrogen has been produced from fossil fuels, biomass, water, and the reform of natural gas; besides, hydrocarbon oxidation, coal gasification, electrolysis of water, and finally dark fermentation of organic substrates (Kumar and Himabindu 2019; Sindhu et al. 2019). Biohydrogen production by dark fermentation to generate hydrogen energy is a friendly environmental alternative to fossil fuels to help meet the needs of carbon emission reduction (Ren et al. 2011). Nevertheless, the quantity of biohydrogen produced via dark fermentation is low (Kumar et al. 2019b).

Nowadays, several advances and tools have been developed to increase the chance of enhancing dark fermentation for biohydrogen production. Recently, an application of nanoparticles (NPs) to enhance bioactivity and metabolite recovery during dark fermentation has gained enormous attention due to the unique surface and quantum size effect. Some examples of inorganic NPs that were used for enhancing biohydrogen production are silver, cobalt, titanium, nickel, and iron; the last one is one of the most promisors because of its versatility and compatibility with other additives (Kumar et al. 2019a). The effect of those nanomaterials could show a positive impact on metabolic key processes.

Yang and Wang (2018) described two mechanisms that enhance hydrogen production during fermentation and were related to a decline in the oxidation-reduction potential in the system, providing a better environment for fermentative bacteria, assisting in the removal of undesired oxygen, thereby contributing to a higher activity of the oxygen-sensitive hydrogenase. Both these mechanisms were studied in zero-valent iron nanoparticles (FeO NPs) supplementation. In this study, it was also proposed that Fe0 NPs could accelerate electron transfer between ferredoxin and hydrogenase and promote the activity of key enzymes by the released Fe^{2+}. The hydrogen yield obtained with Fe0 supplementation (400 mg/L) in this research was 73.1% higher than that of the control group. In 2016, Taherdanak and collaborators, also reported the use of Fe and Ni nanoparticles on dark hydrogen fermentation, specifically Fe0 and Ni0, and they compared them with their equivalents in ion form. Results showed that the order of the hydrogen yield effects was as follows: Ni^{2+} ion (55%) > FeO NPs (37%) > Fe^{2+} ion (15%) > NiO NPs (0.9%) compared with the control without supplementation.

In 2014, Mohanraj and collaborators also reported that an enhancement of ferredoxin oxidoreductase activity in response to NPs addition has been considered to be important to increase the hydrogen production yield during dark fermentation. Thereafter, in 2015, Gadhe and collaborators, showed that an improvement of biohydrogen production with a co-addiction of hematite (Fe_2O_3) plus nickel oxide (NiO) NPs at optimum concentration can be attributed to a higher activity of the ferredoxin oxidoreductase, ferredoxin, and hydrogenase enzymes by surface and quantum size effects of NPs. The hydrogen yield obtained by the co-addiction of Fe_2O_3 and NiO (50 mg/L and 10 mg/L respectively) was 1.2-fold higher than that of the addition of individual nanoparticles. Also, Zhang and collaborators (2018) studied other configurations of iron nanoparticles (ferric oxide/carbon nanoparticles—FOCNPs) for hydrogen production enhancement. Fe_2O_3/C NPs also showed good performance when added to a dark fermentative process based on glucose, reaching 33.7% improvement when FOCNPs were added in a concentration of 200 mg/L.

In 2015, Seelert and collaborators, used magnetite (Fe_3O_4) nanoparticles functionalized with chitosan and alginic acid polyelectrolytes, to promote bacterial attachment (immobilization). They used *Clostridium beijerinckii* with these nanoparticles, and its kinetics resulted in a shorter lag growth phase effect. The greatest hydrogen yield was 2.1 ± 0.7 mol H_2/mol glucose, corresponding to substrate conversion and energy conversion efficiencies of 52 ± 18 and $10 \pm 3\%$, respectively. According to Zhong and collaborators (2020), the addition of magnetite nanoparticles resulted in the formation of electronic conductor chains that enhance the electron transport efficiency and enhance key coenzymes activity in a complex consortium (anaerobic sludge), promoting a relative abundance of ethanol-hydrogen-producing bacteria. Results showed that an addition of 50 mg/L magnetite NPs improved H_2 production by 53.7%.

All these research advances show biohydrogen as one of the most promisor biofuels in the near future. However, there are many bottlenecks in this interesting bioprocess, such as sustainable pretreatments for substrates availability, enhancement stability of key enzymes and coenzymes, better performance in fermentation modes, etc. Thus, the inorganic nanoparticles could be a promising additive in practical application to achieve high hydrogen production, enhancing some of the main challenges that could currently appear in the main steps in bioprocesses.

8.6 Challenges in the Application of Nanotechnology in Biofuels Production

Apart from the advantages of the utilization of nanomaterials in the production of biofuel, several concerns and risks have arisen from the application of nanotechnology. In this regard, the challenges of the use of nanoparticles in biofuel production can be categorized into the following issues.

8.6.1 General Challenges

The nanoparticles could be applied successfully for the development of biofuel production. However, the characterization of many nano-additives studied for biofuel production has not been recognized well. In this regard, physical properties such as particle size, shape, and clustering have been paid less attention (Hossain et al. 2019). More studies should be carried out to solve the problems related to the use of nanomaterials which are accompanied by agglomeration, settling, and erosion. Moreover, little is known about the mechanisms of heat transfer where nanomaterials are applied (Khoo et al. 2020a, b).

On the other hand, enough availability of nanomaterials should be provided for industrial applications since a low quantity of nano-additives is used for laboratory scale. Furthermore, the choice of a proper nanomaterial, scientific approach used for the preparation of nanoparticles for biofuel production should be taken into account to attain the highest production of biofuels (Hossain et al. 2019).

8.6.2 Deleterious Effect of Nanoparticles on the Biofuel Producing Microorganisms

Biofuels are mainly produced by microorganisms. In this context, different yeast, bacteria, and microalgae are exploited for the production of liquid biofuels such as bioethanol, biobutanol, and biodiesel. Furthermore, gaseous biofuels such as biohydrogen as transportation biofuel are produced by microorganisms particularly bacteria (Abdeshahian et al. 2014; Shukor et al. 2014). There is a controversy about the deleterious effect of nanomaterials on microorganisms. It has been reported that carbon nanotubes such as Al_2O_3, CuO, ZnO, and TiO_2 cause toxic effects on the microalgae with oxidative stress, agglomeration, and inappropriate supply of nutrients to algal cells (Khoo et al. 2020a, b). The utilization of nanoparticles in electrodes made for microbial fuel cells (MFC) may cause toxic effects on electrogenic microorganisms including bacteria and fungi, which in turn decreases electricity generation.

8.6.3 The Cost-Effectiveness of Nanomaterials for Biofuel Production

One of the main limitations of the use of nanomaterials is the production costs of biofuel using nanoparticles. In this regard, many nano-materials are relatively expensive which affects their industrial utilization for the economical production of biofuel (Khoo et al. 2020a, b). The exploitation of nanomaterials in the chain of biofuel production consisting of the raw materials to end-product utilization could be analyzed in the aspect of the economic viability of the process. Hence, techno-economical assessment is necessary to evaluate whether the use of nanomaterials for biofuel production is economically variable as the commercialization of biofuel production using nanoparticles is drastically targeted in transportation sectors (Hossain et al. 2019).

8.6.4 Environmental Effect of Nanomaterials

The environmental toxicity of the nanoparticles has been poorly studied. It has been found that nanoparticles have toxic effects on the environment (Khoo et al. 2020a, b). Several nanoparticles are not degradable and can enter the environment and remain for a long time. The nanoparticles settled in the soil can penetrate the deeper layer of the ground and enter the groundwater sources (Engelmann and Hohendorff 2019).

The major concerns are related to the adsorption of the nanoparticles to living organisms which could be accumulated in the cells. In this line, it has been found that due to the low size of nanoparticles, biomolecules such as protein, lipid, and DNA could react with nanoparticles, thereby causing toxic effects on the organism cells. The toxicity of nanomaterials should be studied further in animal models to determine the possible damages to the human cells in the environment (Rai et al. 2016a, b).

8.6.5 Deleterious Effect of Nanomaterials on the Human Body

The nanoparticles could enter the human body through the respiratory system, alimentary canal, and skin injuries. Owing to the small size of the nanoparticles, there is a danger of entering the bloodstream (Engelmann and Hohendorff 2019). Nanoparticles can go to different organs via bloodstreams and enter human cells. They make oxidative reactions in the cells which, in turn, lead to cytotoxic reactions in many tissues. The organs with high metabolism such as the kidney, lung, heart, and liver are at a higher risk of the toxic effects obtained from nanomaterials. Hence, it is necessary to conduct more scientific research to find out the toxicity of the nanomaterials on the human body (Rai et al. 2016a, b).

8.7 Conclusions

It is a well-known fact that the continuous increase in global population and industrialization considerably increases the demand for fossil fuels and looking at limited resources of these fuels, these fuels may be depleted soon. However, environmental concerns like climate change and global warming are the other issues raised due to the burning of fossil fuels. In this context, biofuels are the only alternatives that are reported to mitigate these problems at a significant level. Considering the limitations of conventional approaches commonly used for biofuel production, nanotechnology has come up with the most promising solutions which can make biofuels production easy and economically viable. The direct or indirect use of nanotechnology in general and nanomaterials in particular in the production of various biofuels has been found to be the most effective move which can boost the conventional biorefining industries. Although primary studies conducted so far presented the positive side of nanotechnology in this aspect, there is a constant debate on the use of nanomaterials due to their toxicological concerns. There has been always a difference of opinions from the scientific community about the toxicity of nanomaterials, but we strongly think that further extensive studies are essentially required so that concrete evidence can come out about the toxicity of nanomaterials to the environment and associated living beings.

Acknowledgments The authors would like to thank the Brazilian Federal Agency for the Support and Evaluation of Graduate Education (Award Number 154193/2018-6), Research Council for the State of São Paulo (FAPESP) (Award Number 2014/27055-2; 2016/22086-2; 2016/10636-8; 2018/14095-7; 2016/22086-2; 2018/14095-7; 2020/06323-0). This study was financed in part by the Coordenação de Aperfeiçoamento de Pessoal de Nível Superior—Brasil (CAPES)—Finance Code 001.

References

Abdelhady HH, Elazab HA, Ewais EM, Saber M, El-Deab MS (2020) Efficient catalytic production of biodiesel using nano-sized sugar beet agro-industrial waste. Fuel 261:116481. https://doi.org/10.1016/j.fuel.2019.116481

Abdeshahian P, Al-Shorgani NKN, Salih NKM, Shukor H, Kadier A, Hamid AA, Kalil MS (2014) The production of biohydrogen by a novel strain *Clostridium* sp. YM1 in the dark fermentation process. Int J Hydrog Energy 39:12524–12531

Alaei S, Haghighi M, Toghiani J, Vahid BR (2018) Magnetic and reusable MgO/MgFe2O4 nanocatalyst for biodiesel production from sunflower oil: influence of fuel ratio in combustion synthesis on catalytic properties and performance. Ind Crop Prod 117:322–332. https://doi.org/10.1016/j.indcrop.2018.03.015

Amani T, Haghighi M, Vahid BR (2019) Microwave-assisted combustion design of magnetic Mg–Fe spinel for MgO-based nanocatalyst used in biodiesel production: influence of heating-approach and fuel ratio. J Ind Eng Chem 80:43–52. https://doi.org/10.1016/j.jiec.2019.07.029

ANP (2020) Biodiesel. http://www.anp.gov.br/biocombustiveis/biodiesel

Ashjari M, Garmroodi M, Asl FA, Emampour M, Yousefi M, Lish MP, Habibi Z, Mohammadi M (2020) Application of multicomponent reaction for covalent immobilization of two lipases on aldehyde-functionalized magnetic nanoparticles; production of biodiesel from waste cooking oil. Process Biochem 90:156–167. https://doi.org/10.1016/j.procbio.2019.11.002

Asmat S, Husain Q (2019) A robust nanobiocatalyst based on high performance lipase immobilized to novel synthesised poly(o-toluidine) functionalized magnetic nanocomposite: sterling stability and application. Mater Sci Eng C 99:25–36. https://doi.org/10.1016/j.msec.2019.01.070

Atabani AE, Silitonga AS, Ong HC et al (2013) Non-edible vegetable oils: a critical evaluation of oil extraction, fatty acid compositions, biodiesel production, characteristics, engine performance and emissions production. Renew Sust Energ Rev 18:211–245. https://doi.org/10.1016/j.rser.2012.10.013

Atiroglu V (2020) Lipase immobilization on synthesized hyaluronic acid-coated magnetic nanoparticle-functionalized graphene oxide composites as new biocatalysts: improved reusability, stability, and activity. Int J Biol Macromol 145:456–465. https://doi.org/10.1016/j.ijbiomac.2019.12.233

Avhad MR, Marchetti JM (2015) A review on recent advancement in catalytic materials for biodiesel production. Renew Sust Energ Rev 50:696–718. https://doi.org/10.1016/j.rser.2015.05.038

Badoei-dalfard A, Malekabadi S, Karami Z, Sargazi G (2019) Magnetic cross-linked enzyme aggregates of Km12 lipase: a stable nanobiocatalyst for biodiesel synthesis from waste cooking oil. Renew Energy 141:874–882. https://doi.org/10.1016/j.renene.2019.04.061

Bai C, Colombo P (2018) Processing, properties and applications of highly porous geopolymers: a review. Ceram Int 44(14):16103–16118. https://doi.org/10.1016/j.ceramint.2018.05.219

Banerjee S, Mudliar S, Sen R, Giri B, Satpute D, Chakrabarti T, Pandey RA (2010) Commercializing lignocellulosic bioethanol: technology bottlenecks and possible remedies. Biofuels Bioproducts Biorefining 4(1):77–93

Banerjee S, Rout S, Banerjee S, Atta A, Das D (2019) Fe2O3 Nanocatalyst aided transesterification for biodiesel production from lipid-intact wet microalgal biomass: a biorefinery approach. Energ Convers Manag 195:844–853. https://doi.org/10.1016/j.enconman.2019.05.060

Barbosa MJ, Rocha JMS, Tramper J, Wijffels RH (2001) Acetate as a carbon source for hydrogen production by photosynthetic bacteria. J Biotechnol 85:25–33. https://doi.org/10.1016/S0168-1656(00)00368-0

Baskar G, Soumiya S (2016) Production of biodiesel from Castor oil using Iron (II) doped zinc oxide nanocatalyst. Renew Energy 98:101–107. https://doi.org/10.1016/j.renene.2016.02.068

Baskar G, Selvakumari A, E., Aiswarya R. (2018) Biodiesel production from Castor oil using heterogeneous Ni doped ZnO Nanocatalyst. Bioresour Technol 250:793–798. https://doi.org/10.1016/j.biortech.2017.12.010

Beniwal A, Saini P, Kokkiligadda A et al (2018) Use of silicon dioxide nanoparticles for β-galactosidase immobilization and modulated ethanol production by co-immobilized K. marxianus and S. cerevisiae in deproteinized cheese whey. LWT Food Sci Technol 87:553–561. https://doi.org/10.1016/j.lwt.2017.09.028

Beschkov V (2012) Biogas, biodiesel and bioethanol as multifunctional renewable fuels and raw materials. In: Frontiers in bioenergy and biofuels, p 13

Boboescu IZ, Chemarin F, Beigbeder JB, de Vasconcelos BR, Munirathinam R, Ghislain T, Lavoie JM (2019) Making next-generation biofuels and biocommodities a feasible reality. Curr Opin Green Sustain Chem 20:25–32

Borah MJ, Devi A, Borah R, Deka D (2018) Synthesis and application of co doped ZnO as heterogeneous nanocatalyst for biodiesel production from non-edible oil. Renew Energy 133:512–519. https://doi.org/10.1016/j.renene.2018.10.069

Borah MJ, Das A, Das V, Bhuyan N, Deka D (2019) Transesterification of waste cooking oil for biodiesel production catalyzed by Zn substituted waste egg shell derived CaO nanocatalyst. Fuel 242:345–354. https://doi.org/10.1016/j.fuel.2019.01.060

Botti RF, Innocentini MDM, Faleiros TA, Mello MF, Flumignan DL, Santos LK, Franchin G, Colombo P (2020) Biodiesel processing using sodium and potassium geopolymer powders as heterogeneous catalysts. Molecules 25(12):1–14. https://doi.org/10.3390/molecules25122839

Brethauer S, Wyman CE (2010) Continuous hydrolysis and fermentation for cellulosic ethanol production. Bioresour Technol 101(13):4862–4874

Bugg TD, Ahmad M, Hardiman EM, Rahmanpour R (2011) Pathways for degradation of lignin in bacteria and fungi. Nat Prod Rep 28(12):1883–1896

Callegari A, Bolognesi S, Cecconet D, Capodaglio G (2020) Production technologies, current role, and future prospects of biofuels feedstocks: a state-of-the-art review. Crit Rev Environ Sci Technol 50:384–436. https://doi.org/10.1080/10643389.2019.1629801

Carvalheiro F, Duarte LC, Gírio FM (2008) Hemicellulose biorefineries: a review on biomass pretreatments. J Sci Ind Res 67:849–864. https://doi.org/10.1016/j.talanta.2015.06.045

Cerda A, Mejias L, Rodríguez P, Rodríguez A, Artola A, Font X et al (2019) Valorisation of digestate from biowaste through solid-state fermentation to obtain value added bioproducts: a first approach. Bioresour Technol 271:409–416

Chandel AK, Garlapati VK, Singh AK et al (2018) The path forward for lignocellulose biorefineries: bottlenecks, solutions, and perspective on commercialization. Bioresour Technol 264:370–381. https://doi.org/10.1016/j.biortech.2018.06.004

Chandrasekhar K, Lee YJ, Lee DW (2015) Biohydrogen production: strategies to improve process efficiency through microbial routes. Int J Mol Sci 16:8266–8293. https://doi.org/10.3390/ijms16048266

Chen XY, Vinh-Thang H, Ramirez AA et al (2015) Membrane gas separation technologies for biogas upgrading. RSC Adv 5:24399–24448. https://doi.org/10.1039/c5ra00666j

Cheng S, Logan BE (2007) Sustainable and efficient biohydrogen production via electrohydrogenesis. Proc Natl Acad Sci U S A 104:18871–18873. https://doi.org/10.1073/pnas.0706379104

Cheng JJ, Timilsina GR (2011) Status and barriers of advanced biofuel technologies: a review. Renew Energy 36(12):3541–3549

Cherian E, Dharmendirakumar M, Baskar G (2015) Immobilization of cellulase onto MnO2 nanoparticles for bioethanol production by enhanced hydrolysis of agricultural waste. Chinese J Catal 36(8):1223–1229. https://doi.org/10.1016/s1872-2067(15)60906-8

Cherubini F, Bird ND, Cowie A, Jungmeier G, Schlamadinger B, Woess-Gallasch S (2009) Energy- and greenhouse gas-based LCA of biofuel and bioenergy systems: key issues, ranges and recommendations. Resour Conserv Recycl 53(8):434–447

Chozhavendhan S, Singh MVP, Fransila B et al (2020) A review on influencing parameters of biodiesel production and purification processes. CRGSC 1:1–6. https://doi.org/10.1016/j.rser.2015.05.038

Chua SY, Goh CMH, Tan YH et al (2020) Biodiesel synthesis using natural solid catalyst derived from biomass waste—a review. JIEC 81:41–60. https://doi.org/10.1016/j.jiec.2019.09.022

Costa OYA, Souto BM, Tupinambá DD et al (2015) Microbial diversity in sugarcane ethanol production in a Brazilian distillery using a culture-independent method. J Ind Microbiol Biotechnol 42:73–84. https://doi.org/10.1007/s10295-014-1533-1

Covert T, Greenstone M, Knittel CR (2016) Will we ever stop using fossil fuels? J Econ Perspect 30:117–138

Dadak A, Aghbashlo M, Tabatabaei M, Najafpour G, Younesi H (2016) Sustainability assessment of photobiological hydrogen production using anaerobic bacteria (Rhodospirillum rubrum) via exergy concept: effect of substrate concentrations. Environ Prog Sustain Energy 35 (4):1166–1176

Das D, Veziroğlu TN (2001) Hydrogen production by biological processes: a survey of literature. Int J Hydrog Energy 26:13–28. https://doi.org/10.1016/S0360-3199(00)00058-6

Das RK, Brar SK, Verma M (2016) Potential use of pulp and paper solid waste for the bio-production of fumaric acid through submerged and solid state fermentation. J Clean Prod 112:4435–4444

De A, Boxi SS (2020) Application of Cu impregnated TiO2 as a heterogeneous nanocatalyst for the production of biodiesel from palm oil. Fuel 265:117019. https://doi.org/10.1016/j.fuel.2020.117019

de Gonzalo G, Colpa DI, Habib MH, Fraaije MW (2016) Bacterial enzymes involved in lignin degradation. J Biotechnol 236:110–119

Dhillon GS, Kaur S, Brar SK, Verma M (2012) Potential of apple pomace as a solid substrate for fungal cellulase and hemicellulase bioproduction through solid-state fermentation. Ind Crop Prod 38:6–13

Dotaniya ML, Aparna K, Dotaniya CK, Singh M, Regar KL (2019) Role of soil enzymes in sustainable crop production. In: Enzymes in food biotechnology. Academic, pp 569–589

Duff SJ, Murray WD (1996) Bioconversion of forest products industry waste cellulosics to fuel ethanol: a review. Bioresour Technol 55(1):1–33

Engelmann W, Hohendorff RV (2019) Regulatory challenges in nanotechnology for sustainable production of biofuel in Brazil. In: Rai M, Ingle AP (eds) Sustainable bioenergy. Elsevier, England, pp 367–381

Eryılmaz EB, Dursun D, Dalgıç AC (2016) Multiple optimization and statistical evaluation of astaxanthin production utilizing olive pomace. Biocatal Agric Biotechnol 7:224–227

Fan S, Liu J, Tang X et al (2019) Process operation performance of PDMS membrane pervaporation coupled with fermentation for efficient bioethanol production. Chin J Chem Eng 27 (6):1339–1347. https://doi.org/10.1016/j.cjche.2018.12.005

Feyzi M, Norouzi L (2016) Preparation and kinetic study of magnetic Ca/Fe3O4@SiO2 nanocatalysts for biodiesel production. Renew Energy 94:579–586. https://doi.org/10.1016/j.renene.2016.03.086

Feyzi M, Shahbazi E (2015) Chemical catalytic performance and characterization of Cs—Ca/SiO2—TiO2 nanocatalysts for biodiesel production. J Mol Cat A 404:131–138. https://doi.org/10.1016/j.molcata.2015.04.018

Flumignan DL, Oliveira JE, Hatanaka RR, Sequinel R (2012) Avanços Brasileiros No Desenvolvimento de Normas Técnicas Analíticas Para Certificação e Controle Da Qualidade de Biodiesel. In: Stradiotto NR, Lemos EGM (editors) Bioenergia: Desenvolvimento, Pesquisa e Inovação. São Paulo, pp 319–39. http://www.culturaacademica.com.br/catalogo/bioenergia/

Foust TD, Aden A, Dutta A, Phillips S (2009) An economic and environmental comparison of a biochemical and a thermochemical lignocellulosic ethanol conversion processes. Cellulose 16 (4):547–565

Gaikwad S, Ingle A, da Silva S, Rai M (2018) Immobilized nanoparticles-mediated enzymatic hydrolysis of cellulose for clean sugar production: a novel approach. Curr Nanosci 15:296–303

Gardy J, Rehan M, Hassanpour A, Lai X, Nizami AS (2019) Advances in nano-catalysts based biodiesel production from non-food feedstocks. J Environ Manag 249:109316. https://doi.org/10.1016/j.jenvman.2019.109316

Gaurav N, Sivasankari S, Kiran GS et al (2017) Utilization of bioresources for sustainable biofuels: a review. Renew Sust Energ Rev 73:205–214

Guo XM, Trably E, Latrille E et al (2010) Hydrogen production from agricultural waste by dark fermentation: a review. Int J Hydrog Energy 35:10660–10673. https://doi.org/10.1016/j.ijhydene.2010.03.0

Hahn-Hägerdal B, Karhumaa K, Fonseca C, Spencer-Martins I, Gorwa-Grauslund MF (2007) Towards industrial pentose-fermenting yeast strains. Appl Microbiol Biotechnol 74(5):937–953

Hallenbeck PC (2011) Microbial paths to renewable hydrogen production. Biofuels 2:285–302. https://doi.org/10.4155/bfs.11.6

Hendriks ATWM, Zeeman G (2009) Pretreatments to enhance the digestibility of lignocellulosic biomass. Bioresour Technol 100(1):10–18

Hibino T, Kobayashi K, Ito M et al (2018) Direct electrolysis of waste newspaper for sustainable hydrogen production: an oxygen-functionalized porous carbon anode. Appl Catal B Environ 231:191–199. https://doi.org/10.1016/j.apcatb.2018.03.021

Höök M, Tang X (2013) Depletion of fossil fuels and anthropogenic climate change: a review. Energy Policy 52:797–809

Hossain N, Mahlia TMI, Saidur R (2019) Latest development in microalgae-biofuel production with nano-additives. Biotechnol Biofuels 12(125):1–16

Ingle AP, Chandel AK, Antunes FAF, Rai M, Silva SS (2019a) New trends in application of nanotechnology for the pretreatment of lignocellulosic biomass. Biofuels Bioprod Bioref 13:776–788. https://doi.org/10.1002/bbb.1965

Ingle AP, Philippini RR, da Silva SS (2019b) Pretreatment of sugarcane bagasse using two different acid-functionalized magnetic nanoparticles: a novel approach for high sugar recovery. Renew Energy 150:957–964. https://doi.org/10.1016/j.renene.2019.11.146

Ingle A, Philippini R, Silvério da Silva S (2020a) Pretreatment of sugarcane bagasse using two different acid-functionalized magnetic nanoparticles: a novel approach for high sugar recovery. Renew Energy 150:957–964

Ingle AP, Philippini RR, Melo YCS et al (2020b) Acid-functionalized magnetic nanocatalysts mediated pretreatment of sugarcane straw: an eco-friendly and cost-effective approach. Cellulose. https://doi.org/10.1007/s10570-020-03262-y

Itskos G, Nikolopoulos N, Kourkoumpas D, Koutsianos A, Violidakis I, Drosatos P, Grammelis P (2016) Energy and the environment. In: Environment and development: basic principles, human activities, and environmental implications. Elsevier Inc., pp 363–452

Jain S, Pant K, Kassaye S, Singh S (2014) Nanocatalysts: green fuel and chemical production. In: Dekker encyclopedia of nanoscience and nanotechnology, 3rd edn, pp 1–15. https://doi.org/10.1081/e-enn3-120053518

Johnson FX (2017) Biofuels, bioenergy and the bioeconomy in North and South. Ind Biotechnol 13:289–291

Justine M, Prabu HJ, Johnson I, Raj DMA, Sundaram SJ, Kaviyarasu K (2020) Synthesis and characterizations studies of ZnO and ZnO-SiO2 nanocomposite for biodiesel applications. Mater Today. https://doi.org/10.1016/j.matpr.2020.05.034

Kamani MH, Eş I, Lorenzo JM et al (2019) Advances in plant materials, food by-products, and algae conversion into biofuels: use of environmentally friendly technologies. Green Chem 21:3213–3231. https://doi.org/10.1039/c8gc03860k

Karagiannidis A, Perkoulidis G (2009) A multi-criteria ranking of different technologies for the anaerobic digestion for energy recovery of the organic fraction of municipal solid wastes. Bioresour Technol 100:2355–2360. https://doi.org/10.1016/j.biortech.2008.11.033

Khan I, Saeed K, Khan I (2019) Nanoparticles: properties, applications and toxicities. Arab J Chem 12:908–931

Khoo KS, Chai WY, Tang DYY, Show PL, Chew KW, Chen WH (2020a) Nanomaterials utilization in biomass for biofuel and bioenergy production. Energies 13:892. https://doi.org/10.3390/en13040892

Khoo KS, Chia WY, Tang DYY, Show PL, Chew KW, Chen WH (2020b) Energies. 13, 829, 1–19

Kim Y-K, Park SE, Lee H et al (2014) Enhancement of bioethanol production in syngas fermentation with Clostridium ljungdahlii using nanoparticles. Bioresour Technol 159:446–450. https://doi.org/10.1016/j.biortech.2014.03.046

Konishi M, Yoshida Y, Horiuchi JI (2015) Efficient production of sophorolipids by Starmerella bombicola using a corncob hydrolysate medium. J Biosci Bioeng 119(3):317–322

Kovács KL, Maróti G, Rákhely G (2006) A novel approach for biohydrogen production. Int J Hydrog Energy 31:1460–1468. https://doi.org/10.1016/j.ijhydene.2006.06.011

Kumari D, Singh R (2018) Pretreatment of lignocellulosic wastes for biofuel production: a critical review. Renew Sust Energ Rev 90:877–891

Kumar SS, Himabindu V (2019) Hydrogen production by PEM water electrolysis: a review. Mater Sci Energ Technol 2:442–454

Kumar M, Oyedun AO, Kumar A (2019a) A comparative analysis of hydrogen production from the thermochemical conversion of algal biomass. Int J Hydrogen Energ 44:10384–10397

Kumar S, Sharma S, Thakur S et al (2019b) Bioprospecting of microbes for biohydrogen production: current status and future challenges. Bioproc Biomol Prod 2040:443–471

Kushwaha D, Upadhyay SN, Mishra PK (2018) Nanotechnology in bioethanol/biobutanol production. In: Srivastava N, Srivastava M, Pandey H et al (eds) Green nanotechnology for biofuel production. Springer Nature, Gewerbestrasse, pp 115–127

Lamb CDC, Silva BMZD, de Souza D, Fornasier F, Riça LB, Schneider RDCDS (2018) Bioethanol production from rice hull and evaluation of the final solid residue. Chem Eng Commun 205 (6):833–845

Lazar Z, Liu N, Stephanopoulos G (2018) Holistic approaches in lipid production by Yarrowia lipolytica. Trends Biotechnol 36(11):1157–1170

Leong WH, Lim JW, Lam MK et al (2018) Third generation biofuels: a nutritional perspective in enhancing microbial lipid production. Renew Sust Energ Rev 91:950–961. https://doi.org/10.1016/j.rser.2018.04.066

Limayem A, Ricke SC (2012) Lignocellulosic biomass for bioethanol production: current perspectives, potential issues and future prospects. Prog Energy Combust Sci 38(4):449–467

Liu Y, Zhang P, Fan M, Jiang P (2016) Biodiesel production from soybean oil catalyzed by magnetic nanoparticle MgFe2O4@CaO. Fuel 164:314–321. https://doi.org/10.1016/j.fuel.2015.10.008

López-Domínguez CM, Ramírez-Sucre MO, Rodríguez-Buenfil IM (2019) Enzymatic hydrolysis of Opuntia ficus-indica cladode by Acinetobacter pittii and alcohol fermentation by Kluyveromyces marxianus: pH, temperature and microorganism effect. Biotechnol Rep 24: e00384

Łukajtis R, Hołowacz I, Kucharska K et al (2018) Hydrogen production from biomass using dark fermentation. Renew Sust Energ Rev 91:665–694. https://doi.org/10.1016/j.rser.2018.04.043

Mahdavi M, Abedini E, Darabi AH (2015) Biodiesel synthesis from oleic acid by nanocatalyst (ZrO2/Al2O3) under high voltage conditions. RSC Adv 5:55027–55032. https://doi.org/10.1039/C5RA07081C

Mahlia TMI, Syazmi ZAHS, Mofijur M et al (2020) Patent landscape review on biodiesel production: technology updates. Renew Sust Energ Rev 118:109526. https://doi.org/10.1016/j.rser.2019.109526

Mahmudul HM, Hagos FY, Mamat R et al (2017) Production, characterization and performance of biodiesel as an alternative fuel in diesel engines—a review. Renew Sust Energ Rev 72:497–509. https://doi.org/10.1016/j.rser.2017.01.001

Mapossa AB, Dantas J, Silva MR, Kiminami RHGA, Costa ACFM, Daramola MO (2020) Catalytic performance of NiFe2O4 and Ni0.3Zn0.7Fe2O4 magnetic nanoparticles during biodiesel production. Arab J Chem 13(2):4462–4476. https://doi.org/10.1016/j.arabjc.2019.09.003

Mishra V, Jana AK, Jana MM, Gupta A (2017) Enhancement in multiple lignolytic enzymes production for optimized lignin degradation and selectivity in fungal pretreatment of sweet sorghum bagasse. Bioresour Technol 236:49–59

Mitchell L, Crosman E, Jacques A, Fasoli B, Leclair-Marzolf L, Horel J, Bowling D, Ehleringer J, Lin J (2018) Monitoring of greenhouse gases and pollutants across an urban area using a light-rail public transit platform. Atmos Environ 187:9–23

Mood SH, Golfeshan AH, Tabatabaei M et al (2013) Lignocellulosic biomass to bioethanol, a comprehensive review with a focus on pretreatment. Renew Sust Energ Rev 27:77–93. https://doi.org/10.1016/j.rser.2013.06.033

Moreira FS, Machado RG, Romão BB et al (2017) Improvement of hydrogen production by biological route using repeated batch cycles. Process Biochem 58:60–68

Mudhoo A, Forster-Carneiro T, Sánchez A (2011) Biohydrogen production and bioprocess enhancement: a review. Crit Rev Biotechnol 3:250–263. https://doi.org/10.3109/07388551.2010.525497

NASA. Global Climate Change (2020) COVID-19 mitigation leads to reduced pollution associated with decreased fossil fuel use. https://climate.nasa.gov/climate_resources/224/covid-19-mitigation-leads-to-reduced-pollution-associated-with-decreased-fossil-fuel-use/. Accessed 18 July 2020

Nematian T, Salehi Z, Shakeri A (2020) Conversion of bio-oil extracted from Chlorella Vulgaris micro algae to biodiesel via modified superparamagnetic nano-biocatalyst. Renew Energy 146:1796–1804. https://doi.org/10.1016/j.renene.2019.08.048

Niehus X, Crutz-Le Coq AM, Sandoval G, Nicaud JM, Ledesma-Amaro R (2018) Engineering Yarrowia lipolytica to enhance lipid production from lignocellulosic materials. Biotechnol Biofuels 11(1):1–10

Nobre BP, Villalobos F, Barragán BE et al (2013) A biorefinery from Nannochloropsis sp. microalga—extraction of oils and pigments. Production of biohydrogen from the leftover biomass. Bioresour Technol 135:128–136. https://doi.org/10.1016/j.biortech.2012.11.084

OECD/FAO (2019) OECD-FAO agricultural outlook 2019–2028. OECD Publishing, Paris/Food and Agriculture Organization of the United Nations, Rome. https://doi.org/10.1787/agr_outlook-2019-en

Pandit PR, Fulekar MH (2017) Egg shell waste as heterogeneous nanocatalyst for biodiesel production: optimized by response surface methodology. J Environ Manag 198:319–329. https://doi.org/10.1016/j.jenvman.2017.04.100

Pandit PR, Fulekar MH (2019) Biodiesel production from Scenedesmus Armatus using egg shell waste as nanocatalyst. Mater Today 10:75–86. https://doi.org/10.1016/j.matpr.2019.02.191

Pena L, Ikenberry M, Hohn KL et al (2012) Acid-functionalized nanoparticles for pretreatment of wheat straw. J Biomater Nanobiotechnol 3:342–352. https://doi.org/10.4236/jbnb.2012.33032

Pena L, Xu F, Hohn KL et al (2014) Propyl-sulfonic acid functionalized nanoparticles as catalyst for pretreatment of corn stover. J Biomater Nanobiotechnol 5:8–16. https://doi.org/10.4236/jbnb.2014.51002

Peng W, Hao P, Luo J, Peng B, Han X, Liu H (2020) Guanidine-functionalized amphiphilic silica nanoparticles as a pickering interfacial catalyst for biodiesel production. Ind Eng Chem Res 59:4273–4280

Poorakbar E, Shafiee A, Saboury AA et al (2008) Synthesis of magnetic gold mesoporous silica nanoparticles core-shell for cellulase enzyme immobilization: improvement of enzymatic activity and thermal stability. Process Biochem 71:92–100

Qureshi AS, Khushk I, Ali CH, Chisti Y, Ahmad A, Majeed H (2016) Coproduction of protease and amylase by thermophilic Bacillus sp. BBXS-2 using open solid-state fermentation of lignocellulosic biomass. Biocatal Agric Biotechnol 8:146–151

Rai M, Santos JC, Soler MF et al (2016a) Strategic role of nanotechnology for production of bioethanol and biodiesel. Nanotechnol Rev 5(2):231–250. https://doi.org/10.1515/ntrev-2015-0069

Rai M, dos Santos JC, Soler MF, Marcelino PRF, Brumano LP, Ingle AP, Gaikwad S, Gade A, da Silva SS (2016b) Strategic role of nanotechnology for production of bioethanol and biodiesel. Nanotechnol Rev 5:231–250

Rai M, Ingle AP, Pandit R, Paralikar P, Biswas JK, Silva SS (2019) Emerging role of nanobiocatalysts in hydrolysis of lignocellulosic biomass leading to sustainable bioethanol production. Catal Rev 61:1–26. https://doi.org/10.1080/01614940.2018.1479503

Raj VA, Bharathiraja JB, Vijayakumar B, Arokiyaraj S, Iyyappan J, Kumar RP (2019) Biodiesel production from microalgae Nannochloropsis oculata using heterogeneous poly ethylene glycol (PEG) encapsulated ZnOMn2+ nanocatalyst. Bioresour Technol 282:348–352. https://doi.org/10.1016/j.biortech.2019.03.030

Rastogi M, Shrivastava S (2017) Recent advances in second generation bioethanol production: an insight to pretreatment, saccharification and fermentation processes. Renew Sust Energ Rev 80:330–340

Ren N, Guo W, Liu B et al (2011) Biological hydrogen production by dark fermentation: challenges and prospects towards scaled-up production. Curr Opin Biotechnol 22:365–370

Roberts LG, Patterson TJ (2014) Biofuels. In: Encyclopedia of toxicology, 3rd edn. vol 1, pp 469–475. https://doi.org/10.1016/B978-0-12-386454-3.01054-X
Ruane J, Sonnino A, Agostini A (2010) Bioenergy and the potential contribution of agricultural biotechnologies in developing countries. Biomass Bioenergy 34(10):1427–1439
Rupprecht J, Hankamer B, Mussgnug JH et al (2006) Perspectives and advances of biological H2 production in microorganisms. Appl Microbiol Biotechnol 72:442–449. https://doi.org/10.1007/s00253-006-0528-x
Saha BC (2003) Hemicellulose bioconversion. J Ind Microbiol Biotechnol 30(5):279–291
Sampath PP, Raghava KR, Venkata CR et al (2020) Biohydrogen production from organic waste—a review. Chem Eng Technol 43(7):1240–1248. https://doi.org/10.1002/ceat.20190040
Sanchez OJ, Cardona CA (2008) Trends in biotechnological production of fuel ethanol from different feedstocks. Bioresour Technol 99(13):5270–5295
Sánchez-Ramírez J, Martínez-Hernández JL, Segura-Ceniceros P et al (2016) Cellulases immobilization on chitosan-coated magnetic nanoparticles: application for Agave atrovirens lignocellulosic biomass hydrolysis. Bioprocess Biosyst Eng 40(1):9–22. https://doi.org/10.1007/s00449-016-1670-1
Sanusi IA, Suinyuy TN, Lateef A et al (2020) Effect of nickel oxide nanoparticles on bioethanol production: process optimization, kinetic and metabolic studies. Process Biochem 92:386–400. https://doi.org/10.1016/j.procbio.2020.01.029
Sarno M, Iuliano M (2019) Highly active and stable Fe3O4/Au nanoparticles supporting lipase catalyst for biodiesel production from waste tomato. Appl Surf Sci 474:135–146
Sarrouh BF, da Silva SS (2010) Application of response surface methodology for optimization of xylitol production from lignocellulosic hydrolysate in a fluidized bed reactor. Chem Eng Technol 33(9):1481–1487
Scholl AL, Menegol D, Pitarelo AP, Fontana RC, Zandoná Filho A, Ramos LP et al (2015) Ethanol production from sugars obtained during enzymatic hydrolysis of elephant grass (Pennisetum purpureum, Schum.) pretreated by steam explosion. Bioresour Technol 192:228–237
SCOPUS (2020) TITLE-ABS-KEY (lignocellulosic AND ethanol). https://www.scopus.com/term/analyzer.uri?sid=5ca7d65320dd9964d56eb031ae7bd31d&origin=resultslist&src=s&s=TITLE-ABS-KEY%28lignocellulosic+ethanol%29&sort=plf-f&sdt=b&sot=b&sl=38&count=5578&analyzeResults=Analyze+results&txGid=93b208c49c03b595a6db0ff2787edafc. Accessed 18 July 2020
Sekoai PT, Ouma CNM, du Preez SP et al (2019) Application of nanoparticles in biofuels: an overview. Fuel 237:380–397. https://doi.org/10.1016/j.fuel.2018.10.030
Semwal S, Arora AK, Badoni RP, Tuli DK (2011) Biodiesel production using heterogeneous catalysts. Bioresour Technol 3:2151–2161. https://doi.org/10.1016/j.biortech.2010.10.080
Shanmugam S, Hari A, Pandey A, Mathimani T, Felix LO, Pugazhendhi A (2020) Comprehensive review on the application of inorganic and organic T nanoparticles for enhancing biohydrogen production. Fuel 270:117453. https://doi.org/10.1016/j.fuel.2020.117453
Sharma YC, Singh B, Upadhyay SN (2008) Advancements in development and characterization of biodiesel: a review. Fuel 12:2355–2373. https://doi.org/10.1016/j.fuel.2008.01.014
Shibuya M, Sasaki K, Tanaka Y et al (2017) Development of combined nanofiltration and forward osmosis process for production of ethanol from pretreated rice straw. Bioresour Technol 235:405–410. https://doi.org/10.1016/j.biortech.2017.03.158
Shukor H, Al-Shorgani NKN, Abdeshahian P, Hamid AA, Anuar N, Rahman NA, Kalil MS (2014) Production of butanol by Clostridium saccharoperbutylacetonicum N1-4 from palm kernel cake in acetone-butanol-ethanol fermentation using an empirical model. Bioresour Technol 170:565–573
Sindhu R, Binod P, Pandey A, et al (2019) Biofuel production from biomass: toward sustainable development. In: Current developments in biotechnology and bioengineering: waste treatment processes for energy generation. Elsevier B.V., pp 79–92
Singh S, Jain S, Venkateswaran PS et al (2015) Hydrogen: a sustainable fuel for future of the transport sector. Renew Sust Energ Rev 51:623–633. https://doi.org/10.1016/j.rser.2015.06.040

Singh D, Sharma D, Soni SL et al (2020) A review on feedstocks, production processes, and yield for different generations of biodiesel. Fuel 262:116553. https://doi.org/10.1016/j.fuel.2019.116553

Soam S, Kapoor M, Kumar R, Gupta RP, Puri SK, Ramakumar SSV (2018) Life cycle assessment and life cycle costing of conventional and modified dilute acid pretreatment for fuel ethanol production from rice straw in India. J Clean Prod 197:732–741

Spyridon A, Willem Euverink GJ (2016) Consolidated briefing of biochemical ethanol production from lignocellulosic biomass. Electron J Biotechnol 19(5):44–53

Srivastava N, Singh J, Ramteke PW et al (2015) Improved production of reducing sugars from rice straw using crude cellulase activated with Fe3O4/alginate nanocomposite. Bioresour Technol 183:262–266. https://doi.org/10.1016/j.biortech.2015.02.059

Srivastava RK, Shetti NP, Reddy KR, Aminabhavi TM (2020) Biofuels, biodiesel and biohydrogen production using bioprocesses. A review. Environ Chem Lett. 1–24. https://doi.org/10.1007/s10311-020-00999-7

Sun Y, Cheng J (2002) Hydrolysis of lignocellulosic materials for ethanol production: a review. Bioresour Technol 83(1):1–11

Suo H, Xu L, Xue Y et al (2020) Ionic liquids-modified cellulose coated magnetic nanoparticles for enzyme immobilization: improvement of catalytic performance. Carbohydr Polym 234:115914. https://doi.org/10.1016/j.carbpol.2020.115914

Tan YH, Abdullah MO, Nolasco-Hipolito C, Zauzi NSA (2017) Application of RSM and Taguchi methods for optimizing the transesterification of waste cooking oil catalyzed by solid ostrich and chicken-eggshell derived CaO. Renew Energy 114:437–447. https://doi.org/10.1016/j.renene.2017.07.024

Teo S, Islam A, Chan E, Thomas Choong S, Alharthi N, Taufiq-Yap Y, Awual M (2019) Efficient biodiesel production from Jatropha curcus using CaSO4/Fe2O3-SiO2 core-shell magnetic nanoparticles. J Clean Prod 208:816–826

Vaghari H, Jafarizadeh-Malmiri H, Mohammadlou M et al (2015) Application of magnetic nanoparticles in smart enzyme immobilization. Biotechnol Lett 38(2):223–233. https://doi.org/10.1007/s10529-015-1977-z

Verma ML, Chaudhary R, Tsuzuki T et al (2013) Immobilization of β-glucosidase on a magnetic nanoparticle improves thermostability: application in cellobiose hydrolysis. Bioresour Technol 135:2–6. https://doi.org/10.1016/j.biortech.2013.01.047

Vohra M, Manwar J, Manmode R, Padgilwar S, Patil S (2014) Bioethanol production: feedstock and current technologies. J Environ Chem Eng 2(1):573–584. https://doi.org/10.1016/j.jece.2013.10.013

Wackett LP (2011) Engineering microbes to produce biofuels. Curr Opin Biotechnol 22(3):388–393

Xie W, Huang M (2018) Immobilization of Candida Rugosa lipase onto graphene oxide Fe3O4 nanocomposite: characterization and application for biodiesel production. Energ Convers Manag 159:42–53. https://doi.org/10.1016/j.enconman.2018.01.021

Xie W, Huang M (2020) Fabrication of immobilized Candida Rugosa lipase on magnetic Fe3O4-poly(Glycidyl methacrylate-co-Methacrylic acid) composite as an efficient and recyclable biocatalyst for enzymatic production of biodiesel. Renew Energy 158:474–486. https://doi.org/10.1016/j.renene.2020.05.172

Xu X, Nie S, Ding H, Hou F (2018) Environmental pollution and kidney diseases. Nature Rev Nephrol 14:313–324

Xu C, Nasrollahzadeh M, Sajjadi M, Maham M, Luque R, Puente-Santiago AR (2019) Benign-by-design nature-inspired nanosystems in biofuels production and catalytic applications. Renew Sust Energ Rev 112:195–252. https://doi.org/10.1016/j.rser.2019.03.062

Xue C, Wang Z-X, Du G-Q et al (2016) Integration of ethanol removal using carbon nanotube (CNT)-mixed membrane and ethanol fermentation by self-flocculating yeast for antifouling ethanol recover. Process Biochem 51(9):1140–1146. https://doi.org/10.1016/j.procbio.2016.05.030

Yahmed NB, Carrere H, Marzouki MN, Smaali I (2017) Enhancement of biogas production from Ulva sp. by using solid-state fermentation as biological pretreatment. Algal Res 27:206–214

Yang G, Wang J (2018) Improving mechanisms of biohydrogen production from grass using zero-valent iron nanoparticles. Bioresour Technol 266:413–420. https://doi.org/10.1016/j.biortech.2018.07.004

Zabed H, Sahu JN, Suely A, Boyce AN, Faruq G (2017) Bioethanol production from renewable sources: current perspectives and technological progress. Renew Sust Energ Rev 71:475–501

Zhang XL, Yan S, Tyagi RD, Surampalli RY (2013) Biodiesel production from heterotrophic microalgae through transesterification and nanotechnology application in the production. Renew Sust Energ Rev 26:216–223. https://doi.org/10.1016/j.rser.2013.05.061

Zhong L, Feng Y, Wang G, Wang Z, Bilal M, Lv H, Jia S, Cui J (2020) Production and use of immobilized lipases in/on nanomaterials: a review from the waste to biodiesel production. Int J Biol Macromol 152:207–222. https://doi.org/10.1016/j.ijbiomac.2020.02.258

Zuccaro G, Pirozzi D, Yousuf A (2020) Lignocellulosic biomass to biodiesel. In: Yousuf A, Pirozzi D, Filomena S (eds) Lignocellulosic biomass to liquid biofuels. Academic, Cambridge, pp 127–167

Chapter 9
Role of Metabolic Engineering and Synthetic Biology in the Development of Microbes for Biofuel Production

Raubins Kumar, Anju Mayadevi Nair, and Syed Shams Yazdani

Abstract Limited source of fossil fuel, increasing global need of energy, and environmental concerns due to the use of fossil fuels, such as global warming, pollution, and ozone layer depletion, led to demand for a non-conventional and sustainable source of energy. Biofuels are one such source of energy that are produced from various biological sources. Currently, biofuels are mainly produced from sugars and vegetable oil from sugarcane, corn, soybean, palm, etc. and are catagorized as first-generation biofuels. Considering that these feedstocks are also used for human consumption, the focus has now shifted on second-generation biofuels where non-edible plants and agricultural residues are explored for biofuel production. However, due to the complexity of the feedstock involved, second-generation biofuels need extensive research and development before it can be treated as economically viable. Similarly, third-generation biofuels that are produced from microalgae also need much innovation before it can be commercially realized. In recent years, the rapid development in the field of metabolic engineering and synthetic biology for microorganisms like bacteria, yeast, cyanobacteria, fungi, and algae provides an immense scope for using these techniques and methods for engineering the microbes for higher production of various fuel molecules. In this chapter, we discuss the advancement in the area of metabolic engineering and synthetic biology, along with their applications in enhanced production of biofuels from different microorganisms.

Keywords Biofuels · Metabolic engineering · Synthetic biology · Microorganism

R. Kumar · A. M. Nair · S. S. Yazdani (✉)
Microbial Engineering Group, International Centre for Genetic Engineering and Biotechnology, Aruna Asaf Ali Marg, New Delhi, India
e-mail: shams@icgeb.res.in

J. K. Saini, R. K. Sani (eds.), *Microbial Biotechnology for Renewable and Sustainable Energy*, Clean Energy Production Technologies,
https://doi.org/10.1007/978-981-16-3852-7_9

9.1 Introduction

As the world population increases, the proportional increase in energy demand is inevitable. The limited energy source of conventional fossil fuels and petroleum fuels are unlikely to fulfill the future global energy demand. Further, the partial combustion of these fossil fuels deteriorates the environment and leads to increased CO_2 and CO level in the air, high pollution, global warming, and ozone layer depletion. To overcome these problems, there is a requirement of alternative energy sources, which should ideally be renewable as well as sustainable in nature (Yim et al. 2011).

Biofuel is one of the best alternative sources of energy to meet our global energy demand. Today, biofuel is generally produced from edible plant crops like sugarcane, corn, soybeans, and vegetable oils and is termed as the first-generation biofuel. However, the first-generation biofuel production also sparked energy versus food debate, leading to scientists and policy makers to focus on second-generation biofuel that is produced from non-food feedstock like lignocellulosic biomass, agricultural residues, and other biological wastes. Nonetheless, the current cost of processing to make second-generation biofuel is so high that it cannot be made as a suitable and cheaper source of energy (Dahman et al. 2019). Similar issues exist with the third-generation biofuel as well where microalgae are used as platform to produce lipid. Therefore, microorganisms like bacteria, yeast, fungi, cyanobacteria, and algae are engineered to produce advanced fuel molecules like ethanol, propanol, butanol, 2-methyl-1-butanol, 3-methyl-1-butanol, and isobutanol. Properties such as low hygroscopicity, high energy content, less volatility, and compatibility with existing engines make longer chain alcohols a good substitute for gasoline (Atsumi et al. 2008b).

Various genetic engineering methods have been used to engineer the natural pathways of many microorganisms for biofuel production. The newly emerging fields of metabolic engineering and synthetic biology have boosted the effort to engineer these microbes for the production of various molecules like biofuel, industrial relevant chemicals, and medicinally important compounds. Earlier, one or two gene(s) or enzyme(s) were targeted for engineering the microbes. Now the metabolic engineering provides tools for the integrative investigation of the complete pathway, along with modulation, and optimization of these pathways for increased production of the desired product. Similarly, synthetic biology provides the facility of designing and synthesizing a complete non-natural pathway in a microbial host to facilitate production of molecules of commercial importance (Stephanopoulos 2012).

In this chapter, we will review the definitions of metabolic engineering and synthetic biology, and the synergies and differences which exist between these two fields. Various tools and techniques developed in these fields and their application will also be reviewed. Finally, the successful examples of different microbial hosts like bacteria, yeast, fungi, and algae that have been engineered by the application of metabolic engineering and synthetic biology techniques for the production of biofuel molecules will be highlighted.

9.2 Metabolic Engineering and Synthetic Biology: Definition, Synergy, and Differences

Defining the two terminologies, i.e., metabolic engineering and synthetic biology, distinctly is difficult as most of the time these two terms are used in a closely related context. In this section, we will discuss various definitions that have been attributed to them, and the synergy and the basic difference between these two fields of biological science.

9.2.1 Metabolic Engineering

The field of metabolic engineering foregrounded in the early 1990s when Bailey discussed in his article the development of microorganisms as a platform for producing various renewable chemicals and fuels by using metabolic engineering tools (Bailey 1991). For the enhancement of yield and productivity of metabolites, researchers mainly focus on enzyme overexpression and other modifications in the pathway of that particular product. Stephanopoulos and Vallino described the concept of metabolic flux optimization of the pathway by altering the metabolic rigidity for high product yield after the transformation of the host with desired genes that encode the synthesis of a particular product (Stephanopoulos and Vallino 1991).

Thus, the metabolic engineering is broadly defined as the discipline of engineering that involves genetic modulation and metabolic flux optimization of pathways of living cells or microorganism, for overproduction of desired products like biofuels, chemicals, and pharmaceuticals (Stephanopoulos 2012).

Although the metabolic engineering process uses all the methodologies and techniques of genetic engineering, the distinctions have been made between the two streams. Whereas genetic engineering explores the changes in an individual gene or enzyme, and operon or gene cluster, metabolic engineering includes a comprehensive analysis of integrated metabolic pathways and genetic regulatory networks (Bailey 1991). It is far more complex rather than just arranging the genes together to make a functional pathway (Stephanopoulos 2012).

In fact, metabolic engineering covers all common applications of genetic engineering like gene deletion, gene replacement, the introduction of recombinant DNA cassette, heterologous gene expression of foreign DNA in the non-natural host, etc. (Jullesson et al. 2015). Apart from these molecular biology techniques, it also includes analysis of the metabolic pathways to find target gene for genetic manipulation (Ostergaard et al. 2000; Nielsen and Jewett 2008), redirection of metabolic flux for the pathway modification (Park et al. 2007), alteration of protein level inside the engineered cell, control of fine-tuned gene expression, and also the control of regulators of gene expression (Lee and Lee 2006; Tang and Zhao 2009).

Nielsen suggested following three basic steps of metabolic engineering which can be used or applied for any host engineering: (1) a pathway modeling or design, to target genes for genetic engineering, (2) a pathway construction, in recombinant strains with improved properties, (3) optimization and analysis of recombinant strains, in terms of their performance in comparison with the wild type strain (Nielsen 2001). Metabolic engineering mostly targets the following properties of the host cells for their improvement: (1) establishment of pathways for new products, (2) removal of by-product formation pathways, (3) improvement of overall cellular physiology, (4) enhancement of yield or productivity, (5) heterologous protein expression in a new host, and (6) adaptation of host to various substrate ranges. Various tools of metabolic engineering and their application for biofuel production will be discussed in the later section.

9.2.2 Synthetic Biology

The exact era of the origin of synthetic biology is not very well understood. But some of the researches believe that the emergence of synthetic biology is triggered by the development of cheap and fast method for chemical synthesis of DNA, whereas others think that synthetic biology originated with the development of a genetic oscillatory network (Elowitz 2000) and genetic toggle (an on/off switch having two repressible promoters that inhibit each other) in *E. coli* (Gardner et al. 2000). Furthermore, few support the idea that synthetic biology developed after assembly of full pathway has been achieved with the help of synthetic DNA elements (synthetic DNA which encodes scaffold proteins that direct the intracellular signaling pathway) (Good et al. 2011; Stephanopoulos 2012). Cameron et al. described a timescale-based history of synthetic biology (Cameron et al. 2014). They divided the synthetic biology timeline in three distinct periods—(1) foundation period, (2) intermediate period, and (3) recent period, based on scientific milestones developed in this field. Synthetic biology has distinct relations with metabolic engineering, genetic engineering, and system biology, and it also uses the tools and applications from these fields for further advancements in understanding the functionality of living cells. Metabolic engineering targets on the metabolic pathway alterations and optimization for the maximum production of simple and cost-effective chemicals from natural hosts, but synthetic biology attempts to design synthetic genetic components (promoters, RBS, terminators, and transcription regulators), genetic circuits (toggle switches, oscillators, and repressilators) and assembly of devices for developing model organisms with predicted behaviors of a natural host. Genetic engineering includes the transfer of heterologous gene or operon from one microorganism to another, but synthetic biology includes assembly of new genetic elements or genetic circuits or complete genome, which are well standardized for their function, in a natural host cell or microbe. System biology involves an integrated approach of modeling and simulations like genomics, transcriptomics, and metabolomics to explore the information of a whole biological system and their comparison with the experimental data, whereas synthetic biology focuses on the synthesis of small genetic components to be used in a biological system or synthesis of a whole artificial biological system by using the tools and techniques of system biology (Choi et al. 2019).

Synthetic biology simply defined as—a field of biology and engineering that aims to design synthetic genetic components like promoters, RBS, transcriptional regulators, synthetic DNA circuits, synthetic pathways to modify the biological parts of living cells or even design a completely synthetic organism that does not exist naturally. But there was no proper definition of synthetic biology available until 2014 when the European Commission came up with an operational definition for synthetic biology. European Commission, 2014 defines—"synthetic biology is the application of science, technology, and engineering to facilitate and accelerate the design, manufacture and/or modification of genetic material in living organisms." The definition was further extended by UN Convention of Biodiversity in 2015—"Synthetic biology is a further development and new dimension of modern biotechnology that combines science, technology, and engineering to facilitate and accelerate the understanding, design, redesign, manufacture and/or modification of genetic materials, living organisms and biological systems."

Recently, development of new techniques, for example, different -omics, metagenomics, DNA synthesis, BioBricks (Storch et al. 2015), recombinase technologies, Gibson assembly (Casini et al. 2013), Golden Gate assembly (Potapov et al. 2018), Gap-repair, Lambda-red, MAGE (Multiplex automated genome engineering) (Wang et al. 2012), gTME (global transcription machinery engineering) (Tan et al. 2016), small RNA (sRNA) (Na et al. 2013), CRISPR-Cas9 (Cho et al. 2017), and synthetic scaffold (Lee et al. 2018), has given an immense scope to synthetic biology from the artificial synthesis of codon-optimized genes to assembly of a complete genome (Gibson et al. 2010). Various tools and applications of synthetic biology for higher production of biofuel from different microbes will be discussed in the later section.

9.2.3 Synergies Between Metabolic Engineering and Synthetic Biology

Industrial biotechnology and system metabolic engineering are two fields that extensively exploit the methods and tools of metabolic engineering and synthetic biology. Synergistically, metabolic engineering and synthetic biology facilitate the design and construction of various cell factories with improved and robust properties for the high output of hundreds of chemicals, including biofuels (Atsumi et al. 2008a) and pharmaceuticals (Ajikumar et al. 2010). In industries, a limited number of microbial platforms, for example, *Escherichia coli* and *Saccharomyces cerevisiae*, are being used for a variety of biochemical production. This idea is beneficial in many aspects—first, it is less capital intensive, as there is only need to construct a new pathway on the same platform for producing a new product. When a new pathway is inserted (by use of synthetic biology), initially there is likely to be a

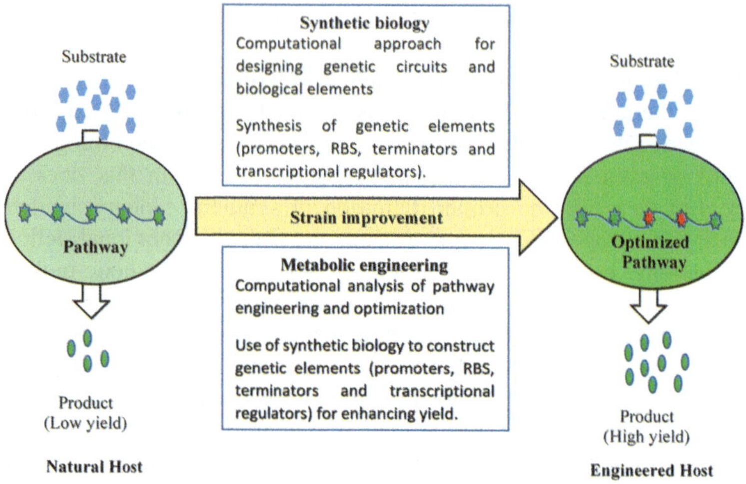

Fig. 9.1 Schematic representation of synergy between metabolic engineering and synthetic biology

low yield of the new product. But by optimization and redirection of metabolic fluxes with the help of metabolic engineering, yield can be improved significantly. Second, more information about the system will be generated by using the same organisms repeatedly. Knowledge about all the interacting components of a biological system can lead to the development of mathematical models. These knowledge base and mathematical models can be applied to predict new models in other organisms to develop new platforms (Nielsen and Keasling 2011).

Computational approaches are utilized for designing metabolic pathways in metabolic engineering and prediction of genetic elements or regulatory circuits in synthetic biology (Holtz and Keasling 2010). Synthetic biology-derived genetic devices or genetic components (promoters, ribosome-binding sites, terminators, and transcriptional regulators) are used in the field of metabolic engineering for controlling biosynthetic pathways in natural hosts (Nielsen et al. 2014). A schematic representation of synergy between these two fields has been shown in Fig. 9.1.

9.2.4 *Differences Between Metabolic Engineering and Synthetic Biology*

Basic aim of these two fields is different. The fundamental objective of metabolic engineering is to develop a biological platform for higher production of the desired chemical. Synthetic biology has the aim to build many biological elements, genetic parts, and modules, which can be applied to manipulate many biological systems. Metabolic engineering is a top-down method (also called as re-engineering) as it involves alteration in already existing pathways or redirection of metabolic flux

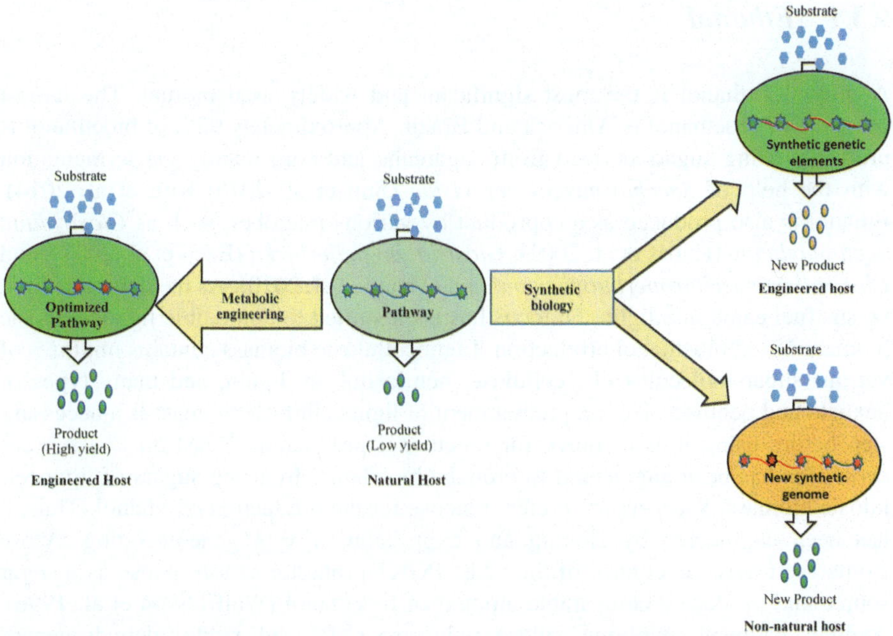

Fig. 9.2 Schematic representation of differences between metabolic engineering and synthetic biology

towards the desired product, whereas synthetic biology is a bottom-up approach (also called as forward engineering) which involves the synthesis of new biological genetic elements or new non-natural hosts (Nielsen and Keasling 2011). Figure 9.2 shows the schematic representation of differences between these two fields.

9.3 Metabolic Engineering and Synthetic Biology Applications for Advancing Research in Microbes for Biofuel Production

A variety of microorganisms, for example, bacteria (*Escherichia coli, Clostridia, Streptococcus, Bacillus,* etc.), cyanobacteria (*Synechococcus, Synechocystis, Nostoc, Spirulina,* etc.), yeast (*Saccharomyces cerevisiae*, *Pichia pastoris, Yarrowia, Rhodotorula*), fungi (*Trichoderma, Aspergillus, Penicillium*, etc.), and microalgae (*Chlorella*, *Parachlorella, Dunaliella*, etc.), have been modified for the production of biofuel molecules like ethanol, propanol, isopropanol, butanol, 3-methyl-1-butanol, isobutanol, biodiesel, and long-chain alkanes, using metabolic engineering and synthetic biology techniques. Here, we will discuss different biofuel molecules produced from the engineered strains of microorganisms.

9.3.1 Ethanol

At present, ethanol is the most significant and widely used biofuel. The largest producer of bioethanol is America and Brazil. Approximately 98% of bioethanol is produced using sugars derived from sugarcane and corn mainly via fermentation with the help of *Saccharomyces cerevisiae* (Sun et al. 2015; Kim et al. 2014). Ethanol is also produced as a coproduct by various microbes, such as *Clostridium acetobutylicum* (Harris et al. 2001), *Clostridium beijerinckii* (Ezeji et al. 2007), and *Clostridium saccharoperbutylacetonicum* (Thang et al. 2010). As the debate of food versus fuel came into light, the focus has been shifted to non-edible lignocellulosic biomass-based bioethanol production. Lignocellulosic biomass contains mixtures of various sugar molecules like cellulose, hemicellulose, lignin, and many types of hexoses and pentoses. Hence, pretreatment of lignocellulosic biomass is a necessary step before using it as a source for bioethanol production. Yeast *Saccharomyces cerevisiae* has been engineered to produce bioethanol by using sugars of lignocellulosic biomass. *S. cerevisiae* prefers glucose for the production of ethanol. Thus, it has been engineered by cloning and expression of *xylA* gene-encoding xylose isomerase under the control of the yeast PGK1 promoter to use xylose as a sugar source and produce a comparable amount of bioethanol (Walfridsson et al. 1996). Another pathway involving xylose reductase (XR) and xylitol dehydrogenase (XDH), where xylose is first converted to xylitol and then oxidized to xylulose, is also commonly used in yeast for fermenting xylose to ethanol (Walfridsson et al. 1997)

The strain was further improved for ethanol production by expressing genes transporter for xylose uptake (Katahira et al. 2008) and cellobiose uptake (Ha et al. 2011). Attempts were also made to develop more effective technique for ethanol production via consolidated bioprocessing (CBP). This method consolidates three steps—secretary expression of cellulase, hydrolysis of cellulose, and production of biofuel in one pot (Lynd et al. 2008).

In recent years, a technique called biomass gasification has got attention due to its potential of converting lignocellulosic biomass into syngas containing hydrogen and CO and CO_2. These oxides of carbon (CO, CO_2) can get assimilated in microbes by the Calvin cycle and the Wood-Ljungdahl pathway. Microbes carrying these pathways have been modified for ethanol production. In photosynthetic plants and some microorganisms, the Calvin cycle has been known to assimilate CO_2 in the form of glucose and other carbohydrates. Calvin cycle yields precursor for glucose formation in three steps—CO_2 fixation, reduction of 3-phosphoglycerate, and ribulose 1,5-bisphosphate recycling. 3-phosphoglycerate can be catalyzed to produce pyruvate and acetyl-CoA, which can further be converted to ethanol. In metabolic engineered cyanobacteria like *Synechococcus elongatus,* flux from pyruvate has been redirected towards ethanol production by expressing pyruvate decarboxylase from *Zymomonas* spp. (Dexter et al. 2015). In acetogens, CO and CO_2 are used as carbon sources via the Wood-Ljungdahl (WL) pathway. The WL pathway consists of two branches—first is the methyl branch, where CO_2 is converted to formate by the enzyme formate dehydrogenase, further the formate produces methylated corrinoid iron-sulfur protein (methyl-CoFeSP) by a cascade of reactions, and second is the carbonyl branch, where carbon monoxide dehydrogenase converts CO_2 to a carbonyl group. The sequential action of acetyl-CoA synthase unite complex methyl-CoFeSP (of the first step) and carbonyl group (of the second step) to generate acetyl-CoA (Jones et al. 2016). Different acetogens like *Clostridium carboxidivorans* have been engineered to convert acetyl-CoA into ethanol via the WL pathway (Cheng et al. 2019).

9.3.2 1-Propanol

1-Propanol or *n*-propanol is a better biofuel than ethanol as it is one carbon atom longer with higher octane value than ethanol. Synthesis of propanol has been achieved through many pathways, for example, the Wood-Werkman pathway, acrylate pathway, threonine pathway, citramalate pathway, succinate pathway, and 1,2-propanediol pathway. Atsumi et al. reported propanol synthesis in *Escherichia coli* by engineering L-threonine pathway. L-threonine was first transformed into 2-ketobutyrate by L-threonine dehydratase (*llvA/TdcB*) and then converted to 1-propanol using 2-ketoacid decarboxylase (*kdc*) (Atsumi et al. 2008b). Further, the group engineered the *E. coli* for heterologous expression of citramalate synthase enzyme from *Methanococcus jannaschii.* This enzyme transforms pyruvate directly into 2-ketobutyrate and hence skips the threonine biosynthesis pathway, making the pathway for 1-propanol synthesis shorter (Atsumi and Liao 2008). Later on, the production of 1-propanol in *E. coli* was enhanced by synergistically combining the threonine pathway and the citramalate pathway. Here, the higher yield (0.15 g/g of glucose) and productivity (0.12 g/L/h) of 1-propanol was achieved in comparison to individual pathways, i.e., the threonine pathway or the citramalate pathway (Shen and Liao 2013).

9.3.3 Isopropanol

Clostridium spp. are native producers of isopropanol. Hanai et al. first reported isopropanol production from engineered *E. coli* by the heterologous expression of a combination of genes from *Clostridium acetobutylicum* (*thl* gene, encodes for acetyl coenzyme A acetyltransferase and *adc* gene, encodes for acetolactate decarboxylase), *E. coli* (*atoAD* gene, encodes for acetoacetyl-CoA transferase), and *C. beijerinckii* (*adh* gene, encodes for alcohol dehydrogenase). This engineered strain of *E. coli* productivity for isopropanol (0.41 g/L/h) is much higher than the native producer *C. beijerinckii* (0.18 g/L/h) (Hanai et al. 2007).

The cyanobacteria *Synechococcus elongatus* has been also genetically modified for production of isopropanol by Kusakabe et al. They made a synthetic pathway with the same genes using same species of bacteria as mentioned above by Hanai et al. and reported 26.5 mg/L of isopropanol after 9 days under anaerobic and dark, optimized conditions (Kusakabe et al. 2013). Soma et al. developed a genetically engineered *E. coli* for production of isopropanol from cellobiose, where cellobiose degrading enzyme β-glucosidase was present on the cell surface of *E. coli*. They have reported 69 mM of isopropanol production in 21 h by cellobiose fermentation, which is 34.6% lower than that of glucose (105.4 mM) (Soma et al. 2012). Isopropanol production pathway was also established in *Cupriavidus necator* strain Re2133, by deleting genes (*phaB1B2B3* and *phaC*) encoding for poly-3-hydroxybutyrate [P(3HB)] and overexpressing two native genes, i.e., *phaA* and *phaC* and two heterologous genes, i.e., *adc* and *adh*. These deletions and expression helped to divert flux from P(3HB) towards isopropanol, producing about 3.44 g/L of isopropanol from fructose (Grousseau et al. 2014).

9.3.4 1-Butanol

Butanol has lower hygroscopicity, lower vapor pressure than ethanol, high octane value, and high energy content. These properties make butanol an alternative biofuel to substitute gasoline. Few species of *Clostridium* naturally produce butanol via the CoA-dependent pathway. Since *Clostridium* is slow growing species, and there is a limitation of available tools of genetic engineering in *Clostridium*, the focus has been shifted towards construction recombinant *E. coli* and *S. cerevisiae* for production of butanol. A set of essential genes *thl* (encodes for acetyl-CoA acetyltransferase), *hbd* (encodes for acetoacetyl-CoA thiolase), *crt* (encodes for 3-hydroxybutyryl-CoA dehydrogenase), *bcd* (butyryl-CoA dehydrogenase), *etf* (electron transfer flavoprotein), and *adhE2* (aldehyde/alcohol dehydrogenase) were transformed into *E. coli*, which produced around 13.9 mg/L of butanol. Further, by deleting competing pathways and optimizing culture media, the butanol production was increased to 552 mg/L (Atsumi et al. 2008a). The final improved strain of *E. coli* produced 30 g/L butanol anaerobically by—modifying NADH level via expression of formate dehydrogenase and disruption of competing NADH consuming pathway, introducing irreversible gene *ter* (*trans*-enoyl-CoA reductase) for stopping the reversible reaction in the pathway, and replacing rate-limiting *Clostridium thl* gene with *atoB* gene of *E. coli* (Shen et al. 2011).

Cyanobacteria *Synechococcus* sp. has also been engineered for the production of butanol by expressing five heterologous genes: *atoB*, (from *E. coli*), *hbd, crt,* and *adhE2*, (from *C. acetobutylicum*) and *ter* (from *Treponema denticola*). The engineered *Synechococcus* strain made 0.01 g/L of butanol during anoxic fermentation performed with 150 μE m^{-2} s^{-1} of light for 7 days (Lan and Liao 2011).

Genetically engineered *E. coli* has been developed to produce *n*-butanol via fatty acid biosynthesis (FASII) pathway. This pathway involves two steps—first production of butyric acid (Jawed et al. 2016) and second conversion of butyric acid into corresponding four carbons chain alcohol, i.e., butanol (Jawed et al. 2020). In the first step, the team has compared three thioesterases from three different bacterial species—*Anaerococcus tetradius* (TesAT), *Bryantella formatexigens* (TesBF), and *Bacteroides thetaiotaomicron* (TesBT). The maximum yield of butyric acid was 1.46 g/L by *E. coli* strain that expressed *tesBT* gene. Process optimization and fed-batch cultivation in phosphorus and carbons limiting source, finally produced 14.3 g/L of butyric acid (Jawed et al. 2016). In second step, butyric acid was converted to butanol. Here, *E. coli* cells expressing native carboxylic acid reductase (*car*) gene and phosphopantetheinyl transferase (*spf*) gene from *Bacillus subtilis* convert butyric acid into butyraldehyde, and subsequently alcohol dehydrogenase (*adh2*) gene from *Saccharomyces cerevisiae* converts butyraldehyde to butanol. Under fed-batch condition, the co-cultivation of butyric acid producing strain from step 1 and butanol producing strain from step 2 yielded 2 g/L butanol titer. In mono-cultivation method under same fed-batch condition, where single strain expressing all the necessary genes to convert glucose into butanol via butyraldehyde, yielded 2.9 g/L of butanol, which was 2.45 fold higher than the co-cultivation approach (Jawed et al. 2020).

9.3.5 Isobutanol

Isobutanol is one of the isomers of n-butanol and has similar properties like butanol (i.e., less corrosive, high energy content, less vapor pressure than ethanol, and less hygroscopic). At the same time, isobutanol is less toxic to the microorganism than linear butanol. Hence, it is a good biofuel candidate to substitute gasoline, as it is also compatible with the existing infrastructure. Atsumi et al. engineered *E. coli* for isobutanol production by introducing only two heterologous genes, i.e., *kivD* from *Lactococcus lactis* and *adh2* from *S. cerevisiae*. *kivD* encodes 2-keto-isovalerate decarboxylase that converts 2-keto-isovalerate to isobutyraldehyde, which got converted to isobutanol by aldehyde dehydrogenase encoded by *adh2* gene. The concentration of isobutanol is dependent on the level of keto-acids. Hence, to increase more amount of keto-acids, the endogenous *ilvCD* gene of *E. coli* and another gene *alsS* from *Bacillus subtilis* that is known to produce more amount of keto-acids were overexpressed. More amount of keto-acids led to more amount of isobutanol production (22 g/L) in *E. coli* (Atsumi et al. 2008b).

Since this method utilizes the intermediates of the amino acid biosynthesis pathway, the accumulation of different keto-acids via this method can produce different alcohol molecules. For example, (1) isoleucine biosynthesis pathway produces 2-ketobutyrate (a precursor for 1-propanol) and 2-keto-3-methyl-valerate (a precursor for 2-methyl-1-butanol), (2) leucine biosynthesis pathway generates 2-keto-4-methyl-pentanoate (a precursor for 3-methyl-1-butanol), (3) phenylalanine biosynthesis pathway generates phenylpyruvate (a precursor for 2-phenylethanol), and (4) norvaline biosynthesis pathway generates precursor for 1-butanol (Atsumi et al. 2008b).

E. coli has also been genetically engineered to produce isobutanol from lignocellulosic biomass having cellobiose. A β-glucosidase gene was expressed on the cell surface of *E. coli* or in extracellular media, which helped in the degradation of cellobiose and cells used it as sole carbon source. Then, the genes involved in isobutanol synthesis were introduced into the system. This system produced isobutanol titer 7.64 g/L at productivity of 0.16 g/L/h (Desai et al. 2014).

S. cerevisiae produces isobutanol as a by-product in fermentation. This isobutanol formation happens as a result of the catabolism of valine in the cytosol via Ehrlich pathway. On the other hand, valine is formed in mitochondria from pyruvate. The spatial separation of valine synthesis and valine degradation in two different compartments of the yeast cell is one of the major restrictions for higher isobutanol production. Wess et al. improved the titer of isobutanol production in yeast by relocating the enzymes of valine synthesis and valine degradation in one cell compartment, i.e., cytosol and also blocking the competing pathways. By this, the team achieved 59.55 mg isobutanol/g of glucose in *S. cerevisiae* at shake flask level (Wess et al. 2019).

9.3.6 Biodiesel

Biodiesel is an environment friendly, non-toxic, and biodegradable alternative energy source. The principal source of industrial production of biodiesel is triacylglyceride-rich vegetable oils. It raises a public concern because a vast land is required for growing seeds for the production of vegetable oils for biodiesel. Hence, focus shifted towards the microbial production of biodiesel as a renewable source of energy. The free fatty acid can be used as a precursor for biofuel synthesis. *E. coli* can produce free fatty acids by expressing gene encoding acyl-acyl carrier protein (acyl-ACP) thioesterase, which breaks the long-chain fatty acyl-ACP to release free fatty acid. Zhang et al. reported the overexpression of the acyl-ACP thioesterase gene from five different heterologous microorganisms in *E. coli.* The acyl-ACP thioesterase gene from *Ricinus communis* and *Jatropha curcas*, produced the maximum free fatty acid with a titer of ~2.0 g/L in 48 h (Zhang et al. 2011). A modular engineering technique has been developed by Xu et al. to remove bottlenecks and improvement in the flux towards fatty acid metabolic pathway. The fatty acid biosynthesis pathway of *E. coli* is categorized into three modules: first, formation of acetyl-CoA, second, activation of acetyl-CoA, and third, fatty acid synthase module. The optimization of these three modules at the level of transcriptional and translation led to a balanced availability of acetyl-CoA and utilization of malonyl-CoA/ACP. The improved strain produced 8.6 g/L of free fatty acid in the fed-batch fermentation (Xu et al. 2013).

9.3.7 Alkanes/Alkenes

Bio-based alkanes or alkenes are important biofuel molecules as they have properties similar to hydrocarbons present in petroleum products (Rahman et al. 2014). Depending upon the number of carbon atoms present, alkane or alkene can be short chain (C_4-C_{12}) and long chain (C_{13}-C_{17}) (Peralta-Yahya et al. 2012). The simplest alkane used as fuel is methane, a major component of natural gas. Methanogens are engineered for overproduction of methane (Lieber et al. 2014). Challenges associated with methane-like production need strict anaerobic condition, difficulties in capture and storage of methane, greenhouse effect of methane, have given attention towards the longer chain alkane or alkene production (Sun et al. 2015). Naturally, cyanobacteria are known to have a pathway for production of alkanes. The pathway consists of two steps—first, acyl-ACP reductase (AAR) converts fatty acyl-ACP into fatty aldehyde and second aldehyde-deformylating oxygenase (ADO) converts them into alkane. The synthesis and secretion of a mixtures of alkanes and alkenes with C_{13} to C_{17} carbon chain length were shown in *E. coli* by heterologous expression of this alkane operon (Schirmer et al. 2010). ADO activity is rate-limiting step for alkane production. Fused expression of AAR-ADO leads to increased production of alkanes by 4.8 fold in comparison to strain expressing these two enzymes separately. The spatial organization of ADO: AAR-binding site on DNA scaffold in 3:1 ratio further increased alkane production by 8.8 fold as compared to control strain (Rahman et al. 2014). Choi et al. developed *E. coli* as platform for production of short-chain alkanes (SCAs), free fatty acids (FFAs), fatty esters, and fatty alcohols. Mutated thioesterase that catalyzes the conversion of short-chain fatty acyl-ACP into their respective free fatty acids is subsequently converted into short-chain alkanes by action of fatty acyl-CoA synthetase (FadD), acyl-ACP reductase (AAR), and aldehyde-deformylating oxygenase (ADO). They also disrupted the *fadE* and *fadR* genes to avoid catalysis of fatty acyl-CoA via β-oxidation pathway, which further enhanced fatty acid biosynthesis. The developed strain of *E. coli* produced 580.8 mg/L of SCA (Choi and Lee 2013). Fatma et al. developed flux balance analysis (FBA)-based metabolic model in *E. coli* for the production of hydrocarbons. Based on in silico predictions, overexpression of *aar, ado,* and *zwf* genes and deletion of *edd, ppsA, ldhA, aceA, poxB*, and *plsX* were done in *E. coli.* The engineered strain produced 425 mg/L alkane and 1506 mg/L fatty alcohols. Finally, fed-batch cultivation of genetically modified strain of *E. coli* shown to produce 2.54 g/L titer of alkane or alkene and 12.5 g/L titer of fatty alcohol (Fatma et al. 2018).

Table 9.1 summarizes the production of different biofuel molecules by application of synthetic biology and metabolic engineering.

Table 9.1 Different biofuel molecules produced from microorganisms by application of synthetic biology and metabolic engineering

Biofuel molecule	Engineered microorganism	References
Ethanol	*Clostridium acetobutylicum*	Harris et al. (2001)
	Clostridium beijerinckii	Ezeji et al. (2007)
	Clostridium saccharoperbutylacetonicum	Thang et al. (2010)
	Clostridium carboxidivorans	Cheng et al. (2019)
	Saccharomyces cerevisiae	Lynd et al. (2008), Kim et al. (2014)
	Synechococcus elongatus	Dexter et al. (2015)
1-Propanol	*Escherichia coli*	Atsumi et al. (2008b), Atsumi and Liao (2008), Shen and Liao (2013)
Isopropanol	*Escherichia coli*	Hanai et al. (2007), Soma et al. (2012)
	Synechococcus elongatus	Kusakabe et al. (2013)
	Cupriavidus necator	Grousseau et al. (2014)
1-Butanol	*Escherichia coli*	Atsumi et al. (2008a), Shen et al. (2011), Jawed et al. (2020)
	Synechococcus sp.	Lan and Liao (2011)
Isobutanol	*Escherichia coli*	Atsumi et al. (2008b), Desai et al. (2014)
	Synechococcus sp.	Wess et al. (2019)
Biodiesel	*Escherichia coli*	Zhang et al. (2011), Xu et al. (2013)
Alka(e)ne	*Escherichia coli*	Rahman et al. (2014), Choi and Lee (2013), Fatma et al. (2018)

9.4 Microbial Engineering of Microorganism for Producing Biofuels

9.4.1 *Bacteria as a Biofuel Producer*

Considering that ethanol is not an ideal fuel due to its high corrosiveness and low hygroscopicity, butanol is thought to be the most feasible biofuel in the existing fuel infrastructure (Stephanopoulos 2007). However, native microorganisms produce butanol in low quantities under conditions not very feasible under industrial condition (Ingram et al. 1999). A synthetic approach was made to express the non-native products in the *E.coli* by altering the already existing metabolic pathways (Spies and Kowalczykowski 2005; Atsumi et al. 2008b). A major source of ethanol and biodiesel are agricultural food resources which are limited in the current situation, but there is a wide unexplored resource of lignocellulosic biomass from agricultural wastes that can be a substrate for the production of biofuel economically and substantially (Blanch et al. 2008). But the production of an efficient and sustainable process for converting lignocellulosic biomass into biofuels has many roadblocks. The major issues are the absence of proper recombinant engineering tools for organisms which has inherent ability to synthesize biological product (non-model organisms) and difficult in balancing the redox state and metabolic pathways in the engineered microbes (Mukhopadhyay et al. 2008). Over the period of time, synthetic biology and metabolic engineering will be able to overcome the roadblocks in engineering strains to efficiently utilize lignocellulosic biomass for the production of biofuel (Lee et al. 2008). Some of the developments in key non-model organisms have been discussed below.

9.4.1.1 *Clostridium*

Clostridium sp. is one of the major producers of biofuel in the industry. *Clostridium acetobutylicum* is known for the biphasic fermentation which leads to butanol and ethanol production. Butanol production via butyryl-CoA pathway seems to more productive than the conventional pathway of acetyl-CoA. Butanol production in *Clostridium acetobutylicum* has been increased to a titer of 18.9 g/L by disrupting the genes coding for phosphotransacetylase and butyrate kinase and simultaneously overexpressing aldehyde/alcohol dehydrogenase gene (Lee et al. 2012b). Successful genetic engineering of *Clostridium ljungdahlii*, which is an acetogenic-anaerobic bacterium, leads to the production of ethanol by using carbon monoxide and carbon dioxide. The genomic database of this species is completely annotated and simple genetic engineering methods can now be used to do the transformation. Thus, *Clostridium ljungdahlii* can be used as an efficient platform for biofuel production industry for synthesis of ethanol using syngas (Kopke et al. 2010). To further increase the ethanol titer, the culture can be grown together/co-cultured with another species *Rhodospirillum rubrum.* High ethanol production is obtained under non-growth conditions of *Rhodospirillum rubrum* via a two-stage cultivation using bioreactor with a high mass transfer (Klasson et al. 1992). To enhance and stabilize the alcohol yield, specific reducing agents like methyl and benzyl viologen is used. These reducing agents increase the ethanol yield by alteration in electron flow (Rao and Mutharasan 1986). These changes contribute to NADH forming out of free hydrogen and direct carbon transfer from acid to alcohol. Butanol pathway from *Clostridium* has also been engineered into *E. coli*. Expression of these genes in *E. coli* helps in enhanced production of butanol in a cost-efficient and large quantities (Bond-Watts et al. 2011; Shen et al. 2011). Microbes open up many indirect fermentation pathways other than direct fermentation for the biofuel production. As discussed earlier, either co-culture of *Rhodospirillum rubrum* and methanogens or by using *Clostridum ljungdahlii* alone, production of ethanol, butanol, and methanol has been demonstrated from syngas (Klasson et al. 1992). For this strategy, syngas is mainly derived from carbon rich feedstock such as coal, natural gas, and biomass.

9.4.2 Cyanobacteria as a Biofuel Producer

Cyanobacteria have been vastly genetically and metabolically engineered for optimum production of biofuel. *Synenchococcus elongatus* PCC 7942, a freshwater cyanobacterium, has been used as host for producing 1-butanol by expressing butyryl-CoA-dependent pathway from *Clostridium* (Lan and Liao 2011). Other well-known cyanobacterium *Synechocystis* sp. PCC6803 has been modified genetically by insertion of genes encoding enzyme pyruvate decarboxylase and alcohol dehydrogenase (ADH) from *Z. mobilis* (Liang et al. 2018). Co-expression of four Calvin-Benson-Bassham (CBB cycle) enzymes, i.e., ribulose-1,5-bisphosphate carboxylase/oxygenase, transketolase, fructose-1,6/sedoheptulose-1,7-bisphosphatase, and aldolase in this cyanobacterium led to increased ethanol and total biomass production. A free-floating filamentous cyanobacterium *Arthrospira platensis* was also used as feedstock for direct-ethanol production by adding lysozyme enzyme for lysis and using recombinant *Saccharomyces cerevisiae* expressing α-amylase and glucoamylase for ethanol fermentation (Aikawa et al. 2018).

9.4.3 Yeast as a Biofuel Producer

Biodiesel is a major alternative for fossil fuels, and it is mostly obtained from soybeans which do not affect the food supply chain, but the major limitations are low yields and the oxidation capacity (Hu and Lu 2015). As compared to oil plants, oleaginous microorganisms, has the capability to accumulate lipids to more than 20% of their dry weight, and are efficient biofuel producers, due to their numerous advantages as higher growth rate and oil productivity, less labor consumption and smaller land coverage, making them potential feedstocks for oil production (Spagnuolo et al. 2019).

9.4.3.1 *Yarrowia lipolytica*

Yarrowia lipolytica, an oleaginous yeast, is an emerging biodiesel producer and acts as an efficient platform for the production of biodiesel due to its inherent lipid synthesis capacity. It is seen that 90% of cell mass gets accumulated with lipid in this oleaginous yeast when the carbon source is glucose (Blazeck et al. 2014). This species can be metabolically engineered easily due to a completely known sequence of the genome (Ledesma-amaro et al. 2016). Using glucose as a sugar source, *Y. lipolytica* has been modified to produce up to 98.9 g/L fatty acid methyl esters (FAME) at 0.269 g/g yield and 1.3 g/L/h productivity in the past few years (Qiao et al. 2017). Normally fatty acids are extracted from the organisms using harsh techniques but recent development made *Y. lipolytica* release fatty acid in the culture. This helped in an increase in the lipid accumulation capacity and reduction in the cost of lipid extraction (Ledesma-amaro et al. 2016). These developments have made this lipid-rich yeast as a major platform for biodiesel production via metabolic engineering. One such effort is to engineer *Y. lipolytica* for production of biodiesel from agricultural wastes. However, the major challenge hear is to make this strain amenable to utilize all the major sugars present in lignocellulosic biomass. Among these sugars, the research is mainly focused on xylose-utilizing *Y. lipolytica* where notable progress has been done over the past few years (Ledesma-amaro et al. 2016). The potentiality of native xylose pathway in *Y. lipolytica* has been exploited and demonstrated in recent studies. However, its native ability to utilize xylose is negligible due to the absence of strong promoters. Two major genes, i.e., genes encoding XKS and XDS, for xylose metabolism have been studied. When these genes were constitutively overexpressed under a strong promoter, higher cell growth, and increased lipid density were obtained. Also, better co-utilization of glucose and xylose and the lipid accumulation were observed when the cells were grown in nitrogen-rich media (Rodriguez et al. 2016). Through metabolic and transcriptomic analysis, native sugar putative pathways in *Y. lipolytica* was studied and found that they have transporters specific for each sugar and specific metabolic enzymes for the assimilation of these sugars. They have a mild carbon catabolite repression (CCR) mechanism which helps in the simultaneous consumption of mixed sugars or co-utilization of sugars (Ryu et al. 2015). Other approaches to obtain more efficient xylose catabolism is by overexpression of the heterologous gene-encoding xylose reductase (XR) and xylose dehydrogenase (XDH) from well-known xylose-utilizing yeast *Scheffersomyces stipitis* (Li and Alper 2016). When XR and XDH were overexpressed in the model organism *S. cerevisiae,* it efficiently utilized xylose, but additional evolutionary engineering needed to be done in *Y. lipolytica* for achieving significant growth (Li and Alper 2016). In most of the case, the lower titer was due to the production of xylitol and citric acid as by-products (Ledesma-amaro et al. 2016). These results suggest that more advanced tools and methods are required to be developed for *Y. lipolytica* for metabilizing xylose.

Several other features useful for using lignocellulosic biomass as feedstock have been developed in *Y. lipolytica*. For example, a number of fungal cellulases were overexpressed in *Y. lipolytica* for hydrolyzing cellulosic biomass and accumulating into lipid for biodiesel (Guo et al. 2017). In addition, *Y. lipolytica* was found to produce extracellular laccase that could eradicate inhibition by lignin-derived generated during pretreatment process (Lee et al. 2012a). Along with the available genomic editing tools such as CRISPR-Cas9, *Y. lipolytica* is likely to be a promising candidate for biodiesel production (Schwartz et al. 2015).

Other wide range of oleaginous yeasts, such as *Rhodosporidium toruloides*, *Cryptococcus curvatus,* and *Lipomyces tetrasporus,* also has the ability to produce lipids from lignocellulosic hydrolysates. A recent study has shown *R. toruloides* as an alternative platform for the production of cellulosic biodiesel via metabolic engineering (Zhang et al. 2016).

9.4.4 Fungi as a Biofuel Producer

For sustainable development of the economy, the source of biofuel production should not be costly or economically inconvenient. Agricultural wastes contain a large amount of lignocellulose that can serve as a feedstock for biofuel production. However, the high cost and less efficient lignocellulosic enzymes affect the release of fermentable sugars from agricultural wastes.

9.4.4.1 *Trichoderma reesei*

Filamentous fungi, such as *Trichoderma reesei*, are known to be very efficient producers of copious amount of biomass saccharifying enzymes, such as cellulases, hemicellulases, and ligninases. It is highly desirous to increase the efficiency of enzyme production and efficacy of enzyme composition for successful valorization of lignocellulosic biomass. Various approaches have been implemented to achieve these objectives. The magnetic nano-particle has been used to immobilize cellulose obtained from filamentous fungi *Trichoderma longibrachiatum*, which was further used to hydrolyze *Sesbania bispinosa* biomass and ferment to ethanol (Baskar et al. 2016).

The presence of inducers in cellulase production is one of the rate determining factors in the synthesis. The cost-effective production of *T. reesei* cellulase depends on the efficient and low-cost cellulase inducers. Many recent reports establish that the presence of inducers show a significant change in the production of cellulase. Usage of pulp and paper sludge as substrate has shown to have increased cellulase production (Lai et al. 2017). When substrate used for cellulase production from *T. reesei* is bead milled straw, the cellulase activity in the extracellular medium increased by 1.5 times compared to the wild type. There is also a 2–3 times increase in β-glucosidase activity (Zheng et al. 2017). *T. reesei* is known for higher cellulase enzyme yield, but it produces less amount of β-glucosidase. The beta-glucosidase is one of the major enzymes for producing fermentable sugars. So, the above-mentioned work shows a significant increase in beta-glucosidase which is a notable milestone. Soluble inducers can be used for cellulase production conveniently in *T. reesei.* Efforts were made to produce cheaper inducers for cellulase production by using glucose-disaccharide mixed sugar along with transglycosiding β-glucosidase, leading to improved cellulase production by several times (Li et al. 2016; Srivastava et al. 2018).

9.4.5 Microalgae as a Source of Biofuel

The third-generation, algae-derived, biofuels are candidate for fulfilling future energy demand. But their slow-growth rate and the extensive energy requirement for algal cultivation have been the major obstacles preventing the commercialization

Table 9.2 Microalgae as biofuel source: chemical composition (% of dry matter)

Strain	Lipid content (%)	Protein content (%)	Carbohydrate content (%)	References
Chlamydomonas reinhardtii	21	48	17	Mata et al. (2010)
Chlorella vulgaris	14–22	51–58	12–17	Mata et al. (2010)
Dunaliella salina	6	57	32	Gouveia and Oliveria (2009)
Haematococcus pluvialis	25	–	–	Satyanarayana et al. (2011)
Spirulina maxima	6–7	60–71	13–16	Satyanarayana et al. (2011)
Euglena gracilis	4–20	39–61	14–18	Mata et al. (2010)

of microalgal biofuels (Borowitzka 2013). Even then there are many algal species which are being explored for having relatively higher growth rate, efficient photosynthetic machinery, and higher lipid content (Peng et al. 2016). Microalgae has the ability to convert carbon dioxide and sunlight into a wide range of products such as carbohydrates, vitamins, proteins, lipids, and many compounds of pharmaceutical significance (Ugwu et al. 2008). The major advantage of using microalgae in the production for biofuels is overall reduced carbon dioxide emission, which is the main greenhouse gas, thereby reducing adverse climate impact (Costa and de Morais 2011). Carbohydrates, lipids, and protein ratios vary from microalgal species (Table 9.2). Different species of algae produce different proportion of carbohydrates, proteins, and lipids. Major portion of lipid in microalgae in accumulated as triacyl glycerol (TAG), a precursor for biodiesel synthesis, which makes microalgae a good alternative for fossil fuels (Johnson 2009). There is a need to increase the microalgal lipid content for better economic impact. One way to achieve this goal is to minor manipulation of chemical composition in their culture medium to modulate algal metabolism (Chisti 2007). Microalgae such as *Botryococcus* and *Chlorella* have a lipid content of 60–80% which is sufficient for the production of biofuel (Costa and de Morais 2011). *Tetraselmis* sp. and *Dunaliella* sp., which are marine microalgae, have lipid content of 56% and 50% of their dry cell weight, respectively (Peng et al. 2019). Suitable growth conditions lead to a doubling of biomass in less than 24 h. Selection of microalgal species, thus, are on the basis of traits such as faster doubling rate, higher lipid content, and ease in cultivation (Zhao et al. 2010). In the following section, we will be discussing about a eukaryotic green microalga *Scenedesmus obliquus* and its biofuel application.

9.4.5.1 *Scenedesmus obliquus*

Scenedesmus obliquus belonging to Phylum Chlorophyta, class Chlorophyceae, order Chlorococcales, and family Scenedesmaceae, is a freshwater alga commonly known as a cosmopolitan green alga. The colony morphology changes significantly according to the medium used for the cell growth. In a growth medium containing low salt or phosphorous concentration, it grows as single, unicellular long elliptical cells of size 10 μm. It has the ability to grow either in light or dark, photoautotrophically, and heterotrophically at 30°C (Funes et al. 2002). This is a pioneer species to be investigated by biologists due to its ease of cultivation and maintenance (Wünschiers and Lindblad 2002). These factors paved the way for the discovery of hydrogen metabolism. Under anaerobic conditions, if the air is replaced by nitrogen it releases hydrogen 10 times faster than in the presence of light (Gaffron and Rubin 1942). The TAG content in *Scenedesmus obliquus* is estimated at 40–55%.

An efficient genetic engineered *S. obliquus* has been developed by transferring the type 2 diacylglycerolacyltransferase (DGTT1) gene from a green algae *Chlamydomonas reinhardtii* to increase the lipid content. The results revealed that expression of the DGTT1 gene in *S. obliquus* has stimulated growth rate of microalga as well as its oil production capacity to 234.3 mg/L/day and 18.1%, respectively. The rate of biomass production and oil content were 125 mg/L/ day and 12.3%, respectively, in a 40 L tubular photobioreactor. Thus, genetic engineering of *S. obliquus* by insertion of DGTT1 seems to have enhanced both the lipid content as well as productivity (Chen et al. 2016).

The presence of saturated fatty acid-palmitate and monounsaturated oleate as lipid makes *S. obliquus* a relevant feedstock for the production of biofuel (Mandal and Mallick 2009). Many strategies of genetic engineering have been documented in this species. A strain of *Scenedesmus*, named R-16, which can grow in high glucose (up to 100 g/L) and a wide range of pH (4.0–11.0), was isolated and characterized. By varying the carbon and nitrogen source they found that optimum glucose (10 g/L) and nitrogen (0.6 g/L of sodium nitrate) leads to the lipid accumulation of 43.4% in heterotrophic condition. And, the biomass production was increased to 3.46 g/L (Ren et al. 2013). Above this, when the *Scenedesmus* undergoes nitrogen-deprived situation, the accumulation of lipid is increased to 52.6% of their dry cell weight. UV mutagenesis of *Scenedesmus obliquus* leads to the first starchless mutants which have shown an enhanced TAG content without compromising cell growth rate. The study shows that the mutants can grow in dark-light illumination in nitrogen depletion condition and gives a lipid content of 49.4% of cell's dry weight (De Jaeger et al. 2014). When the mutant strain was grown in outdoor condition, the maximum yield of TAG was increased to 0.217 ± 0.011 g TAG/mol photon.

Depending on the partial hydrogen pressure, *Scenedesmus* can either take hydrogen or release hydrogen. In both light and dark conditions, hydrogen gas is released anaerobically by *Scenedesmus*. As oxygen's pressure drops below the partial pressure, photosynthetic organisms undergo dark fermentation (Gaffron and Rubin 1942), where oxygen and hydrogen are produced simultaneously in anaerobic conditions (Wünschiers and Lindblad 2002). In the light, oxygen is produced by photosynthesis, leading to a halt in hydrogen production. So, algae produce hydrogen only when it is grown in anaerobic conditions. This may be a limiting factor for growing these algae in bioreactors or open ponds. But anaerobic conditions can be adapted for microalgae via two different methods: flushing nitrogen gas for 3 h continuously or growing algae in dark conditions for overnight. In the first condition, it creates an effect similar to fermentation where oxygen is replaced by nitrogen and the cell is forced to undergo fermentation. In the second method, oxygen is used by respiration. The addition of sodium-thionate will remove the oxygen produced by photosynthesis in the algal cell and remove the remaining oxygen during experiments. It is hypothesized that the redox environment activates the inactive hydrogenase thereby increasing the hydrogen production.

As demands for an economic, pollution-free, sustainable energy increase, microalgae are the major alternative for fossil fuels. This third-generation fuel doesn't depend on the feedstocks or lignocellulosic biomasses; it depends on the algal biomasses. Therefore, genetic engineering of these microalgal strains and developing mutant strains having the ability to produce higher lipid content is an essential step. There should be efficient mechanisms for extracting lipids and by-products effectively and in large quantities. In addition, the availability of lots of sunlight and favorable tropical temperature will facilitate the production of hydrogen in a closed low-cost photobioreactor (Shumbulo and Ki 2018).

Table 9.2 summarized various strains of microalgae engineered for biofuel production.

9.5 Conclusion and Future Perspective

In this chapter, we have discussed the origin and definition of metabolic engineering and synthetic biology. Synergy and differences between these two fields are also discussed. We have also discussed the engineering of microbes for the production of biofuels molecules like bioethanol, 1-propanol, isopropanol, 1-butanol, isobutanol, and biodiesel by the application of various tools and techniques developed by metabolic engineering and synthetic biology. Also discussed here about how metabolic engineered and synthetic biology advances the research for developing new microorganisms or living cells as cell factories. Many examples of bacteria, cyanobacteria, yeast, fungi, and microalgae have been given which are engineered for the production of biofuels.

Recent developments in rapid DNA assembly, rapid, and high fidelity DNA synthesis, omics studies resulted in the development of various mathematical models that can predict the different domains of cell function and behavior. In the near future, the combined approach of metabolic engineering and synthetic biology can lead to the creation of standardized and optimized microbes or living cells whose function and behavior are completely known. System metabolic engineering is emerging as a new field that brings metabolic engineering and synthetic biology together and will be extremely useful in developing microbial platform for efficient production of biofuels.

References

Aikawa S, Inokuma K, Wakai S et al (2018) Biotechnology for biofuels direct and highly productive conversion of cyanobacteria *Arthrospira platensis* to ethanol with – $CaCl_2$ addition. Biotechnol Biofuels 1–9. https://doi.org/10.1186/s13068-018-1050-y

Ajikumar PK, Xiao WH, Tyo KEJ et al (2010) Isoprenoid pathway optimization for Taxol precursor overproduction in *Escherichia coli*. Science 330(6000):70–74. https://doi.org/10.1126/science.1191652

Atsumi S, Liao JC (2008) Directed evolution of *Methanococcus jannaschii* citramalate synthase for biosynthesis of 1-propanol and 1-butanol by *Escherichia coli*. Appl Environ Microbiol 74: 7802–7808. https://doi.org/10.1128/AEM.02046-08

Atsumi S, Cann AF, Connor MR et al (2008a) Metabolic engineering of *Escherichia coli* for 1-butanol production. Metab Eng 10:305–311. https://doi.org/10.1016/j.ymben.2007.08.003

Atsumi S, Hanai T, Liao JC (2008b) Non-fermentative pathways for synthesis of branched-chain higher alcohols as biofuels. Nature 451:86–89. https://doi.org/10.1038/nature06450

Bailey JE (1991) Toward a science of metabolic engineering. Science 252(5013):1668–1675. https://doi.org/10.1126/science.2047876

Baskar G, Naveen Kumar R, Heronimus Melvin X et al (2016) *Sesbania aculeate* biomass hydrolysis using magnetic nanobiocomposite of cellulase for bioethanol production. Renew Energy 98:23–28. https://doi.org/10.1016/j.renene.2016.04.035

Blanch et al (2008) Addressing the need for alternative transportation fuels: the Joint BioEnergy Institute. J Chem Inf Model 53:1689–1699. https://doi.org/10.1017/CBO9781107415324.004

Blazeck J, Hill A, Liu L et al (2014) Harnessing *Yarrowia lipolytica* lipogenesis to create a platform for lipid and biofuel production. Nat Commun 5:1–10. https://doi.org/10.1038/ncomms4131

Bond-Watts BB, Bellerose RJ, Chang MCY (2011) Enzyme mechanism as a kinetic control element for designing synthetic biofuel pathways. Nat Chem Biol 7:222–227. https://doi.org/10.1038/nchembio.537

Borowitzka MA (2013) High-value products from microalgae — their development and commercialisation. J Appl Phycol. https://doi.org/10.1007/s10811-013-9983-9

Cameron DE, Bashor CJ, Collins JJ (2014) A brief history of synthetic biology. Nat Rev Microbiol 12:381–390. https://doi.org/10.1038/nrmicro3239

Casini A, MacDonald JT, De JJ et al (2013) One-pot DNA construction for synthetic biology: the modular overlap-directed assembly with linkers (MODAL) strategy. Nucleic Acids Res 42:1–13. https://doi.org/10.1093/nar/gkt915

Chen CY, Kao AL, Tsai ZC et al (2016) Expression of type 2 diacylglycerol acyltransferse gene DGTT1 from *Chlamydomonas reinhardtii* enhances lipid production in *Scenedesmus obliquus*. Biotechnol J 11:336–344. https://doi.org/10.1002/biot.201500272

Cheng C, Li W, Lin M, Yang ST (2019) Metabolic engineering of *Clostridium carboxidivorans* for enhanced ethanol and butanol production from syngas and glucose. Bioresour Technol 284: 415–423. https://doi.org/10.1016/j.biortech.2019.03.145

Chisti Y (2007) Biodiesel from microalgae. Biotechnol Adv 25:294–306. https://doi.org/10.1016/j.biotechadv.2007.02.001

Cho JS, Choi KR, Prabowo CPS et al (2017) CRISPR/Cas9-coupled recombineering for metabolic engineering of *Corynebacterium glutamicum*. Metab Eng 42:157–167. https://doi.org/10.1016/j.ymben.2017.06.010

Choi YJ, Lee SY (2013) Microbial production of short-chain alkanes. Nature 502:571–574. https://doi.org/10.1038/nature12536

Choi KR, Jang WD, Yang D et al (2019) Systems metabolic engineering strategies: integrating systems and synthetic biology with metabolic engineering. Trends Biotechnol 37:817–837. https://doi.org/10.1016/j.tibtech.2019.01.003

Costa JAV, de Morais MG (2011) The role of biochemical engineering in the production of biofuels from microalgae. Bioresour Technol 102:2–9. https://doi.org/10.1016/j.biortech.2010.06.014

Dahman Y, Syed K, Begum S, et al (2019) Biofuels: their characteristics and analysis. Elsevier Ltd. https://doi.org/10.1016/B978-0-08-102426-3.00014-X

De Jaeger L, Verbeek REM, Draaisma RB et al (2014) Superior triacylglycerol (TAG) accumulation in starchless mutants of *Scenedesmus obliquus*: (I) mutant generation and characterization. Biotechnol Biofuels 7:1–11. https://doi.org/10.1186/1754-6834-7-69

Desai SH, Rabinovitch-Deere CA, Tashiro Y, Atsumi S (2014) Isobutanol production from cellobiose in *Escherichia coli*. Appl Microbiol Biotechnol 98:3727–3736. https://doi.org/10.1007/s00253-013-5504-7

Dexter J, Armshaw P, Sheahan C, Pembroke JT (2015) The state of autotrophic ethanol production in cyanobacteria. J Appl Microbiol 119:11–24. https://doi.org/10.1111/jam.12821

Elowitz (2000) A synthetic oscillatory network repressilator. Nature 403:335–338. https://doi.org/10.1038/35002125

Ezeji TC, Qureshi N, Blaschek HP (2007) Production of acetone butanol (AB) from liquefied corn starch, a commercial substrate, using *Clostridium beijerinckii* coupled with product recovery by gas stripping. J Ind Microbiol Biotechnol 34:771–777. https://doi.org/10.1007/s10295-007-0253-1

Fatma Z, Hartman H, Poolman MG et al (2018) Model-assisted metabolic engineering of *Escherichia coli* for long chain alkane and alcohol production. Metab Eng 46:1–12. https://doi.org/10.1016/j.ymben.2018.01.002

Funes S, Davidson E, Reyes-Prieto A et al (2002) A green algal apicoplast ancestor. Science 298(5601):2155. https://doi.org/10.1126/science.1076003

Gaffron H, Rubin J (1942) Fermentative and photochemical production of hydrogen in algae. J Gen Physiol 26:219–240. https://doi.org/10.1085/jgp.26.2.219

Gardner TS, Cantor CR, Collins JJ (2000) Construction of a genetic toggle switch in *Escherichia coli*. Nature 403:339–342. https://doi.org/10.1038/35002131

Gibson DG, Glass JI, Lartigue C et al (2010) Creation of a bacterial cell controlled by a chemically synthesized genome. Science 329(5987):52–56. https://doi.org/10.1126/science.1190719

Good MC, Zalatan JG, Lim WA (2011) Scaffold proteins: hubs for controlling the flow of cellular information. Science 332(6030):680–686. https://doi.org/10.1126/science.1198701

Gouveia L, Oliveira AC (2009) Microalgae as a raw material for biofuels production. J Ind Microbiol Biotechnol 36(2):269–274. https://doi.org/10.1007/s10295-008-0495-6

Grousseau E, Lu J, Gorret N et al (2014) Isopropanol production with engineered Cupriavidus necator as bioproduction platform. Appl Microbiol Biotechnol 98:4277–4290. https://doi.org/10.1007/s00253-014-5591-0

Guo Z, Duquesne S, Bozonnet S et al (2017) Biotechnology for biofuels conferring cellulose – degrading ability to *Yarrowia lipolytica* to facilitate a consolidated bioprocessing approach. Biotechnol Biofuels 10(132):1–13. https://doi.org/10.1186/s13068-017-0819-8

Ha SJ, Galazka JM, Kim SR et al (2011) Engineered *Saccharomyces cerevisiae* capable of simultaneous cellobiose and xylose fermentation. Proc Natl Acad Sci U S A 108:504–509. https://doi.org/10.1073/pnas.1010456108

Hanai T, Atsumi S, Liao JC (2007) Engineered synthetic pathway for isopropanol production in *Escherichia coli*. Appl Environ Microbiol 73:7814–7818. https://doi.org/10.1128/AEM.01140-07

Harris L, Blank L, Desai RP et al (2001) Fermentation characterization and flux analysis of recombinant strains of *Clostridium acetobutylicum* with an inactivated soIR gene. J Ind Microbiol Biotechnol 27:322–328. https://doi.org/10.1038/sj.jim.7000191

Holtz WJ, Keasling JD (2010) Engineering static and dynamic control of synthetic pathways. Cell 140:19–23. https://doi.org/10.1016/j.cell.2009.12.029

Hu C, Lu W (2015) Insight into yeast: a study model of lipid metabolism and terpenoid biosynthesis. Biotechnol Appl Biochem 62:323–328. https://doi.org/10.1002/bab.1272

Ingram LO, Aldrich HC, Borges ACC et al (1999) Enteric bacterial catalysts for fuel ethanol production. Biotechnol Prog 15:855–866. https://doi.org/10.1021/bp9901062

Jawed K, Mattam AJ, Fatma Z et al (2016) Engineered production of short chain fatty acid in *Escherichia coli* using fatty acid synthesis pathway. PLoS One 11:1–20. https://doi.org/10.1371/journal.pone.0160035

Jawed K, Abdelaal AS, Koffas MAG, Yazdani SS (2020) Improved butanol production using FASII pathway in *E. coli*. ACS Synth Biol. https://doi.org/10.1021/acssynbio.0c00154

Johnson WZ (2009) Microalgae as a feedstock for biofuel production. Virginia Coop Ext 11:442–886. https://doi.org/10.1016/j.nbt.2009.06.630

Jones SW, Fast AG, Carlson ED et al (2016) CO_2 fixation by anaerobic non-photosynthetic mixotrophy for improved carbon conversion. Nat Commun 7:1–9. https://doi.org/10.1038/ncomms12800

Jullesson D, David F, Pfleger B, Nielsen J (2015) Impact of synthetic biology and metabolic engineering on industrial production of fine chemicals. Biotechnol Adv 33:1395–1402. https://doi.org/10.1016/j.biotechadv.2015.02.011

Katahira S, Ito M, Takema H et al (2008) Improvement of ethanol productivity during xylose and glucose co-fermentation by xylose-assimilating *S. cerevisiae* via expression of glucose transporter Sut1. Enzym Microb Technol 43:115–119. https://doi.org/10.1016/j.enzmictec.2008.03.001

Kim JH, Ryu J, Huh IY et al (2014) Ethanol production from galactose by a newly isolated *Saccharomyces cerevisiae* KL17. Bioprocess Biosyst Eng 37:1871–1878. https://doi.org/10.1007/s00449-014-1161-1

Klasson KT, Ackerson MD, Clausen EC, Gaddy JL (1992) Bioconversion of synthesis gas into liquid or gaseous fuels. Enzym Microb Technol 14:602–608. https://doi.org/10.1016/0141-0229(92)90033-K

Köpke M, Held C, Hujer S et al (2010) *Clostridium ljungdahlii* represents a microbial production platform based on syngas. Proc Natl Acad Sci U S A 107(29):13087–13092. https://doi.org/10.1073/pnas.1004716107

Kusakabe T, Tatsuke T, Tsuruno K et al (2013) Engineering a synthetic pathway in cyanobacteria for isopropanol production directly from carbon dioxide and light. Metab Eng 20:101–108. https://doi.org/10.1016/j.ymben.2013.09.007

Lai TT, Pham TTH, Adjallé K et al (2017) Production of *Trichoderma reesei* RUT C-30 lignocellulolytic enzymes using paper sludge as fermentation substrate: an approach for on-site manufacturing of enzymes for biorefineries. Waste Biomass Valor 8:1081–1088. https://doi.org/10.1007/s12649-016-9686-y

Lan EI, Liao JC (2011) Metabolic engineering of cyanobacteria for 1-butanol production from carbon dioxide. Metab Eng 13:353–363. https://doi.org/10.1016/j.ymben.2011.04.004

Ledesma-amaro R, Lazar Z, Rakicka M et al (2016) Metabolic engineering of *Yarrowia lipolytica* to produce chemicals and fuels from xylose. Metab Eng 38:115–124. https://doi.org/10.1016/j.ymben.2016.07.001

Lee SJ, Lee SY (2006) Genome-based metabolic engineering of *Mannheimia succiniciproducens* for succinic acid production. Appl Environ Microbiol 72:1939–1948. https://doi.org/10.1128/AEM.72.3.1939

Lee SK, Chou H, Ham TS et al (2008) Metabolic engineering of microorganisms for biofuels production: from bugs to synthetic biology to fuels. Curr Opin Biotechnol 19:556–563. https://doi.org/10.1016/j.copbio.2008.10.014

Lee K, Kalyani D, Kumar M et al (2012a) Bioresource technology enhanced enzymatic hydrolysis of rice straw by removal of phenolic compounds using a novel laccase from yeast *Yarrowia lipolytica*. Bioresour Technol 123:636–645. https://doi.org/10.1016/j.biortech.2012.07.066

Lee SY, Jang YS, Lee JY et al (2012b) Enhanced butanol production obtained by reinforcing the direct butanol-forming route in *Clostridium acetobutylicum*. MBio 3(5):e00314–e00312. https://doi.org/10.1128/mBio.00314-12

Lee MJ, Mantell J, Hodgson L et al (2018) Engineered synthetic scaffolds for organizing proteins within the bacterial cytoplasm. Nat Chem Biol 14:142–147. https://doi.org/10.1038/nchembio.2535

Li H, Alper HS (2016) Enabling xylose utilization in *Yarrowia lipolytica* for lipid production. Biotechnol J 11:1230–1240. https://doi.org/10.1002/biot.201600210

Li Y, Liu C, Bai F, Zhao X (2016) Overproduction of cellulase by *Trichoderma reesei* RUT C30 through batch-feeding of synthesized low-cost sugar mixture. Bioresour Technol 216:503–510. https://doi.org/10.1016/j.biortech.2016.05.108

Liang F, Englund E, Lindberg P, Lindblad P (2018) Engineered cyanobacteria with enhanced growth show increased ethanol production and higher biofuel to biomass ratio. Metab Eng 46: 51–59. https://doi.org/10.1016/j.ymben.2018.02.006

Lieber DJ, Catlett J, Madayiputhiya N et al (2014) A multienzyme complex channels substrates and electrons through acetyl-coA and methane biosynthesis pathways in *Methanosarcina*. PLoS One 9:1–8. https://doi.org/10.1371/journal.pone.0107563

Lynd LR, Laser MS, Bransby D et al (2008) How biotech can transform biofuels. Nat Biotechnol 26:169–172. https://doi.org/10.1038/nbt0208-169

Mandal S, Mallick N (2009) Microalga *Scenedesmus obliquus* as a potential source for biodiesel production. Appl Microbiol Biotechnol 84:281–291. https://doi.org/10.1007/s00253-009-1935-6

Mata TM, Martins AA, Caetano NS (2010) Microalgae for biodiesel production and other applications: a review. Renew Sust Energ Rev 14:217–232. https://doi.org/10.1016/j.rser.2009.07.020

Mukhopadhyay A, Redding AM, Rutherford BJ, Keasling JD (2008) Importance of systems biology in engineering microbes for biofuel production. Curr Opin Biotechnol 19:228–234. https://doi.org/10.1016/j.copbio.2008.05.003

Na D, Yoo SM, Chung H et al (2013) Metabolic engineering of *Escherichia coli* using synthetic small regulatory RNAs. Nat Biotechnol 31:170–174. https://doi.org/10.1038/nbt.2461

Nielsen J (2001) Metabolic engineering. Appl Microbiol Biotechnol 55:263–283. https://doi.org/10.1007/s002530000511

Nielsen J, Jewett MC (2008) Impact of systems biology on metabolic engineering of *Saccharomyces cerevisiae*. FEMS Yeast Res 8:122–131. https://doi.org/10.1111/j.1567-1364.2007.00302.x

Nielsen J, Keasling JD (2011) Synergies between synthetic biology and metabolic engineering. Nat Biotechnol 29:693–695. https://doi.org/10.1038/nbt.1937

Nielsen J, Fussenegger M, Keasling J et al (2014) Engineering synergy in biotechnology. Nat Chem Biol 10:319–322. https://doi.org/10.1038/nchembio.1519

Ostergaard S, Olsson L, Nielsen J (2000) Metabolic engineering of *Saccharomyces cerevisiae*. Microbiol Mol Biol Rev 64:34–50. https://doi.org/10.1128/MMBR.64.1.34-50.2000

Park JH, Lee KH, Kim TY, Lee SY (2007) Metabolic engineering of *Escherichia coli for* the production of L-valine based on transcriptome analysis and in silicogene knockout simulation. Proc Natl Acad Sci U S A 104:7797–7802

Peng L, Zhang Z, Cheng P et al (2016) Cultivation of *Neochloris oleoabundans* in bubble column photobioreactor with or without localized deoxygenation. Bioresour Technol 206:255–263. https://doi.org/10.1016/j.biortech.2016.01.081

Peng L, Fu D, Chu H et al (2019) Biofuel production from microalgae: a review. Environ Chem Lett 18:285–297. https://doi.org/10.1007/s10311-019-00939-0

Peralta-Yahya PP, Zhang F, Del Cardayre SB, Keasling JD (2012) Microbial engineering for the production of advanced biofuels. Nature 488:320–328. https://doi.org/10.1038/nature11478

Potapov V, Ong JL, Kucera RB et al (2018) Comprehensive profiling of four base overhang ligation fidelity by T4 DNA ligase and application to DNA assembly. ACS Synth Biol 7:2665–2674. https://doi.org/10.1021/acssynbio.8b00333

Qiao K, Wasylenko TM, Zhou K et al (2017) Lipid production in *Yarrowia lipolytica* is maximized by engineering cytosolic redox metabolism. Nat Biotechnol 35:173–177. https://doi.org/10.1038/nbt.3763

Rahman Z, Sung BH, Yi JY et al (2014) Enhanced production of n-alkanes in *Escherichia coli* by spatial organization of biosynthetic pathway enzymes. J Biotechnol 192:187–191. https://doi.org/10.1016/j.jbiotec.2014.10.014

Rao G, Mutharasan R (1986) Alcohol production by *Clostridium acetobutylicum* induced by methyl viologen. Biotechnol Lett 8:893–896. https://doi.org/10.1007/BF01078655

Ren HY, Liu BF, Ma C et al (2013) A new lipid-rich microalga *Scenedesmus* sp. strain R-16 isolated using Nile red staining: effects of carbon and nitrogen sources and initial pH on the biomass and lipid production. Biotechnol Biofuels 6:1–10. https://doi.org/10.1186/1754-6834-6-143

Rodriguez GM, Hussain MS, Gambill L et al (2016) Engineering xylose utilization in *Yarrowia lipolytica* by understanding its cryptic xylose pathway. Biotechnol Biofuels 9:1–15. https://doi.org/10.1186/s13068-016-0562-6

Ryu S, Hipp J, Trinh CT (2015) Activating and elucidating complex sugar metabolism in *Yarrowia lipolytica*. Appl Environ Microbiol 82:1334–1345. https://doi.org/10.1128/AEM.03582-15

Satyanarayana KG, Mariano AB, Vargas JVC (2011) A review on microalgae, a versatile source for sustainable energy and materials. Int J Energy Res 35:291–311. https://doi.org/10.1002/er.1695

Schirmer A, Rude MA, Li X et al (2010) Microbial biosynthesis of alkanes. Science 329(5991): 559–562. https://doi.org/10.1126/science.1187936

Schwartz CM, Hussain MS, Blenner M, Wheeldon I (2015) Synthetic RNA polymerase III promoters facilitate high efficiency CRISPR-Cas9 mediated genome editing in *Yarrowia lipolytica*. ACS Syn Bio 5(4):356–359. https://doi.org/10.1021/acssynbio.5b00162

Shen CR, Liao JC (2013) Synergy as design principle for metabolic engineering of 1-propanol production in *Escherichia coli*. Metab Eng 17:12–22. https://doi.org/10.1016/j.ymben.2013.01.008

Shen CR, Lan EI, Dekishima Y et al (2011) Driving forces enable high-titer anaerobic 1-butanol synthesis in *Escherichia coli*. Appl Environ Microbiol 77:2905–2915. https://doi.org/10.1128/AEM.03034-10

Shumbulo E, Ki D (2018) Microalgae to biofuels: 'promising ' alternative and renewable energy, review. Renew Sust Energ Rev 81:743–755. https://doi.org/10.1016/j.rser.2017.08.042

Soma Y, Inokuma K, Tanaka T et al (2012) Direct isopropanol production from cellobiose by engineered *Escherichia coli* using a synthetic pathway and a cell surface display system. J Biosci Bioeng 114:80–85. https://doi.org/10.1016/j.jbiosc.2012.02.019

Spagnuolo M, Yaguchi A, Blenner M (2019) Oleaginous yeast for biofuel and oleochemical production. Curr Opin Biotechnol 57:73–81. https://doi.org/10.1016/j.copbio.2019.02.011

Spies M, Kowalczykowski SC (2005) Homologous recombination by the RecBCD and RecF pathways. Bact Chromosom:389–403. https://doi.org/10.1128/9781555817640.ch21

Srivastava N, Srivastava M, Mishra PK et al (2018) Applications of fungal cellulases in biofuel production: advances and limitations. Renew Sust Energ Rev 82:2379–2386. https://doi.org/10.1016/j.rser.2017.08.074

Stephanopoulos G (2007) Challenges in engineering microbes for biofuels production. Science 315(5813):801–804. https://doi.org/10.1126/science.1139612

Stephanopoulos G (2012) Synthetic biology and metabolic engineering. ACS Synth Biol 1:514–525. https://doi.org/10.1021/sb300094q

Stephanopoulos G, Vallino JJ (1991) Network rigidity and metabolic engineering in metabolite overproduction. Science 252(5013):1675–1681. https://doi.org/10.1126/science.1904627

Storch M, Casini A, Mackrow B et al (2015) BASIC: a new biopart assembly standard for idempotent cloning provides accurate, single-tier DNA assembly for synthetic biology. ACS Synth Biol 4:781–787. https://doi.org/10.1021/sb500356d

Sun X, Shen X, Jain R et al (2015) Synthesis of chemicals by metabolic engineering of microbes. Chem Soc Rev 44:3760–3785. https://doi.org/10.1039/c5cs00159e

Tan F, Wu B, Dai L et al (2016) Using global transcription machinery engineering (gTME) to improve ethanol tolerance of *Zymomonas mobilis*. Microb Cell Factories 15:1–9. https://doi.org/10.1186/s12934-015-0398-y

Tang WL, Zhao H (2009) Industrial biotechnology: tools and applications. Biotechnol J 4:1725–1739. https://doi.org/10.1002/biot.200900127

Thang VH, Kanda K, Kobayashi G (2010) Production of acetone-butanol-ethanol (ABE) in direct fermentation of cassava by *Clostridium saccharoperbutylacetonicum* N1-4. Appl Biochem Biotechnol 161:157–170. https://doi.org/10.1007/s12010-009-8770-1

Ugwu CU, Aoyagi H, Uchiyama H (2008) Photobioreactors for mass cultivation of algae. Bioresour Technol 99:4021–4028. https://doi.org/10.1016/j.biortech.2007.01.046

Walfridsson M, Bao X, Anderlund M et al (1996) Ethanolic fermentation of xylose with *Saccharomyces cerevisiae* harboring the *Thermus thermophilus* xylA gene, which expresses an active xylose (glucose) isomerase. Appl Environ Microbiol 62:4648–4651. https://doi.org/10.1128/aem.62.12.4648-4651.1996

Walfridsson M, Anderlund M, Bao X, Hahn-Hagerdal B (1997) Expression of different levels of enzymes from the *Pichia stipitis* XYL1 and XYL2 genes in *Saccharomyces cerevisiae* and its effects on product formation during xylose utilisation. Appl Microbiol Biotechnol 218–224. https://doi.org/10.1007/s002530051041

Wang HH, Kim H, Cong L et al (2012) Genome-scale promoter engineering by coselection MAGE. Nat Methods 9:591–593. https://doi.org/10.1038/nmeth.1971

Wess J, Brinek M, Boles E (2019) Improving isobutanol production with the yeast *Saccharomyces cerevisiae* by successively blocking competing metabolic pathways as well as ethanol and glycerol formation. Biotechnol Biofuels 12:1–15. https://doi.org/10.1186/s13068-019-1486-8

Wünschiers R, Lindblad P (2002) Hydrogen in education – a biological approach. Int J Hydrog Energy 27:1131–1140. https://doi.org/10.1016/S0360-3199(02)00098-8

Xu P, Gu Q, Wang W et al (2013) Modular optimization of multi-gene pathways for fatty acids production in *E. coli*. Nat Commun 4:1408–1409. https://doi.org/10.1038/ncomms2425

Yim H, Haselbeck R, Niu W et al (2011) Metabolic engineering of *Escherichia coli* for direct production of 1,4-butanediol. Nat Chem Biol 7:445–452. https://doi.org/10.1038/nchembio.580

Zhang X, Li M, Agrawal A, San KY (2011) Efficient free fatty acid production in *Escherichia coli* using plant acyl-ACP thioesterases. Metab Eng 13:713–722. https://doi.org/10.1016/j.ymben.2011.09.007

Zhang S, Skerker JM, Rutter CD et al (2016) Engineering Rhodosporidium toruloides for increased lipid production. Biotechnol Bioeng 113:1056–1066. https://doi.org/10.1002/bit.25864

Zhao T, Liu Z, Zhao C, Mao C (2010) Energy consumption and carbon calculations of microalgae biodiesel. World Acad Sci Eng Technol 65:963–968. https://doi.org/10.5281/zenodo.1077621

Zheng W, Chen X, Xue Y et al (2017) The influence of soluble polysaccharides derived from rice straw upon cellulase production by *Trichoderma reesei*. Process Biochem 61:130–136. https://doi.org/10.1016/j.procbio.2017.06.007

Chapter 10
Microbial Biotechnology for Renewable and Sustainable Energy: The Current Status of Biogas, Biodiesel, and Bioethanol in Brazil

F. A. F. Antunes, A. P. Ingle, T. M. Rocha, A. Hernandez-Perez, R. R. Philippini, S. E. Martiniano, S. Sánchez-Muñoz, C. A. Pradro, A. V. Paula, D. L. Flumignan, L. K. Santos, D. D. V. Silva, K. J. Dussán, J. C. Santos, and S. S. da Silva

Abstract The study of new approaches, as well as the straightness of the current environment-friendly process for biofuels production, are pivotal keys for the sustainable development of the world. In this context, the present chapter mainly focuses on an overview of the current status of biogas, biodiesel, and bioethanol in Brazil, based on microbial biotechnology for renewable and sustainable energy. Brazil has a large territorial land as well as a healthy climate for different and abundant crops, enabling it to provide potential different renewable substrates to bioprocesses. Firstly, the process of biogas production will be elucidated based on the feasibility for production and the possibility to scale up in Brazil. Then, different processes to obtain bioethanol, such as the consolidated first-generation production (from sugarcane juice); the second generation approach and its challenges (from lignocellulosic biomass), and the third generation possibility (from algae) will be discussed. In the last subject presentation, the fundaments of biodiesel production faced in the currently available and feasible substrates will be pointed. Prospects will be taken into account according to processes technology feasibility verified in the light of literature.

F. A. F. Antunes (✉) · A. P. Ingle · T. M. Rocha · A. Hernandez-Perez · R. R. Philippini · S. E. Martiniano · S. Sánchez-Muñoz · C. A. Pradro · J. C. Santos · S. S. da Silva
Department of Biotechnology, Engineering School of Lorena, University of Sao Paulo (USP), Lorena, Brazil

A. V. Paula · D. L. Flumignan · L. K. Santos
Department of Biochemistry and Organic Chemistry, Institute of Chemistry, São Paulo State University (UNESP), Araraquara, SP, Brazil

D. D. V. Silva · K. J. Dussán
Department of Engineering, Physics and Mathematics, Institute of Chemistry, São Paulo State University (UNESP), Araraquara, SP, Brazil

J. K. Saini, R. K. Sani (eds.), *Microbial Biotechnology for Renewable and Sustainable Energy*, Clean Energy Production Technologies,
https://doi.org/10.1007/978-981-16-3852-7_10

Keywords Bioenergy · Biomass · Biogas · Bioethanol · Biodiesel

10.1 Introduction

The consumption of fossil fuel energy sources, such as petroleum has led to the emission of greenhouse gases worldwide, thereby causing adverse effects upon the environment including massive pollution and ecological imbalance (Krajick 2001). Additionally, the scarcity of those resources is a matter of concern due to the limited stock available globally. To address this issue, advances in bioprocess technologies have contributed to the development of different means to synthesize biofuels from a variety of renewable feedstocks (e.g., crops, lignocellulosic materials, vegetal and microbial-derived oils, dairy manure, microalgae, etc.) which represent reliable alternatives for substituting the conventional combustible fossil sources owing to their eco-friendly status, renewability, and lower degree of gases emission (Mussgnug et al. 2010; Mathur et al. 2017; Winquist et al. 2019; Yew et al. 2019).

In this respect, biofuels can be synthesized in three different manners termed as first-, second-and third-generation processes, whereby food crops (e.g., sugarcane and corn), biomasses (e.g., lignocellulose), and algae biomass, respectively are used. The primary generation (1G) can be achieved simply by microbial fermentation using the entrained carbohydrates in food crops as carbon source; however, this process may generate competition between food supply and biofuels negatively affecting the agricultural lands and food prices (Boboescu et al. 2019; Lazar et al. 2018). The second-generation (2G) comprises nonedible plant biomass feedstock, such as lignocellulose and biowastes. Finally, third-generation (3G) is an extent of the others and is based primarily on the utilization of algae biomass as feedstock (Behera et al. 2015).

For instance, Brazil has a considerably vast climate potential nationwide and has inexhaustibly favorable land to cultivate different types of crops. Furthermore, agribusiness represents one of the major commercial guidelines for Brazilian economics. In 2019, Brazil exported more than USD 90 billion in agro-products (Agrostat 2020). Meat (from cattle, poultry birds, and pigs) production has also a notable influence on the economy contributing to 25 tons of total meat production (Bolfe 2018). Within this context, it is known that tons of agricultural residues and animal manures are generated from harvesting those commodities which represent a valuable opportunity to increasingly enhance its utilization in biorefineries in order to form value-added and profitable products such as biofuels.

For many years, stalk juice from sugarcane has been the main substrate for bioethanol production conferring approximately 79% of the total ethanol titer in Brazil (Costa et al. 2015). As mentioned above, this pathway may create competition with the food supply chain. Moreover, the seasonal availability of sugarcane is another relevant constraint factor that may stimulate the search for diverse and alternative feedstocks, e.g. microalgae and lignocellulosic biomasses. The latter has great potential to be exploited and can be divided into several groups, such as perennial grasses, aquatic plants, forest materials, and finally agricultural residues (Zabed et al. 2017). The structure of lignocellulosic materials is composed of three major molecules: hemicellulose (pentoses and hexoses), cellulose (hexoses), and recalcitrant macromolecule lignin, whereas the respective portion of each may vary according to cultivation conditions, type, and crop hybrids (Zabed et al. 2017).

To facilitate microbial access to those fermentable sugars lignin must be removed by pretreatment process in order to decrease the degree of toxicity for microbial digestion and ameliorate productivity (Mahmood et al. 2019). Thus, meticulous methods have been concretized for pretreating those materials and their main objectives are to reduce the crystallinity and the degree of polymerization of cellulose as well as increase porosity of the fibers by disrupting its compactness structure, thereby releasing soluble entrained sugars for the fermentative phase (Rastogi and Shrivastava 2017). Nevertheless, microalgae biomass also represents a valuable carbon source for the generation of bio-based fuels and consists in the cultivation and harvesting of microalgae cellular biomass. Its attractiveness relies largely on the fact of the easier digestibility due to the absence of lignin (Rizza et al. 2017). Its cultivation may be mediated with low-cost nutrients such as municipal waste and besides it might assist in the management of industrial waste and wastewater treatment (Zhou et al. 2013).

Likewise, biodiesel can be obtained through first- and second-generations processes and is synthesized commonly from animal fats, vegetal oils, recycled greases, and the attractive arising choice of microbial derived-oils (Drozdzynska et al. 2011; Lazar et al. 2018; Spagnuolo et al. 2019). The chemical structure of biodiesel is precisely characterized as long-chain alkyl esters formed by transesterification reaction whereby triacylglyceride reacts with alcohol in the presence of a catalyst mostly assigned as NaOH or KOH (Mitrea et al. 2017). The overall conversion generates biodiesel and glycerol as a by-product which can be further purified and used as a carbon source for a series of various microbially mediated bio-based products such as citric acid (Garlapati et al. 2016; Rywińska et al. 2013).

To contextualize, in 2013 over 2.8 billion liters of biodiesel were produced by several Brazilian biodiesel facilities where main feedstock such as soybean oil, animal tallow, and cottonseed oil was utilized (Brasil et al. 2017). On the other hand, the current emerging technologies demonstrate a promising pathway in the near future to substitute crops and vegetal oils for oleaginous yeast derived-oils. There are advantageous aspects in the lipid production by oleaginous microorganisms such as the utilization of inexpensive substrates that are further metabolized under nitrogen limitation to considerably accumulate fatty acids in the form of triglycerides (Zhu et al. 2012). Furthermore, the advances in genetic engineering tools have provided an optimist perspective for scientists to tailor engineered microorganisms to over-accumulate lipids and significantly increase microbial derived-oil synthesis capability thereby enhancing the productivity of biodiesel (Bhutada et al. 2017; Niehus et al. 2018).

Another sustainable energy generation approach available in the Brazilian market scenario relies on biogas utilization. Early and concurrently inexhaustible efforts have been exerted in the field of biogas production (Zhong et al. 2012; Elniski et al.

2019; Sangeetha et al. 2011; Carotenuto et al. 2016). This combustible gaseous substance is composed of two major components: carbon dioxide (40–45% v/v) and methane (55–60% v/v) plus minor traces of other gases and organic acids are assessed as well (Lopes et al. 2004; Kalia and Singh 2001). The fermentative process for biogas synthesis occurs through anaerobic digestion in a synergistic consortium composed of bacterial and archaeal diversity. Its attractiveness may be attributed to the broad range of digestible feedstocks commonly used to produce biogas, comprising microalgae biomass, food wastes, animal manures, municipal solid wastes, agricultural wastes, etc. (Zabed et al. 2020; Elalami et al. 2019).

To gain insight, it is noteworthy to underline the abundance of such types of feedstock in the Brazilian agricultural land which notably presents 172.3 million hectares under pasture and 76.7 million hectares under the cultivation of food crops (de Oliveira et al. 2018), thus a strategy for more sustainable management of the residues generated in these tillable lands is of significant benefits; however, several drawbacks are inevitable and each substrate imposes different hurdles to be overcome in order to have higher productivity of biogas. Therefore, this chapter attempts to guide the Brazilian market to a more ecological approach and to address the valuable utilization of the low-cost feedstocks available in the predominant local agribusiness. In alignment, the recent advances, challenges, and future economic prospects regarding biogas, bioethanol, and biodiesel production are disclosed throughout this work to prompt alternative green products realization.

10.2 Ethanol

10.2.1 First Generation Ethanol

Fermentation of sugars to produce fuel ethanol in Brazil predates the 1920s and this industry is based on sugarcane (Soccol et al. 2005; Andersen 2015). According to Eaglin (2019), in the beginning, ethanol was a by-product of sugar production, and from the 1970s onwards it became an important part of the Brazilian energy sector.

As can be seen in Fig. 10.1, which highlights the main events in the history of ethanol production in Brazil, in the early 1930s, to stabilize and modernize the sugar

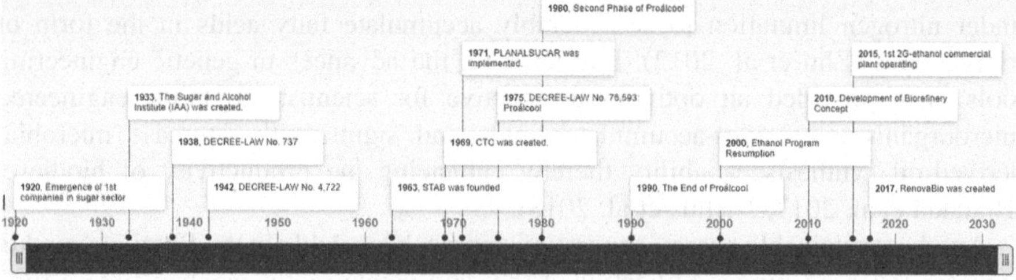

Fig. 10.1 Timeline: Highlights in Brazilian Ethanol Production History

economy, the Sugar and Alcohol Institute (IAA) was created by president Getulio Vargas.

Shortly thereafter, the government made it mandatory to blend anhydrous ethanol with gasoline (Decree-Law No. 737). In 1942, Decree-Law No. 4722 declared the alcohol industry of national interest and established price guarantees for alcohol and the raw material intended for its manufacture. In the 1960s, the Society of Alcoholic Sugar Technicians of Brazil (STAB) was created to promote scientific and cultural exchange between various regions producing sugarcane, alcohol, and derivatives, not only in Brazil but also abroad, and the Sugarcane Technology Center (CTC) was created to invest in the development of more productive varieties and add quality to the production of sugar and alcohol. In 1971 the National Sugarcane Ethanol Plan (Planalsucar) was implemented with the aim of developing new varieties, improving productivity, and modernizing agricultural and industrial parks. Finally, in 1975, the National Alcohol Program (Proálcool—DECREE-LAW No. 76593) was established with the main objectives to stimulate the production of alcohol, providing the needs of the domestic and foreign market and the automotive fuel policy (Morgera et al. 2009; Cruz et al. 2016; Stolf and Oliveira 2020).

The following years (Second Phase of Proálcool) were a period when alcohol production grew substantially, with emphasis on the 1985–1986 harvest. Production started to be also in autonomous distilleries, dedicated exclusively to the production of alcohol, without sugar production. It was the beginning of the production of hydrated alcohol, which would allow its widespread use in alcohol cars. The 1990s were marked by the end of Proálcool. During this period, Proálcool formally ceased to exist as a government incentive program to produce fuel alcohol. However, policies to support the production of sugarcane and the use of fuel alcohol were continued, given the increase in the production of alcohol vehicles by the automobile industry. It was also the time of the extinction of the Sugar and Alcohol Institute (IAA). The 2000s can be considered the ethanol program resumption period when it was observed that the introduction of flex-fuel technology (flex-fuel vehicle) in Brazil, associated with a new geopolitical strategy aimed at Agroenergy (ethanol, bioelectricity, biorefineries, and biodiesel) boosted this agro-industrial sector. In the late 2000s, integration of the 1G ethanol process was proposed with possible 2-G routes through biochemistry and thermochemistry with the energy cogeneration. Also, studies have highlighted the recovery of 3G biofuels, from algae. In this period, the new company Petrobras Biocombustíveis was also created, boosting research in the ethanol area cellulosic, biodiesel, other biofuels, and second-generation biochemicals (Leite and Cortez 2008; Cruz et al. 2016; Stolf and Oliveira 2020).

The Brazilian sugar-alcohol sector has shown great organization and increased productivity and competitiveness. Most of the plants, with autonomous distilleries, can produce ethanol and sugar in variable proportions, within certain limits established for reasons technical, commercial, and political, mainly adaptable to the changes in the prices of these products due to competitive market and the impact that the price of gasoline has on the sugar and alcohol sector (Moraes and Bacchi 2014; Melo and Sampaio 2016; Gilio and Castro 2017).

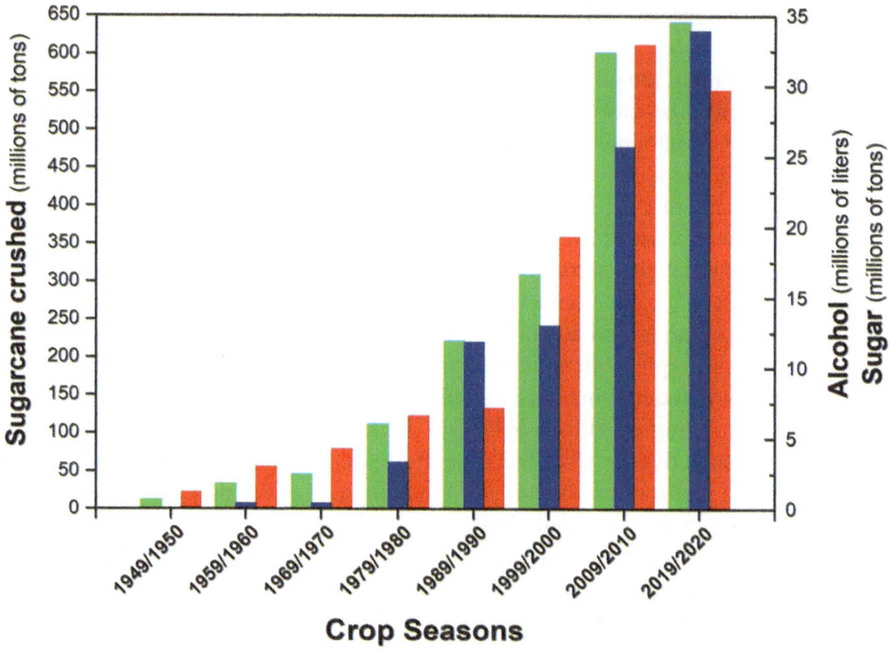

Fig. 10.2 Production of sugarcane crushed (green bar), alcohol (blue bar) and sugar (red bar) in different crop seasons in Brazil (data from ANP 2020a; CEPEA—Centro de Estudos Avançados em Economia Aplicada 2020)

As can be verified in Fig. 10.2, the amount of crushed sugarcane has increased in different crop seasons, consequently, there is also an increase in the production of the two main products of the agro-industrial sector (sugar and alcohol). According to Moraes and Bacchi (2014) and Melo and Sampaio (2016), the main factors that greatly interfered with that were the introduction of flex vehicles from 2003, mandatory addition of anhydrous ethanol to gasoline, petroleum price, and international price of sugar.

As stated earlier, ethanol production is strongly impacted by international sugar market prices and the sugar and energy sector takes advantage of its industrial production flexibility to adjust to market fluctuations. In the 1989/90 and 2019/20 crop seasons (Fig. 10.2), in which the dynamics of the sugar market were reversed and, together with the increase in the price of gasoline to the consumer, the destination of total recoverable sugar (TRS) was boosted from cane for ethanol (ANP 2020a; CEPEA—Centro de Estudos Avançados em Economia Aplicada 2020; EPE 2019).

As ethanol production is subject to the harvest and off-season cycles, influenced by market and opportunity laws and imbalance between supply and demand, different factors contribute to a fluctuation in prices, as shown in Fig. 10.3, which shows the variation in the prices of ethanol and sugar over the last 15 years in Brazil.

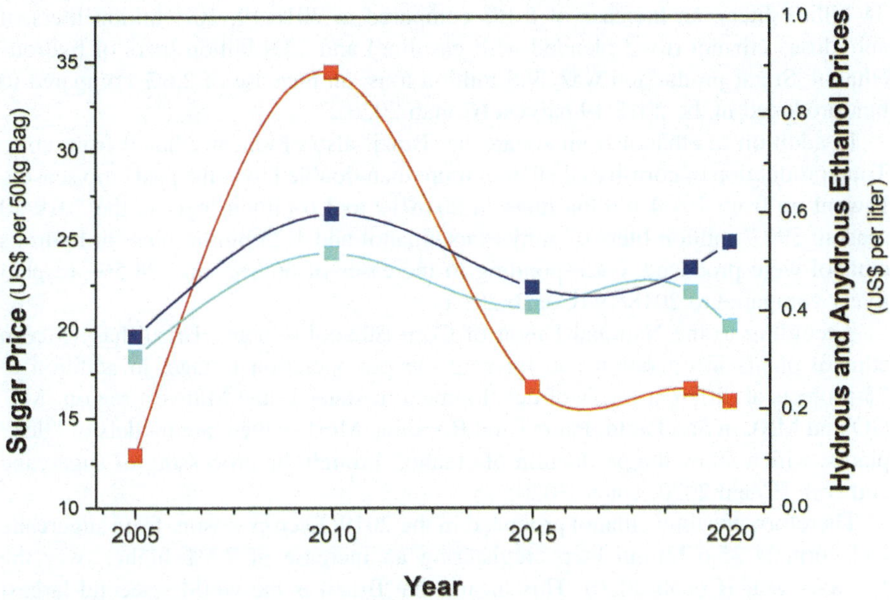

Fig. 10.3 Variation in prices (without taxes) of sugar (red symbol), hydrous ethanol (light blue symbol), and anydrous ethanol (dark blue symbol) in the last 15 years in Brazil (data from ANP 2020a; CEPEA—Centro de Estudos Avançados em Economia Aplicada 2020)

Figure 10.3 also shows the price differences between anhydrous and hydrous ethanol. Hydrous ethanol is the common ethanol (minimum content of 92.6% and maximum content of 93.8% of ethanol), the one used to supply cars with alcohol or flex-fuel, while anhydrous ethanol (minimum content of 99.3% and maximum content of 99.6% ethanol) is blended into gasoline (Almeida et al. 2017; ANP 2020a). In the analysis of Fig. 10.3, it is also highlighted that the demand for fuels is intricately linked to the growth of the economy and income. In 2010, Brazil recorded, according to the IBGE, a great economic development, registering a significant increase in Gross Domestic Product—GDP (7.5%), among other factors that favored the prices of the sugar and alcohol industry products (Torquato 2011; Fronzaglia and Torquato 2007; Torquato 2011).

The growth of the flex-fuel vehicle fleet in the Brazilian market associated with government actions, direct and indirect, such as the increase in the anhydrous ethanol mandatory blending requirement for gasoline from 20% [E20 blend] to 25% [E25 blend] (MAPA 2013) and to 27% [E27 blend] (MAPA 2015), has been providing conditions for the growth of the sugar and alcohol sector.

Currently, Brazil has more than 360 active sugarcane alcohol plants and distilleries, most of them are in the state of São Paulo (Novacana 2020a; UDOP 2020).

According to the National Supply Company (Conab), for the 2019/20 crop season, more than 642.7 million tons of sugarcane were harvested, representing an increase of 3.6% compared to 2018/19. The estimative of ethanol production was

34 billion liters, an increase of 5.1% compared to 2018/19, 10.1 billion liters of anhydrous ethanol (used blended with gasoline) and 23.9 billion liters of hydrous ethanol. Sugar production was 29.8 million tons, an increase of 2.6% compared to that produced in the 2018/19 harvest (Conab 2020).

In addition to ethanol from sugarcane, Brazil also produces ethanol from corn. Total production of corn-based ethanol more than doubled over the past crop season. It went up from 791.4 million liters in 2018/19 to 1.6 billion liters in the 2019/20 season: 390.7 million liters of anhydrous ethanol and 1.25 billion liters of hydrous ethanol were produced, corresponding to increases of 66.8% and 124.5%, respectively compared to 2018/19 (Conab 2020).

According to the National Union of Corn Ethanol (Unem), Brazil has 15 corn ethanol plants in operation and three in the pre-operational stage, in addition to 23 projects at different levels of development in states in the Midwest region (MT, GO and MS), in São Paulo, Paraná and Roraima. Most of them are models of "flex" plants, which allow the production of ethanol through the processing of sugarcane and corn (Canal 2020; Unem 2020).

Therefore, the total ethanol produced in the 2019/20 crop season, from sugarcane and corn, is 35.6 billion liters, registering an increase of 7.5% higher over the previous year (Conab 2020). This means that Brazil is the world's second-largest producer of ethanol, after the USA, which has almost twice the production by Brazil (Garside 2019).

The prospect of replacing gasoline with biofuel and advancing policies to decarbonize transport matrix, as well as a scenario of expansion of domestic demand, is increasing in several countries, which with greater interest for blending ethanol-gasoline contribute to increasing Brazilian ethanol exports because flex-fuel vehicles can run on any proportion of gasoline and ethanol, together with the mandatory E27 blend throughout the country; these factors stimulate Brazilian ethanol production in the coming years (Magalhães and Braunbeck 2014; EPE 2019; Ramos 2019). According to the Brazilian Sugarcane Industry Association (Unica), in the 2018/19 period, there was a significant increase (about 9%) in sales of ethanol in the domestic market (Unica 2020). In addition, analysis of the trade balance indicates that sales to the foreign market should reach 1.6 billion liters in the 2019/2020 crop season and that imports should total 1.25 billion liters in this cycle (Albuquerque 2019; OECD/FAO 2019).

At the domestic level, the federal government created RenovaBio, the National Biofuel Policy instituted by Law No. 13576/2017, a program to stimulate the production of biofuels. RenovaBio seeks the energy efficiency of the entire production system with consequent cost reduction in the production of biofuels and rewards the agents with the highest energy productivity and the lowest CO_2 emissions, which should encourage new investments in this sector of renewable fuels in Brazil (FIESP 2018; ANP 2020b). Unica's estimates indicate broad adherence to the sugar sector program. Mills already certified or in the process of certification accounted for 85% of domestic sales of biofuel in 2019 (Unica 2020).

Recent estimates project growth of the sugar and alcohol sector, with sugarcane production of 817.6 million tons (an increase of 36%), sugar production of 42.8

million tons (an increase of 39%), and ethanol 41.9 billion liters (an increase of 43%) for the period 2028/29. There are also estimates for expansion of corn ethanol production, expecting a rise from the current 4–20% in 2028 (FIESP 2018; Unem 2020).

However, due to the Covid-19 pandemic, the entire sector has been suffering in different ways. According to data from Conab (2020), a 2% reduction in sugarcane crushing in the 2020/2021 crop season is estimated compared to the previous period. Such reduction is directly related to social distancing measures (reducing and staff reallocation), new practices to minimize the impacts of the health crisis, the possibility of interruptions in planting and harvesting, in processing, in addition to unavailability in its supply and service chain (EPE 2020a).

Another important factor affecting the transport sector is the decrease in travel and economic activities, which reduced the consumption of fuels such as ethanol (EPE 2020a). It is estimated that there will be a reduction in the total production of biofuels by around 13% in 2020, with a 15% reduction in the production of ethanol (IEA 2020a, b). According to Unica (2020), during the period January–April, the volume of ethanol sold totaled 6.35 billion liters, 11.3% lower than that registered in the same period of 2019.

There are predictions that 25% of sugar and alcohol plants in operation in the country are in danger of closing their doors by the end of the year, especially those that are only distilleries, that is they produce only ethanol as they lack the flexibility to make sugar, and are exposed to drop in both consumption and the ethanol price (InfoMoney 2020).

For some more capitalized groups, the alternative is to change the industry mix, starting to produce more sugar to pass the most acute moment of the crisis. With this, it is expected a gain of sugar participation in the production mix, which, despite the fall in the international sugar quotation, with the devaluation of the Real against the U.S. dollar favors exports (EPE 2020a; InfoMoney 2020).

10.2.2 Second-Generation Ethanol

The integral energy utilization of sugarcane, in addition to ethanol and sugar production usually obtained in the sugar-alcohol sector in a conventional way, is linked to the current concept of a biorefinery that allows the use of lignocellulosic biomass as raw material for the production of different chemicals and/or biofuels (Júnior et al. 2020; Renó et al. 2014). Over the last years, researchers from different sectors of academia and industry have been making efforts to develop technologies that increase the efficiency and sustainability (economic and environmental) of the second-generation ethanol process. The importance of the use of lignocellulosic biomass as a raw material for the production of second-generation ethanol and other bioproducts is evident, as it is the largest source of natural carbohydrates, with an estimated 50% of the biomass in the world (Singhvi and Gokhale 2019; Claassen et al. 1999).

These studies are based on the development of bioethanol production processes from a variety of vegetal biomasses, though mainly from the sugarcane bagasse, due to its large quantity in the sugar-alcohol industry despite its usual use for energy cogeneration (Hossain et al. 2017; Balat 2011; Quintero et al. 2011; Cardona et al. 2010). It is considered that for each ton of processed cane, 280 kg of straw and 280 kg of bagasse are generated. In addition, it is increasingly evident that to enable the production of second-generation ethanol, this technology must be coupled with the already established technologies of the first-generation ethanol process.

According to National Agency for Petroleum, Brazil produces first-generation (1G) and second-generation (2G) ethanol. In 2019, 99.20% of ethanol was produced from sugarcane as raw material and 0.80% from other raw materials, with only 0.98% from other raw materials such as bagasse or straw, indicating that 2G ethanol production is still incipient (ANP 2020a, b). This production comes from two commercial 2G ethanol plants (Granbio and Raízen) with a nominal production capacity of 60 and 40 million liters per year, respectively. There is also an experimental plant at the Sugarcane Technology Center (CTC 2020; EPE 2020b; GranBio 2020a; Raizen 2020).

During the 2018/19 crop season, Raízen's 2G ethanol production was 16.5 million liters, a record production according to the company (Raizen 2019). GranBio, on the other hand, declared that it was capable of breaking its daily production records on several occasions and that the daily production volume was equivalent to 40% of the capacity authorized by the ANP; however, the company has accumulated losses and often chooses to stop ethanol production in benefit of electricity generation with the same raw material (GranBio 2020b).

The projection of ethanol supply in Brazil considers a series of premises, including the technological stage of second-generation ethanol—cellulosic or 2G (EPE 2020b). As reported by Yusuf et al. (2019) and Milanez et al. (2015) in 2015 the country had an installed production capacity of around 140 million liters of 2G ethanol per year, of which 82 million liters were produced at Bioflex 1-GranBio, the first plant with a commercial scale 2G ethanol production capacity in the country.

However, according to Milanez et al. (2017), the current commercial 2G ethanol plants tend to be considered experimental units, due to the associated technological uncertainties that still exist. There are still barriers such as breaking down the lignocellulosic matrix that need to overcome in order to ensure the availability of monomeric sugars, e.g., C6 (glucose) and C5 (xylose and arabinose) for the fermentation by microorganisms to ethanol. One of the difficulties for the use of biomass as raw material is inherent to the nature of plant cell wall which consists of cellulose (homopolymer of glucose), hemicellulose (a mixture of several polymerized monosaccharides, mainly xylose and a smaller amount of arabinose, glucose, galactose, and mannose) and lignin (polyphenolic macromolecule) (Rezania et al. 2020; Balat 2011; Cardona et al. 2010; Sánchez and Cardona 2008). Moreover, the deconstruction process of the lignocellulosic matrix is not trivial due to (1) the complex interactions between hemicellulose and cellulose present in the cell wall of plants and between these polysaccharides and lignin, (2) the crystalline nature of cellulose, (3) the physical barrier formed by lignin around the cellulosic fibers and, and (4) the difficulty of operating the equipment stably and continuously. For this reason, the biomass must be pre-treated to separate the lignin, reduce the degree of crystallinity of the cellulose and allow consistent hemicellulose hydrolysis, improving the efficiency of the chemical or enzymatic attack process (Loow et al. 2016; Brethauer and Wyman 2010; Wyman et al. 2004). Besides, engineering solutions that apply to other fiber processing industries, such as cellulose and paper, could not be successful in processing large quantities of cane straw, which present excessive levels of moisture, impurities, abrasiveness, heterogeneity, and low density, requiring solutions not yet developed.

The enzymatic hydrolysis process of cellulose, which has been extensively studied, involves the synergistic action of cellulolytic enzymes that present a high cost associated with low production (Hu et al. 2018). Several studies have been carried out in the search for enzymes capable of hydrolyzing cellulose more effectively, either by the optimization of fermentative processes, by the combination of enzymes to obtain more efficient cellulosic complexes, or by the improvement of species through genetic engineering methods (Bala and Singh 2019). The large-scale industrial production of cellulases requires an understanding and control of the parameters involved in the growth of the microorganism and its capacity to produce the enzyme (Singhania et al. 2010). The use of cellulolytic enzymes in lignocellulosic hydrolysis has been studied extensively for the liberation of fermentable sugars from the cellulose matrix (Houfani et al. 2020). Although there are several studies on the production of cellulases by microorganisms, the development of new projects is still necessary, since natural reserves have the great biotechnological potential for the exploration of new wild strains not yet studied.

One aspect of the investments is in the production of enzyme cocktails with high efficiency with a range of diverse enzymes: endoglucanases, exoglucanases, beta glucosidases, lytic polysaccharide monooxygenases (LPMOs), and xylanases, acting in a complementary way to make even more feasible production of 2G-ethanol from sugarcane straw and bagasse and boost this type of process in Brazil (Embrapa 2020; Novacana 2020b). On the other hand, it is necessary to obtain enzymes more economically because of the high cost of commercial enzymes. For example, the cost of producing one gallon of ethanol without the use of enzymes is about US$ 2, while in the enzymatic process this value rises to US$ 2.00–3.60 due to the cost of producing these proteins. However, it is estimated that in an efficient and optimized process of enzymatic hydrolysis from lignocellulosic biomass this value can be reduced (Carrigan 2016).

Some projections indicate the integration of cellulosic ethanol production with conventional methods makes the production process more economical and competitive, which should lead plants that already have cogeneration and are interested in producing cellulosic ethanol to assess the availability and diversity of raw material, as well as the efficiency of the production process (exchange of boilers and turbines and electrification of equipment). The 2G-ethanol production should use a small amount of bagasse and straw and should increase gradually, reaching around 722 million liters in 2029 (yield factor 300 ethanol liters/dry metric tonne of

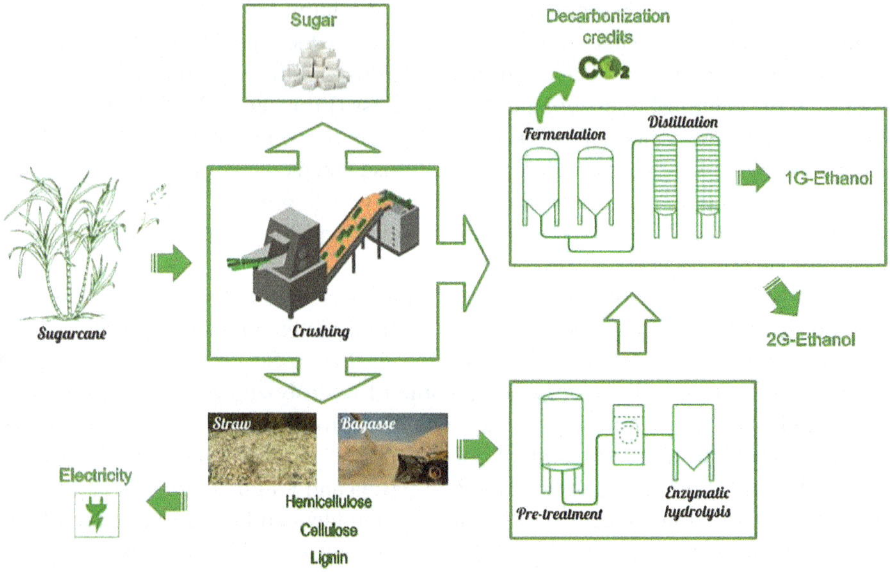

Fig. 10.4 Today's Sugarcane Mills

sugarcane bagasse), with an estimated investment of approximately U$ 22 billion between 2010 and 2029 (Araújo 2016; EPE 2020b).

Finally, it is important to highlight new economic opportunities that have emerged in the sugar and alcohol industry from the use of materials that were considered as residues, such as bagasse, stillage, and other by-products. One of the most effective and inexpensive alternatives to the total use of cane biomass is the co-generation of energy using bagasse, tops, and leaves, becoming not only an extra source of revenue in a short term of investment but the third main product of the sugar-energy sector (Rosillo-Calle and Cortez 1998; EPE 2020a). The energy usage of residual biomass generated in the industrial processing of sugarcane, both in the production of heat and electricity, is intended for self-consumption and the production of surplus electricity, exported to the Brazilian National Interconnected System—SIN (EPE 2020b).

Therefore, it is believed that it is possible to establish synergy between sugar-energy products, with bioelectricity from sugarcane bagasse considered as another asset in this sector, which involves the sale of four products (Fig. 10.4): sugar, ethanol (1G and 2G), electricity and, more recently, decarbonization credits. With the full implementation of the RenovaBio program, it may induce an increase in the efficiency of the sugar-energy producing units, increasing the score for environmental energy efficiency, favoring a sustainable and environmentally friendly bioenergy production chain (Niphadkar et al. 2018; EPE 2020b).

10.2.3 *Third Generation*

Currently, most fuels are from fossil origin. Studies on renewable alternatives have been increasing. In Brazil, there are consolidated techniques in the first and second generation for the production of bioethanol from sugarcane biomass (Dragone et al. 2010; Darda et al. 2019). The first generation of ethanol production is produced from the fermentation of glucose present in the sugarcane juice, using yeasts such as *Saccharomyces cerevisiae*. In other countries, other sugar-rich crops used for the production of bioethanol are potatoes, beets, cassava, wheat, yams, and barley (Mussatto et al. 2010; Silva and da Silva 2019). In the case of second-generation ethanol, residual crop material (e.g., sugarcane bagasse in Brazil) is used as the main source for this process (Carneiro et al. 2017). And, ethanol production using microalgae is classified as third-generation (Jambo et al. 2019).

In addition, third-generation biofuels are composed of high hydrogen content thanks to proteins and chlorophyll. When compared to many sources for the production of biofuels, the third generation has a higher caloric power, low density and viscosity, which make it more suitable for the production of biofuels with 1G origin (Miao et al. 2011; Suganya et al. 2016).

Ethanol represents a product with great impact, due to its demand in the market, and because current production is limited, making ethanol from the fermentation of algae biomass is a possible alternative for fuel demand (Silva and da Silva 2019). The advantages of the third-generation process are:

1. Its raw material does not compete with land for food cultivation since algae can be found in the seas, and
2. It can be created artificially in pools and farms. For example, in the production of 1G and 2G bioethanol, the raw material promotes competition in the use of agricultural land for food production, which can stimulate social issues, thereby increasing their cost (Balat et al. 2008; Gouveia and Oliveira 2009; Borines et al. 2013; Mesa et al. 2016) in addition to the cost of the 1G and 2G production process, which has up to 40% of its total cost from raw materials (Farrell et al. 2006; Silva and da Silva 2019).

As algae are responsible for covering more than 70% of the Earth's surface, they are disposed of in oceans. Being an abundant raw material and easy to grow, it does not require an agricultural area or drinking water. Furthermore, there are more than 10,000 species and their growth is greater than terrestrial cultures (Torzillo et al. 1986; Goh and Lee 2010). Brazil possesses a large tropical coastal area that has approximately 12% of the world's freshwater supply and contains approximately 34% of the total cataloged species in the world (Forzza et al. 2012). Worldwide cultivation of algae has grown by 10% over the last 10 years (Mesa et al. 2016; Silva and da Silva 2019). Brazil produces 30% of total world microalgae and just 700 tons (<0.1%) of macroalgae annually (EMBRAPA 2020).

An alga is an aquatic photosynthetic microorganism that develops in different environments such as saltwater, wastewater in urban areas, or on land unsuitable for agriculture (Chen et al. 2011; Suganya et al. 2016). Microalgae are microscopic organisms (Kose 2017). There are several species of microalgae and macroalgae suitable for the production of biofuels (Brennan and Owende 2010; Pablo et al. 2019). Indeed, the amount of lipids and carbohydrates, such as the presence of starch, produced in its interior, is highlighted. In addition, some characteristics are important to make the biofuel production process possible. For example, the carbohydrates rate of macroalgae and microalgae is up to 50% and 70%, respectively of its weight or favors the production of ethanol (Wi et al. 2009; Behera et al. 2015; Rizza et al. 2017). Algae also have lower recalcitrance, being more amenable for hydrolysis procedures, because of lack of lignin compared with lignocellulosic materials (Hargreaves et al. 2013). For example, Smachetti et al. (2020), identified useful microalgal strains and provided optimized conditions for sucrose production using seawater, and modeled efficient conversion into ethanol by mild methods with a productivity of 4200 L ethanol·ha^{-1}·year^{-1}.

According to Brennan and Owende (2010), another important characteristic is that the microalgae can tolerate shear stresses found in photo bioreactors; be dominant over contaminating microorganisms; absorb large amounts of CO_2; tolerate temperature variations, which occur due to seasonality; high photosynthetic efficiency; can self-flocculate by forming cellular aggregates, and facilitate the recovery of microalgae biomass. The species *Chorella vulgaris*, for example, is widely studied for the production of biofuels because it is easy to grow and sensitive to different process conditions (Jambo et al. 2016; Pablo et al. 2019).

These parameters sought in species of micro and macro algae are essential because the production of bioethanol goes through the following stages:

1. First, the cultivation of micro algae is carried out using solar energy. In this stage, several types of reactors such as open, covered lagoons or closed photobioreactors can be used (these can be tubular, flat plate, or other models) (Klein 2013; Jambo et al. 2016).
2. In the second stage, the biomass needs to be concentrated at least 30 times to perform starch (carbohydrate) extraction. This extraction can be physical or biological (carried out by enzymes).
3. In the third stage, the starch is broken down enzymatically with the aid of amylase enzymes.
4. In the fourth stage, the fermentation of simple sugars is promoted by the fungus species *S. Cerevisiae* which promotes alcoholic fermentation (Dragone et al. 2010).
5. In the last stage, the alcohol broth is pumped into a holding tank to feed a distillation unit, finalizing the production of ethanol as biofuel. In this context, some studies seek to find methods that contribute to the release of more glucose from these algae (Yoon et al. 2010; Suganya et al. 2016) in order to improve ethanol production through alcoholic fermentation (Mussatto et al. 2010; Klein 2013; Darda et al. 2019).

10.2.3.1 Current Status

Recurrently, different ways of increasing 3G ethanol are being studied, some studies show that it is possible to release a greater amount of glucose from specific species of algae (Yoon et al. 2010). In addition, instead of extraction, some species of algae are capable of conducting automatic fermentation (de Souza et al. 2012; Carneiro et al. 2017).

In Brazil, studies with 3G ethanol advance with research in the field of cultivation of microalgae species, where the concentration of CO_2 in the medium of innovative sources of nutrients such as N ($NaNO_3$) and pH of cultures are studied (Klein 2013; Carneiro et al. 2017), in addition to the reuse of water in the cultivation process, luminous flux and the impact on the production of carbohydrates and 3G ethanol (Pablo et al. 2019). Photosynthetic efficiency and variation of the use of luminous flux by microalgae species affect directly the carbohydrate yield and production of a greater amount of 3G fuel produced from the microalgae (Klein 2013). Another way to optimize production is through design, which enables the reuse of process inputs, such as solvents and catalysts. At the end of the process, waste can be reused as fertilizers. The fundamental of a project that aims to the best performance and maximum use of the system, is to proceed with microalgae species that grow fast (Silva and da Silva 2019).

10.2.3.2 Market and Challenges

From the financial comparison, it can be concluded that the production cost of third-generation bioethanol is quite higher than of the first and second generations (Jambo et al. 2016). The productivity rate of algal feedstock such as seaweed is very high until it has become one of the sources of income for the coastal population by increasing the employment opportunities through seaweed farming or cultivation (Noraini et al. 2014). In recent years, the total global production of aquatic plants (macro and microalgae) has increased exponentially and reached 30.1 million tons of wet weight in 2018 with a value of more than US$ 1.5 billion (current biomass prices average about US$ 50 per dry ton) (Bjerregaard et al. 2016; Pablo et al. 2019).

Annual seaweed harvesting could produce about 1.25 billion megawatt-hours' worth of methane or liquid fuel. The world used about 85 billion megawatt-hours of energy from fossil fuels in 2012, so energy production from these seaweed products equates to roughly 1.5% of current energy use from fossil fuels (IEA 2014). In the case of microalgae, this microorganism could produce approximately 5000–15,000 gal of ethanol per acre annually which is more reliable than the first-generation bioethanol feedstock (707.4 gal/acre from sugarcane in Brazil) (Goldemberg and Guardabassi 2010; Nguyen 2012).

However, the commercial viability of algal bioethanol is still in the doubt, in spite of the promising opportunities envisioned for microalgae-integrated biorefineries, the economic viability of large-scale algal biomass cultivation for low-value products (e.g., biofuels, bulk chemicals, and biomaterials) has not been achieved yet. And one of the main factors for its commercialization is the lack of an efficient and reliable established technology, this in addition to the costs for microalgae biofuel production two-fold higher than for its fossil-based counterparts (Jambo et al. 2016; Brasil et al. 2017). Another challenge is the concerns regarding water scarcity, food availability and competition with human being feed, and land/soil degradation are increasing, and these issues are far from being independent of bioenergy production. The competing need for land and water resources in food and bioenergy production has been at the forefront of policy debates (Benites-Lazaro et al. 2020).

Brazilian microalgae production chain has considerable potential for development, the largest barriers to the Brazilian microalgae production chain are the cost of raw material production (US$ 0.37–3.80 per kg) and the lack of organizations representative of the sector. Currently, only a few companies (~17, when compared with top algae country producers) produce or commercialize microalgae in Brazil, demonstrating that the Brazilian microalgae production chain is still under development and has yet to demonstrate itself as a commercially viable industry (Andrade et al. 2020).

10.3 Biogas

Biogas is a biofuel produced by some microorganisms during the anaerobic digestion of several biomasses and is mainly composed of methane (CH_4) and carbon dioxide (CO_2) (Santos et al. 2018). The technological strategies in biogas are present in the whole manufacturing process comprising: feedstocks, production, treatment, logistics, and uses, but the biotechnological approaches are more focused on production and treatment (Oliveira and Negro 2019).

Several organic materials can be used as substrate for the anaerobic digestion, which comprises the following steps: (1) hydrolysis, (2) acidogenesis, (3) acetogenesis, and (4) methanogenesis (Yang and Ge 2016). Methane and carbon dioxide contribute about 50–75% and 25–50%, respectively, of the final biogas composition (Da Costa-Gomez 2013). However, other common components and some impurities such as H_2O, O_2, H_2S, NH_3, N_2, chlorines, and siloxane (Da Costa-Gomez 2013; Yang and Ge 2016) are present in low concentrations.

The process is generally performed at mesophilic temperatures (35–37 °C) for about 30–35 days (Wang et al. 2018; Yılmaz and Şahan 2020), though it can vary according to each biomass, microbial load, and environmental conditions. After the biogas production, a liquid effluent (bioslurry) and solid residues (sludge and scum) are also generated, though in low amounts (FAO 2013). The inoculum is obtained from anaerobic biodigesters such as slurry and sludge and consists of a diversified microbial community (Holliger et al. 2016; Juárez et al. 2018; Wang et al. 2018). Microorganisms involved in hydrolysis and acidogenesis processes consist of obligated and facultative anaerobic bacteria, while methanogenesis is performed by methanogens archaea (Neshat et al. 2017). Some of the most common

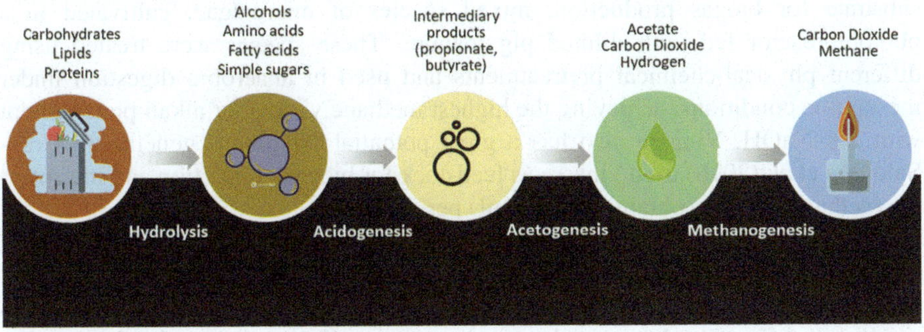

Fig. 10.5 Biomass digestion for biogas production

microorganisms in bioslurry include bacteria belonging to the genus *Clostridium*, *Methanosaeta*, *Terrisporobacter*, *Methanobacterium*, and *Methanosarcina* (Zuo et al. 2019). Figure 10.5 shows the digestion steps for biogas production.

The feedstocks used as a substrate for biogas can be divided into three main types: (1) agro-industrial wastes and animal manure, (2) urban wastes, and (3) industrial effluents (Da Costa-Gomez 2013); they can be used alone or in association for a co-digestion process. From a biogas production perspective, the utilization of food waste is a better substrate than sludge, as promotes a weak alkaline environment and avoids the excessive acidification effect, gradually increasing the digestion reaction (Liu et al. 2016a).

These biomasses are rich in carbohydrates, proteins, and/or fats, and the biogas composition and purity are directly influenced by the type of organic biomass (Yang and Ge 2016; Freitas et al. 2019a, b).

Animal manure contains several nutrients and has traditionally been used as a fertilizer for crops, but this practice is declining over the last decades due to an increase of synthetic fertilizers and the use of new cultivation and logistics techniques (Neshat et al. 2017). The growing amount of manure generated daily and its nutrient composition make this residue a great feedstock for biogas production through anaerobic digestion. In a study performed by Yılmaz and Şahan (2020), the authors evaluated the anaerobic digestion of poultry manure and obtained the maximum cumulative biogas production of 8965.87 mL, composed of 71.3% methane and percentage of chemical oxygen demand (COD %), removal of 68.47% under optimum conditions after 35 days. Wang et al. (2018) investigated the co-digestion of fresh deer manure and mushroom residue, added with biogas slurry, and obtained up to 24,346 mL of cumulative biogas production after 31 days.

Agro-industrial wastes can be used alone or along with animal manure in the anaerobic digestion co-digestion after a pre-treatment process. Ning et al. (2019) studied the co-digestion of corn straw and pig manure; corn straw was pre-treated with 5% NaOH solution at room temperature for 5 days before using, obtaining biogas with 62.17% methane content. Zuo et al. (2019) used rice straw pretreated with 2% ammonia for 3 days but obtained a slow decrease in the biogas production after 10 days. Juárez et al. (2018) used microalgae biomass grown in pig manure as a substrate for biogas production, mixed species of microalgae, cultivated in a photobioreactor fed with diluted pig manure. These species were treated using different physical-chemical pretreatments and used in anaerobic digestion under mesophilic conditions, achieving the highest methane yield after alkali pretreatment with 5% NaOH. Vinasse also has a great potential for biogas generation, Cruz-Salomón et al. (2016) used vinasse as feedstock for biogas production and improved the methane content, obtaining 2140 mL per day.

The municipal wastes, rural wastes, and landfills are rich in organic matter and good feedstocks for biogas production. Lavergne et al. (2019) compared the digestion of sewage sludge and its co-digestion with pig manure and observed no significant differences between these two processes. In another study, the concentrated primary sludge was anaerobically digested and the authors observed that alkaline conditions (pH $\geq$8) have increased the methane production and decreased the hydrogen sulfide content (Zhao et al. 2020). Villanueva-Estrada et al. (2019) studied the biogas emission from a landfill and reported 75% of biogas generation with about 55% of methane content from wells.

Brazil predominates in the use of biodigesters with low solid matter feedstocks, most related to pig farming, and placed close to the residues sources lands (ABBM 2016; Freitas et al. 2019a, b). The anaerobic digestion plants are divided into high-solids and low-solids (LS) systems. Low-solids anaerobic digestion (LSAD) systems are reported by presenting 5–10% of solids in the feedstock, as the high-solids anaerobic digestion (HSAD) systems present up to 40% solids (Guendouz et al. 2008). Even though LSAD systems are still mostly used, this process still presents many limitations such as waste management, including water wasting, inconvenient operation, low energy recovery, and low fermentation slurry concentration (Chen et al. 2015). Therefore, there is a fast growth of HSAD systems, as they are easier to handle, presenting lower energy and heating requirements and smaller operational reactor volumes. (Guendouz et al. 2008; Liao and Li 2015). The drawback presented in the utilization of high-solids feedstock is the generation of high concentrated intermediates, which correspondingly increases the risk of system imbalance, resulting in the accumulation of inhibitory intermediates (Liao et al. 2014; Liu et al. 2016b). In high-solid systems, there are two primary technologies: plug-flow and dry fermentation. A plug-flow reactor is a reactor type that presents separated acidogenic and methanogenic phases through the reactor path. This separation could improve reactor stability and pretreatment efficiency (Namsree et al. 2012). The dry fermentation technology presents some advantages related to the lower feedstock pretreatment, such as organic waste on size reduction, removal of inert materials such as plastic (which accumulates inside the digesters or clogging pipes and pumps). Dry anaerobic digestion also does not present foaming, sedimenting, and crust surface formation, and requires no energy for stirring, allowing similar biogas/energy production when compared to the wet digestion process (Chiumenti et al. 2018). It is observed that both low and high-solid technologies are interesting as can be used for producing energy or renewable biogas, contributing to the greenhouse gas emission reduction and the preferred operational model to be implemented must be based on waste type and desired biogas volume output.

10.3.1 Current Status

Biogas production has been growing in Brazil in the last two decades (Fig. 10.6), based on its large potential to improve sustainability and diversity of energetic matrix both at the national and local and decentralized level, to represent a viable technology for waste treatment, and to promote the creation of income sources (EPE 2019a, b; Freitas et al. 2019a, b). The progressive improvement based on its economic and technical viability has promoted the emergence of small, midsize, and large plants in the country since 2003 (Fig. 10.6a), which offers biogas for various applications (Fig. 10.6b) and use diverse waste and byproducts as substrates (Fig. 10.6c).

According to the database *Biogásmap* (CIBiogás 2020), 486 biogas plans were operating in Brazil in 2019, of which almost 76% were small plants (less than 2.500 Nm^3/day), 18% were midsize plants (2500–12,500 Nm^3/day) and 6% were large plants (higher than 12,500 Nm^3/day). Most of the plants are located in states from the south, southeast, and midwest regions, where plants implementation started sooner (since 2003), while the first plants appeared in the northeast region in 2010 and the northern one just in 2017. For comparison, world biogas production is led by Germany (more than 10,000 plants) and the United Kingdom (approximately 1000 plants), whereas Brazilian production is similar to Korea, Denmark, and The Netherlands, and higher than other European countries (IEA 2020a, b).

The main application of biogas generated in Brazil is for electrical energy (76%) followed by thermal energy (22%), as shown in Fig. 10.2b. Regarding substrates, the highest proportion (approximately 70%) of biogas plants in Brazil in 2019 were based on the use of waste from pig farming (Fig. 10.6c), which were also the oldest ones (since 2003) and mostly small and midsize plants. Wastes from the food and beverage industry (FBI, 11%), cattle (Ct, 7%), and the landfill (Lf, 5%) were in second, third, and fourth places (Fig. 10.6c), respectively, which mainly include midsize and large plants in the case of FBI and Lf, and small plants for Ct. Other sources for biogas productions are sewage sludge (3.7%), waste co-digestion (1.6%), laying or cutting poultry (0.8%), sugar-and-energy industry (0.6%), and poultry and swine slaughterhouse (0.4%), which in most cases correspond to small and midsize plants.

The continuous increase in biogas production in Brazil has been driven by public policies, private initiatives, and research. Regarding the last one, it is noteworthy that Brazilian scientific documents in the Scopus database with the keyword *biogas* started to be published in increasing numbers since 2004–2005, as shown in Fig. 10.7b, and with a growth tendency coherent with the world's research trends (Fig. 10.7a). Until June 2020, 978 documents had been published (77% are papers, 5.6% reviews, and 3.2% chapters, among others), but patents were not found. Brazil's scientific production on biogas represents 3.2% of the global production (30,378 documents) and is in eighth place (Fig. 10.7c). Worldwide scientific research is largely dominated by China, United States, Germany, India, and other European countries (Fig. 10.7c). Nevertheless, it is interesting to notice that

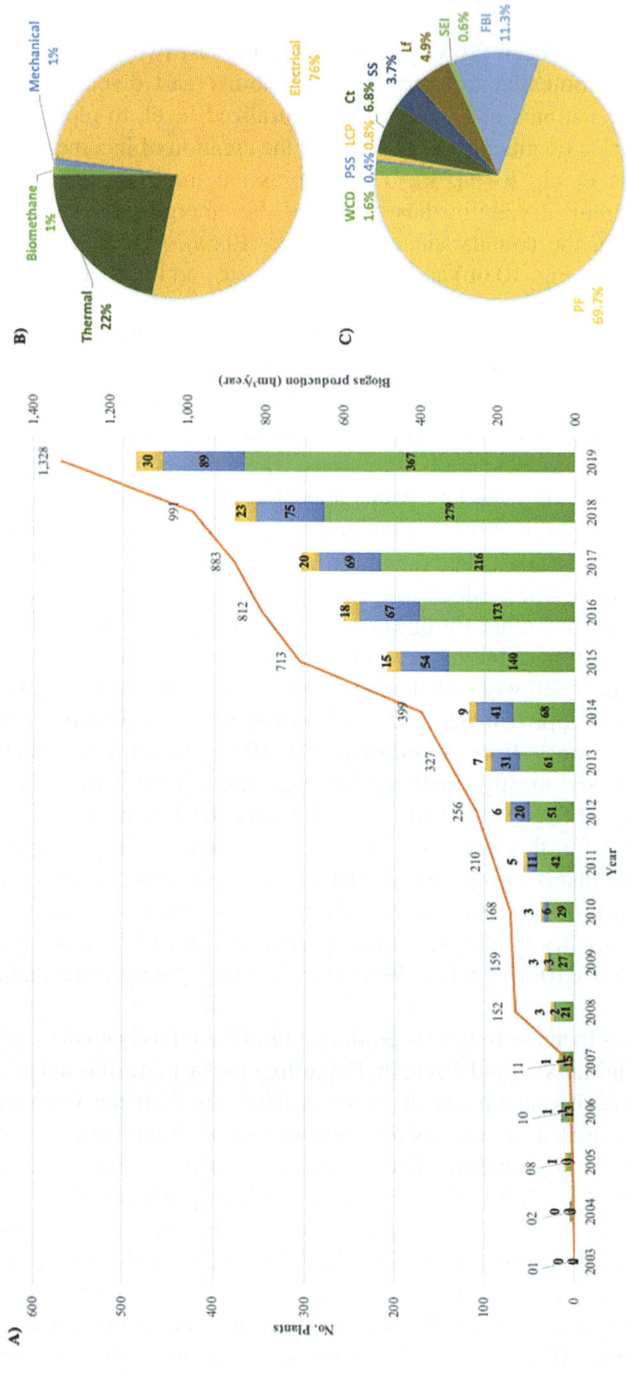

Fig. 10.6 Biogas production in Brazil in the period 2003–2019 (**a**), main applications of biogas generated, (**b**) substrates for biogas production, (**c**) Constructed with data from https://mapbiogas.cibiogas.org

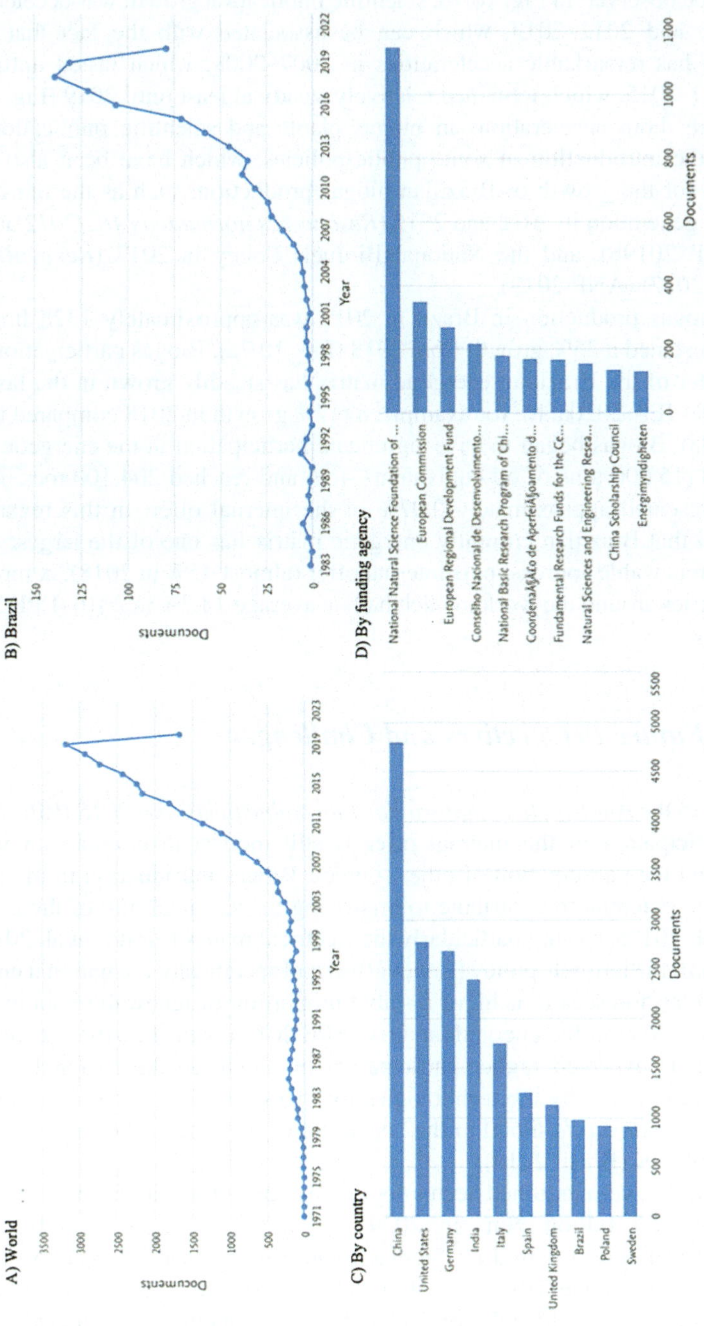

Fig. 10.7 Research trends on Biogas worldwide **(a)**, particularly in Brazil **(b)**, participations of countries **(c)** and of funding agencies **(d)**

Brazilian funding agencies, such CNPq and CAPES, occupy positions 4 th and 6 th worldwide (Fig. 10.7d), suggesting the federal commitment to this issue.

As can be observed in Fig. 10.7b, scientific publication growth was accelerated in 2004–2005 and 2012–2013, which can be associated with the fact that biogas production has remarkable accelerations in 2007–2008, which lasted until 2014, and in 2014–2015, which remained relatively steady at least until 2019 (Fig. 10.7a). Furthermore, both accelerations in biogas plants and scientific publication were related to the introduction of some public policies, which have been also one of the drivers for the growth of Brazilian biogas production, such as the ones on the distributed generation in 2012 and 2015 (*Resoluções normativas 482/2012* and *687/2015*) (EPE 2019b), and the National Biofuels Policy in 2017 (*RenovaBio, Lei 13.576 de 2017*) (ANP 2019).

Total biogas production in Brazil in 2019 was approximately 1328 hm^3/year, which represented a 25% growth from 2018 (Fig. 10.7a). Biogas participation in the internal offer of the Brazilian energetic matrix has steadily grown in the last years (EPE 2019b; IEA 2020a, b), for example, a 6.7% growth in 2018 compared to 2017 (EPE 2019a). Biogas began to have significant participation in the energetic matrix since 2010 (15,000 tons of oil equivalent—toe) and reached 204,100 toes in 2018, which represented approximately 0.07% of the internal offer. In this regard, it is highlighted that Brazilian currently energetic matrix has one of the largest participations of renewable sources in its internal offer (almost 45% in 2018), compared to other countries around the world, which had on average 13.7% in 2016 (EPE 2019a).

10.3.2 Future Perspectives and Challenges

According to the *Análise de cojuntura dos biocombustiveis ano 2018* (EPE 2019b), biogas participation in the internal offer is still modest in comparison with its potential and the participation of other sources. Biogas introduction in the internal offer has the potential to contribute to improving the sustainability of the energetic matrix (EPE 2019a, b), and particularly the electrical matrix (Freitas et al. 2019a, b). In this regard, the biofuels policy (RenovaBio) and specifically the national energetic expansion plan foresees to achieve a stable proportion of renewables sources of at least 48% by 2029 in the energetic matrix (EPE 2019a; Brasil 2019). According to this plan, it is expected higher participation of biogas in the national energetic matrix, driven mainly by the technologies for digestion of the by-products vinasse and "filter pie" (*torta de filtro)* in the sugar-and-energy sector, besides animal and municipal wastes (Brasil 2019).

From the above-mentioned technologies, the potential biogas production can reach up to 7.2 billion Nm^3 in 2029, which can account for 11% of the thermoelectrical energy based on biomass, and can represent also almost 3.9 billion Nm^3 of biomethane, (Brasil 2019) which may have a large potential to cover a high proportion (approximately 40–50%) of the total natural gas demand (Schmid et al. 2019). It is expected that the biogas offer for the electrical matrix should be at least 30 MW/year since 2023 (Brasil 2019).

Despite the existence of federal policies and plans for biogas promotion, authors argued that there are still political, economic, and social challenges to allow a higher growth on the implementation of biogas generation and waste treatment (Freitas et al. 2019a, b; Nadaleti et al. 2020; Ribeiro et al. 2020). Since biogas production technologies are already considered technically viable, the main bottlenecks are the need for robust public policies and regulation to promote financial aids to make them economically viable both in rural and urban areas (Freitas et al. 2019a, b). Renewable energy can represent a higher initial investment than traditional technologies, but that the potential economic, environmental, and social gains can be also larger (Nadaleti et al. 2020; Ribeiro et al. 2020; Campello et al. 2020). Furthermore, a higher promotion of research and national and international cooperation programs is necessary to provide and keep suitable equipment, storage, distribution grids, human resources for technical support and negotiation of energy offers, available for all the potential partners in the country (Freitas et al. 2019a, b; Nadaleti et al. 2020; Ribeiro et al. 2020).

Regarding technical challenges that can be addressed by each particular biogas production plant, factors that affect the anaerobic digestion are substrate and organic matter concentration, temperature, pH, carbon-nitrogen ratio, macro and micronutrients, and presence of heavy metals (Campello et al. 2020; Ribeiro et al. 2020). Ribeiro et al. (2020) also highlighted the importance of some operations factors, such as residence time, water excess, and possible contamination from washing and disinfection. The challenges already mentioned are especially important regarding biogas production from vinasses (Nadaleti et al. 2020). According to these authors, not only biomethane production from vinasses in sugar and ethanol plants seemed to be technically and economically viable, but also bio digested vinasses had better properties to be used for fertirrigation.

10.4 Biodiesel

10.4.1 Fundaments

Biodiesel is composed of alkyl esters of fatty acids, can be synthesized by chemical or enzymatic catalysis, from renewable feedstocks such as vegetable oil, non-edible vegetable oil, algae oil, waste frying/cooking oil, and animal fats.

Historically, the production technology in Brazil began in the 1980s with the first pilot plant by Expedito Parente in northeastern Brazil (Suarez et al. 2016). Since then, in the world, biodiesel has attracted attention because it is a renewable, biodegradable, and non-toxic fuel, enabling the development of a sustainable energy source.

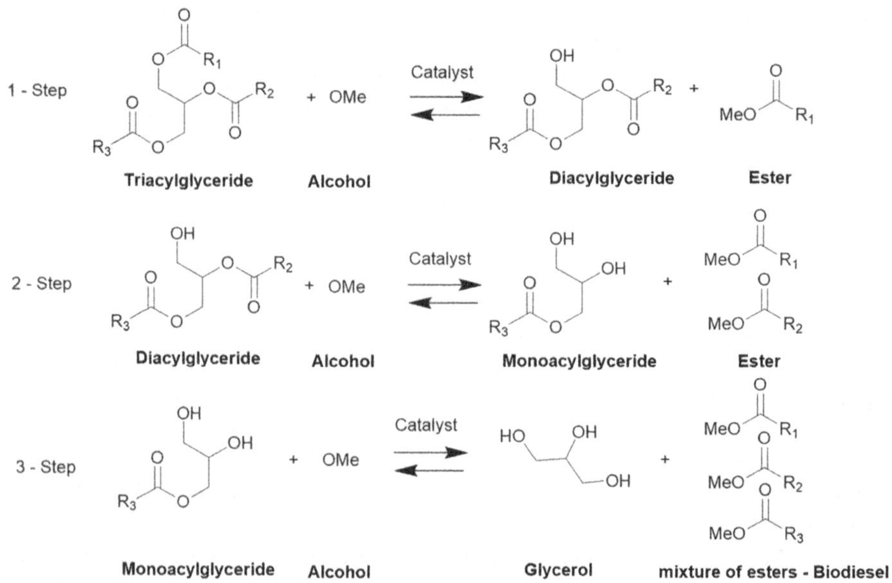

Fig. 10.8 Transesterification reaction of a triacylglyceride

Biodiesel is miscible and physically-chemically similar to diesel oil and usable in diesel engines without the need for significant or numerous adaptations (Knothe et al. 2010).

The main biodiesel production route is the transesterification reaction by homogeneous base catalysis. The term transesterification, in general, is used to describe an important class of organic reactions in which one ester is transformed into another, through the exchange of alkoxide groups in the presence of a catalyst producing a mixture of monoalkyl esters and glycerol (Fig. 10.8).

The biodiesel synthesis is a sequence of three consecutive and reversible reactions in which the triacylglycerides are gradually converted into diacylglycerides and monoacylglycerides as intermediates, one mole of the ester is released at each stage.

The transesterification reaction can be accelerated by chemical catalysts (homogeneous or heterogeneous), biocatalysts (free or immobilized), or non-catalytic (supercritical), as shown in Table 10.1. The advantages and disadvantages of each process are also detailed. Beyond that, biodiesel can be produced by other routes, such as esterification, hydroesterification, or non-catalytic processes.

In the history of biodiesel production in Brazil, the industry opted for the use of transesterification via alkaline catalysis using mainly sodium methoxide as a catalyst for the production of monoalkyl esters, as it had a low cost. However, there are disadvantages in this method mainly due to the occurrence of secondary reactions (Fig. 10.9), which reduce the yield in monoalkyl esters productions. In addition, it is a route that requires large amounts of energy, wastewater treatment, and high-quality raw material requirements (low levels of water and free fatty acids), limiting the use of various raw materials, increasing thus the price of the production of esters.

Table 10.1 Comparison of types of catalysts for biodiesel production (Robles-Medina et al. 2009; Aransiola et al. 2014; De Lima et al. 2016; Norjannah et al. 2016)

Catalyst type	Advantages	Disadvantages
Homogeneous	The very fast reaction rate	Low-quality glycerol produced thus requires a lengthy distillation process for purification
	High conversion (≥99%)	Hydroscopic nature of catalysts (NaOH, KOH)
	Feedstock must be of high quality in order to maintain the yield of the process: free of FFAs (<0.5%) and water (<0.3%)	Basic homogeneous catalyst—FFAs react with the catalyst and form soap
	Sometimes, esterification and transesterification occur in conjugated processes	Acid homogeneous catalysts are very harmful, very corrosive to the reactor and pipeline, and require careful handling
	Relatively low-cost basic catalyst and available (NaOH and KOH)	The water promotes hydrolysis of the alkyl esters to FFAs
	A preferred method for high-grade feedstock	The high-quality feedstock represents 70–95% of the final cost of biodiesel
	1% of catalyst based on the mass of oil	The purification stage of the biodiesel process is relatively difficult and requires a huge amount of water
		The catalyst can not be reused
Heterogeneous	High catalytic stability against leaching	Converts triglycerides at a relatively slower rate
	Easy separation of the catalyst from the product	Complicated catalyst synthesis procedures lead to higher cost
	The separation of the glycerol and catalyst from biodiesel is much easier	Effectively active only on surface atoms
	Economic because of its reusable nature	
Biocatalyst	FFAs are converted into biodiesel, without loss of raw material	Long process time due to very slow reaction rate (8–72 h)
	Low-price feedstock can be employed	Sensitive to alcohol, that can deactivate the enzyme
	High possibility to reuse and regenerate the catalyst	High cost of biocatalyst
	Only a simple purification step is required	
	Easy separation of biodiesel and biocatalyst by filtration	
	Easy separation of biodiesel and glycerol by decanting	
	Glycerol is of high quality and has a high sale value	
	Enzymes are biodegradable	
Non-catalytic (supercritical)	Low-quality feedstock could be transformed easily into biodiesel	

(continued)

Table 10.1 (continued)

Catalyst type	Advantages	Disadvantages
		More energy is required by the reaction step especially in the heating step as high power consumption is involved
	High-quality glycerin is generated as a coproduct	Possible generation of thermal degradation products
	Simpler separation and purification steps involved	High temperature and pressure required
	High conversion (98%)	High alcohol to oil ratio is needed
	Short reaction time (7–15 min)	
	No catalyst cost	

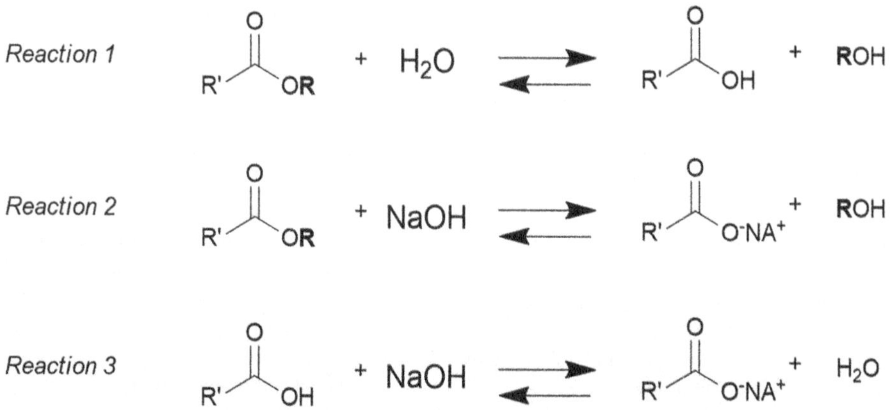

Fig. 10.9 Secondary reactions that may occur during the transesterification of vegetable oils: (1) hydrolysis; (2) saponification and (3) neutralization

In this context, interest in the use of biocatalysts has increased as it is a viable interdisciplinary technology that provides a more sustainable production of biodiesel production. Thus, the difficulties presented by conventional catalysts (Table 10.1) can be eliminated.

10.4.2 Biocatalysts in Biodiesel Production

The use of enzymes for the production of biodiesel has presented many advantages. Among biocatalysts, the application of lipases is the class of enzymes most widely used in organic synthesis, because it has excellent catalytic activity, high specificity, and can catalyze different reactions.

Lipases (triacylglycerol acyl hydrolases-EC 3.1.1.3) are enzymes generated through microorganisms (Table 10.2), animals and/or plants. (Christopher et al.

Table 10.2 Different sources of lipase (Christopher et al. 2014)

Fungi	Bacteria	Yeasts
Alternaria brassicicola	*Achromobacter lipolyticum*	*Candida deformans*
Aspergillus niger	*Aeromonas hydrophila*	*Candida parapsilosis*
Candida antarctica	*Bacillus subtilis*	*Candida rugose*
Mucor miehei	*Burkholderia glumae*	*Candida quercitrusa*
Rhizomucor miehei	*Chromobacterium viscosum*	*Pichia burtonii*
Rhizopus chinensis	*Pseudomonas aeruginosa*	*Pichia sivicola*
Rhizopus oryzae	*Pseudomonas cepacia*	*Pichia xylosa*
Streptomyces exfoliates	*Staphylococcus aureus*	*Saccharomyces lipolytica*
Thermomyces lanuginosus	*Staphylococcus carnosus*	*Geotrichum candidum*

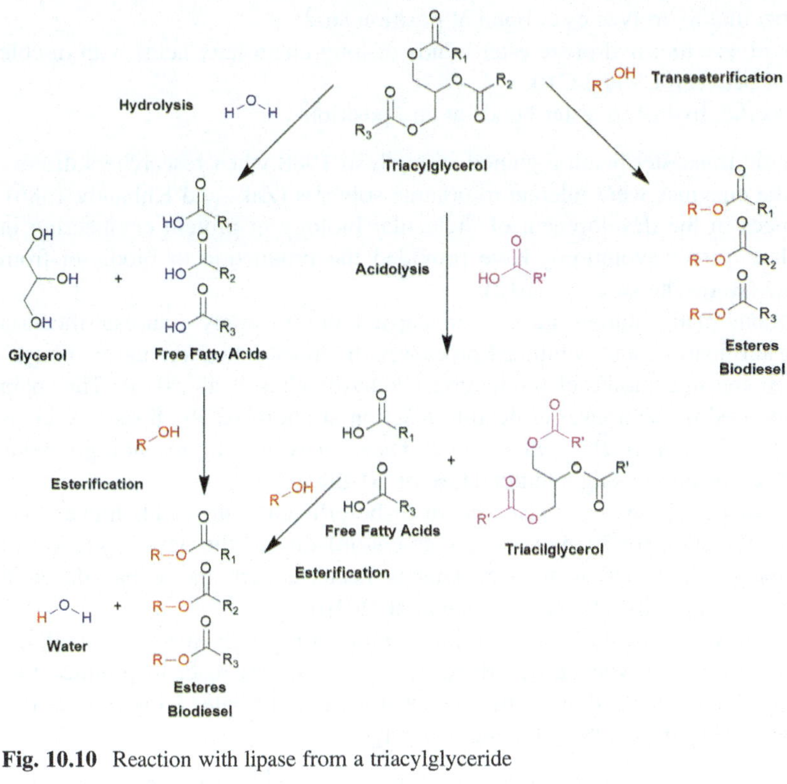

Fig. 10.10 Reaction with lipase from a triacylglyceride

2014; Filho et al. 2019). They belong to the class of hydrolases, capable of hydrolyzing the insoluble triacylglycerols at the interface between the substrate and the water. In addition to hydrolysis, they are capable of carrying out various types of reactions, depending on the composition of the reaction system such as esterification, transesterification, interesterification, alcoholysis and acidolysis (Fig. 10.10). All of these processes will generate monoalkyl esters (biodiesel).

Moreover, all co-products generated and obtained in all processes, in special glycerol, are of high purity and quality.

Lipases could be classified as extracellular and intracellular. Extracellular cells are obtained mainly from the purification of living products produced by microorganisms. The main extracellular microorganisms are *Mucor miehei, R. oryzae*, *C. antarctica* and *P. cepacia* (Fjerbaek et al. 2009; Norjannah et al. 2016). Intracellular lipases are present within the cell or in the cell-producing wall. In most cases, intracellular lipases are found in immobilized form.

The lipases possess different regioselectivity, specificity, and catalytic activity. In terms of regioselectivity, lipases can be divided into four groups (Gog et al. 2012; Guldhe et al. 2015):

1. Sn-1,3-specific: hydrolyze ester bonds at position sn-1 and sn-3;
2. Sn-2-specific: hydrolyze ester bond at position sn-2;
3. Fatty acid-specific: hydrolyze ester bonds of long-chain fatty acids with double bonds in between C9 and C10;
4. Non-specific: hydrolyze ester bonds at any position.

Enzymatic transesterification gained strength in 1988 when researchers discovered that the enzymes were tolerant to organic solvents (Zaks and Klibanov 1986). The advances in the development of molecular biology of protein engineering in 1990, called "direct evolution", have provided the production of biodiesel from biocatalysts (Bornscheuer et al. 2012).

Since then, many studies have been carried out to apply transesterification reaction, esterification, and combined processes (hydroesterification and two-steps) aiming at increasing biodiesel production (Pourzolfaghar et al. 2016). The main biocatalysts used in the transesterification reaction are non-specific lipases, such as *C. antarctica*, *C. rugosa*, *P. cepacia* and *P. fluorescens*, which present high yields above 99% conversion using a temperature of 30–50 ° C.

Several Sn-1,3 lipases have also shown to be efficient with yields higher than theoretical efficiency (66%, maximum conversion) due to the acyl migration of position due to the reaction system. This is because each lipase has different specificity about its substrates (Norjannah et al. 2016).

Lipase catalyst transesterification follows ping–pong bi–bi mechanism. Ping–pong bi–bi mechanism can be described as two substrates react to produce two products through the formation of enzyme-substrate intermediates (Gog et al. 2012).

The literature proposed three kinetic pathways:

1. direct alcoholysis of glycerides (triglycerides, diglycerides and monoglycerides) into alkyl esters;
2. two consecutive steps which consist of hydrolysis (conversion of glycerides into free fatty acids) and followed by esterification (conversion of free fatty acids into alkyl esters);
3. simultaneous reactions of both alcoholysis and hydrolysis followed by esterification.

Table 10.3 Comparison between free lipase versus immobilized lipase (Ranganathan et al. 2008; Hanefeld et al. 2009; Sheldon and Van Pelt 2013)

Lipase catalyst	Advantages	Disadvantages
Free lipase	More efficient catalytic activity	Inactivation of lipase by alcohol
	Highly selective	Unstable lipase with a small change in the reaction medium
		High cost
		Unable to recover
Immobilized lipase	Better stability, especially towards organic solvents and higher temperatures	High resistance to mass transfer
	High operational stability	Lower reaction rate
	Co-immobilization with other enzymes is possible	Centrifugation and filtration required for the separation
	Use of fixed bed or batch reactors without the need of membrane to isolate enzyme from product	Enzyme activity decreased during the immobilization process
	Recycle	Reduced costs of downstream processing

In addition, it is important to examine enzyme kinetics, because they help in the determination of the optimum reaction parameters, in special, studies involving lipase type, lipase immobilization, solvent type, the effect of temperature, reactant concentrations, and mass transfer limitations; which are very important to scaling-up the process and to reactor design.

The use of immobilized lipase is more preferred than free lipase, as it reduces the high cost of the biocatalyst, which is the biggest obstacle to the use of lipases. Moreover, immobilized lipase promotes easy recovery and allows the reuse of the enzyme. The advantages and disadvantages of each type of lipase are illustrated in Table 10.3 (Ranganathan et al. 2008; Hanefeld et al. 2009; Sheldon and Van Pelt 2013).

On the other hand, enzyme immobilization can affect the enzyme activity, specificity, and selectivity of the enzyme and also change its structural form. These changes are not always beneficial. Someone can cause possible effects such as temperature stability, solvents (alcohol) stabilization of the hyperactivated form of the enzyme, dispersion of the enzyme on the support surface, and denaturation, while others can lead to negative effects such as the inhibition of lipase.

The immobilization method and the support material are essential for good immobilization. The most used methods for immobilization are adsorption, entrapment, encapsulation, and crosslinking, new area of promising immobilization methods are cross-linked enzyme aggregates (CLEAs) and protein-coated microcrystals (PCMCs), cross-linked PCMC (CLPCMC), magnetic particles carrier, and electrospun nano-fibers (Hanefeld et al. 2009; Sheldon and Van Pelt 2013).

The commercial immobilized lipase widely used to produce biodiesel by transesterification is Novozym 435 (*Candida antarctica*) which is immobilized by the method of adsorbed on a macroporous resin called Lewatit VP OC 1600. According to the information given by the Novozymes Co. on their website, this lipase produces biodiesel with above 97% (w/w) of monoalkyl esters (José et al. 2013).

Currently, Novozym 435 (*Candida antarctica*) is the commercial lipase that surpasses the high cost of the biocatalyst, as it presents the best biodiesel production yield when compared to other immobilized lipases. However, it is not tolerant to high concentrations of methanol and water, which promotes its denaturation, a negative effect that is easily contemporary because the biodiesel industry operates observing the water levels and fractionally applying alcohol. Finally, the efficient recycling of the biocatalyst and the control of final product composition promote a better advantage to the commercial process.

10.4.3 Current Status: Biodiesel Plant in Brazil

Brazil is among the three largest biodiesel producers in the world. The production of commercial biodiesel in Brazil started in 2005. Figure 10.11 shows the growing production of biodiesel in Brazil until June 2020 with 58 production plants in the country.

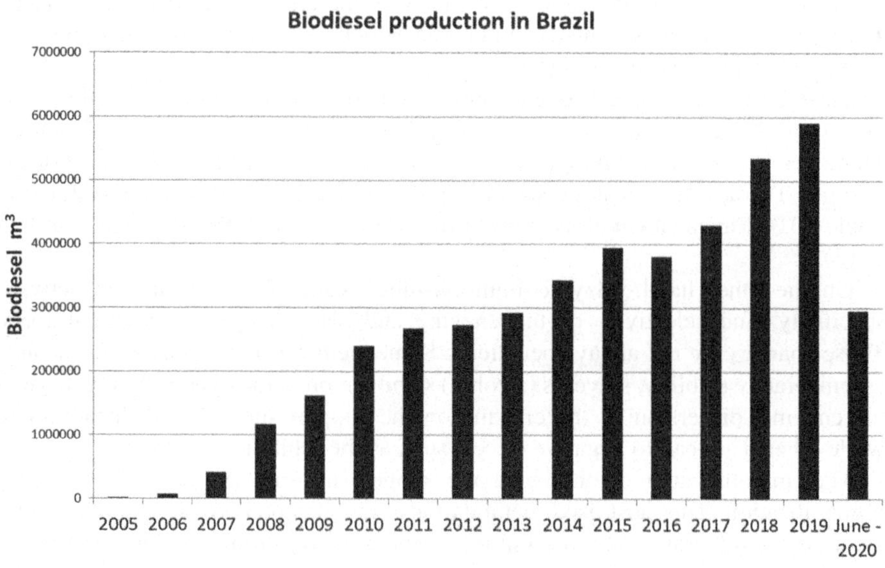

Fig. 10.11 Brazil: Biodiesel Production. B100 years: 2005–2020 (m^3). (Source: ANP (National Agency of Petroleum Natural Gas and biofuels)—Data updated on July 23, 2020)

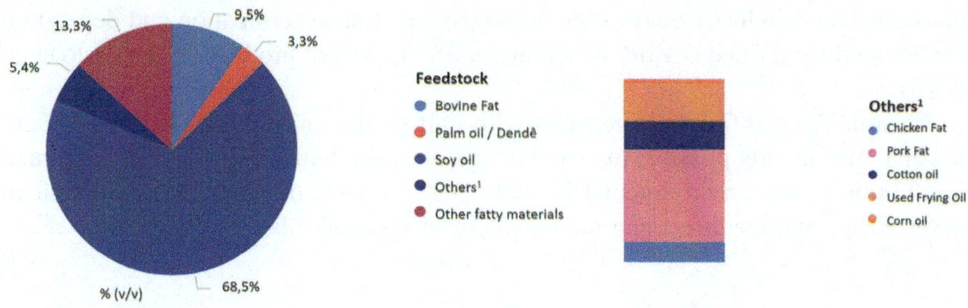

Fig. 10.12 The Feedstocks that were used for the production of biodiesel in Brazil in the first half of 2020. (Source: ANP (National Agency of Petroleum Natural Gas and biofuels)—Data updated on July 23, 2020)

Most of the plants are located in the south and midwest of the country (Kuss et al. 2015). Binary mixtures of biodiesel and diesel petroleum oil are designated worldwide by the BX nomenclature, where X is the percentage by volume of biodiesel added to diesel oil. For example, B2, B5, B20 and B100 are fuels with a concentration of 2%, 5%, 20% and 100% biodiesel, respectively.

In 2008, the mixture of B2 became mandatory regulated by the ANP (National Agency of Petroleum Natural Gas and biofuels). And an increase plan as stipulated by the Brazilian Law "Lei 13.263" (Brazil 2016). The mix in 2020 is B11 and the prospect is that in 2030 it will reach B100 since biodiesel is a strong ally against climate change.

However, it is necessary to develop technology, 68% of the production uses refined soybean oil as feedstock (Fig. 10.12). This characteristic limits a large part of the Brazilian production potential, especially the productions that necessitate low-quality feedstock, due to its high water content and free fatty acids (Suarez et al. 2016).

Several Brazilian biodiesel plants operate using the conventional transesterification route. Commercially, there are no plants in Brazil producing biodiesel from biocatalysts. However, there are research groups that have concentrated efforts on the development of efficient biocatalysts for biodiesel production using low-quality feedstocks, such as crude and residual oils, in addition to studies in bioreactors to minimize costs with the enzyme.

10.4.4 Commercial Prospects

Biodiesel is the main biofuel used in the heavy transport sector. The increase in Binary mixtures will stimulate demand for raw materials.

Brazil suffers from an idle capacity from a very promising market, which allows replacing the use of diesel oil. A large part of this capacity is not used because industries are unable to work with various feedstocks, especially low-quality ones. It occurs because industries use basic homogeneous transesterification and depend on the demand for refined soy oil, which increases the cost of production (Dágosto et al. 2015).

Encarnação (2007) explained that with half of the animal fats, fatty acids, and residual oils already possible to meet 80% (v/v) of the demand for biodiesel in Brazil in addition to the environmental benefits that biodiesel presents as a reduction of gases with a sponge effect and the recycling of residual oils.

10.4.5 Challenges

In the last 20 years, the objective of the chemical industry has been on implementing sustainable processes, the use of biotechnology is essential in this progress. Biocatalysts for biodiesel production have an advantage over other routes, which are (a) the control of the final composition of the product, and (b) the co-product (glycerin) with more high purity and, consequently, more economic value (Dágosto et al. 2015).

However, the production cost of biocatalysts remains a significant challenge. The development of research on immobilized enzymes and reactors will help to leverage the production of biodiesel using biocatalysts, which present technological advantages that point to a bright future.

Acknowledgments The authors would like to thank the Brazilian Federal Agency for the Support and Evaluation of Graduate Education (Award Number 154193/2018-6), Research Council for the State of São Paulo (FAPESP) (Award Number 2014/27055-2; 2016/22086-2; 2016/10636-8; 2016/22086-2). This study was financed in part by the Coordenação de Aperfeiçoamento de Pessoal de Nível Superior—Brasil (CAPES)—Finance Code 001.

References

ABBM (2016) RenovaBio Biocombustíveis 2030. Associação Brasileira de Biogás e Metano, Santa Cruz do Sul. http://www.mme.gov.br/c/document_library/get_file?uuid=868dc2f2-d486-14dc-2b3b-3aa0d0007f4d&groupId=36224

AGROSTAT: statistics of the external market in Brazilian agribusiness. http://sistemasweb.agricultura.gov.br/pages/AGROSTAT.html. Accessed 5 July 2020

Albuquerque F (2019) Produção de etanol na safra 2019/2020 deve aumentar 7,1%, diz Única. Agência Brasil, São Paulo. https://agenciabrasil.ebc.com.br/economia/noticia/2019-12/producao-de-etanol-na-safra-20192020-deve-aumentar-71-diz-unica#. Accessed 24 June 2020

Almeida VP, Longhi GM, dos Santos LR (2017) Etanol: 40 anos de evolução do mercado de combustíveis e automóveis no Brasil. Revista Teoria e Evidência Econômica, 23(49). http://seer.upf.br/index.php/rtee/article/view/8269

Andersen AD (2015) A functions approach to innovation system building in the south: the pre-Proálcool evolution of the sugarcane and biofuel sector in Brazil. Innov Dev 5(1):1–21. https://doi.org/10.1080/2157930X.2014.996855

Andrade DS, Telles TS, Castro GHL (2020) The Brazilian microalgae production chain and alternatives for its consolidation. J Clean Prod 250:119526
ANP (2019) RenovaBio. Agência Nacional de Petróleo, Gás Natural e Biocombustíveis, Rio de Janeiro. http://www.anp.gov.br/producao-de-biocombustiveis/renovabio
ANP (2020a) Ethanol. http://www.anp.gov.br/producao-de-biocombustiveis/etanol. Accessed 18 June 2020
ANP (2020b) RenovaBio. http://www.anp.gov.br/producao-de-biocombustiveis/renovabio. Accessed 23 June 2020
Aransiola EF, Ojumu TV, Oyekola OO et al (2014) A review of current technology for biodiesel production: state of the art. Biomass Bioenergy 61:276–297. https://doi.org/10.1016/j.biombioe.2013.11.014
Araújo WA (2016) Ethanol industry: surpassing uncertainties and looking forward. In: Salles-Filho SLM, et al. (eds.), Global Bioethanology. Academic, pp 1–33, isbn:9780128031414. https://doi.org/10.1016/B978-0-12-803141-4.00001-0
Bala A, Singh B (2019) Cellulolytic and xylanolytic enzymes of thermophiles for the production of renewable biofuels. Renew Energy 136:1231–1244
Balat M (2011) Production of bioethanol from lignocellulosic materials via the biochemical pathway: a review. Energy Convers Manag 52(2):858–875
Balat M, Balat H, Öz C (2008) Progress in bioethanol processing. Progress Energy Combust Sci 34(5):551–573
Behera S, Singh R, Arora R, Sharma NK, Shukla M, Kumar S (2015) Scope of algae as third generation biofuels. Front Bioeng Biotechnol 2:90
Benites-Lazaro LL, Giatti LL, Junior WS, Giarolla A (2020) Land-water-food nexus of biofuels: discourse and policy debates in Brazil. Environ Dev 33:100491
Bhutada G, Kavšček M, Ledesma-Amaro R, Thomas S, Rechberger GN, Nicaud JM, Natter K (2017) Sugar versus fat: elimination of glycogen storage improves lipid accumulation in Yarrowia lipolytica. FEMS Yeast Res 17(3):fox020
Bjerregaard, R, Valderrama D, Radulovich R, Diana, J, Capron, M, McKinnie, CA, ..., Forster J (2016) Seaweed aquaculture for food security, income generation and environmental health in tropical developing countries (No. 107147). The World Bank, pp 1–16. https://documents.worldbank.org/en/publication/documents-reports/documentdetail/947831469090666344/seaweed-aquaculture-for-food-security-income-generation-and-environmental-health-in-tropical-developing-countries. Accessed 13 Aug 2020
Boboescu IZ, Chemarin F, Beigbeder JB, de Vasconcelos BR, Munirathinam R, Ghislain T, Lavoie JM (2019) Making next-generation biofuels and bio commodities a feasible reality. Curr Opin Green Sust Chem 20:25–32
Bolfe L (2018) Visão 2030: O futuro da agricultura brasileira (Coord.). Embrapa, Brasília
Borines MG, de Leon RL, Cuello JL (2013) Bioethanol production from the macroalgae Sargassum spp. Bioresour Technol 138:22–29
Bornscheuer UT, Huisman GW, Kazlauskas RJ et al (2012) Engineering the third wave of biocatalysis. Nature 485(7397):185–194. https://doi.org/10.1038/nature11117
BRASIL. Presidência da República. Casa Civil. Lei n° 13.263. Altera a Lei n° 13.033, de 24 de setembro de 2014, para dispor sobre os percentuais de adição de biodiesel ao óleo diesel comercializado no território nacional, Diário Oficial da União, Brasília, DF, 23 mar. 2016. Disponível em: http://www.planalto.gov.br/ccivil_03/_ato2015-2018/2016/lei/L13263.htm. Accessed 23 July 2020
Brasil (2019) Plano Decenal de Expansão de Energia 2029. Ministério de Minas e Energia, Empresa de Pesquisa Energética. https://www.epe.gov.br/pt/publicacoes-dados-abertos/publicacoes/plano-decenal-de-expansao-de-energia-2029
Brasil BSAF, Silva FCP, Siqueira FG (2017) Microalgae biorefineries: the Brazilian scenario in perspective. New Biotechnol 39:90–98

Brennan L, Owende P (2010) Biofuels from microalgae—a review of technologies for production, processing, and extractions of biofuels and co-products. Renew Sustain Energy Rev 14(2):557–577

Brethauer S, Wyman CE (2010) Review: continuous hydrolysis and fermentation for cellulosic ethanol production. Bioresour Technol 101(13):4862–4874

Campello LD, Barros RG, Filho GLT, Santos IFS (2020) Analysis of the economic viability of the use of biogas produced in wastewater treatment plants to generate electrical energy. Environ Dev Sustain. https://doi.org/10.1007/s10668-020-00689-y

Canal—Jornal da Bioenergia. Etanol de milho: um futuro promissor. 2020. https://www.canalbioenergia.com.br/etanol-de-milho-um-futuro-promissor/. Accessed 19 June 2020

Cardona CA, Quintero JÁ, Paz IC (2010) Production of bioethanol from sugarcane bagasse: status and perspectives. Bioresour Technol 101(13):4754–4766

Carneiro MLN, Pradelle F, Braga SL, Gomes MSP, Martins ARF, Turkovics F, Pradelle RN (2017) Potential of biofuels from algae: comparison with fossil fuels, ethanol and biodiesel in Europe and Brazil through life cycle assessment (LCA). Renew Sustain Energy Rev 73:632–653

Carotenuto C, Guarino G, Minale M, Morrone B (2016) Biogas production from anaerobic digestion of manure at different operative conditions. Int J Heat Technol 34(4):623–629

Carrigan J (2016) Chapter 15: applications of cellulase in biofuel industry. In: Gupta VK (ed) New and future developments in microbial biotechnology and bioengineering. microbial cellulase system properties and applications. Elsevier, Amsterdam

CEPEA—Centro de Estudos Avançados em Economia Aplicada (2020) Consultas ao banco de dados do site. https://www.cepea.esalq.usp.br/br/consultas-ao-banco-de-dados-do-site.aspx. Accessed 10 June 2020

Chen CY, Yeh KL, Aisyah R, Lee DJ, Chang JS (2011) Cultivation, photobioreactor design and harvesting of microalgae for biodiesel production: a critical review. Bioresour Technol 102(1):71–81

Chen C, Zheng D, Liu G et al (2015) Continuous dry fermentation of swine manure for biogas production. Waste Manag 38:436–442. https://doi.org/10.1016/j.wasman.2014.12.024

Chiumenti A, da Borso F, Limina S (2018) Dry anaerobic digestion of cow manure and agricultural products in a full-scale plant: efficiency and comparison with wet fermentation. Waste Manag 71:704–710. https://doi.org/10.1016/j.wasman.2017.03.046

Christopher LP, Kumar H, Zambare VP (2014) Enzymatic biodiesel: challenges and opportunities. Appl Energy 119:497–520. https://doi.org/10.1016/j.apenergy.2014.01.017

CIBiogás (2020) Biogásmap. https://mapbiogas.cibiogas.org

Claassen PAM, Lier JBV, Lopez Contreras AM, Niel EJWV, Sijtsma L, Stams AJM, Vries SSD, Weusthuis RA (1999) Utilisation of biomass for the supply of energy carriers. Appl Microb Biotechnol 52:741–755

CONAB (2020) Companhia Nacional de Abastecimento. Acompanhamento da safra brasileira de cana-de-açúcar 2019/2020. https://www.conab.gov.br/info-agro/safras/cana. Accessed 10 June 2020

Costa OYA, Souto BM, Tupinambá DD, Bergmann JC, Kyaw CM, Kruger RH et al (2015) Microbial diversity in sugarcane ethanol production in a Brazilian distillery using a culture-independent method. J Indus Microb Biotechnol 42(1):73–84

Cruz CHB … [et al.], Cortez, LAB (org.). Universidades e empresas: 40 anos de ciência e tecnologia para o etanol brasileiro—São Paulo: Blucher, 2016. 224 p. isbn:978-85-212-1062-7

Cruz-Salomón A, Meza-Gordillo R, Rosales-Quintero A et al (2016) Biogas production from a native beverage vinasse using a modified UASB bioreactor. Fuel 198:170–174. https://doi.org/10.1016/j.fuel.2016.11.046

CTC—Centro de Tecnologia Canavieira. Etanol Celulósico. https://ctc.com.br/etanol-celulosico/. Accessed 1 July 2020

Da Costa-Gomez C (2013) Biogas as an energy option: an overview. In: Wellinger A, Murphy J, Baxter D (eds) The biogas handbook. Woodhead Publishing Elsevier, Sawston, pp 1–16

Dágosto MDA, Vieira Da Silva MA, De Oliveira CM et al (2015) Evaluating the potential of the use of biodiesel for power generation in Brazil. Renew Sust Energ Rev 43:807–817. https://doi.org/10.1016/j.rser.2014.11.055

Darda S, Papalas T, Zabaniotou A (2019) Biofuels journey in Europe: currently the way to low carbon economy sustainability is still a challenge. J Clean Prod 208:575–588

De Lima AL, Ronconi CM, Mota CJA (2016) Heterogeneous basic catalysts for biodiesel production. Catalysis Sci Technol 6:2877–2891. https://doi.org/10.1039/c5cy01989c

de Oliveira BR, Carvalho JLN, Lal R, de Figueiredo EB, de Oliveira BG, La Scala N (2018) Sustainability of sugarcane production in Brazil. A review. Agron Sustain Dev 38(2):13

de Souza MP, Bjerk TR, Gressler PD, de Souza Schneider RDC, Corbellini VA, Moraes MSA (2012) As Microalgas Como Uma Alternativa Para a Produção De Biocombustíveis Parte I: Bioetanol. Tecno-Lógica 16(2):108–116

Dragone G, Fernandes BD, Vicente AA, Teixeira JA (2010) Third generation biofuels from microalgae. In: Méndez-Vilas A (ed), Current research, technology and education topics in applied microbiology and microbial biotechnology, pp 1355–1366

Drozdzynska A, Leja K, Czaczyk K (2011) Biotechnological production of 1, 3-propanediol from crude glycerol. J Biotech Comput Biol Bionanotechnol 92(1):92–100

Eaglin J (2019) The demise of the Brazilian ethanol program: environmental and economic shocks. Environ Hist 24:104–129. https://doi.org/10.1093/envhis/emy056

Elalami D, Carrère H, Monlau F, Abdelouahdi K, Oukarroum A, Barakat A (2019) Pretreatment and co-digestion of wastewater sludge for biogas production: recent research advances and trends. Renew Sustain Energy Rev 114:109287

Elniski AR, Chatterjee, SG, Mondal C, Doelle K, (2019) Effects of substrate to inoculum ration on biogas production from anaerobic co-digestion of office paper and cow manure. J Energy Renew Rev 1–15

EMBRAPA (2020) Empresa Brasileira de Pesquisa Agropecuária. Mistura de enzimas mostrou alto desempenho para gerar etanol a partir de bagaço de cana. https://www.embrapa.br/busca-de-noticias/-/noticia/52639673/mistura-de-enzimas-mostrou-alto-desempenho-para-gerar-etanol-a-partir-de-bagaco-de-cana. Accessed 3 July 2020

Encarnação APG (2007) Geração de Biodiesel Pelos Processos de Transesterificação e Hidroesterificação, Uma Avaliação Econômica. Dissertação (Mestrado Em Tecnologia de Processos Químicos e Bioquímicos)—Universidade Federal Do Rio de Janeiro—UFRJ, Escola de Química—EQ, 144

EPE (2019) Empresa de Pesquisa Energética. Cenários de Oferta de Etanol e Demanda do Ciclo Otto: 2020–2030. https://www.epe.gov.br/sites-pt/publicacoes-dados-abertos/publicacoes/PublicacoesArquivos/publicacao-255/topico-503/EPE_DPG_SDB_Bios_NT_02_2019_r0_Cenarios_de_Oferta_de_Etanol.pdf. Accessed 24 June 2020

EPE (2019a) Balanço Energético Nacional 2019: Ano-base 2018. Empresa de Pesquisa Energética, Rio de Janeiro. https://www.epe.gov.br/pt/publicacoes-dados-abertos/publicacoes/balanco-energetico-nacional-2019

EPE (2019b) Análise de Conjuntura dos Biocombustíveis: Ano 2018. Empresa de Pesquisa Energética, Rio de Janeiro. https://www.epe.gov.br/sites-pt/publicacoes-dados-abertos/publicacoes/PublicacoesArquivos/publicacao-402/Análise_de_Conjuntura_Ano%202018.pdf

EPE (2020a) Empresa de Pesquisa Energética. Nota Técnica Impactos da pandemia de Covid-19 no mercado brasileiro de combustíveis. https://www.epe.gov.br/sites-pt/publicacoes-dados-abertos/publicacoes/PublicacoesArquivos/publicacao-485/NT-DPG-SDB-2020-02_Impactos_da_COVID-19_no_mercado_brasileiro_de_combustiveis.pdf. Accessed 25 June 2020

EPE (2020b) Empresa de Pesquisa Energética. Plano Decenal de Expansão de Energia 2029. https://www.epe.gov.br/pt/publicacoes-dados-abertos/publicacoes/plano-decenal-de-expansao-de-energia-2029. Accessed 30 June 2020

FAO (2013) Bioslurry = Brown Gold? A review of scientific literature on the co-product of biogas production. Food and Agriculture Organization of the United Nations, Rome. http://www.fao.org/3/a-i3441e.pdf Accessed 17 July 2020

Farrell AE, Plevin RJ, Turner BT, Jones AD, O'Hare M, Kammen DM (2006) Ethanol can contribute to energy and environmental goals. Science 311(5760):506–508

FIESP, Departamento do Agronegócio (2018) Outlook Fiesp 2028: projeções para o agronegócio brasileiro/FIESP. FIESP, São Paulo. 86 p: isbn:978-85-7201-035-1. http://hotsite.fiesp.com.br/outlookbrasil/2028/files/assets/basic-html/page-1.html#

Filho DG, Silva AG, Guidini CZ (2019) Lipases: sources, immobilization methods, and industrial applications. Appl Microbiol Biotechnol 103:7399–7423. https://doi.org/10.1007/s00253-019-10027-6

Fjerbaek L, Christensen KV, Norddahl B (2009) A review of the current state of biodiesel production using enzymatic transesterification. Biotechnol Bioeng 102:1298–1315. https://doi.org/10.1002/bit.22256

Forzza RC, Baumgratz JFA, Bicudo CEM, Canhos DA, Carvalho AA Jr, Coelho MA et al (2012) New Brazilian floristic list highlights conservation challenges. BioScience 62(1):39–45

Freitas FF, De Souza SS, Ferreira LRA et al (2019a) The Brazilian market of distributed biogas generation: overview, technological development and case study. Renew Sustain Energy Rev 101:146–157. https://doi.org/10.1016/j.rser.2018.11.007

Freitas FF, De Souza SS, Ferreira LRA, Otto RB, Alessio FJ, De Souza SNM, Venturini OJ, Ando Junior OH (2019b) The Brazilian market of distributed biogas generation: overview, T technological development and case study. Renew Sustain Energy Rev 101:146–157. https://doi.org/10.1016/j.rser.2018.11.007

Fronzaglia T, Torquato SA (2007) Mercado de Álcool: desajustes e excesso de expectativas. Análise e Indicadores do Agronegócio, São Paulo, 2:8. http://www.iea.sp.gov.br. Accessed 18 June 2020

Garlapati VK, Shankar U, Budhiraja A (2016) Bioconversion technologies of crude glycerol to value added industrial products. Biotech Rep 9:9–14

Garside M (2019) Ethanol fuel production in top countries. https://www.statista.com/statistics/281606/ethanol-production-in-selected-countries/. Accessed 19 June 2020

Gilio L, Castro NR (2017) Avaliação de aspectos limitantes ao crescimento do etanol e o setor sucroenergético no Brasil. Revista Eletrônica de Energia 6(1) https://revistas.unifacs.br/index.php/ree/article/view/4503

Gouveia L, Oliveira AC (2009) Microalgae as a raw material for biofuels production. J Ind Microbiol Biotechnol 36(2):269–274

Gog A, Roman M, Toşa M et al (2012) Biodiesel production using enzymatic transesterification—current state and perspectives. Renew Energy 39:10–16. https://doi.org/10.1016/j.renene.2011.08.007

Goh CS, Lee KT (2010) A visionary and conceptual macroalgae-based third-generation bioethanol (TGB) biorefinery in Sabah, Malaysia as an underlay for renewable and sustainable development. Renew Sustain Energy Rev 14(2):842–848

Goldemberg J, Guardabassi P (2010) The potential for first-generation ethanol production from sugarcane. Biofuels Bioproducts Biorefining 4(1):17–24

GranBio (2020a) BioFlex® I: Produção de Biocombustível. http://www.granbio.com.br/conteudos/biocombustiveis/. Accessed 1 July 2020

GranBio (2020b) Press releases. http://www.granbio.com.br/press-releases/para-presidente-da-granbio-e2g-esta-consolidado-e-empresa-esta-pronta-para-nova-etapa/. Accessed 1 July 2020

Guendouz J, Buffière P, Cacho J, Carrère M, Delgenes J-P (2008) High-solids anaerobic digestion: comparison of three pilot scales. Water Sci Technol 58(9):1757–1763. https://doi.org/10.2166/wst.2008.521

Guldhe A, Singh B, Mutanda T, et al (2015) Advances in synthesis of biodiesel via enzyme catalysis: novel and sustainable approaches. Renew Sust Energ Rev. 41

Hanefeld U, Gardossi L, Magner E (2009) Understanding enzyme immobilisation. Chem Soc Rev 38:453–468. https://doi.org/10.1039/b711564b

Hargreaves PI, Barcelos CA, da Costa ACA, Pereira N Jr (2013) Production of ethanol 3G from Kappaphycus alvarezii: evaluation of different process strategies. Bioresour Technol 134:257–263
Holliger C, Alves M, Andrade D et al (2016) Towards a standardization of biomethane potential tests. Water Sci Technol 74(11):2515–2522. https://doi.org/10.2166/wst.2016.336
Hossain N, Zaini JH, Mahlia TMI (2017) A review of bioethanol production from plant-based waste biomass by yeast fermentation. Int J Technol 8(1):5–18
Houfani AA, Anders N, Spiess AC, Baldrian P, Benallaoua S (2020) Insights from enzymatic degradation of cellulose and hemicellulose to fermentable sugars—a review. Biomass Bioenergy 134:105481
Hu J, Zhai R, Tian D, Saddler JN (2018) Substrate factors that influence cellulase accessibility and catalytic activity during the enzymatic hydrolysis of lignocellulosic biomass. In: Fungal cellulolytic enzymes. Springer, Singapore, pp 239–256
IEA (2014) Key world energy statistics 2014. International Energy Agency, Paris
IEA (2020a) IEA bioenergy task 37—country reports summaries 2019. http://task37.ieabioenergy.com/
IEA (2020b) Renewable energy market update—Outlook for 2020 and 2021. https://www.iea.org/reports/renewable-energy-market-update. Accessed 25 June 2020
InfoMoney (2020) Um quarto das usinas do país pode fechar as portas. https://www.infomoney.com.br/negocios/um-quarto-das-usinas-do-pais-pode-fechar-as-portas/. Accessed 25 June 2020
Jambo AS, Abdulla R, Azhar SHM, Marbawi H, Gansau JÁ, Ravindra P (2016) A review on third-generation bioethanol feedstock. Renew Sustain Energy Rev 65:756–769
Jambo SA, Abdulla R, Marbawi H, Gansau JA (2019) Response surface optimization of bioethanol production from third-generation feedstock-Eucheuma cottonii. Renew Energy 132:1–10
José C, Austic GB, Bonetto RD et al (2013) Investigation of the stability of Novozym® 435 in the production of biodiesel. Catal Today 213:73–80. https://doi.org/10.1016/j.cattod.2013.02.013
Juárez JM, Pastor ER, Sevilla JMF et al (2018) Effect of pretreatments on biogas production from microalgae biomass grown in pig manure treatment plants. Bioresour Technol 257:30–38. https://doi.org/10.1016/j.biortech.2018.02.063
Júnior JCF, Palacio JCE, Leme RC, Lora EES, da Costa JEL, Reyes AMM, del Olmo OA (2020) Biorefineries productive alternatives optimization in the Brazilian sugar and alcohol industry. Appl Energy 259:113092
Kalia AK, Singh SP (2001) Effect of mixing digested slurry on the rate of biogas production from dairy manure in the batch fermenter. Energy Sources 23(8):711–715
Klein B (2013) Cultivo de microalgas para produção de bioetanol de terceira geração. Universidade de Campinas
Knothe G, Krahl J, Van Gerpen J (2010) Biodiesel handbook, 2nd edn
Kose OSS (2017) Algae as a promising resource for biofuel industry: facts and challenges. Int J Energy Res 41(7):924–951
Krajick K (2001) Long-term data show lingering effects from acid rain, pp 195–196
Kuss VV, Kuss AV, Da Rosa RG et al (2015) Potential of biodiesel production from palm oil at Brazilian Amazon. Renew Sust Energ Rev 50:1013–1020. https://doi.org/10.1016/j.rser.2015.05.055
Lavergne C, Bovio-Winkler P, Etchebehere C et al (2019) Towards centralized biogas plants: co-digestion of sewage sludge and pig manure maintains process performance and active microbiome diversity. Bioresour Technol 297:122442. https://doi.org/10.1016/j.biortech.2019.122442
Lazar Z, Liu N, Stephanopoulos G (2018) Holistic approaches in lipid production by Yarrowia lipolytica. Trends Biotechnol 36(11):1157–1170
Leite RCC, Cortez LAB (2008) O etanol combustível no Brasil. In: Biocombustíveis no Brasil: Realidades e Perspectivas. Ministério das Relações Exteriores, pp 60–75. http://www.agencia.cnptia.embrapa.br/Repositorio/etanol3_000g7gq2cz702wx5ok0wtedt3xdrmftk.pdf

Liao X, Li H (2015) Biogas production from low-organic-content sludge using a high-solids anaerobic digester with improved agitation. Appl Energy 148:252–259. https://doi.org/10.1016/j.apenergy.2015.03.082
Liao X, Li H, Cheng Y et al (2014) Process performance of high solids batch anaerobic digestion of sewage sludge. Environ Technol 35:2652–2659. https://doi.org/10.1080/09593330.2014.916756
Liu C, Li H, Zhang Y et al (2016a) Improve biogas production from low-organic-content sludge through high-solids anaerobic co-digestion with food waste. Biotechnology 219:252–260. https://doi.org/10.1016/j.biortech.2016.07.130
Liu C, Li H, Zhang Y et al (2016b) Characterization of methanogenic activity during high-solids anaerobic digestion of sewage sludge. Biochem Eng J 109:96–100. https://doi.org/10.1016/j.bej.2016.01.010
Loow YL, Wu TY, Jahim JM, Mohammad AW, Teoh WH (2016) Typical conversion of lignocellulosic biomass into reducing sugars using dilute acid hydrolysis and alkaline pretreatment. Cellulose 23(3):1491–1520
Lopes WS, Leite VD, Prasad S (2004) Influence of inoculum on performance of anaerobic reactors for treating municipal solid waste. Biotechnology 94(3):261–266
Magalhães PSG, Braunbeck OA (2014) Sugarcane and straw harvesting for ethanol production. In: Luis Augusto Barbosa Cortez (Coord.). Sugarcane bioethanol—R&D for productivity and sustainability, Editora Edgard Blücher, São Paulo, pp 465–476. https://doi.org/10.5151/BlucherOA-Sugarcane-SUGARCANEBIOETHANOL_43
Mahmood H, Moniruzzaman M, Iqbal T, Khan MJ (2019) Recent advances in the pretreatment of lignocellulosic biomass for biofuels and value-added products. Curr Opin Green Sustain Chem 20:18–24
Mathur S, Umakanth AV, Tonapi VA, Sharma R, Sharma MK (2017) Sweet sorghum as biofuel feedstock: recent advances and available resources. Biotechnol Biofuels 10(1):146
Melo AS, Sampaio YSB (2016) Uma Nota Sobre o Impacto do Preço do Açúcar, do Etanol e da Gasolina na Produção do Setor Sucroalcooleiro. Rev Bras Econ 70(1):61–69. https://doi.org/10.5935/0034-7140.20160004
Mesa L, López N, Cara C, Castro E, González E, Mussatto SI (2016) Techno-economic evaluation of strategies based on two steps organosolv pretreatment and enzymatic hydrolysis of sugarcane bagasse for ethanol production. Renew Energy 86:270–279
Miao Z, Grift TE, Hansen AC, Ting KC (2011) Energy requirement for comminution of biomass in relation to particle physical properties. Ind Crop Prod 33(2):504–513
Milanez AY et al (2015) De promessa a realidade: como o etanol celulósico pode revolucionar a indústria da cana-de-açúcar—uma avaliação do potencial competitivo e sugestões de política pública. BNDES Setorial 41:237–294. https://web.bndes.gov.br/bib/jspui/handle/1408/4283
Milanez AY et al (2017) O acordo de Paris e a transição para o setor de transportes de baixo carbono: o papel da plataforma para o biofuturo. BNDES Setorial 45:285–340. https://web.bndes.gov.br/bib/jspui/handle/1408/11756
Mitrea L, Trif M, Cătoi AF, Vodnar DC (2017) Utilization of biodiesel derived-glycerol for 1, 3-PD and citric acid production. Microb Cell Factories 16(1):190
Moraes M, Bacchi M (2014) Etanol: do início às fases atuais de produção. Revista de Política Agrícola. https://seer.sede.embrapa.br/index.php/RPA/article/view/950. Accessed 15 June 2020
Morgera E, Kulovesi, K, Gobena A (2009) Case studies on bioenergy policy and law: options for sustainability. FAO Legislative Study. Rome. isbn:978-92-5-10-6455-9. http://www.fao.org/3/i1285e/i1285e00.htm. Accessed 19 June 2020
Mussatto SI, Dragone G, Guimarães PM, Silva JPA, Carneiro LM, Roberto IC et al (2010) Technological trends, global market, and challenges of bio-ethanol production. Biotechnol Adv 28(6):817–830
Mussgnug JH, Klassen V, Schlüter A, Kruse O (2010) Microalgae as substrates for fermentative biogas production in a combined biorefinery concept. J Biotechnol 150(1):51–56

Nadaleti WC, Lourenço VA, Filho PB, Santos GB, Przybyla G (2020) National potential production of methane and electrical energy from T sugarcane vinasse in Brazil: a thermo-economic analysis. J Environ Chem Eng 8:103422. https://doi.org/10.1016/j.jece.2019.103422

Namsree P, Suvajittanont W, Puttanlek C et al (2012) Anaerobic digestion of pineapple pulp and peel in a plug-flow reactor. J Environ Manag 110:40–47. https://doi.org/10.1016/j.jenvman.2012.05.017

Neshat SA, Mohammadi M, Najafpour GD et al (2017) Anaerobic co-digestion of animal manures and lignocellulosic residues as a potent approach for sustainable biogas production. Renew Sustain Energy Rev 79:308–322. https://doi.org/10.1016/j.rser.2017.05.137

Nguyen THM (2012) Bioethanol production from marine algae biomass: prospect and troubles. J Vietnam Environ 3(1):25–29

Niehus X, Crutz-Le Coq AM, Sandoval G, Nicaud JM, Ledesma-Amaro R (2018) Engineering Yarrowia lipolytica to enhance lipid production from lignocellulosic materials. Biotechnol Biofuels 11(1):1–10

Ning J, Zhou M, Pan X et al (2019) Simultaneous biogas and biogas slurry production from co-digestion of pig manure and corn straw: performance optimization and microbial community shift. Bioresour Technol 282:37–47. https://doi.org/10.1016/j.biortech.2019.02.122

Niphadkar S, Bagade P, Ahmed S (2018) Bioethanol production: insight into past, present and future perspectives. Biofuels 9(2):229–238. https://doi.org/10.1080/17597269.2017.1334338

Noraini MY, Ong HC, Badrul MJ, Chong WT (2014) A review on potential enzymatic reaction for biofuel production from algae. Renew Sustain Energy Rev 39:24–34

Norjannah B, Ong HC, Masjuki HH, et al (2016) Enzymatic transesterification for biodiesel production: a comprehensive review. RSC Adv

Novacana (2020a) As usinas de Açúcar e Etanol do Brasil. https://www.novacana.com/usinas_brasil. Accessed 22 June 2020

Novacana (2020b) Laboratório Nacional de Biorrenováveis desenvolve coquetel enzimático para E2G. https://www.novacana.com/n/etanol/2-geracao-celulose/laboratorio-nacional-biorrenovaveis-coquetel-enzimatico-producao-e2g-170620. Accessed 3 July 2020

OECD/FAO (2019) OECD-FAO Agricultural Outlook 2019–2028, OECD Publishing, Paris/Food and Agriculture Organization of the United Nations, Rome. https://doi.org/10.1787/agr_outlook-2019-en. Accessed 24 June 2020

Oliveira LGS, Negro S (2019) Contextual structures and interaction dynamics in the Brazilian biogas innovation system. Renew Sust Energ Rev 107:462–481. https://doi.org/10.1016/j.rser.2019.02.030

Pablo G, Domínguez E, Domínguez VD, Romaní A, Domingues L, Garrote G (2019) Third-generation bioethanol from invasive macroalgae Sargassum muticum using autohydrolysis pretreatment as first step of a biorefinery. Renew Energy 141:728–735

Pourzolfaghar H, Abnisa F, Daud WMAW, Aroua MK (2016) A review of the enzymatic hydroesterification process for biodiesel production. Renew Sust Energ Rev 61:245–257. https://doi.org/10.1016/j.rser.2016.03.048

Quintero JA, Rincón LE, Cardona CA (2011) Production of bioethanol from agroindustrial residues as feedstocks. In: Biofuels. Academic, Amsterdam, pp 251–285

Raizen (2019) Relatório Anual 2018–2019. https://www.raizen.com.br/relatorioanual/1819/pt/. Accessed 1 July 2020

Raizen (2020) Etanol. https://www.raizen.com.br/nossos-negocios/etanol#b2b. Accessed 1 July 2020

Ramos CS (2019) RenovaBio dá esperança a indústrias de base. Valor Econômico, São Paulo. https://valor.globo.com/agronegocios/noticia/2019/12/27/renovabio-da-esperanca-a-industrias-de-base.ghtml. Accessed 24 June 2020

Ranganathan SV, Narasimhan SL, Muthukumar K (2008) An overview of enzymatic production of biodiesel. Bioresour Technol 99:3975–3981. https://doi.org/10.1016/j.biortech.2007.04.060

Rosillo-Calle F, Cortez LAB (1998) Towards Pro Alcool II—a review of the Brazilian bioethanol programme. Biomass Bioenergy 14(2):115–124. https://doi.org/10.1016/S0961-9534(97)10020-4

Rastogi M, Shrivastava S (2017) Recent advances in second generation bioethanol production: an insight to pretreatment, saccharification and fermentation processes. Renew Sustain Energy Rev 80:330–340

Renó MLG, del Olmo OA, Palacio JCE, Lora EES, Venturini OJ (2014) Sugarcane biorefineries: case studies applied to the Brazilian sugar–alcohol industry. Energy Convers Manag 86:981–991

Rezania S, Oryani B, Cho J, Talaiekhozani A, Sabbagh F, Hashemi B et al (2020) Different pretreatment technologies of lignocellulosic biomass for bioethanol production: an overview. Energy 199:117457

Ribeiro MCP, Nadal CP, Junior WFR, Fragoso RMS, Lindino CA (2020) Institutional and legal framework of the Brazilian energy market: biomass as a sustainable alternative for Brazilian agribusiness. Sustainability 12:1554. https://doi.org/10.3390/su12041554

Rizza LS, Smachetti MES, Do Nascimento M, Salerno GL, Curatti L (2017) Bioprospecting for native microalgae as an alternative source of sugars for the production of bioethanol. Algal Res 22:140–147

Robles-Medina A, González-Moreno PA, Esteban-Cerdán L, Molina-Grima E (2009) Biocatalysis: towards ever greener biodiesel production. Biotechnol Adv 27:398–408. https://doi.org/10.1016/j.biotechadv.2008.10.008

Rywińska A, Juszczyk P, Wojtatowicz M, Robak M, Lazar Z, Tomaszewska L, Rymowicz W (2013) Glycerol as a promising substrate for Yarrowia lipolytica biotechnological applications. Biomass Bioenergy 48:148–166

Sánchez OJ, Cardona CA (2008) Trends in biotechnological production of fuel ethanol from different feedstocks. Bioresour Technol 99(13):5270–5295

Sangeetha P, Babu S, Rangasamy R (2011) Potential of green alga Chaetomorpha litorea (Harvey) for biogas production. Int J Curr Sci 1:24–29

Santos IFS, Vieira NDB, Nóbrega LGB et al (2018) Assessment of potential biogas production from multiple organic wastes in Brazil: impact on energy generation, use, and emissions abatement. Resour Conserv Recycling 131:54–63. https://doi.org/10.1016/j.resconrec.2017.12.012

Schmid C, Horschig T, Pfeiffer A, Szarka N, Thrän D (2019) Biogas upgrading: a review of National Biomethane Strategies and support policies in selected countries. Energies 12:3803. https://doi.org/10.3390/en12193803

Sheldon RA, Van Pelt S (2013) Enzyme immobilisation in biocatalysis: why what and how. Chem Soc Rev 42:6223–6235. https://doi.org/10.1039/c3cs60075k

Silva BM, da Silva WSD (2019) Um panorama da implantação do etanol de 3ª geração como uma fonte de energia sustentável. Engevista 21(1):176–192

Singhania RR, Sukumaran RK, Patel AK, Larroche C, Pandey A (2010) Advancement and comparative profiles in the production technologies using solid-state and submerged fermentation for microbial cellulases. Enzyme Microbial Technol 46(7):541–549

Singhvi MS, Gokhale DV (2019) Lignocellulosic biomass: hurdles and challenges in its valorization. Appl Microb Biotechnol 103(23–24):9305–9320

Smachetti MES, Coronel CD, Salerno GL, Curatti L (2020) Sucrose-to-ethanol microalgae-based platform using seawater. Algal Res 45:101733

Soccol C et al (2005) Brazilian biofuel program: an overview. J Sci Indus Res 64:897–904

Spagnuolo M, Yaguchi A, Blenner M (2019) Oleaginous yeast for biofuel and oleochemical production. Curr Opin Biotechnol 57:73–81

Stolf R, Oliveira APR (2020) The success of the Brazilian alcohol program (PROÁLCOOL)—a decade-by-decade brief history of ethanol in Brazil. Eng Agríc 40(2):243–248. https://doi.org/10.1590/1809-4430-eng.agric.v40n2p243-248/2020

Suarez PAZ, Da Cruz RS, Meneghetti SMP (2016) Biodiesel mo Brasil: Impulso tecnológico. In: Menezes RS (ed) vol 1. Lavras, UFLA, MG, p 244

Suganya T, Varman M, Masjuki HH, Renganathan S (2016) Macroalgae and microalgae as a potential source for commercial applications along with biofuels production: a biorefinery approach. Renew Sustain Energy Rev 55:909–941

Torquato AS (2011) Conflitos e propostas de soluções: o comportamento do preço do álcool na entressafra canavieira. Análises e Indicadores do Agronegócio, v. 6, n. 5, maio 2011. issn:1980-0711. http://www.iea.sp.gov.br/out/TerTexto.php?codTexto=12137

Torzillo G, Pushparaj B, Bocci F, Balloni W, Materassi R, Florenzano G (1986) Production of spirulina biomass in closed photobioreactors. Biomass 11(1):61–74

UDOP—União Nacional da Bioenergia (2020). https://www.udop.com.br/. Accessed 22 June 2020

UNEM—União Nacional do Etanol de Milho (2020). http://etanoldemilho.com.br/. Accessed 19 June 2020

Unica—União da Indústria de Cana-de-Açúcar (2020). https://unica.com.br/comunicacao/noticias/. Accessed 19 June 2020

Villanueva-Estrada RE, Rocha-Miller R, Arvizu-Fernández JL (2019) Energy production from biogas in a closed landfill: a case study of Prados de la Montaña, Mexico City. Sustain Energy Technol 31:236–244. https://doi.org/10.1016/j.seta.2018.12.005

Wang H, Xu J, Sheng L et al (2018) Effect of addition of biogas slurry for anaerobic fermentation of deer manure on biogas production. Energy 165:411–418. https://doi.org/10.1016/j.energy.2018.09.196

Wi SG, Kim HJ, Mahadevan SA, Yang DJ, Bae HJ (2009) The potential value of the seaweed Ceylon moss (*Gelidium amansii*) as an alternative bioenergy resource. Bioresour Technol 100(24):6658–6660

Winquist E, Rikkonen P, Pyysiäinen J, Varho V (2019) Is biogas an energy or a sustainability product?—business opportunities in the Finnish biogas branch. J Clean Prod 233:1344–1354

Wyman CE, Decker SR, Himmel ME, Brady JW, Skopec CE, Viikari L (2004) Hydrolysis of cellulose and hemicellulose. In: Polysaccharides. CRC Press, Boca Raton

Yang L, Ge X (2016) Biogas and syngas upgrading. In: Li Y, Ge X (eds) Advances in bioenergy, vol 1. Academic, Cambridge, pp 125–188

Yew GY, Lee SY, Show PL, Tao Y, Law CL, Nguyen TTC, Chang JS (2019) Recent advances in algae biodiesel production: from upstream cultivation to downstream processing. Bioresour Technol Rep 7:100227

Yılmaz Ş, Şahan T (2020) Utilization of pumice for improving biogas production from poultry manure by anaerobic digestion: a modeling and process optimization study using response surface methodology. Biomass Bioenergy 138:10560. https://doi.org/10.1016/j.biombioe.2020.105601

Yoon JJ, Kim YJ, Kim SH, Ryu HJ, Choi JY, Kim GS, Shin MK (2010) Production of polysaccharides and corresponding sugars from red seaweed. In: Advanced materials Research, vol 93. Trans Tech Publications Ltd, pp 463–466

Yusuf KA, et al (2019) Potencial da Agroindústria Canavieira no Brasil: Etanol de Segunda Geração e Bioeletricidade. In: XXII Encontro de Economia da Região Sul Associação Nacional dos Centros de Pós-Graduação em Economia. https://www.anpec.org.br/sul/2019/submissao/files_I/i4-d4a8e1c382a485918f926d12b81df610.pdf. Accessed 30 June 2020

Zabed H, Sahu JN, Suely A, Boyce AN, Faruq G (2017) Bioethanol production from renewable sources: current perspectives and technological progress. Renew Sustain Energy Rev 71:475–501

Zabed HM, Akter S, Yun J, Zhang G, Zhang Y, Qi X (2020) Biogas from microalgae: technologies, challenges and opportunities. Renew Sust Energ Rev 117:109503

Zaks A, Klibanov AM (1986) Substrate specificity of enzymes in organic solvents vs water is reversed. J Am Chem Soc 10:2767–2768. https://doi.org/10.1021/ja00270a053

Zhao J, Hou T, Lei Z et al (2020) Effect of biogas recirculation strategy on biogas upgrading and process stability of anaerobic digestion of sewage sludge under slightly alkaline condition. Bioresour Technol 308:123293. https://doi.org/10.1016/j.biortech.2020.123293

Zhong W, Zhang Z, Luo Y, Qiao W, Xiao M, Zhang M (2012) Biogas productivity by co-digesting Taihu blue algae with corn straw as an external carbon source. Biotechnology 114:281–286

Zhou Y, Schideman L, Yu G, Zhang Y (2013) A synergistic combination of algal wastewater treatment and hydrothermal biofuel production maximized by nutrient and carbon recycling. Energy Environ Sci 6(12):3765–3779

Zhu Z, Zhang S, Liu H, Shen H, Lin X, Yang F et al (2012) A multi-omic map of the lipid-producing yeast Rhodosporidium toruloides. Nat Commun 3(1):1–12

Zuo X, Yuan H, Wachemo AC et al (2019) The relationships among sCOD, VFAs, microbial community, and biogas production during anaerobic digestion of rice straw pretreated with ammonia. J Chem Eng 28:286–292. https://doi.org/10.1016/j.cjche.2019.07.015

Chapter 11
Enzymatic Saccharification Technologies for Biofuel Production: Challenges and Prospects

Priyadharshini Ramachandran, J. Beslin Joshi, Lakshmi Kasirajan, Julie A. Maupin-Furlow, and Sivakumar Uthandi

Abstract Significant advancement has been made in biomass valorization, especially in the twenty-first century. Reasons for these advancements include population growth, depletion in petroleum and fossil fuels, and growing demand for fuels, lignin derivatives, and petrochemicals. The energy demand is increasing tremendously, and today's energy needs can be met by producing fuels and chemicals from renewable feedstocks. Agricultural by-products and other lignocellulosic biomass (LCB) are abundant feedstocks for this purpose. A plethora of biocatalysts are available for biomass conversion, and the discovery of new and efficient enzymes is ever increasing. The significant challenges faced in this area are bridging the efficient utilization of biomass and developing enzyme cocktails with improved saccharification efficiency in a cost-effective manner. Overcoming the inhibitors generation during pretreatment, understanding biomass complexity, enhancing biocatalyst efficiency, optimizing saccharification, and reducing operating costs are challenging needs. This chapter provides a comprehensive review of biomass feedstocks, the enzymes available for the conversion and saccharification of these renewable substrates, the challenges for optimized conversion, and the production of

P. Ramachandran · S. Uthandi (✉)
Biocatalysts Laboratory, Department of Agricultural Microbiology, Directorate of Natural Resource Management, Tamil Nadu Agricultural University, Coimbatore, Tamil Nadu, India

J. B. Joshi
Department of Plant Biotechnology, Centre for Plant Molecular Biology and Biotechnology, Tamil Nadu Agricultural University, Coimbatore, Tamil Nadu, India

L. Kasirajan
Indian Council of Agricultural Research-Sugarcane Breeding Institute, Coimbatore, Tamil Nadu, India

J. A. Maupin-Furlow
Department of Microbiology and Cell Science, Institute of Food and Agricultural Sciences, University of Florida, Gainesville, FL, USA

Genetics Institute, University of Florida, Gainesville, FL, USA

J. K. Saini, R. K. Sani (eds.), *Microbial Biotechnology for Renewable and Sustainable Energy*, Clean Energy Production Technologies,
https://doi.org/10.1007/978-981-16-3852-7_11

platform chemicals that can serve as substrates for generating other high-value products.

Keywords Biomass conversion · Saccharification · Glycosyl hydrolases · Thermozymes · Platform chemicals

11.1 Introduction

Most biorefineries rely upon the production of biogas, bioethanol, and/or biodiesel from lignocellulosic biomass (LCB). Biogas fuel is generated by breaking down biomass in anaerobic environments using methanogens and acidogenic microbes that produce a biogas mixture of 40% carbon dioxide and 60% methane. Bioethanol is a renewable and ecofriendly liquid fuel that is recovered from the fermentation of sugars released from LCB pretreated with physical, chemical, or biological hydrolysis techniques. Biodiesel is produced by trans-esterification processes that employ feedstocks such as oilseeds and can be used to replace fossil diesel (Nikkhah et al. 2020). Platform chemicals produced from LCB are also gaining attraction owing to the dwindling supply of fossil reserves, fluctuating crude oil prices, and environmental concerns. Whether the finished products are fuels or chemicals, LCB-based bioconversion is of global interest to strengthen economies, minimize climate change, conserve energy, and maximize food security (Limayem and Ricke 2012). In the process of bioethanol production, the cost for LCB saccharification is still extraordinarily high, owing to primarily the cost of cellulase to saccharify the cellulose. The presently available industrial cellulases are not optimal for harsh conditions and lack sufficient enzymes for complete hydrolysis. LCB-based biorefineries opt for cellulases that possess temperature tolerance and wider range of stability to pH, metal ions, and solvents. Additionally, the saccharification process to achieve higher sugar recovery needs to be optimized. Hence, the present chapter focuses on the biocatalysts, their suitability for enhancing the saccharification process, and the issues and challenges about biomass conversion.

11.1.1 History of Feedstock

Historical transitions have occurred in the type of feedstock supplies used for bioenergy production. Accordingly, the first-generation biofuel production plants rely on edible food crops like grains, starchy, and sugar-rich feedstocks, and they are competing with the food supply. The second-generation (2G) biofuel production utilizes non-edible biomass such as energy crops and waste residues from forestry and agricultural processes. Also, the LCBs abundant in nature are highly feasible for use as a substrate in bioenergy. Second-generation biofuels might not affect food security and the environment than first-generation biofuels. Thus, 2G biofuels crops could be grown on marginal lands without competing with the land used for food industries. The third generation utilize algal sources such as microalgal (*Cholera vulgaris*) and macroalgal (*Ulva sp.*) biomass as a major substitute for biofuel production. The third-generation biofuels present the best possibility for alternative fuels as they show a rich nutritional profile with high lipids and carbohydrates and are easily cultivated in an aquatic environment. However, there are still some limitations in making them economically feasible. The fourth-generation biofuels are derived from genetically modified algae to enhance biofuel production (Raud et al. 2019). However, the potential environmental and health-related risks such as modified algal systems are yet to be studied.

11.1.2 Composition of Lignocellulosic Biomass

LCB remains a sustainable material for use as feedstock in biofuels and bioproducts. LCB contains a carbohydrate fraction of cellulose and hemicelluloses and a non-carbohydrate fraction of lignin, proteins, and extractives (Yoo et al. 2020). Lignin is found at 15–40% of the LCB material. Lignin is a complex macromolecule composed of monomeric units of para-coumaryl, coniferyl, and sinapyl alcohols that are cross-linked via stable covalent bonds between the polysaccharides and lignin polymer. This lignin-based cross-linked matrix complicates the degradation of biomass using microorganisms (Dragone et al. 2020).

The lignocellulosic residues comprise an abundance of complex carbon components derived from plant sources after harvesting or processing. Lignin forms the protective layer for the hemicellulose and cellulose matrix. The cellulose polymer consists of glucose units linked via β-1-4 glycosidic bonds and forms a linear crystalline structure. Cellulose requires three types of enzymes for efficient degradation, including (1) cellobiohydrolases (CBHs) to cleave the cellobiose from the reducing and non-reducing ends of cellulose chains, (2) endo-β-glucanases to cleave the glycosidic bonds, and iii) β-glucosidases for hydrolysis of the free cellobiose and cellodextrin fractions; hemicellulose is located between the cellulose and lignin. It is a complex polysaccharide mainly composed of arabinoxylan with branched heteropolymers of D-glucose, D-galactose, D-mannose, and D-xylan. Xylan and lignin are covalently bound together by cinnamic acids. Hydrolysis of the xylan component of hemicellulose requires endoxylanases and accessory enzymes like β-xylosidases, α-L-arabinofuranosidases, 4-O-methyl-D-glucuronidases, and acetyl xylan esterases (Cintra et al. 2020). Lignin is the most recalcitrant molecule to degrade among the LCB constituents, as it is a complex polymer of phenolic, amorphous, and hydrophobic nature due to its varied precursor components (Pereira et al. 2016).

Lignin acts as a physical barrier and hinders the conversion of biomass to biofuel. It affects both the pretreatment and enzymatic hydrolysis process due to its resistant nature and cross-linked networks. Thus, lignin has been targeted by various pretreatment methods like alkaline, alcohol-based organosolv, ionic liquid pretreatment, and biological methods such as enzymes and microbes (Yoo et al.

2020). Ligninolytic enzymes are produced by certain fungi and bacterial strains in large amounts. The efficient and beneficial characteristics of these strains are highly preferable for biocatalyst development for biofuel production, biopulping, textile industries, and platform chemicals (Gaur et al. 2018). The on-site enzyme production and tailor-made enzyme cocktail formulation can be used to pretreat LCB effectively. However, these added enzymes are of high-cost commodity due to the production cost, including nutrient costs, operational and capital cost, formulation, transport cost, and enzyme activity. Moreover, the performance of the enzyme is another limitation that usually differs based on the lignocellulosic substrates (Dragone et al. 2020).

11.2 Renewable Lignocellulosic Feedstocks

Agricultural residues (e.g., rice straw, sugarcane bagasse, corn stover, stalks, and other secondary products), energy crops, forestry residues, and industrially processed residues can be used as feedstocks for the production of biofuels and chemicals (Raud et al. 2019). These plant-based feedstocks are sustainable and have the potential to be generated under harsh conditions like saline, drought, and hot climates. Sweet sorghum is a highly feasible lignocellulosic crop containing both soluble and insoluble sugars to improve the sugar yield for further conversion to biofuel. The sorghum biomass produces (1.26–1.80 t $acre^{-1}$) bioethanol comparatively higher than any other feedstocks (Dar et al. 2018). On the other hand, the biomass from sugarcane, cassava, and plant seeds can also be used as renewable feedstocks based on the availability of bioenergy production resources (Adewuyi 2020). Several crop wastes are also considered for biofuel production. The availability of banana peduncle is 1% compared to that of sugarcane. One in five parts of sorghum is considered waste and has been employed for bioethanol using commercial fermentation using yeast and biogas production (Pazmiño-Hernandez et al. 2019). Alternative biomass for biodiesel production includes plant-based oils (e.g., olive oil, rapeseed oil, and palm oil), waste cooking oil, and crude tall oil derived as a by-product of pulping woody residues.

11.2.1 Industrial and Municipal Solid Waste

Industrial and municipal waste can be used for renewable fuels and chemical production. Municipal wastes, including animal waste, rotten vegetables and fruits, and tubers, have been used for bioethanol production (Adewuyi 2020). Paper mill sludge (PMS) materials from paper and milling industries can also be used as renewable feedstocks for biofuel production using feasible biological conversion approaches. As the PMS materials are obtained from the woody biomass, an increased amount of cellulose and other components like hemicelluloses and lignin with minimal quantity can be effectively utilized as a feedstock (Tawalbeh et al. 2021).

11.2.2 Macroalgal and Microalgal Sources

Macroalgae and microalgae are useful feedstocks with numerous beneficiary bioproducts. Microalgal oil, seaweeds, and natural algae are large-scale and renewable feedstocks used for biofuel production. Algal varieties with high lipid content, fast growth rate, reduced nutritional requirement, and biological traits amenable to pretreatment methods that reduce production cost include *Chlorella vulgaris* and mixed cultures like *Chlorophyceae* sp., *Cyanophyceae* sp., *Euglenophyceae* sp., *Bacillariophyceae* sp., and *Nannochloropsis* sp. (Japar et al. 2017; Thirugnanasambantham et al. 2020). Industrial effluents, often a menace to the environment, can also be useful resources for bioenergy production. The meat processing industry is one such manufacturing unit where the effluent is often organic rich and amenable for use as a bio-based feedstock for algal cultivation. Techniques like thermal, physicochemical, and biochemical methods are preferred for algal biomass conversion (Okoro et al. 2017). Macroalgae, such as *Ulva* sp., predominantly known as seaweeds, that has high sugars (at least 50%) can be used in biofuel production (Margareta et al. 2020; Nagarajan et al. 2020). Microalgae produces several different kinds of renewable biofuel, such as (a) anaerobic digestion of the algal biomass produces methane, (b) biodiesel from microalgal oil, and (c) biohydrogen through photobiological mechanism (Rajkumar et al. 2014).

11.3 Challenges in Biomass Processing

Biorefinery process designs must consider multiple factors to ensure the system is economically viable. Biomass processing requires optimized conditions like pH, temperature, inoculum, agitation rate, biocatalyst, and the concentration of the final product for efficient conversion. The nature and complexity of the biomass used as feedstock can dictate the combination of pretreatment techniques needed to ensure efficient bioconversion. These factors can lead to technical complications that could render the system economically unprofitable. The utilization of waste resources, while renewable and of limited impact on food security, can harbor undesired variables as the biomass can be diverse and include drastic fluctuations in pathogenic, organic, and moisture content (Okoro et al. 2017).

11.3.1 Consideration of Pretreatment Versus Inhibitors Generated

Pretreatment of biomass is an essential step for overcoming the recalcitrant nature of lignocelluloses and enabling access to the sugars for fermentation. The degradation products produced from pretreatment of lignocellulose depends on both the biomass and the pretreatment conditions, including temperature, duration, pressure, pH, redox conditions, and presence of catalysts (Klinke et al. 2004). Fermentation inhibitors are generated as by-products during pretreatment that interferes with the metabolism of microorganisms during bioconversion and further fermentation. Short-chain aliphatic acids (formic acid, acetic acid, and levulinic acid) are reported as inhibitors (Zhang et al. 2011a, 2016). The concentration and composition of inhibitors generated depend on the raw materials and the pretreatment method (Bellido et al. 2011). The choice of pretreatment often affects inhibitor formation. Acid-based pretreatments often generate aliphatic carboxylic acids, phenolic compounds, furans, and other related by-products. Likewise, hydrothermal processing produces acetic acid and furan aldehydes. Mild alkaline pretreatments methods are considered to be slow processes and may produce several acids and phenolic compounds that can inhibit biocatalysis. Similarly, oxidative methods produce aldonic and aldaric acids, furoic acid, phenolic acids, and acetic acid. Contrary to these methods, ammonia fiber explosion produces inhibitors such as ferulic acid that attack the biofuel process (Chundawat et al. 2010; Jönsson and Martín 2016; Piotrowski et al. 2014).

During ethanol fermentation, acetic acid affects the growth of *Saccharomyces cerevisiae* by a prolonged lag phase (Pampulha and Loureiro-Dias 2000; Zhang et al. 2011b). Similarly, several compounds of phenols, furans, ionic liquids, and other types of inhibitors are generated during pretreatment when harsh processes are employed. The presence of furan aldehydes in the fermentation media during ethanol production can decrease the specific growth rate and ethanol yield. Inhibition problems are increased due to the by-products accumulation during water recirculation and the high solid loads that are used to obtain more amount of sugar (Jönsson and Martín 2016).

The inhibitors generated after pretreatment include dehydrated sugar monomers (furans), degraded lignin polymers (phenols), and small organic acids). The major degradation products of glucose and xylose are 5-(hydroxymethyl)-2-furaldehyde (5-HMF) and furan-2-carbaldehyde (furfural), respectively (Damião Xavier et al. 2018; Rasmussen et al. 2014). 5-HMF may result from the dehydration of hexoses and furfural, resulting from the dehydration of pentoses during pretreatment. Pretreatments involving high temperatures and high acid concentrations for lignin removal result in undesirable compounds such as furans (Kabel et al. 2007). It was observed that there was a significant decrease in ethanol yield and productivity due to the synergistic combination of acetic acid, furfural, and lignin derivatives than due to the combined inhibition of individual compounds (Nigam 2001). In ethanol fermentations, furfural is more toxic than HMF, promoting the inhibition of enzymes acting on carbon catalysis, including acetaldehyde dehydrogenase, alcohol dehydrogenase, aldehyde dehydrogenase, glyceraldehyde 3-phosphate dehydrogenase, and pyruvate dehydrogenase (Guo et al. 2008).

The phenolic compounds generation depends on the molecular weight, polarity, and side-chain characteristics of the lignin structure and pretreatment method applied. Phenolic compounds inhibit cellulases and increase the pretreatment severity with liquid hot water, resulting in the solubilization of phenolic compounds (Michelin et al. 2016; Ximenes et al. 2011). Phenolic compounds affect the integrity in biological membranes, cell growth, ability of cell membrane to serve as barriers and enzymatic matrices, decrease the cellular assimilation of sugars, and inhibit protein synthesis. Low-molecular-weight phenolics or salts are more toxic by penetrating the cell membranes, whereas fermentation inhibitors with high molecular weight affect the transporters of sugar and ion (Kang et al. 2012; Klinke et al. 2004).

11.3.2 Lignin Complexity

Lignin in LCB acts as solid adhesive to cellulose and hemicellulose and contributes for the compactness and integrity of the structure. Lignin contains diverse phenolic acids such as p-coumaryl, coniferyl, guaiacyl, syringyl, and sinapyl, which is one of the dominant compounds that can release various inhibitory by-products during the pretreatment (Kim 2018). Pretreatment is the primary step in producing biofuel production from LCB, followed by saccharification or hydrolysis of the biomass. The removal of lignin enables efficient access to the cellulosic biomass for enzymatic hydrolysis. The saccharification process is the rate-limiting process since the utilization of all sugar in the biomass is vital to achieve the maximum end product. The high bulk lignin content in softwood might be responsible for strong inhibitory effect. Removing bulk lignin can improve enzymatic hydrolysis (Yoo et al. 2020). The inhibitory role of lignin on enzymatic hydrolysis revealed that the type of lignin and molecular weight influenced the inhibition. Similarly, kraft pine lignin precipitated on the cellulose surface, preventing it from contacting with the enzyme. The low-molecular-weight lignin could bind enzyme non-productively, and when the molecular weight increased, the steric repulsion was caused by lignin deposition on cellulose. The lignin structural features like functional groups and syringyl/guaiacyl ratio affected the behaviors of lignin in enzymatic hydrolysis. The high aliphatic hydroxyl groups and low carboxylic groups lead to high surface hydrophobicity, increasing the adsorption between lignin and enzyme. In addition, substrate reactivity is also an essential factor that affects enzymatic hydrolysis (Li and Zheng 2017). The extent of lignin inhibition on enzymatic hydrolysis is closely related to how lignin undergoes non-productive binding and physical blocking of the enzyme biocatalyst (Kumar et al. 2012; Li and Zheng 2017). It has been shown that the bulk lignin can be more inhibitory than the extractable lignin owing to differences in the physicochemical properties and condensed subunit content of these lignin fractions. Milled wood lignin possess a higher enzyme adsorption capacity, leading to the stronger inhibitory effects of residual lignin during enzymatic hydrolysis, as compared to extractable lignin. Milled wood lignin from softwood exhibits a stronger inhibitory effect on enzymatic hydrolysis of Avicel than pretreated sweetgum (Lai et al. 2015, 2017).

Of the several pretreatment strategies, biological-based methods are promising, as they minimize inhibitor formation, consume less energy, and are eco-friendly. A combination of more than one pretreatment method is also found to enhance delignification efficiency (Wang et al. 2012). Recently, coupling hydrodynamic cavitation with laccase was successful in LCB pretreatment (Thangavelu et al. 2018). The cavitation effect of degrading lignin moieties generates highly reactive radicals (-H and -OH) (Davis et al. 2016). In this hydrodynamic cavitation reactor (HCR)—laccase process, phenoxy radicals are released, eliminating recalcitrant portions of LCB and improving delignification. Coupling a multi-copper oxidase (LccH) from the hyper laccase-producing fungus *Hexagonia hirta* MSF2 in a HCR was also found to be successful for delignification of corn cob and wood biomass (Kandasamy et al. 2016).

11.3.3 Economics of Enzyme Production

Enzyme production is the most costly process in converting LCB to bioethanol, which covers about 40% of the total cost of the conversion process (Du et al. 2010) (Kabel et al. 2007). Finding cheaper methods of producing cellulase and hemicellulase fractions to use as substrates is another challenge for meeting the economics of biofuel production. Improved means of enzyme production and commercially economic enzyme on a large scale are some of the most pressing needs of the industry. Discovering new thermostable enzymes and optimizing methods to produce enzymes from natural polymers through solid-state fermentation (SSF) is envisioned as cost cutting and efficient bioconversion approaches. While doing so, the simultaneous saccharification and enzymes of cellulase and hemicellulase productions are gaining momentum. In this regard, a thermotolerant enzyme cocktail that includes a novel GH family 13 enzymes from the thermophilic fungi *Chaetomium thermophilum* EDWF1 has registered endoglucanase (EGL) activity of 484.10 $IU.mL^{-1}$ under SSF along with xylanase activity (Saranya 2017).

11.3.4 Biomass Size, Complexity, and Utilization Factors

Enzymatic saccharification of LCB is affected by various inhibitors that limit enzyme activity. In order to attain effective conversion of cellulosic substrates, the factors negatively affecting saccharification productivity must be overcome (Su et al. 2017). The main factors influencing enzymatic hydrolysis are the type of substrate and enzyme-related factors. In general, the two main chemical and physical parameters that affect substrate saccharification using cellulases are (1) the cellulose crystallinity and its degree of polymerization and (2) the complexity of the lignin-cellulose structure that acts as a physical barrier that blocks the enzymes from reaching the cellulose (Cateto et al. 2011; Fockink et al. 2016; Zhang and Lynd 2004). The lignin and hemicellulose content, the particle size, and the accessible surface area of the substrate also affect the saccharification efficiency.

Furthermore, cellulase-mediated hydrolysis includes three major steps: (1) cellulase adsorption to substrate surface, (2) fermentable sugar production, and (3) desorption of the cellulase. However, the substrate content, enzyme level, and reaction condition influence the above steps. The biomass particle sizes influencing the sugar recovery were studied using biomass with different sizes from 0.5 to 2.5 cm. The particle size of 1.0 and 0.5 cm gave 99.6% glucan and 67% xylan recovery, while the particle size of 2.5 cm yielded the maximum sugar conversion (100% for glucan and 83% for xylan). With the particle size increase, the surface area of pretreated biomass significantly increased with a decreased crystallinity index of pretreated biomass resulting in maximum hydrolysis and sugar conversion. The large particle size of corn stover biomass also helped in better mixing during steam explosion pretreatment (Liu et al. 2013). Therefore, conditions including the size of the biomass must be optimized to achieve maximal sugar recovery.

11.3.5 Product Inhibition During Saccharification

Metal ions are reported to act as potentiators or inhibitors of the enzymatic saccharification of LCB. Metal ions that potentiate or inhibit cellulases and hemicellulase activity include Co^{2+}, Mg^{2+}, Fe^{2+}, Fe^{3+}, Mn^{2+}, Ni^{2+}, Cu^{2+}, Hg^{2+}, Zn^{2+}, Ca^{2+}, Na^{+}, and K^{+} (Mandels and Reese 1965). Metal ions association with the enzyme catalyst alters enzyme activity and the formation of various complexes. The ions interact with the carboxyl and amino groups and affect the enzyme structure (Pereira et al. 2016). Metal ions formed during the acidic processing of biomass may corrode equipment and release metal ions, such as copper, nickel, chromium, and iron, and can be inhibitory to fermenting microorganisms(Watson et al. 1984). Other cations, viz., Na, Ca, and Mg, may result from the chemicals used in pretreatment or pH adjustment (Jönsson and Martín 2016). Biofuel end products themselves are inhibitory. Ethanol and isobutanol produced during saccharification can act as end-product inhibitors that reduce enzyme activity. Implementing ethanol-tolerant microbes for fermentation can address this latter issue.

11.4 Biomass Hydrolyzing Enzymes

Biomass hydrolyzing enzymes require synergistic action of many enzymes, and there are different classes of enzyme with unique functionality. The complexity of biomass varies enormously, and the enzymes for its hydrolysis also vary considerably. Plants have unique cell walls composed of (1) middle lamella, (2) primary cell wall, and (3) secondary cell wall structures. In general, the plant cell wall composition, including lignin content, varies among monocots, dicots, softwood, and hardwood (Rytioja et al. 2014; Vogel 2008). The major polysaccharides of the plant cell walls are cellulose, hemicellulose, and pectin, and its complexion with lignin makes the plant cell wall recalcitrant. The depolymerization of LCB requires the synergistic action of numerous oxidative, hydrolytic, and non-hydrolytic enzymes (Sistakameshwar and Qin 2018). According to the CAZy database, plant biomass polysaccharide-degrading enzymes and their subunits can be divided into six major families: (1) glycoside hydrolases (GHs), (2) glycosyl transferases (GTs), (3) polysaccharide lyases (PLs), (4) carbohydrate esterases (CEs), (5) carbohydrate-binding modules (CBMs), and (6) auxiliary activities (AAs) based on structural or sequence similarities (Lombard et al. 2014).

The group of enzymes involved in cellulose hydrolysis are classified into cellulases, hemicellulases, lignin-modifying enzymes, and non-hydrolytic proteins. In general, redox enzymes catalyze the auxiliary activities (AAs) that can assist and work simultaneously with other GHs to saccharify LCB. Cellulose decomposition was thought to be mediated primarily through the hydrolytic action of cellulases. Later, polysaccharide degradation was discovered to be mediated by oxidative reactions catalyzed by CBM33s (chitin-binding proteins in bacteria) and GH61s (EGs in fungi) (Vaaje-Kolstad et al. 2010). These are called lytic polysaccharide monooxygenases (LPMOs) and are reclassified as AA families 10 and 9, respectively, in the CAZy database (Levasseur et al. 2013).

The non-hydrolytic proteins that take part in the amorphogenesis of cellulose include swollenin (SWO1), which resembles plant expansins can degrade crystalline cellulose. *Trichoderma reesei,* SWO1s possess close amino acid sequence similarity to the plant expansins (Arantes and Saddler 2010; Gourlay et al. 2012). Similar to the expansins, SWO1s with no catalytic activity appear to disrupt the structure of cellulose microfibers, possibly by breaking hydrogen bonds (Saloheimo et al. 2002). SWO1 synergistically enhances endoxylanase and then endoglucanase or cellobiohydrolase activities during enzymatic hydrolysis of pretreated corn stover (Gourlay et al. 2013). The proposed mode of action of SWO1 is that the protein renders the xylan portion of LCB more accessible for degradation by xylanases and thereby indirectly promotes the action of cellulases. The two proteins CIP1 and CIP2 (cellulose-induced protein), which are induced along with most of the cellulases (Brown et al. 2003), are shown to be essential to degrade lignocellulose efficiently (Banerjee et al. 2010). CIP1 has synergistic activity with swollenins, while CIP2 cleaves hemicellulose-lignin cross links. CIP1 consists of a GH family 1 CBM connected via a linker region to a domain with yet unknown function. Though

CIP1 lacks lyase activity, it shows structural similarities with lyases (Jacobson et al. 2013). CIP2 is a glucuronoyl esterase of the carbohydrate esterase family 15. The glucuronoyl esterase could separate the lignin from hemicelluloses by hydrolysis of the ester bond between 4-O-methyl-D-glucuronic acid moieties of glucuronoxylans and aromatic alcohols of lignin (Pokkuluri et al. 2011). Expansins cell wall loosening action weakens the lignocellulose structure and enhances cellulose hydrolysis by cellulases (Baker et al. 2000).

11.5 Glycosyl Hydrolases (GHs)

11.5.1 Cellulases

The discovery of *T. reesei* (then *T. viride*) for its astonishing extracellular cellulases producing potential is exploited by many industries. Predominant biorefineries use *T. reesei* enzymes to saccharify lignocellulose from renewable plant biomass in order to produce bio-based fuels and chemicals (Bischof et al. 2016). Among 14,000 molds screened for cellulase, *Trichoderma* sp. QM6a was found to display the ability to degrade native crystalline cellulose. This strain was regarded as the *T. reesei* reference strain, and most of the mutants used in industry today have been derived from this strain. Subsequently, a 20-fold increase in the extracellular protein produced by the original strain QM6a was achieved through mutagenesis, which opened its industrial applicability (Bischof et al. 2016). By the end of the 1990s, *Hypocrea jecorina*, the sexual form of *T. reesei*, was discovered. Since then, numerous cellulolytic microorganisms have been discovered, and their cellulases have been characterized.

Cellulolytic microorganisms have developed two major cellulase strategies: discrete non-complexed cellulases and complexed cellulases (Lynd et al. 2002; Zhang and Lynd 2004) (Fig. 11.1). Most aerobic cellulolytic microorganisms degrade cellulose by secreting a set of individual cellulases, which possess a CBM linked N-terminus or C-terminus to the catalytic module. In contrast, most anaerobic microorganisms produce large (> one million Da molecular mass) multienzyme complexes, called cellulosomes, which are attached to the cell surface of the microorganisms (Bayer et al. 2004). Only a few of the enzymes in cellulosomes contain a CBM, but most of them are attached to the scaffolding protein that contains a CBM. Certain anaerobic bacteria produce both cellulosomes and free cellulases.

The whole process of cellulose bioconversion to glucose occurs in two steps. The first step is catalyzed by exoglucanases and endoglucanases that reduce the degree of polymerization in the liquefaction stage, releasing cellobiose; the second step is performed by β-glucosidase that cleaves cellobiose to glucose. Synergism has been observed between endo- and exo-β-glucanases as well as among exo-β-glucanases that act from the reducing and non-reducing ends. Four different types of synergism exist among these enzymes as proposed by Teeri (1997): (1) endo–exo synergy between endoglucanases and exoglucanases, (2) exo–exo synergy between

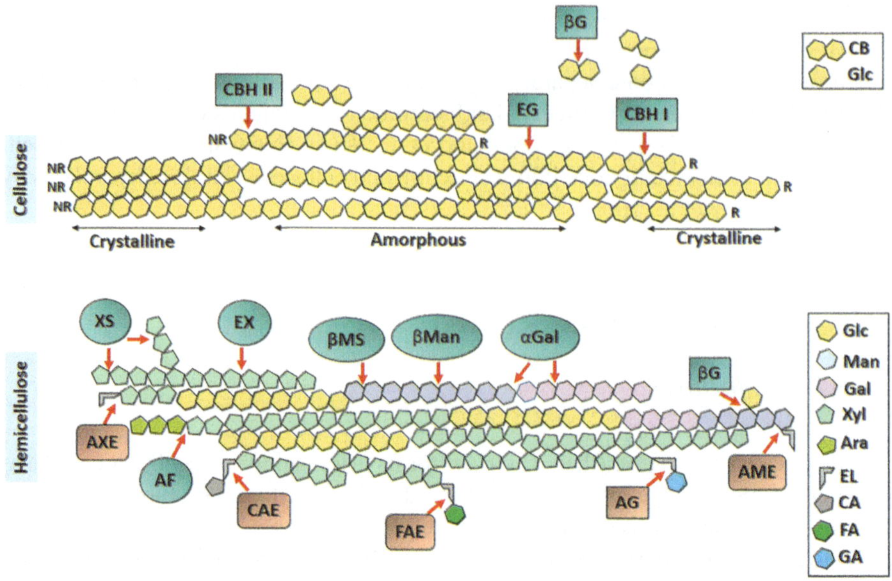

Fig. 11.1 Schematic portrayal showing enzymatic depolymerization of cellulose and hemicellulose. CBH I & II hydrolyze the cellulase chain from the non-reducing end (NR) and the reducing ends (R) of the cellulose chain, respectively, liberating glucose or cellobiose. EG hydrolyze the cellulose chain randomly in the amorphous region of the cellulose. βG acts on the cellobiose to produce glucose units. Hemicellulose is a branched polymer consisting of many different sugars. The complete hydrolysis of hemicellulose requires the concerted action of many enzymes. The enzymes that participate in xylan biomass hydrolysis include endo-1,4-β-xylanase (EX), exo-1,4-β-xylosidase or β-xylosidase (XS), β-mannanases (βMan), β-mannosidase (βMS), α-D-galactosidase (αGal), α-L-arabinofuranosidase (AF), α-D-glucuronidase (AG), acetyl xylan esterase (AXE), ferulic acid esterase (FAE), para-coumaroyl esterase (CAE), and acetyl mannan esterase (AME); βG also act on glucose and mannose linked units to liberate free sugars. *CBH* cellobiohydrolase, *EG* endoglucanase, *βG* β-glucosidase, *Glc* glucose, *CB* cellobiose, *Man* mannan, *Gal* galactose, *Xyl* xylose, *Ara* arabinose, *EL* ester linkage, *CA* para-coumaric acid, *FA* ferulic acid, *GA* glucuronic acid. Red arrows represent the enzyme action on glycosidic bonds or ester linkages present in the biomass component

reducing-end exoglucanases and non–reducing-end exoglucanases, (3) synergy between exoglucanases and β-glucosidase, and (4) intramolecular synergy between CBMs and catalytic modules. The CBMs aid in disrupting the cellulose fibers as well as helping the cellulases bind to the cellulose (Zhang and Zhang 2013).

Cellulases of bacteria are ideal compared to the fungal enzymes owing to the fast multiplication, various genetic diversity, and ease of genetic manipulation (Chandel et al. 2010). Many bacteria produce endoglucanases that can hydrolyze amorphous celluloses viz. carboxymethyl cellulose (CMC) but can be limited in the efficient hydrolysis of crystalline cellulose (Wilson 2011). Only few *Bacillus* spp. produce microcrystalline cellulose (Avicel)-degrading endoglucanases (Han et al. 1995). Furthermore, thermotolerant bacteria identified to synthesize cellulases with β-glucosidase activity can overcome the rate-limiting steps of the saccharification process leading to increased glucose yield (Bhalla et al. 2013). However, successful biomass hydrolysis and synergistic action of cellulase rely mainly on an optimum pretreatment process.

11.5.2 Endoglucanase, Exoglucanase, and β-Glucosidase

Endoglucanase, 1,4-β-D-glucan-4-glucanohydrolase, or carboxymethylcellulases (CMCases) (EC 3.2.1.4) are found to cut randomly at the β-1,4-bonds of cellulose chains, generating new ends. EGLs hydrolyze cellulose at the amorphous regions and produce accessible free chain ends for the further action of CBH (Fig. 11.1). In general, fungal EGLs possess a catalytic module with or without a CBM, while bacterial EGLs may have multiple catalytic modules, CBMs, and other modules with unknown functions. The catalytic modules of most EGLs possess a cleft/grove-shaped active site, which allows the endoglucanases to bind and cleave the cellulose chain that generates glucose, soluble cellodextrins, or insoluble cellulose fragments. Certain EGLs act "processively," to hydrolyze crystalline cellulose and produce the major products as cellobiose or longer cellodextrins (Cohen et al. 2005; Medve et al. 1998).

Exoglucanases, 1,4-β-D-glucan glucohydrolases (EC. 3.2.1.74), or CBHs acts on the reducing or non-reducing ends of cellulose chains, releasing either cellobiose or glucose as major products. CBHs join with the ends of cellulose microfibrils and then processively slide down the strands and cleave off cellobiose. The processive nature of CBHs is mediated by tunnel-like active sites, which can only accept a substrate chain via its terminal regions. These exo-acting CBH enzymes function by threading the cellulose chain through the tunnel, removing cellobiose units in a sequential manner (Kurašin and Väljamäe 2011; Yeoman et al. 2010). The CBHs also act on swollen, partly degraded amorphous substrates and cellodextrins but do not hydrolyze soluble derivatives of cellulose like carboxymethyl cellulose and hydroxyethyl cellulose (Sajith et al. 2016).

β-Glucosidase is also called as cellobiase (EC 3.2.1.21) that completes the process of cellulose hydrolysis by cleaving cellobiose and removing glucose from the non-reducing end of oligosaccharides. The β-glucosidases hydrolyze β-glucosidic linkages present in disaccharides, oligosaccharides, or conjugated glucosides. Based on substrate specificity, β-glucosidases are divided into three groups: aryl-β-glucosidases, cellobiases, and broad-specificity β-glucosidases. Aryl-β-glucosidases prefer hydrolysis of aryl-β-glucosides, whereas cellobiases only hydrolyze cello-oligosaccharides and cellobiose. Broad-specificity β-glucosidases show significant activity on both substrate types and represent the most commonly observed group in cellulolytic microbes (Bhatia et al. 2002). β-Glucosidase is the rate-limiting enzyme because it hydrolyzes the final step of lignocellulose breakdown in which cellobiose and short cellodextrins are converted into glucose.

11.5.3 Hemicellulases

Hemicellulose is complex and heterogeneous, and the complete hydrolysis of hemicellulose requires the interactive action of several hydrolytic enzymes (Beg et al. 2001). In hemicelluloses, xylanase is involved in the enzymatic hydrolysis of xylan. Based on the mode of action on the substrate, endo-1,4-β-xylanase or endoxylanases (EC 3.2.1.8) and exo-1,4-β-xylosidase or β-xylosidase or xylobiase (EC 3.2.1.37) hydrolyze the hemicellulose. The xylan hydrolysis demands the use of endo-β-1,4-xylanases, acting randomly on the internal bond of xylan to release a diverse range of products, such as xylobiose, xylotriose, xylotetraose, and longer and/or branched xylooligomers (Collins et al. 2005). Reducing-end xylose-releasing exooligoxylanases are called Rexs (EC 3.2.1.156). Rexs hydrolyze the xylan backbone or xylo-oligosaccharide (XOS) from the reducing end producing short XOSs and xylose (Malgas et al. 2019). β-xylosidase hydrolyzes the non-reducing ends of xylose chains, xylobiose, and xylo-oligomers to release xylose but do not hydrolyze xylan (Huy et al. 2015; Knob et al. 2010; Yan et al. 2008). Several supplementary enzymes, such as α-L-arabinofuranosidase (EC 3.2.1.55), α-D-glucuronidase (EC 3.2.1.139), α-D-galactosidase (EC 3.2.1.22), acetyl xylan esterase (EC 3.1.1.72), and feruloyl esterase (EC 3.1.1.73), participate in xylan biomass hydrolysis (Fig. 11.1).

Hemicellulose in softwood has mannan as the major component. Mannan is primarily composed of mannose residues. This polysaccharide is known as glucomannan when combined with glucose residues, galactomannan when combined with galactose, and galactoglucomannan with all three sugar units are present. β-mannanases or endo-β-1,4-mannanase (EC 3.2.1.78) hydrolyze mannan linkages via cleaving β-1,4 bonds and producing new reducing and non-reducing ends. Most of the β-mannanases are active on oligosaccharides containing three or four monomers. β-mannanases hydrolyze mannan with the help of β-mannosidase or exo-β-1,4-mannosidases (EC 3.2.1.25) and produce the terminal, non-reducing β-D-mannose residues. β-glucosidases can cleave the bond between one mannose and one glucose residue during glucomannan degradation. In softwood, endomannanases also catalyze internal linkages in mannan chains, constituting galactoglucomannans and glucomannans (Andlar et al. 2018). Acetyl mannan esterase (AME) (EC 3.1.1.6) plays a key role in removing side-chain acetyl substituents attached at various points on the mannan structure.

Generally, debranching enzymes can remove side groups linked to the main chain of the polysaccharides or oligomers. α-L-arabinofuranosidases cleaves arabinose residues from arabinan, arabinoxylan, or pectin. This activity facilitates the debranching and degradation of xylan and disrupts the lignin-carbohydrate complex. Similalry, α-glucuronidases catalyze the release of glucuronic acid or 4-O-methylglucuronic acid from xylan, showing a synergistic effect with endoxylanases. α-D-Galactosidases are involved in the cleavage of terminal α-1,6-linked galactose residues of galactomannans, galactoglucomannans, and oligosaccharides (Ademark et al. 2001; Lei et al. 2016).

Carbohydrate esterases act synergistically for efficient hemicellulose degradation. These accessorial enzymes are acetyl xylan esterase (AXE) (EC 3.1.1.72), feruloyl esterase (FAE) (EC 3.1.1.73), para-coumaroyl esterase (CAE) (EC 3.1.1.B10), exo-acting α-L-arabinofuranosidase (EC 3.2.1.55), endo-acting arabinofuranosidase (EC 3.2.1.99), xylan α-1,2-glucuronosidase (EC 3.2.1.131), and α-glucuronidase (EC 3.2.1.139). The esterases are considered as hemicellulases since they hydrolyze the ester bonds between hemicellulose and other components (Andlar et al. 2018; Malgas and Pletschke 2019; Zhang et al. 2011b). AXEs are involved in the liberation of acetic acid from acetylated polysaccharides by hydrolysis of ester bonds, thereby the main chain is accessible to GHs. FAEs cleave ester bonds between a hydroxyl-cinnamate and acetyl xylan, liberating phenolic acids including ferulic acid or *p*-coumaric acid (Wong et al. 2013). Glucuronoyl esterases (EC3.1.1.B11) cleave ester bonds between lignin-aliphatic alcohols and the 4-O-methyl-D-glucuronic acid substituents of glucuronoxylans (Arnling Bååth et al. 2016). Ferulic and para-coumaric acid esterases hydrolyze ester bonds between hydroxycinnamic acids and sugars and release ferulic acid and para-coumaric acid from these polymers. α-Glucuronidase catalyzes the hydrolysis of xylan into glucuronic acid or 4-O-methyl-glucuronic acid. The action of esterases can enhance the accessibility of the cellulose fibers and be used to produce bioactive chemicals and biofuels (Polizeli et al. 2005). Pectinases (EC 3.2.1.15) depolymerize (hydrolases and lyases) and deesterify (esterases) pectic substances present in the plant cell wall.

11.6 Thermophilic Biocatalysts Hydrolyzing Plant Biomass

The industrial conversion of LCB necessitates a pretreatment step that facilitates the subsequent enzymatic saccharification. This step is often characterized by a combination of extremely harsh conditions (high temperatures, pressures, and pH). Thermozymes are enzymes that works under high temperatures. These highly stable enzymes offer advantages during pretreatment steps to minimize the cost and complication of varying process conditions, including enzymatic hydrolysis steps. Of the extremozymes, polyextremophilic enzymes simultaneously withstand a combination of more than one harsh condition such as high temperature and pressure (thermopiezophilic), low temperature and high pressure (psychropiezophilic), or high temperature and low pH (thermoacidophilic). These enzymes allow saccharification at higher temperatures, shortens the reaction time, and avoids contamination (Guerriero et al. 2015).

Physical, chemical, and biological pretreatment processes can be customized based on the nature of the LCB. Laccases, also called green catalysts, hold a critical role in biological pretreatment processes and provide flexibility to the pretreatment process when these enzymes are expressed at high levels in a stable form. One such example is the laccase of the halophilic archaeon *Haloferax volcanii* (LccA). LccA is secreted at high levels into the culture supernatant of *H. volcanii* US02 with peak laccase activity detected at the stationary phase, thus, finding application in biorefineries. LccA is tolerant to high salt, mixed organosolvents, and high temperatures, with a half-life of inactivation at 50 °C of 1.3 days (Uthandi et al. 2010, 2012; Hepowit et al. 2012). A hyper laccase-producing white-rot fungus, *Hexagonia hirta* MSF2 (1944.44 U.mL^{-1}), is also found to hold promise in pretreatment strategies as it delignifies wood and corncob biomass to a level of 28.6 and 16.5%, respectively (Kandasamy et al. 2016). HCR coupled with *H. hirta* laccase pretreatment shows 47% delignification efficiency in corn cob in 1 h (Thangavelu et al. 2018). As inhibitors are typically not generated using biological pretreatments, robust enzymes are needed to develop economic and efficient LCB bioconversion processes. Xylitol was produced from the pretreated corncob biomass (Ariyan and Uthandi 2019; Yamunasri et al. 2021).

Thermophilic bacteria are bioprospected for LCB-modifying enzymes. *Bacillus* spp. including *Bacillus tequilensis, Bacillus subtilis,* and *Bacillus licheniformis* were isolated for this purpose by in situ enrichment methods from the hot springs of Manikaran (~95 °C), Kalath (~50 °C), and Vasist (~65 °C), The Himalayas, India (Thangappan et al. 2017). Cellulases and xylanases identified by this approach are found tolerant of temperatures up to 80 °C and pH 7. The identified endoglucanases also exhibit high-level activity in the presence of calcium and potassium ions (Thankappan et al. 2018). Under submerged conditions, the thermophilic bacterium *B. aerius* CMCPS1, isolated from paddy straw compost, showed maximum activity of FPAse of 4.36 IU mL^{-1} and endoglucanase of 2.98 IU mL^{-1} at 44 h (Ganesan et al. 2020). The GHs encoding genes from thermophilic fungi engineered in a suitable yeast-based vector system are also a feasible technology for the optimal and sustainable production of GHs from thermophilic fungi. While bioprospecting endophytes for biomass conversion, perennial grasses are also unique sources of GHs. Endophytes from a C4 perennial grass *Neyraudia reynaudiana* L viz., *Bacillus tequilensis* BT5 and *Alcaligenes faecalis* B12, show FPAase, β-glucosidase, and xylanase activities (Vegnesh et al. 2019).

11.7 Accelerated Saccharification

Multifunctional cellulases are showing high-temperature tolerance, work at harsh conditions, and accelerate saccharification (Bhalla et al. 2013). The cellulase with high catalytic efficiency would reduce the viscosity of the medium and simultaneously increasing the diffusion of simple sugars from complex polysaccharides. Thus, screening diverse cellulases suitable for industrial requirements is an important goal (Krahe et al. 1996; Mozhaev 1993).

Multi-functional cellulases of the *Bacillus subtilis* CMCPS1 recorded a saccharification efficiency of 55% at 50 °C and pH 5.0 (Ganesan et al. 2020). Similarly, thermophilic fungi are more efficient than bacteria, as they produce good yields of GHs. However, the maintenance of thermophilic fungi under laboratory conditions is challenging (Saranya and Uthandi 2017). The thermophilic fungus *Chaetomium thermophilum* EDWF1 was isolated from elephant dung and produces thermotolerant and alkali-tolerant cellulases, endoglucanases, and beta-glucosidase (Saranya and Uthandi 2017).

A novel one-pot enzyme technology that comprises laccase, cellulase, and β-glucosidase have been co-immobilized to facilitate bioethanol production from *Typha angustifolia, Arundo donax, Saccharum arundinaceum,* and *Ipomoea carnea.* The co-immobilized enzyme system is more stable at different temperatures compared to free enzymes. Enzymatic saccharification of *S. arundinaceum* recorded the highest reducing sugar of 205 mg/g and the highest bioethanol yield of 63% with *I. carnea* among the LCB (Sankar et al. 2018).

The enzymes involved in cellulose degradation are not produced at an optimal level in a single microbe, and cellulases from a single organism may not be hydrolyzing different feedstocks. The enzyme-producing firms make cocktails of cellulase by enzyme assembly (multienzyme mixtures) or use of engineered microorganisms to express the desired combination of enzymes. Enzyme cocktails are also often produced from the co-fermentation of several microorganisms. The most productive major source of cellulases comes from the filamentous fungi and mutant strains of *Trichoderma* (*T. viride, T. reesei,* and *T. longibrachiatum*). The two leading companies that supply commercial cellulases are Novozymes and Genencor, supported by the US Department of Energy. Genencor has launched four new blends: Accelerase®1500, Accelerase®XP, Accelerase®XC, and Accelerase®BG. Each of these enzyme blends includes two or more enzymes. Accelerase®1500 includes exoglucanase, endoglucanase, hemicellulase, and β-glucosidase. Accelerase®XP improves both xylan and glucan conversion. Accelerase®XC comprises of hemicellulase and cellulase activities. Accellerase®Duet has exoglucanase, endoglucanase, β-glucosidase, and xylanase enzymes and can hydrolyze LCB into fermentable monosaccharides such as glucose and xylose (Genencos 2010). In contrast, Accelerase®BG includes only β-glucosidase enzyme designed as an accessory product to supplement whole cellulases deficient in beta-glucosidase.

Similarly, Cellic CTec in combination with Cellic HTec produced by Novozymes can be helpful for the conversion of the carbohydrates in biomass materials into simple sugars using a wide variety of pretreated feedstocks, such as sugarcane bagasse, corn cob, corn fiber, and wood pulp. Most of the commercial cellulases are optimally active at 50 °C and pH of 4.0–5.0. Similarly, enzyme mixtures produced Biocellulase A and Cellulase AP 30K produced by Quest Intl. (Sarasota, Fl) and Amano Enzyme Inc., respectively can work at higher temperatures from 50 to 60 °C (Verardi et al. 2012).

11.8 Conclusion and Perspective

Many times, the combination of more than one pretreatment method is helpful in effective delignification and deconstruction, resulting in maximal saccharification efficiency. Regardless of the pretreatment method, it should aim for the recovery of monomers without the generation of inhibitors. Another approach for effective and economical conversion of biomass is to find suitable multi-functional thermophilic GHs. Applications of GHs in biorefineries to produce sugars and concomitant fermentation products can be accelerated by using enzyme that possess multi-stability of pH, temperature, metal ions, and organic solvents. Therefore, methods to enhance saccharification efficiency are urgently needed and may be solved by staggered enzyme loading, assembly of enzyme cocktails, and optimizing the conditions of monosaccharide generation. Additionally, candidate microbial strains that produce multi-functional GHs should possess cellulase activity in the presence of hydrophobic solvents at thermo-alkali conditions. Such strains are more potent in terms of activity and stability and, thereby, make the strain a cost-efficient resource. High-value commodity chemicals from biomass can be produced using the optimized process of in-house thermophilic GHs production through submerged and solid-state fermentations. While producing GHs through SSF, cheaply available renewable biomass materials, such as corn cob and *Erianthus*, may be used as substrates.

Valorization of lignin has gained importance in recent years for the production of low-molecular-weight value-added products in industries because of the abundance and aromatic polymeric structure of lignin. The integrated biorefinery approach of catalytic depolymerization of lignin using enzymatically pretreated LCB seems to be a viable technology for lignin routed high-value commodities. However, this combined and sequential process needs to be perfected for the recovery of high-value platform chemicals. Hence, catalytic and biocatalytic approaches of deconstructing LCB for lignin-derived platform chemicals holds promise.

Acknowledgements The authors acknowledge the financial support provided to SU by DBT, GoI, through Indo-Russian collaboration (No. DBT/IC2/Indo-Russia/2014-16) and SERB (No. EEQ/2020/000583). Authors also thank DBT-BIOCARe for the financial support to RP and SU through the project grant No. BT/PR18134/BIC/101/795/2016 and for the support to JMF through the U.S. Department of Energy, Office of Basic Energy Sciences, Division of Chemical Sciences, Geosciences and Biosciences, Physical Biosciences Program (grant number DOE DE-FG02-05ER15650) to advance microbial bioenergy.

References

Ademark P, De Vries RP, Hägglund P, Stålbrand H, Visser J (2001) Cloning and characterization of *Aspergillus niger* genes encoding an alpha-galactosidase and a beta-mannosidase involved in galactomannan degradation. Eur J Biochem 268(10):2982–2990

Adewuyi A (2020) Challenges and prospects of renewable energy in Nigeria: a case of bioethanol and biodiesel production. J Energy Rep 6:77–88

Andlar M, Rezić T, Marđetko N, Kracher D, Ludwig R, Šantek B (2018) Lignocellulose degradation: an overview of fungi and fungal enzymes involved in lignocellulose degradation. J Eng Life Sci 18(11):768–778

Arantes V, Saddler JN (2010) Access to cellulose limits the efficiency of enzymatic hydrolysis: the role of amorphogenesis. Biotechnol Biofuels 3(1):1–11

Ariyan M, Uthandi S (2019) Xylitol production by xylose reductase over producing recombinant *Escherichia coli* M15. Madras Agric J 106(1–3):1

Arnling Bååth J, Giummarella N, Klaubauf S, Lawoko M, Olsson L (2016) A Glucuronoyl esterase from *Acremonium alcalophilum* cleaves native lignin-carbohydrate ester bonds. FEBS Lett 590 (16):2611–2618

Baker JO, King MR, Adney WS, Decker SR, Vinzant TB, Lantz SE, Nieves RE, Thomas SR, Li LC, Cosgrove D (2000) Investigation of the cell-wall loosening protein expansin as a possible additive in the enzymatic saccharification of lignocellulosic biomass. In: Twenty-first symposium on biotechnology for fuels and chemicals. Springer, pp 217–223

Banerjee G, Car S, Scott-Craig JS, Borrusch MS, Aslam N, Walton JD (2010) Synthetic enzyme mixtures for biomass deconstruction: production and optimization of a Core set. Biotechnol Bioeng 106(5):707–720

Bayer EA, Belaich J-P, Shoham Y, Lamed R (2004) The cellulosomes: multienzyme machines for degradation of plant cell wall polysaccharides. Annu Rev Microbiol 58:521–554

Beg Q, Kapoor M, Mahajan L, Hoondal G (2001) Microbial xylanases and their industrial applications: a review. Appl Microbiol Biotechnol 56(3):326–338

Bellido C, Bolado S, Coca M, Lucas S, González-Benito G, García-Cubero MT (2011) Effect of inhibitors formed during wheat straw pretreatment on ethanol fermentation by *Pichia stipitis*. Bioresour Technol 102(23):10868–10874

Bhalla A, Bansal N, Kumar S, Bischoff KM, Sani RK (2013) Improved lignocellulose conversion to biofuels with thermophilic bacteria and thermostable enzymes. Bioresour Technol 128:751–759

Bhatia Y, Mishra S, Bisaria V (2002) Microbial beta-glucosidases: cloning, properties, and applications. Crit Rev Biotechnol 22(4):375–407

Bischof RH, Ramoni J, Seiboth B (2016) Cellulases and beyond: the first 70 years of the enzyme producer *Trichoderma reesei*. Microb Cell Factories 15(1):1–13

Brown D, Dean R, Diener S, Houfek TD, Mitchell T, Foreman PK, Dankmeyer L, Dunn-Coleman NS, Goedegebuur F, England G (2003) Transcriptional regulation of biomass-degrading enzymes in the filamentous fungus *Trichoderma reesei*. J Biol Chem 278(34):31988–31997

Cateto C, Hu G, Ragauskas A (2011) Enzymatic hydrolysis of Organosolv Kanlow switchgrass and its impact on cellulose crystallinity and degree of polymerization. Energy Environ Sci 4(4):1516–1521

Chandel AK, Singh OV, Chandrasekhar G, Rao LV, Narasu ML (2010) Key drivers influencing the commercialization of ethanol-based biorefineries. J Commer Biotechnol 16(3):239–257

Chundawat SP, Vismeh R, Sharma LN, Humpula JF, Da Costa Sousa L, Chambliss CK, Jones AD, Balan V, Dale BE (2010) Multifaceted characterization of cell wall decomposition products formed during ammonia fiber expansion (Afex) and dilute acid based pretreatments. Bioresour Technol 101(21):8429–8438

Cintra LC, Da Costa IC, De Oliveira ICM, Fernandes AG, Faria SP, Jesuíno RSA, Ravanal MC, Eyzaguirre J, Ramos LP, De Faria FP (2020) The boosting effect of recombinant hemicellulases on the enzymatic hydrolysis of steam-treated sugarcane bagasse. Enzyme Microb Technol 133: 109447

Cohen R, Suzuki MR, Hammel KE (2005) Processive endoglucanase active in crystalline cellulose hydrolysis by the Brown rot basidiomycete *Gloeophyllum trabeum*. Appl Environ Microbiol 71 (5):2412–2417

Collins T, Gerday C, Feller G (2005) Xylanases, xylanase families and extremophilic xylanases. FEMS Microbiol Rev 29(1):3–23

Damião Xavier F, Santos Bezerra G, Florentino Melo Santos S, Sousa Conrado Oliveira L, Luiz Honorato Silva F, Joice Oliveira Silva A, Maria Conceição M (2018) Evaluation of the simultaneous production of xylitol and ethanol from sisal fiber. Biomol Ther 8(1):2

Dar RA, Dar EA, Kaur A, Phutela UG (2018) Sweet sorghum-A promising alternative feedstock for biofuel production. J Renew Sustain Energy Rev 82:4070–4090

Davis KM, Rover M, Brown RC, Bai X, Wen Z, Jarboe LR (2016) Recovery and utilization of lignin monomers as part of the biorefinery approach. Energies 9(10):808

Dragone G, Kerssemakers AA, Driessen JL, Yamakawa CK, Brumano LP, Mussatto SI (2020) Innovation and strategic orientations for the development of advanced biorefineries. Bioresour Technol 302:122847

Du B, Sharma LN, Becker C, Chen SF, Mowery RA, Van Walsum GP, Chambliss CK (2010) Effect of varying feedstock–pretreatment chemistry combinations on the formation and accumulation of potentially inhibitory degradation products in biomass hydrolysates. Biotechnol Bioeng 107(3):430–440

Fockink DH, Urio MB, Chiarello LM, Sánchez JH, Ramos LP (2016) Principles and challenges involved in the enzymatic hydrolysis of cellulosic materials at high total solids. In: Green fuels technology, pp 147–173

Ganesan M, Vinayakamoorthy RM, Thankappan S, Muniraj I, Uthandi S (2020) Thermotolerant glycosyl hydrolases-producing *Bacillus aerius* CMCPS1 and its saccharification efficiency on HCR-laccase (LccH)-pretreated corncob biomass. Biotechnol Biofuels 13(1):1–14

Gaur N, Narasimhulu K, Pydisetty Y (2018) Biochemical and kinetic characterization of laccase and manganese peroxidase from novel *Klebsiella pneumoniae* strains and their application in bioethanol production. RSC Adv 8(27):15044–15055

Genencos (2010) A trademark product introduced in 2010 by Accellerase Genencos

Gourlay K, Arantes V, Saddler JN (2012) Use of substructure-specific carbohydrate binding modules to track changes in cellulose accessibility and surface morphology during the amorphogenesis step of enzymatic hydrolysis. Biotechnol Biofuels 5(1):1–14

Gourlay K, Hu J, Arantes V, Andberg M, Saloheimo M, Penttilä M, Saddler J (2013) Swollenin aids in the amorphogenesis step during the enzymatic hydrolysis of pretreated biomass. Bioresour Technol 142:498–503

Guerriero G, Hausman J-F, Strauss J, Ertan H, Siddiqui KS (2015) Destructuring plant biomass: focus on fungal and extremophilic cell wall hydrolases. Plant Sci 234:180–193

Guo G-L, Chen W-H, Chen W-H, Men L-C, Hwang W-S (2008) Characterization of dilute acid pretreatment of Silvergrass for ethanol production. Bioresour Technol 99(14):6046–6053

Han SJ, Yoo YJ, Kang HS (1995) Characterization of a bifunctional cellulase and its structural gene: the Cel gene of *Bacillus* sp. D04 has exo-and endoglucanase activity. J Biol Chem 270 (43):26012–26019

Hepowit NL, Uthandi S, Miranda HV, Toniutti M, Prunetti L, Olivarez O, De Vera IM, Fanucci GE, Chen S, Maupin-Furlow JA (2012) Archaeal jab 1/Mpn/Mov 34 metalloenzyme (Hvjamm 1) cleaves ubiquitin-like small archaeal modifier proteins (Samp S) from protein-conjugates. Mol Microbiol 86(4):971–987

Huy ND, Le Nguyen C, Seo J-W, Kim D-H, Park S-M (2015) Putative endoglucanase Pcgh5 from *Phanerochaete chrysosporium* is a beta-xylosidase that cleaves xylans in synergistic action with endo-xylanase. J Biosci Bioeng 119(4):416–420

Jacobson F, Karkehabadi S, Hansson H, Goedegebuur F, Wallace L, Mitchinson C, Piens K, Stals I, Sandgren M (2013) The crystal structure of the Core domain of a cellulose induced protein (Cip1) from *Hypocrea jecorina*, at 1.5 Å resolution. PLoS One 8(9):E70562

Japar AS, Takriff MS, Yasin NH (2017) Harvesting microalgal biomass and lipid extraction for potential biofuel production: a review. J Environ Chem Eng 5(1):555–563

Jönsson LJ, Martín C (2016) Pretreatment of lignocellulose: formation of inhibitory by-products and strategies for minimizing their effects. Bioresour Technol 199:103–112

Kabel MA, Bos G, Zeevalking J, Voragen AG, Schols HA (2007) Effect of pretreatment severity on xylan solubility and enzymatic breakdown of the remaining cellulose from wheat straw. Bioresour Technol 98(10):2034–2042

Kandasamy S, Muniraj IK, Purushothaman N, Sekar A, Sharmila D, Kumarasamy R, Uthandi S (2016) High level secretion of laccase (Lcch) from a newly isolated white-rot basidiomycete, *Hexagonia hirta* MSF2. Front Microbiol 7:707

Kang L, Lee YY, Yoon S-H, Smith AJ, Krishnagopalan GA (2012) Ethanol production from the mixture of hemicellulose prehydrolysate and paper sludge. Bioresources 7(3):3607–3626

Kim D (2018) Physico-chemical conversion of lignocellulose: inhibitor effects and detoxification strategies: a mini review. Molecules 23(2):309

Klinke HB, Thomsen A, Ahring BK (2004) Inhibition of ethanol-producing yeast and bacteria by degradation products produced during pretreatment of biomass. Appl Microbiol Biotechnol 66(1):10–26

Knob A, Terrasan CF, Carmona E (2010) Β-xylosidases from filamentous fungi: an overview. World J Microbiol Biotechnol 26(3):389–407

Krahe M, Antranikian G, Märkl H (1996) Fermentation of Extremophilic microorganisms. FEMS Microbiol Rev 18(2–3):271–285

Kumar L, Arantes V, Chandra R, Saddler J (2012) The lignin present in steam pretreated softwood binds enzymes and limits cellulose accessibility. Bioresour Technol 103(1):201–208

Kurašin M, Väljamäe P (2011) Processivity of cellobiohydrolases is limited by the substrate. J Biol Chem 286(1):169–177

Lai C, Tu M, Yong Q, Yu S (2015) Disparate roles of solvent extractable lignin and residual bulk lignin in enzymatic hydrolysis of pretreated sweetgum. RSC Adv 5(119):97966–97974

Lai C, Tu M, Xia C, Shi Z, Sun S, Yong Q, Yu S (2017) Lignin alkylation enhances enzymatic hydrolysis of lignocellulosic biomass. Energy Environ Sci 31(11):12317–12326

Lei Z, Shao Y, Yin X, Yin D, Guo Y, Yuan J (2016) Combination of xylanase and debranching enzymes specific to wheat Arabinoxylan improve the growth performance and gut health of broilers. J Agric Food Chem 64(24):4932–4942

Levasseur A, Drula E, Lombard V, Coutinho PM, Henrissat B (2013) Expansion of the enzymatic repertoire of the Cazy database to integrate auxiliary redox enzymes. Biotechnol Biofuels 6(1):1–14

Li X, Zheng Y (2017) Lignin-enzyme interaction: mechanism, mitigation approach, modeling, and research prospects. Biotechnol Adv 35(4):466–489

Limayem A, Ricke SC (2012) Lignocellulosic biomass for bioethanol production: current perspectives, potential issues and future prospects. Progress Energy Combust Sci 38(4):449–467

Liu Z-H, Qin L, Pang F, Jin M-J, Li B-Z, Kang Y, Dale BE, Yuan Y-J (2013) Effects of biomass particle size on steam explosion pretreatment performance for improving the enzyme digestibility of corn Stover. Ind Crop Prod 44:176–184

Lombard V, Golaconda Ramulu H, Drula E, Coutinho PM, Henrissat B (2014) The carbohydrate-active enzymes database (Cazy) in 2013. Nucleic Acids Res 42(D1):D490–D495

Lynd LR, Weimer PJ, Van Zyl WH, Pretorius IS (2002) Microbial cellulose utilization: fundamentals and biotechnology. Microbiol Mol Biol Rev 66(3):506–577

Malgas S, Pletschke BI (2019) The effect of an oligosaccharide reducing-end xylanase, Bhrex8a, on the synergistic degradation of Xylan backbones by an optimised Xylanolytic enzyme cocktail. Microbial Enzyme Technol 122:74–81

Malgas S, Mafa MS, Mkabayi L, Pletschke BI (2019) A mini review of xylanolytic enzymes with regards to their synergistic interactions during hetero-xylan degradation. World J Microbiol Biotechnol 35(12):1–13

Mandels M, Reese ET (1965) Inhibition of cellulases and beta-glucosidases. In: Advances in enzymic hydrolysis of cellulose and related materials pergamon, pp 115–157

Margareta W, Nagarajan D, Chang J-S, Lee D-J (2020) Dark fermentative hydrogen production using macroalgae (*Ulva* sp.) as the renewable feedstock. Appl Energy 262:114574

Medve J, Karlsson J, Lee D, Tjerneld F (1998) Hydrolysis of microcrystalline cellulose by cellobiohydrolase I and endoglucanase ii from *Trichoderma reesei*: adsorption, sugar production pattern, and synergism of the enzymes. Biotechnol Bioeng 59(5):621–634

Michelin M, Ximenes E, De Moraes MD, Ladisch MR (2016) Effect of phenolic compounds from pretreated sugarcane bagasse on cellulolytic and hemicellulolytic activities. Bioresour Technol 199:275–278

Mozhaev VV (1993) Mechanism-based strategies for protein thermostabilization. Trends Biotechnol 11(3):88–95

Nagarajan D, Nandini A, Dong C-D, Lee D-J, Chang J-S (2020) Lactic acid production from renewable feedstocks using poly (vinyl alcohol)-immobilized *Lactobacillus plantarum* 23. Ind Eng Chem Res 59(39):17156–17164

Nigam J (2001) Ethanol production from wheat straw hemicellulose hydrolysate by *Pichia stipitis*. J Biotechnol 87(1):17–27

Nikkhah A, Assad ME, Rosentrater KA, Ghnimi S, Van Haute S (2020) Comparative review of three approaches to biofuel production from energy crops as feedstock in a developing country. Bioresour Technol Rep 10:100412

Okoro OV, Sun Z, Birch J (2017) Meat processing waste as a potential feedstock for biochemicals and biofuels—a review of possible conversion technologies. J Clean Prod 142:1583–1608

Pampulha ME, Loureiro-Dias MC (2000) Energetics of the effect of acetic acid on growth of *Saccharomyces cerevisiae*. FEMS Microbiol Lett 184(1):69–72

Pazmiño-Hernandez M, Moreira CM, Pullammanappallil P (2019) Feasibility assessment of waste banana peduncle as feedstock for biofuel production. Biofuels 10(4):473–484

Pereira JH, Heins RA, Gall DL, Mcandrew RP, Deng K, Holland KC, Donohue TJ, Noguera DR, Simmons BA, Sale KL (2016) Structural and biochemical characterization of the early and late enzymes in the lignin B-aryl ether cleavage pathway from *Sphingobium* sp. SYK-6. J Biol Chem 291(19):10228–10238

Piotrowski JS, Zhang Y, Sato T, Ong I, Keating D, Bates D, Landick R (2014) Death by a thousand cuts: the challenges and diverse landscape of lignocellulosic hydrolysate inhibitors. Front Microbiol 5:90

Pokkuluri PR, Duke N, Wood SJ, Cotta MA, Li XL, Biely P, Schiffer M (2011) Structure of the catalytic domain of glucuronoyl esterase Cip2 from *Hypocrea jecorina*. Proteins 79(8):2588–2592

Polizeli M, Rizzatti A, Monti R, Terenzi H, Jorge JA, Amorim D (2005) Xylanases from fungi: properties and industrial applications. Appl Microbiol Biotechnol 67(5):577–591

Rajkumar R, Yaakob Z, Takriff MS (2014) Potential of micro and macro algae for biofuel production: a brief review. Bioresources 9(1):1606–1633

Rasmussen H, Sørensen HR, Meyer AS (2014) Formation of degradation compounds from lignocellulosic biomass in the biorefinery: sugar reaction mechanisms. Carbohydr Res 385: 45–57

Raud M, Kikas T, Sippula O, Shurpali N (2019) Potentials and challenges in lignocellulosic biofuel production technology. Renew Sustain Energy Rev 111:44–56

Rytioja J, Hildén K, Yuzon J, Hatakka A, De Vries RP, Mäkelä MR (2014) Plant-polysaccharide-degrading enzymes from basidiomycetes. Microbiol Mol Biol Rev 78(4):614–649

Sajith S, Priji P, Sreedevi S, Benjamin S (2016) An overview on fungal cellulases with an industrial perspective. J Nutr Food Sci 6(1):461

Saloheimo M, Paloheimo M, Hakola S, Pere J, Swanson B, Nyyssönen E, Bhatia A, Ward M, Penttilä M (2002) Swollenin, a *Trichoderma reesei* protein with sequence similarity to the plant expansins, exhibits disruption activity on cellulosic materials. Eur J Biochem 269(17):4202–4211

Sankar MK, Ravikumar R, Kumar MN, Sivakumar U (2018) Development of co-immobilized tri-enzyme biocatalytic system for one-pot pretreatment of four different perennial lignocellulosic biomass and evaluation of their bioethanol production potential. Bioresour Technol 269: 227–236

Sistakameshwar A, Qin W (2018) Comparative study of genome-wide plant biomass-degrading Cazymes in white rot, Brown rot and Soft rot fungi. Mycology 9(2):93–105

Saranya S, Uthandi S (2017) Thermophilic cellulolytic fungi: cellulase production, characterization and biomass conversion. Thesis submitted to Tamil Nadu Agricultural University, pp 1–101

Su L-H, Zhao S, Jiang S-X, Liao X-Z, Duan C-J, Feng J-X (2017) Cellulase with high B-glucosidase activity by *Penicillium oxalicum* under solid-state fermentation and its use in hydrolysis of cassava residue. World J Microbiol Biotechnol Adv 33(2):37

Tawalbeh M, Rajangam AS, Salameh T, Al-Othman A, Alkasrawi M (2021) Characterization of paper mill sludge as a renewable feedstock for sustainable hydrogen and biofuels production. Int J Hydrog Energy 46(6):4761–4775

Teeri TT (1997) Crystalline cellulose degradation: new insight into the function of cellobiohydrolases. Trends Biotechnol 15(5):160–167

Thangappan S, Kandasamy S, Uthandi S (2017) Deciphering thermostable xylanases from hot springs: the heritage of Himachal Pradesh for efficient biomass deconstruction. Madras Agric J 104:282

Thangavelu K, Desikan R, Taran OP, Uthandi S (2018) Delignification of corncob via combined hydrodynamic cavitation and enzymatic pretreatment: process optimization by response surface methodology. Biotechnol Biofuels 11(1):203. https://doi.org/10.1186/S13068-018-1204-Y

Thankappan S, Kandasamy S, Joshi B, Sorokina KN, Taran OP, Uthandi S (2018) Bioprospecting thermophilic glycosyl hydrolases, from hot springs of Himachal Pradesh, for biomass valorization. AMB Express 8(1):1–15

Thirugnanasambantham R, Elango T, Elangovan K (2020) *Chlorella vulgaris* microalgae as a feedstock for biofuel. Mater Today 33:3182–3185

Uthandi S, Saad B, Humbard MA, Maupin-Furlow JA (2010) LccA, an archaeal laccase secreted as a highly stable glycoprotein into the extracellular medium by *Haloferax volcanii*. Appl Environ Microbiol 76(3):733–743

Uthandi S, Prunetti L, De Vera IM, Fanucci GE, Angerhofer A, Maupin-Furlow JA (2012) Enhanced archaeal laccase production in recombinant Escherichia Coli by modification of N-terminal propeptide and twin arginine translocation motifs. J Ind Microbiol 39(10):1523–1532

Vaaje-Kolstad G, Westereng B, Horn SJ, Liu Z, Zhai H, Sørlie M, Eijsink VG (2010) An oxidative enzyme boosting the enzymatic conversion of recalcitrant polysaccharides. Science 330(6001):219–222

Vegnesh R, Thankappan S, Singh BP, Kennedy ZJ, Ratul Saikia Z, Uthandi S (2019) Glycosyl hydrolases producing bacterial endophytes from Perennial Grass Species (*Neyraudia reynaudiana* L.) for biomass deconstruction. Madras Agric J 106(1–3):1

Verardi A, De Bari I, Ricca E, Calabrò V (2012) Hydrolysis of lignocellulosic biomass: current status of processes and technologies and future perspectives. In: Bioethanol, vol 2012. Intech Rijeka, pp 95–122

Vogel J (2008) Unique aspects of the grass cell wall. Curr Opin Plant Biol 11(3):301–307

Wang C, Sun J, Liu X, Sudo A, Endo T (2012) Synthesis and copolymerization of fully bio-based benzoxazines from guaiacol, furfurylamine and stearylamine. Green Chem 14(10):2799–2806

Watson NE, Prior BA, Lategan PM, Lussi M (1984) Factors in acid treated bagasse inhibiting ethanol production from D-xylose by *Pachysolen tannophilus*. Enzym Microb Technol 6 (10):451–456

Wilson DB (2011) Microbial diversity of cellulose hydrolysis. Curr Opin Microbiol 14(3):259–263

Wong DW, Chan VJ, Liao H, Zidwick MJ (2013) Cloning of a novel feruloyl esterase gene from rumen microbial metagenome and enzyme characterization in synergism with endoxylanases. J Indus Microbiol Biotechnol 40(3–4):287–295

Ximenes E, Kim Y, Mosier N, Dien B, Ladisch M (2011) Deactivation of cellulases by phenols. Enzyme Microb Technol 48(1):54–60

Yamunasri P, Priyadharshini R, Uthandi S (2021) Evaluation of efficient transformation method for xylose reductase gene integration in *Pichia pastoris* GS115. Madras Agric J 107(10–12):1

Yan Q, Wang L, Jiang Z, Yang S, Zhu H, Li L (2008) A xylose-tolerant B-xylosidase from *Paecilomyces thermophila*: characterization and its co-action with the endogenous xylanase. Bioresour Technol 99(13):5402–5410

Yeoman CJ, Han Y, Dodd D, Schroeder CM, Mackie RI, Cann IK (2010) Thermostable enzymes as biocatalysts in the biofuel industry. Adv Appl Microbiol 70:1–55

Yoo CG, Meng X, Pu Y, Ragauskas AJ (2020) The critical role of lignin in lignocellulosic biomass conversion and recent pretreatment strategies: a comprehensive review. Bioresour Technol 301: 122784

Zhang YH, Lynd LR (2004) Toward an aggregated understanding of enzymatic hydrolysis of cellulose: noncomplexed cellulase systems. Biotechnol Bioeng 88(7):797–824

Zhang X-Z, Zhang YH (2013) Cellulases: characteristics, sources, production, and applications. In: Bioprocessing technologies in biorefinery for sustainable production of fuels, chemicals, polymers, vol 1, pp 131–146

Zhang C-M, Jiang L, Mao Z-G, Zhang J-H, Tang L (2011a) Effects of propionic acid and pH on ethanol fermentation by *Saccharomyces cerevisiae* in cassava mash. Appl Biochem Biotechnol 165(3):883–891

Zhang J, Siika-Aho M, Tenkanen M, Viikari L (2011b) The role of acetyl Xylan esterase in the solubilization of xylan and enzymatic hydrolysis of wheat straw and giant reed. Biotechnol Biofuels 4(1):1–10

Zhang L, Li X, Yong Q, Yang S-T, Ouyang J, Yu S (2016) Impacts of lignocellulose-derived inhibitors on L-lactic acid fermentation by *Rhizopus oryzae*. Bioresour Technol 203:173–180

Chapter 12
New Trends and Commercial Aspects of Enzymatic Saccharification of Lignocellulosic Biomass

Nathiely Ramírez-Guzmán, Erick M. Peña-Lucio, Orlando de la Rosa, Jorge Angulo-López, Salvador Saldaña-Mendoza, Sandra Pacios, Leidy Johana Valencia-Hernández, Laihsa Rodriguez, and Cristóbal N. Aguilar

Abstract Currently, society is looking for new alternative energy sources, cleaner and less harmful to the environment, and an example of this is the depletion of fossil fuels and the search for biofuels from various renewable materials, which can be classified as first and second generation. The most common in the industry is the first-generation biofuel obtained through edible oils or vegetable sugars, mainly corn and sugar calla, and the second-generation biofuel obtained from the exploitation of residual raw material residues from food industries, forest residues among others. On the other hand, the third-generation biofuels are obtained from non-food species by using molecular biology techniques in which microalgae currently stand out, and finally, in a similar way, the fourth-generation biofuels are manufactured from non-arable land. However, unlike third-generation biofuels, it does not require the destruction of biomass. The relationship between the different types of biofuels is the search for the saccharification process, which is a process in which a polysaccharide is transformed into fermentable sugar. Enzymatic hydrolysis is a type of saccharification in which the process is catalyzed by a group of enzymes generically called cellulases, which are a mixture of different enzymatic activities whose combined action degrades cellulose. During enzymatic hydrolysis, cellulose is degraded by cellulases to reducing sugars that can be fermented by yeast or bacteria to ethanol. This chapter will address the recent developments in the enzymatic saccharification

N. Ramírez-Guzmán (✉)
Center for Interdisciplinary Studies and Research (CEII-UAdeC), Universidad Autónoma de Coahuila, Saltillo, Mexico
e-mail: nathiely.ramirez@uadec.edu.mx

E. M. Peña-Lucio · O. de la Rosa · J. Angulo-López · S. Saldaña-Mendoza · S. Pacios · L. J. Valencia-Hernández · L. Rodriguez · C. N. Aguilar (✉)
Bioprocesses and Bioproducts Research Group (BBG-DIA), Food Research Department, School of Chemistry, Universidad Autónoma de Coahuila, Saltillo, Mexico
e-mail: cristobal.aguilar@uadec.edu.mx

J. K. Saini, R. K. Sani (eds.), *Microbial Biotechnology for Renewable and Sustainable Energy*, Clean Energy Production Technologies,
https://doi.org/10.1007/978-981-16-3852-7_12

(ES) technologies for biofuel production in their current state, challenges, products in the market, and prospects.

Keywords Saccharification · Enzyme · Biofuel · Cellulase · Lignocellulose

12.1 Biofuels

Mainly, the energy demand has been covered by fossil fuels; however, the increase in the population has generated the development of different energy supplies to improve the quality of life (Rodionova et al. 2016). The use of fossil fuels presents some disadvantages such as the generation of pollutants and the emission of greenhouse gases (Beig et al. 2021). Recently, other energy generating alternatives have been investigated, for example, biofuels, which are energy-rich chemicals generated through biological processes or derived from the biomass of living organisms such as microalgae, plants, or bacteria (Rodionova et al. 2016). The production of biofuels has been studied in some species of bacteria such as *Escherichia coli* and *Bacillus subtilis* (Hasunuma et al. 2013). *Saccharomyces cerevisiae* is an another microorganism that has been utilized for the efficient production of ethanol by a fermentative process (Rodionova et al. 2016). It has been found that some species of algae can produce biofuels; *Botryococcus braunii* and *Chlorella protothecoides* contain high amounts of terpenoid hydrocarbons and glyceryl lipid, which can be transformed into shorter hydrocarbons to produce bioethanol, triterpenic hydrocarbons, isobutyraldehyde, and isobutanol (Rodionova et al. 2016).

Biofuels can be used in a wide variety of ways – liquid fuels (long-chain alcohols, bioethanol, biodiesel, and biobutanol) and gaseous products (methane and hydrogen) (Beig et al. 2021). They are composed of ethanol, 1-butanol, isobutanol, isobutene, isoprene, and farnesene. To obtain, it requires the fermentation of sugars derived from biomass such as corn, sugar cane, and vegetable oil (Choi et al. 2020). Biofuels have been classified into three different generations, the first generation which is derived from food biomass (Immethun et al. 2016); the second generation which is derived from non-edible biomasses, or lignocellulosic biomasses (Saladini et al. 2016), and the third generation which is derived from photosynthetic microorganisms as microalgae (Alaswad et al. 2015).

12.1.1 Conventional Biofuel Production

The application of the biomass resource that is produced in vicinity of the site of production of the biofuel has some advantages in its procurement and less cost in transportation (Grisolia et al. 2020). Biofuel production requires some characteristics, for example, easy availability, technical and environmental feasibility, and economic competitivity (Grisolia et al. 2020). Commonly, biofuel production is realized by the biomass from organisms and plants, such as firewood, wood chips, pellets, animal waste, forest and crop residues, and landfill gas. These materials are composed of ethanol, alcohols, triglycerides, fatty acids, lipids, carbohydrates, and cellulose, which are considered as the major biofuels sources (Rodionova et al. 2016). Bioethanol and biomethane are produced through the fermentation of starch or sugars; biodiesel is obtained by the transesterification of oil crops and the hydrogen from microalgae and microbes (Dragone et al. 2010). Lignocellulosic biomass is an important source of sugars for the production of bioethanol. Currently, lignocellulosic biomass from rice straw or cane has been used for the biofuels production; also, some plants with a high content of starch such as maize has been applied. Bioethanol is produced by distillation, hydrolysis, and subsequent fermentation (Dias et al. 2009). Lipids are commonly accumulated in cell biomass which can be converted into multiple products (Dong et al. 2016). Biofuels are also produced by the oleaginous microorganisms; obtention of biological lipids is favoured because direct lipid extraction from the wet cell biomass eliminates the need for costly dehydration (Dong et al. 2016).

12.1.2 Use of Enzymes in the Production of Biofuels

In the production of second-generation biofuels, lignocellulosic materials are employed. These materials come from a wide variety of sources and their composition can vary. In general, lignocellulosic materials are composed of cellulose, hemicellulose, and lignin. The process involves several steps: fermentation, pretreatment, and enzymatic saccharification (ES) (Binod et al. 2019).

The pretreatment is considered to be a very crucial initial step as it needs to be well chosen to ensure the hemicellulose and lignin removal from the biomass to improve the enzyme contact with the matrix (Guo et al. 2018). Different kinds of pretreatments can be applied to the lignocellulosic biomass such as chemical, physical, and biological and their severity can deteriorate the biomass to release polymeric sugars (Guo et al. 2018; Nargotra et al. 2018; Rattanaporn et al. 2018). When choosing the pretreatment method, it should be considered that some pretreatments can release inhibitors for fermentation (Guo et al. 2018).

Oxidation pretreatments have also been reported to be successful for lignocellulosic materials. A recent study (Xiao et al. 2017) evaluated two oxidation pretreatments (Fenton reagent and peroxyacetic acid) for biofuel production employing sugarcane bagasse, *Eichhornia crassipes*, and *Metasequoia glyptostroboides*. Peroxyacetic acid resulted in the improved lignin removal for sugar cane bagasse (reaching carbohydrate content up to 90.63%) and *Metasequoia glyptostroboides* (up to 93.73% of carbohydrates). On the other hand, the Fenton reagent displayed better performance on *Eichhornia crassipes*. Also, the results showed higher porosity and improved surface area for the action of enzymes. This highlights the influence of different pretreatments methods on the different types of materials.

To perform the enzymatic saccharification, the enzymes to be used need to be properly chosen due to the complexity of the biomass. One single enzyme is not sufficient to perform the enzymatic hydrolysis; so a pool of enzymes is often selected to carry out the process. These enzymes are grouped as cellulases, xylanases, peroxidases, and laccases (Binod et al. 2019; Guo et al. 2018; Siqueira et al. 2020).

The cellulases are composed of endoglucanases EC 3.2.1.4 (which randomly hydrolyze internal β-1,4-glucosidic bonds), cellobiohydrolases EC 3.2.1.91 (which can produce cellobiose by hydrolyzing β-1,4-glycosidic linkages at the reducing and non-reducing ends), and β-glucosidases EC 3.2.1.21 (which may act on cellobiose degrading it into glucose) (Binod et al. 2019; Guo et al. 2018; Gupta et al. 2016; Siqueira et al. 2020). Lignocellulosic enzymes are also composed of xylan which can be degraded by xylanases. These englobe endo-1,4-β-xylanase (EC 3.2.1.8), β-xylosidase (EC 3.2.1.37), and α-arabinofuranosidases (EC 3.3.1.55) (these enzymes release xylooligosaccharides that are further degraded to xylobiose and sequentially to xylose) (Biely et al. 2016; Binod et al. 2019; Cui and Zhao 2012; Thomas et al. 2013). Peroxidases such as lignin peroxidase enzyme (LiP, EC 1.11.1.7) and laccases (EC 1.10.3.2) help cellulases in hydrolyzing lignocellulose (Gupta et al. 2016).

12.1.3 Enzymatic Saccharification and its Use for the Production of Biofuels

For efficient saccharification, the process parameters need to be well defined. These parameters involve pH, optimal enzyme concentration, temperature, and time (Bala and Singh 2019; Faizal et al. 2020; Rattanaporn et al. 2018).

A recent study (Narra et al. 2020) employed response surface methodology to optimize the culture conditions for four hydrolytic enzymes from the fungi *Aspergillus tubingensis* M7. Among those, the optimized parameters were incubation time, inoculum size, moisture content, and substrate (g%). The results showed a high saccharification efficiency up to 86.02%.

An optimized saccharification process for the bioethanol production was reported by (Faizal et al. 2020) utilizing four species of duckweeds, *Lemnaaequinoctialis*, *Landoltia punctata*, *Spirodelapolyrrhiza*, and *Wolffia arrhiza*. Best starch conversion to sugar was achieved after 24 h at 50 °C with a 2: 1 (v/v) of α-amylase and amyloglucosidase. Sugar conversion was further carried out obtaining 0.16–0.19 g of ethanol/g of dry biomass.

Enhancement in the enzymatic saccharification yield is highly influenced by the pretreatment; several authors focus on pretreatment to reach more efficiency by their enzymatic methods. (Nargotra et al. 2018) reports an improved enzymatic digestibility (163.42 mg sugars/g biomass) followed by an alkali (NaOH) and ionic liquid 1-butyl-3-methyl imidazolium chloride pretreatment. (Rattanaporn et al. 2018) reports a chemical pretreatment composed of organic acids (acetic acid, oxalic acid, and citric acid) in which enhancement of enzymatic saccharification was observed to be around 2.3 times higher sugar yield compared to the untreated biomass.

12.2 Challenges for Research and Development

Biofuel production from lignocellulosic residues is one of the most common alternatives (Beig et al. 2021). However, there are many challenges in obtaining biofuels from lignocellulosic biomass. Biomass is a highly oxygenated and highly functionalized material, so it is necessary to increase energy density and reduce reactivity when generating a biofuel (Alonso et al. 2012). The conversion of lignocellulosic materials into fermentable sugars as fuel precursors such as ethanol (Lin et al. 2019) has several stages that start with a pretreatment step (Yang et al. 2020; Wang and Lü 2021), followed by enzymatic hydrolysis, and ends with the fermentation of the obtained sugars (Yang et al. 2020; Wang and Lü 2021). The pretreatment applied to biomass represents 20–30% of the total costs associated with biofuel production (Axelsson et al. 2012; Beig et al. 2021). Hydrolysis of hemicellulose and cellulose into reducing sugars is a critical point in the conversion process (Alonso et al. 2012). Thus far, biofuel production from lignocellulosic material has been limited (Lin et al. 2019). This is mainly attributed to the hydrolysis stage, where some barriers are present, which affect the viability of the process. Some of them are the cost associated with the enzymes in charge of hydrolyzing the biomass (Lin et al. 2019), the low hydrolysis efficiency, and the high production costs (Wang et al. 2020). On the other hand, there is a need for low-cost feedstocks that can be effectively digested by hydrolytic enzymes (Lin et al. 2019) with low processing costs(Sandesh and Ujwal 2021).

Hydrolyzing the lignocellulose into monosaccharides remains a technical challenge due to the indigestibility of the cellulose structure (Khaire et al. 2021). Pretreatment of the lignocellulosic biomass is paramount to improve the cellulose accessibility for the enzymes to release fermentable sugars at the hydrolysis stage and reduce the enzyme usage (Marulanda et al. 2019). Several studies have focused on determining the process conditions that allow better yield at a lesser cost.

Pretreatments applied to lignocellulosic biomass before enzymatic hydrolysis that require high energy demands (Beig et al. 2021) can be biological, physical, chemical, and physicochemical (Shafiei et al. 2015; Lin et al. 2019; Houfani et al. 2020). These are applied independently or in combination to improve the enzyme efficiency during saccharification (Jamaldheen et al. 2018). The combination of methods has greater advantages than a single pretreatment method as it favors the monosaccharide production, reduces the inhibitor formation at high concentration, and reduces the effects of extreme pretreatment conditions. The combination of pretreatments results in higher productivity. The most effective pretreatments for lignin and hemicellulose removal are the combination of dilute acid with a steam explosion, alkaline pretreatments, and microwave-assisted alkaline pretreatments (Lu et al. 2009).

Accordingly, the current challenge lies in the development of an efficient pretreatment process that meets the following requirements:

1. To be energy efficient without compromising production.
2. Minimize the loss of compounds and products, particularly sugars.
3. Avoid the application of products that may act as inhibitors to the reactions. Avoid washing or neutralization steps that increase the cost of operation.
4. Synchronize subsequent operations to increase the overall efficiency of the process.
5. Include a subsequent sugar preconcentration step to improve the efficiency of the process.
6. Optimize the fermentation time between 3 and 4 days (Houghton 2006; Axelsson et al. 2012; Beig et al. 2021).

Other challenges for the practice of large-scale enzymatic saccharification are the low enzymatic activity and the high costs (Chen and Fu 2016; Guo et al. 2018).

12.3 Marketing and Products in the Market

The development of biofuels from the renewable sources is an essential issue for the conservation of the planet's fossil resources. Different raw materials have been reported as substrates to produce biofuels. Biomasses such as food crop, non-food lignocellulosic biomass, microalgae, forest, agricultural residues, and agri-food residues (Raud et al. 2019) have gained global importance due to their environmental impact on the ecosystem (Torres-Valenzuela et al. 2020). An important case is that of agro-industrial waste, which could generate up to five billion tons of waste per year globally (Naidu et al. 2018). The use of this wastes in the production of biofuels requires an initial stage of enzymatic hydrolysis or saccharification which is decisive for the viability of the process with a contribution of 25% of the operational costs (Valdivia et al. 2016). The saccharifying enzymes are used as the complex carbohydrate degraders in biofuel production and play an important role in the optimization of the process conditions.

As analyzed by the Business Communication Company (BCC), in 2023, total world-wide industrial enzyme market should reach \$7.0 billion and the estimated compound annual growth rate is 4.9% from 2018 to 2023. By 2021, market studies predict an increase in the production of technical enzymes including those used to produce biofuels which is directly related to the creation of new production processes.

The countries with the highest demand for enzymes are North America, Western Europe, Japan, and Canada. By 2021, it is estimated that the global enzyme market could increase by 6.8–7.9% for the North American and Asia-Pacific regions. Other markets such as Eastern Europe, Middle East, and Africa have also been highlighted.

In 2016, the global technical enzymes market was dominated by Europe, the Middle East, and Africa, accounting for approximately 35%; however, the market may grow further in North America and the Asia-Pacific region in the coming years (Dewan 2014). For the biofuel enzyme market, sales of more than $300 were predicted in the Europe and North America regions by 2020 (BBC Research 2015).

New enzymatic technologies have made it possible to overcome the problems of converting recalcitrant biomasses or lignocellulosic materials. Currently, saccharolytic enzymes are produced in the market by making blends of enzymes (Lange 2017; Lange et al. 2021). Many enzymes are available on the market mainly from Novozymes (Denmark), Danisco/Dupont (US), BASF (Germany), DSM (Netherlands), and Abengoa representing an important segment of total enzyme production (Dewan 2014). Other important companies are Denykem (UK), Megazyme (Ireland), Advanced Enzymes Technologies (India), and MetGen (Finland). Novozymes released an annual report in 2019, estimating that it comprises approximately 48% of the global enzyme market; it also reported that sales during the same year increased moderately with predominant growth in India and a weakening of the market in China and emerging markets. Currently, Novozymes has launched the product Fortiva®, composed of alpha-amylase, to increase ethanol production yields by 1%. It should also be noted that this company has focused its efforts in the production of yeasts to produce the first-generation biofuels, under the name Innova® yeast technology. According to its annual report, sales in the bioenergy sector are expected to grow 1–5%.

Commercially available enzymes for biofuel production from different feedstocks can be grouped into cellulases, amylases, β-glucosidases, xylanases, proteases, lipases, keratinases, laccases, lignin peroxidase, and manganese peroxidase (de Pereira Scarpa et al. 2019). The applications of these enzymes can vary according to the type of fermentation such as solid-state (SSF) or submerged (SmF) fermentation (de Castro and de Castro 2012). Among the saccharifying enzymes that dominate the market are cellulases that are popular due to the wide range of industrial applications; other enzymes such as lipases, catalase, and xylanase are being investigated based on catalytic activity (Chapman et al. 2018). In the market, these enzymes can be prepared as cocktails which contain different enzymes with specific properties and other substances such as secondary metabolites produced by microbial strains (Álvarez et al. 2016). The main option in the biofuel enzymes market is Spirizyme®, portfolio launched by Novozymes (Table 12.1), which contains eight gluco-amylases for saccharification. The most outstanding enzyme is trehalase which has allowed to increase the starch ethanol production yields with reducing fermentation times (Novozyme 2021).

Market studies have shown that the price of the enzyme should stabilize at $0.4/gallon; however, this cost may increase in a commercial presentation. By 2020, the total cost of enzymes for biofuel has been estimated at $1.0 billion (BBC Research 2015; Lopes et al. 2018)). Previous studies report that the enzyme costs higher than 30% in the bioethanol production (Solarte-Toro et al. 2019). Stabilization of the enzyme cocktail costs requires increasing demand and competition. The design of

Table 12.1 Commercially available enzymes in the biofuel market

Trade names	Companies	Enzyme type	Reference
Spirizyme®	Novozymes	Glucoamylases	Novozyme (2021)
Celluclast®, Cellic CTec2 Cellic CTec3	Novozymes	Cellulase	Khare et al. (2015); Scott et al. (2016); Brar et al. (2019)
HTec3®	Novozymes	Cellulase	Sharma et al. (2016)
Termamyl1®	Novozymes	Amylase	Fasim et al. (2021)
AMG1®	Novozymes		Fasim et al. (2021)
Viscozyme L®	Novozymes	Multienzyme	Gama et al. (2015)
Novozyme 188®	Novozymes	Glucosidase	Khare et al. (2015)
BrewZymeLP®	Danisco	β-Glucanase	Sharma et al. (2016)
Boli GA-150®	Boli bioproducts	Glucoamylase	Sharma et al. (2016)
Spezyme®	Genencor	Cellulase	Khare et al. (2015)
Accelarase 1500®	Genencor	Cellulase	Khare et al. (2015)
Optimax L-1000®	Genencor	Pullulanase	Sharma et al. (2016)

new enzymes or preparations from lignocellulosic biomass is necessary to ensure the stability of this market segment (Valdivia et al. 2016).

Genetic engineering has played a crucial role in the design of new enzymes or enzyme preparations. Currently, there is an increasing interest in improving the thermal stability of the enzyme, the temperature being a determining factor in the viability of the plant material. The enzyme production yield and its catalytic efficiency, as well as the reduction of protein production costs, and inhibition of the final product are some of the issues under study (Elleuche et al. 2014; Valdivia et al. 2016).

The production of technical enzymes is affected by the Research & Development (R&D) activities and the environmental policies and legislation in each country. The Paris Climate Agreement encourages the production of fuels from renewable sources to reduce greenhouse gases, a situation that has favoured the demand for saccharifying enzymes (Dewan 2014). However, the cost of technical enzymes remains an important factor in the growth of the market.

12.4 Success and Failure Stories and New Trends in Enzymatic Saccharification Technologies

12.4.1 Success and Failure Stories

The enzymatic saccharification is an efficient and environment friendly process to enhance the reducing sugars from polymeric sugars in the lignocellulosic biomass (Tan et al. 2016; Manisha 2017).

The lignocellulosic material was decomposed to monosaccharides using acid-catalyzed or alkali-catalyzed hydrolysis. Acid hydrolysis can be performed using dilute acid or concentrated acid. Alkaline hydrolysis results in efficient lignin removal and low inhibitor formation, but this technique is expensive and results in alteration of the lignin structure (Saeed and Saleem 2018). However, it causes corrosion of the gas equipment and produces by-products that inhibits further fermentation.

The enzymatic hydrolysis is of great interest because it could overcome the disadvantages of acid and alkali catalyzed hydrolysis. However, there are still some downsides such as slow reaction rate and limited enzymatic accessibility to polysaccharides. Pretreatment is necessary to open the biomass cell wall structure which would increase the enzymatic accessibility during enzymatic hydrolysis (Saeed and Saleem 2018).

The obstacles for carrying out enzymatic saccharification on a larger scale are increased costs and little profitability (Chen and Fu 2016). Genetic engineering has been one of the solution tools for enzyme technology. However, it suffers from various drawbacks such as posttranslational modifications, inclusion bodies, costs, tediouness, time-consumption, and expertise requirement. Immobilization has been the foremost enzyme technology being used due to its simplicity, decreased labor, and cost-efficacy. It leads to physical confinement or localization of enzymes in a specifically defined region of space with retention of their catalytic activities and less sensitivity towards their environment with insistent usability (Dwevedi and Kayastha 2011).

Since several obstacles are encountered in the process, it was first necessary to use extensive pretreatment processes. Since the objective is to fractionate the cellulose, hemicellulose and lignin of the biomass, to later hydrolyze it using selected enzymes at reduced doses. (Zhang et al. 2016). Pretreatment with an organic solvent has been studied. (Li et al. 2016). However, the selection of the solvent should meet several requirements such as low risk to health, production of cellulose for the subsequent phases, and low cost.

The reducing sugars obtained in the saccharification stage will be fermented in the next phase and will be able to produce some biofuel. Separate saccharification and fermentation (SHF) is a standard practice; however, its optimal conditions are generally different (Guo et al. 2018). Therefore, there is strong research interest to seek combined processes to increase general enzymatic saccharification and subsequent fermentation of yeast that include simultaneous saccharification and fermentation (SSF), presaccharification and simultaneous saccharification and fermentation (PSSF), and consolidated bioprocesses (CBP) (Loaces et al. 2017; Hilares et al. 2017).

12.4.2 *New Trends in Enzymatic Saccharification Technologies*

As mentioned above, the enzymatic saccharification processes have been accompanied by the genetic recombination processes of microorganisms to obtain higher enzyme titers and enzyme recovery for reuse in search of an increase in the production yields (Guo et al. 2018). Nowadays, research on the production of biofuels continues to make use of the enzymatic saccharification processes. The main difficulties derive from the crystallinity and degree of polymerization of cellulose, the accessibility to the substrate surface, and mainly from the presence of lignin. The last one prevents the swelling of fibers and produces non-productive adsorption of cellulases (Sheng et al. 2021). Novel pretreatments accompany the enzymatic saccharification processes to contribute the delignification of the lignocellulosic biomass.

The hybrid pretreatment of ultrasound and organic solvents consists of solvents synergistic action with free radicals production. These radicals are produced through the sonochemical effect to exert an attack on the biomass components and reduce the cellulose crystallinity by rearrangement of molecules through mechanoacoustic development (Lee et al. 2020). Research has been carried out on the use of lytic polysaccharide monooxygenases (LPMOs) in conjunction with cellulase enzymes to improve saccharification yields through the oxidation of substrates surface, facilitating the access of hydrolytic enzymes (Velasco et al. 2021).

In recent years, the use of deep eutectic solvents (DES) as a pretreatment in the enzymatic saccharification has been reported. DES has properties similar to the ionic liquids, although they stand out for being simple to synthesize, biodegradable, and have a low cost (Ling et al. 2020). As lignin is an essential component of lignocellulosic materials, its revaluation in saccharification processes could contribute to biofuel production's profitability (Huang et al. 2021). Research has recently been conducted using DES with lignin derivatives. In 2020, the first report of DES prepared with *p*-hydroxybenzoic acid (derived from lignin) and choline chloride for the pretreatment of woody biomass improved the percentage of delignification enzymatic hydrolysis, also achieving a sustainable process by recycling the DES used (Wang et al. 2020). Similarly, Huang et al. (2021) reported that using a DES pretreatment consisting of choline chloride/guaicol (derived from lignin) with traces of $AlCl_3$ contributed significantly to the degradation of hemicellulose and lignin, resulting in complete enzymatic hydrolysis from wheat straw.

Use of the alkaline hydrogen peroxide has been reported to increase the enzymatic digestibility of corn stubble leading to the breaking of the hydrogen bonds of cellulose and hemicellulose and the elimination of lignin, reducing the non-productive adsorption of cellulases to the biomass (Yang et al. 2021). The use of a novel hybrid pretreatment has recently been reported by Tang et al. (2021). They combined an organic surfactant (humic acid) with dilute sulfuric acid, achieving an increase in the percentage of lignin and hemicellulose removal and the ES of wheat straw, reaching a saccharification percentage of 92.9% (Tang et al. 2021). Similarly, a pretreatment effect based on ozonolysis with a subsequent washing with sulfuric acid under mild conditions before enzymatic saccharification has been evaluated with good results in cane bagasse (Perrone et al. 2021).

There is currently a trend toward the use of enzymatic saccharification using the macroalgal biomass. The use of macroalgal biomass is because, in contrast to the terrestrial lignocellulosic biomass sources, they do not requireland for agriculture or fertilizers, water, or pesticides. In some cases, they respond better to the thermal pretreatments, increasing the percentage of saccharification compared to that achieved with some terrestrial biomass (Thygesen et al. 2020).

12.5 Conclusions

In this chapter, after reviewing, it was concluded that the need for replacement and overexploitation of fossil fuels is imminent due to the consequences and risks they represent, for which it is very important to continue research and studies of new sources of renewable energies such as biofuels. It is essential to invest time and efforts to improve their production and yields and investigate new technologies that offer greater benefits. Enzymatic saccharification is an option, as we observed in this chapter, is quite interesting and attractive for its use and that has already been used with active products in the market and with very interesting success stories; however, there is still a long way to go, starting with research focused on this issue since there are still complications that have not been studied and it is important to consider better production processes at the industrial level.

References

Alaswad A, Dassisti M, Prescott T, Olabi AG (2015) Technologies and developments of third generation biofuel production. Renew Sust Energ Rev 51:1446–1460. https://doi.org/10.1016/j.rser.2015.07.058

Alonso DM, Wettstein SG, Dumesic JA (2012) Bimetallic catalysts for upgrading of biomass to fuels and chemicals. Chem Soc Rev 41(24):8075–8098. https://doi.org/10.1039/c2cs35188a

Álvarez C, Reyes-Sosa FM, Díez B (2016) Enzymatic hydrolysis of biomass from wood. Microb Biotechnol 9(2):149–156. https://doi.org/10.1111/1751-7915.12346

Axelsson L, Franzén M, Ostwald M, Berndes G, Lakshmi G, Ravindranath NH (2012) Perspective: Jatropha cultivation in southern India: assessing farmers' experiences. Biofuels Bioprod Biorefin, 6(3), 246–256. https://doi.org/https://doi.org/10.1002/bbb.1324

Bala A, Singh B (2019) Development of an environmental-benign process for efficient pretreatment and saccharification of Saccharum biomasses for bioethanol production. Renew Energy 130:12–24

BBC Research (2015) Global markets and technologies for biofuel enzymes—EGY099B. BBC Research, Wellesley. https://www.bccresearch.com/market-research/energy-and-resources/biofuels-enzymes-global-markets-technologies-report-egy099b.html. Accessed 28 Jan 2021

Beig B, Riaz M, Raza S, Hassan M, Zheng Z, Karimi K, Million M (2021) Current challenges and innovative developments in pretreatment of lignocellulosic residues for biofuel production: a review original equipment manufacturer. Fuel 287:119670. https://doi.org/10.1016/j.fuel.2020.119670

Biely P, Singh S, Puchart V (2016) Towards enzymatic breakdown of complex plant xylan structures: state of the art. Biotechnol Adv 34:1260–1274

Binod P, Gnansounou E, Sindhu R, Pandey A (2019) Enzymes for second generation biofuels: recent developments and future perspectives. Bioresour Technol Rep 5:317–325

Brar KK et al (2019) Evaluating novel fungal secretomes for efficient saccharification and fermentation of composite sugars derived from hydrolysate and molasses into ethanol. Bioresour Technol 273:114–121. https://doi.org/10.1016/j.biortech.2018.11.004

Chapman J, Ismail A, Dinu C (2018) Industrial applications of enzymes: recent advances, techniques, and outlooks. Catalysts 8(6):238. https://doi.org/10.3390/catal8060238

Chen H, Fu X (2016) Industrial technologies for bioethanol production from lignocellulosic biomass. Renew Sust Energ Rev 57:468–478. https://doi.org/10.1016/j.rser.2015.12.069

Choi KR, Jiao S, Lee SY (2020) Metabolic engineering strategies toward production of biofuels. Curr Opin Chem Biol 59:1–14. https://doi.org/10.1016/j.cbpa.2020.02.009

Cui F, Zhao L (2012) Optimization of xylanase production from Penicillium sp.WX-Z1 by a two-step statistical strategy: Plackett-Burman and Box-Behnken experimental design. Int J Mol Sci 13:10630–10646

de Castro SM, de Castro AM (2012) Assessment of the Brazilian potential for the production of enzymes for biofuels from agroindustrial materials. Biomass Convers Biorefin 2(1):87–107. https://doi.org/10.1007/s13399-012-0031-9

de Pereira Scarpa JC et al (2019) Saccharification of pretreated sugarcane bagasse using enzymes solution from Pycnoporussanguineus MCA 16 and cellulosic ethanol production. Indus Crops Prod 141:111795. https://doi.org/10.1016/j.indcrop.2019.111795

Dewan SS (2014) Global markets for enzymes in industrial applications. BCC Research, Wellesley

Dias MOS, Ensinas AV, Nebra SA, Filho RM, Rossell CEV, Maciel MRW (2009) Production of bioethanol and other bio-based materials from sugarcane bagasse: integration to conventional bioethanol production process. Chem Eng Res Des 87:120616

Dong T, Knoshaug EP, Pienkos PT, Laurens LML (2016) Lipid recovery from wet oleaginous microbial biomass for biofuel production: a critical review. Appl Energy 177:879–895. https://doi.org/10.1016/j.apenergy.2016.06.002

Dragone G, Fernande B, Vicente AA, Teixeira JA (2010) Third generation biofuels from microalgae. In: Mendez-Vilas A, editor. Current research, technology and education topics in applied microbiology and microbial biotechnology. Formatex, p 135566

Dwevedi A, Kayastha AM (2011) Enzyme immobilization: a breakthrough in enzyme technology and boon to enzyme based industries. In: Protein structure, (April), pp 31–50

Elleuche S et al (2014) Extremozymes-biocatalysts with unique properties from extremophilic microorganisms. Curr Opin Biotechnol 29:116–123. https://doi.org/10.1016/j.copbio.2014.04.003

Faizal A, Sembada AA, Priharto N (2020) Production of bioethanol from four species of duckweeds (Landoltia punctata, Lemnaaequinoctialis, Spirodelapolyrrhiza, and Wolffia arrhiza) through optimization of saccharification process and fermentation with Saccharomyces cerevisiae. Saudi J Biol Sci 28:294–301

Fasim A, More VS, More SS (2021) Large-scale production of enzymes for biotechnology uses. Curr Opin Biotechnol 69:68–76. https://doi.org/10.1016/j.copbio.2020.12.002

Gama R, Van Dyk JS, Pletschke BI (2015) Optimisation of enzymatic hydrolysis of apple pomace for production of biofuel and biorefinery chemicals using commercial enzymes. 3 Biotech 5(6):1075–1087. https://doi.org/10.1007/s13205-015-0312-7

Grisolia G, Fino D, Lucia U (2020) Thermodynamic optimisation of the biofuel production based on mutualism. Energy Rep 6:1561–1571. https://doi.org/10.1016/j.egyr.2020.06.014

Guo H, Chang Y, Lee DJ (2018) Enzymatic saccharification of lignocellulosic biorefinery: research focuses. Bioresour Technol 252(December):198–215. https://doi.org/10.1016/j.biortech.2017.12.062

Gupta VK, Kubicek CP, Berrin JG, Wilson DW, Couturier M, Berlin A, Filho EXF, Ezeji T (2016) Fungal enzymes for bio-products from sustainable and waste biomass. Trends Biochem Sci 41: 633–645

Hasunuma T, Okazaki F, Okai N, Hara KY, Ishii J, Kondo A (2013) A review of enzymes and microbes for lignocellulosic biorefinery and the possibility of their application to consolidated bioprocessing technology. Bioresour Technol 135:51322

Hilares RT et al (2017) Ethanol production in a simultaneous saccharification and fermentation process with interconnected reactors employing hydrodynamic cavitation-pretreated sugarcane bagasse as raw material. Bioresour Technol 243:652–659. https://doi.org/10.1016/j.biortech.2017.06.159

Houfani AA, Anders N, Spiess AC, Baldrian P, Benallaoua S (2020) Insights from enzymatic degradation of cellulose and hemicellulose to fermentable sugars—a review. Biomass Bioenergy 134:105481. https://doi.org/10.1016/j.biombioe.2020.105481

Houghton J (2006) Breaking the biological barriers to cellulosic ethanol: a joint research agenda. EERE Publication and Product

Huang C et al. (2021) Facilitating enzymatic hydrolysis with a novel guaiacol-based deep eutectic solvent pretreatment. In: Bioresource technology, p 124696. https://doi.org/10.1016/j.biortech.2021.124696

Immethun CM, Henson WR, Wang X, Nielsen DR, Moon TS (2016) Engineering central metabolism for production of higher alcohol-based biofuels. In: Biotechnology for biofuel production and optimization, pp 1–34. https://doi.org/10.1016/B978-0-444-63475-7.00001-7

Jamaldheen SB, Sharma K, Rani A, Moholkar VS, Goyal A (2018) Comparative analysis of pretreatment methods on Sorghum (Sorghum durra) stalk agrowaste for holocellulose content. Prep Biochem Biotechnol 48(6):457–464. https://doi.org/10.1080/10826068.2018.1466148

Khaire KC, Moholkar VS, Goyal A (2021) Bioconversion of sugarcane tops to bioethanol and other value added products: an overview. Mater Sci Energy Technol 4:54–68. https://doi.org/10.1016/j.mset.2020.12.004

Khare SK, Pandey A, Larroche C (2015) Current perspectives in enzymatic saccharification of lignocellulosic biomass. Biochem Eng J 102:38–44. https://doi.org/10.1016/j.bej.2015.02.033

Lange L (2017) Fungal enzymes and yeasts for conversion of plant biomass to bioenergy and high-value products. In: The fungal kingdom. ASM Press, Washington, DC, pp 1027–1048. https://doi.org/10.1128/9781555819583.ch51

Lange L et al (2021) Developing a sustainable and circular bio-based economy in EU: by partnering across sectors, upscaling and using new knowledge faster, and for the benefit of climate, environment biodiversity, and people business. Front Bioeng Biotechnol 8:1456. https://doi.org/10.3389/fbioe.2020.619066

Lee KM et al (2020) Synergistic ultrasound-assisted organosolv pretreatment of oil palm empty fruit bunches for enhanced enzymatic saccharification: an optimization study using artificial neural networks. Biomass Bioenergy 139(May):105621. https://doi.org/10.1016/j.biombioe.2020.105621

Li MF et al (2016) Recent advances in alcohol and organic acid fractionation of lignocellulosic biomass. Bioresour Technol 200:971–980. https://doi.org/10.1016/j.biortech.2015.10.004

Lin W, Chen D, Yong Q, Huang C, Huang S (2019) Improving enzymatic hydrolysis of acid-pretreated bamboo residues using amphiphilic surfactant derived from dehydroabietic acid. Bioresour Technol 293(August):122055. https://doi.org/10.1016/j.biortech.2019.122055

Ling Z et al (2020) Deconstruction of oriented crystalline cellulose by novel levulinic acid based deep eutectic solvents pretreatment for improved enzymatic accessibility. Bioresour Technol 305(February):123025. https://doi.org/10.1016/j.biortech.2020.123025

Loaces I et al (2017) Ethanol production by Escherichia coli from Arundo donax biomass under SSF, SHF or CBP process configurations and in situ production of a multifunctional glucanase and xylanase. Bioresour Technol 224:307–313. https://doi.org/10.1016/j.biortech.2016.10.075. PMID: 27815044.

Lopes AM, Ferreira Filho EX, Moreira LRS (2018) An update on enzymatic cocktails for lignocellulose breakdown. J Appl Microbiol 125(3):632–645. https://doi.org/10.1111/jam.13923

Lu X, Zhang Y, Angelidaki I (2009) Optimization of H2SO4-catalyzed hydrothermal pretreatment of rapeseed straw for bioconversion to ethanol: focusing on pretreatment at high solids content. Bioresour Technol 100(12):3048–3053

Manisha YSK (2017) Technological advances and applications of hydrolytic enzymes for vaporization of lignocellulosic biomass. Bioresour Technol 245:1727–1739. https://doi.org/10.1016/j.biortech.2017.05.066. Epub 2017 May 15. PMID: 28552567.

Marulanda VA, Gutiérrez CDB, Álzate CAC (2019) Thermochemical, biological, biochemical, and hybrid conversion methods of bio-derived molecules into renewable fuels. In: Advanced bioprocessing for alternative fuels. Woodhead Publishing, pp 59–81

Naidu DS, Hlangothi SP, John MJ (2018) Bio-based products from xylan: a review. Carbohydr Polym 179:28–41. https://doi.org/10.1016/j.carbpol.2017.09.064

Nargotra P, Sharma V, Gupta M, Kour S, Bajaj BK (2018) Application of ionic liquid and alkali pretreatment for enhancing saccharification of sunflower stalk biomass for potential biofuel-ethanol production. Bioresour Technol 267:560–568

Narra M, Rudakiya DM, Macwan K, Patel N (2020) Black liquor: a potential moistening agent for production of cost-effective hydrolytic enzymes by a newly isolated cellulo-xylano fungal strain Aspergillus tubingensis and its role in higher saccharification efficiency. Bioresour Technol 306:123149

Novozyme L (2021) Saccharification [en línea]. https://www.novozymes.com/en/advance-your-business/bioenergy/saccharification. Accessed 28 Jan 2021

Perrone OM et al (2021) Improving cellulosic ethanol production using ozonolysis and acid as a sugarcane biomass pretreatment in mild conditions. Bioresour Technol Rep 13:100628. https://doi.org/10.1016/j.biteb.2021.100628

Rattanaporn K, Tantayotai P, Phusantisampan T, Pornwongthong P, Sriariyanun M (2018) Organic acid pretreatment of oil palm trunk: effect on enzymatic saccharification and ethanol production. Bioprocess Biosyst Eng 41:467–477

Raud M et al (2019) Potentials and challenges in lignocellulosic biofuel production technology. Renew Sustain Energy Rev 111:44–56. https://doi.org/10.1016/j.rser.2019.05.020

Rodionova MV, Poudyal RS, Tiwari I, Voloshin RA (2016) Biofuel production: challenges and opportunities. Int J Hydrog Energy 42(12):8450–8461. https://doi.org/10.1016/j.ijhydene.2016.11.125

Saeed S, Saleem M (2018) Novel pretreatment methods to improve enzymatic saccharification of sugarcane bagasse: a report. Iran J Chem Chem Eng 37(5):225–234

Saladini F, Patrizi N, Pulselli FM, Marchettini N, Bastianoni S (2016) Guidelines for energy evaluation of first, second and third generation biofuels. Renew Sustain Energy Rev 66:221–227. https://doi.org/10.1016/j.rser.2016.07.073

Sandesh K, Ujwal P (2021) Trends and perspectives of liquid biofuel—process and industrial viability. Energy Convers Manag 10:100075. https://doi.org/10.1016/j.ecmx.2020.100075

Scott BR et al (2016) Catalase improves saccharification of lignocellulose by reducing lytic polysaccharide monooxygenase-associated enzyme inactivation. Biotechnol Lett 38(3):425–434. https://doi.org/10.1007/s10529-015-1989-8

Shafiei M, Kumar R, Karimi K (2015) Pretreatment of lignocellulosic biomass lignocellulose-based bioproducts. Springer, Cham, pp 85–154

Sharma A, Parashar D, Satyanarayana T (2016) Acidophilic microbes: biology and applications. Springer, Cham, pp 215–241. https://doi.org/10.1007/978-3-319-13521-2_7

Sheng Y et al (2021) Enzymatic conversion of pretreated lignocellulosic biomass: a review on influence of structural changes of lignin. Bioresour Technol 324:124631. https://doi.org/10.1016/j.biortech.2020.124631

Siqueira JGW, Rodrigues C, de Souza Vandenberghe LP, Woiciechowski AL, Soccol CR (2020) Current advances in on-site cellulase production and application on lignocellulosic biomass conversion to biofuels: a review. Biomass Bioenergy 132:105419

Solarte-Toro JC et al (2019) Acid pretreatment of lignocellulosic biomass for energy vectors production: a review focused on operational conditions and techno-economic assessment for bioethanol production. Renew Sustain Energy Rev 107:587–601. https://doi.org/10.1016/j.rser.2019.02.024

Tan HT et al (2016) Emerging technologies for the production of renewable liquid transport fuels from biomass sources enriched in plant cell walls. Front Plant Sci 7:1854. https://doi.org/10.3389/fpls.2016.01854. PMCID: PMC5161040. PMID: 28018390.

Tang W et al (2021) Natural surfactant-aided dilute sulfuric acid pretreatment of waste wheat straw to enhance enzymatic hydrolysis efficiency. Bioresour Technol 324:124651. https://doi.org/10.1016/j.biortech.2020.124651

Thomas L, Larroche C, Pandey A (2013) Current developments in solid-state fermentation. Biochem Eng J 81:146–161

Thygesen A et al (2020) Microstructural and carbohydrate compositional changes induced by enzymatic saccharification of green seaweed from West Africa. Algal Res 47:101894. https://doi.org/10.1016/j.algal.2020.101894

Torres-Valenzuela LS, Ballesteros-Gómez A, Rubio S (2020) Green solvents for the extraction of high added-value compounds from agri-food waste. Food Eng Rev 12:1–18. https://doi.org/10.1007/s12393-019-09206-y

Valdivia M et al (2016) Biofuels 2020: biorefineries based on lignocellulosic materials. Microb Biotechnol 9(5):585–594. https://doi.org/10.1111/1751-7915.12387

Velasco J et al (2021) Comparative analysis of two recombinant LPMOs from Aspergillus fumigatus and their effects on sugarcane bagasse saccharification. Enzyme Microb Technol 144:109746. https://doi.org/10.1016/j.enzmictec.2021.109746

Tao Wang, Xin Lü. (2021). Chapter 8: Advances in 2nd generation of bioethanol production. In: Overcome saccharification barrier: advances in hydrolysis technology. Woodhead Publishing, pp 137–159. https://doi.org/10.1016/B978-0-12-818862-0.00005-4

Wang Y et al (2020) Investigation of a lignin-based deep eutectic solvent using p -Hydroxybenzoic acid for efficient Woody biomass conversion. ACS Sustain Chem Eng 8(33):12542–12553. https://doi.org/10.1021/acssuschemeng.0c03533

Xiao K, Yuan Y, Xiao N, Zhou W, Yang Q, Feng W, Zhu D (2017) Enzymatic saccharification responses of Eichhornia crassipes, sugarcane bagasse and Metasequoia glyptostroboides to two oxidation pretreatments for biofuel production. Ind Crop Prod 107:22–29

Yang H, Jin Y, Shi Z, Wang D, Zhao P, Yang J (2020) Effect of hydrothermal pretreated bamboo lignin on cellulose saccharification for bioethanol production. Ind Crop Prod 156:112865. https://doi.org/10.1016/j.indcrop.2020.112865

Yang L et al (2021) Features correlated to improved enzymatic digestibility of corn stover subjected to alkaline hydrogen peroxide pretreatment. Bioresour Technol 325:124688. https://doi.org/10.1016/j.biortech.2021.124688

Zhang ZY et al (2016) Organosolv pretreatment of plant biomass for enhanced enzymatic saccharification. Green Chem 18:360–381. https://doi.org/10.1039/C5GC02034D

Chapter 13
Yeasts for Single Cell Oil Production from Non-conventional Bioresources

Sagia Sajish, Surender Singh, and Lata Nain

Abstract Oleaginous microorganisms accumulating more than 20% of their dry weight biomass as lipids are used for the production of microbial lipid, also called as single cell oil (SCO). SCO from oleaginous yeasts with a fatty acid profile comparable to that of vegetable oil can be a potential feedstock for biodiesel production. Biodiesel is a renewable biofuel, alternative to petroleum fuels. Due to increasing energy demand and depletion of existing fossil fuel reserves, intensive research has been focused on sustainable biodiesel production. Oleaginous yeasts are more advantageous compared to other oleaginous microorganisms because of their fast duplication rate, shorter life cycle, easier to scale up, and amenability to genetic modifications. Production of microbial lipid with oleaginous yeasts from nonedible and abundant lignocellulosic biomass has been viewed as a novel potential technology to fulfill the increasing energy demand. But lignocellulosic biomass being recalcitrant requires pretreatment step and hydrolysis for the conversion of complex polymers into their respective monomers like glucose that can be assimilated into lipids by oleaginous yeasts. These pretreatment methods also generate various degradation products that inhibit enzyme hydrolysis and subsequent fermentation. Understanding the mechanism of lipid accumulation, improvement of strains for high lipid yield from lignocellulosic hydrolysate is necessary for sustainable biodiesel production. In this chapter, we discuss the importance of lignocellulosic biomass as a raw material for sustainable single cell oil production from oleaginous yeasts.

Keywords Single cell oil (SCO) · Bioresources · Oleaginous yeasts · Enzyme hydrolysis · Lignocellulose

---

S. Sajish · L. Nain
Division of Microbiology, ICAR-Indian Agricultural Research Institute, New Delhi, India

S. Singh (✉)
Division of Microbiology, ICAR-Indian Agricultural Research Institute, New Delhi, India

Department of Microbiology, Central University of Haryana, Mahendargarh, India

J. K. Saini, R. K. Sani (eds.), *Microbial Biotechnology for Renewable and Sustainable Energy*, Clean Energy Production Technologies,
https://doi.org/10.1007/978-981-16-3852-7_13

13.1 Introduction

The growing energy demand accompanied with limited reserves of fossil fuels and global environmental degradation associated with the use of fossil fuels propelled the worldwide attention in alternative clean and renewable energy sources. In this respect, biofuels produced from renewable resources are of utmost importance as these are produced directly or indirectly from organic material including plant and animal wastes. Thus, biofuels can serve as a feasible alternate to fossil fuels for easing the world energy crisis and also for mitigating the greenhouse gases emission. The two most common types of biofuels are biodiesel and bioethanol. Biodiesel is a type of biofuel made from methyl esters of fatty acids that are derived from renewable resources. Burning of biodiesel results in lesser emission of carbon monoxide, hydrocarbons, sulfate oxides, and further toxic compounds than that after burning fossil fuels (Lotero et al. 2006).

Vegetable oils like soybean oil, palm oil, and rapeseed oil are used mainly as the triglycerides feedstocks for biodiesel production. Such biodiesel made from vegetable oils as prominent feedstock is termed as first-generation biodiesel. However, due to competition in the food chain leading to the "food vs fuel" controversy, there is a necessity to look for other non-edible eco-friendly renewable oil sources. This has given rise to second-generation biodiesel that is derived from non-edible oil resources like jatropha, jojoba, animal fats, and grease as well as waste oils from cooking. But these non-edible oils are not available in abundance to meet the global needs for biofuel generation. Further, biodiesel from animal fats do not perform well in cold weather. Moreover, using vegetable oils and animal fats as substrate covers 70–85% of the total cost of production and thus unsuitable to substitute the fossil fuels. Microbial sources can be used for biolipid production throughout the year, unlike plants. The above-mentioned limitations in first- and second-generation biodiesel has led to the growth of third-generation biodiesel from microbial resources (oleaginous microbes); the so-called single cell oil (SCO) seems to be an attractive substitute for the plant, animal, and crude oil feedstock for biodiesel. The term single cell oil is used analogously to single cell protein to represent oils of microbial origin. The composition of microbial oil closely resembles that of vegetable oils (Li et al. 2008) and thus makes it appropriate as biodiesel feedstock (Karatay and Dönmez 2010). Oleaginous microorganisms are those species that accumulate more than 20% of their biomass as lipids. These classes of microorganisms include bacteria, algae, fungi, and yeasts that utilize an organic carbon source to synthesize lipids in their intracellular compartment. Oleaginous microorganisms have better productivity than oil-producing crops, with higher lipid yield, lack of any seasonal and climatic changes, less labor intensive, easier to scale up, and amenable to genetic modification.

The use of synthetic media makes the microbial oil economically uncompetitive; therefore, the use of low-cost carbon substrates for microbial lipid synthesis will be of great significance. Among the various low-cost substrates, non-edible biomass like lignocellulosic biomass seems to be a suitable option because of its availability and low cost. Lignocellulose is a recalcitrant biopolymer and is composed of several classes of polymers including cellulose, hemicellulose, pectin, and lignin. The recalcitrance of lignocellulosic biomass makes its hydrolysis into desired monomers ineffective. Therefore, pretreatment is needed to degrade the crystalline structure and to separate these polymers, thus making each of the polymers available for enzymatic hydrolysis. Several methods including physical, physicochemical, and biological pretreatments have been developed which are suitable for different types of plant materials (Saritha et al. 2012). However, the pretreatment process leads to the formation of some inhibitors like neutral and acidic phenolics, hydroxymethyl furfural, furfural, and acetic acid necessitating detoxification and also the selection of microbial strains that can tolerate such inhibitors during fermentation (Almeida et al. 2009). The lignocellulosic material will be saccharified after pretreatment using microbial cellulases to release monosaccharides and oligo-saccharides. In nature, hydrolytic enzymes are secreted by microbes such as bacteria and fungi. At the industrial scale, commercially available cellulase cocktails comprising cellulases, hemicellulases, and pectinases are utilized for the deconstruction of lignocellulosic biomass. Following saccharification, a mixture of monosaccharides like glucose and xylose is obtained, out of which glucose is fermented to bioethanol by yeast leaving behind the xylose part. Therefore, oleaginous yeasts that can co-metabolize both glucose and xylose available in saccharification hydrolysates will be better suited for single cell oil production.

13.2 Oleaginous Yeasts

For thousands of years, yeasts have been used commercially for several biotechnological applications including the production of recombinant proteins. Recent years of research in yeasts are dedicated to the sustainable and renewable production of fuels and value-added chemicals. *Torula pulcherrima* was found with the ability to accumulate intracellular fat by Lindner in 1899 (Woodbine 1959). In 1915, Lindner observed that lipid accumulation in *Endomycopsis vernalis* (currently, *Guehomyces pullulans*) occurs in the medium under nitrogen-limited conditions and used the term "Fetthefe" (yeast fat in German) for the lipids accumulated by oleaginous yeasts. Sulfite waste liquor as carbon source was used for industrial fat production by *Endomycopsis vernalis* (Lundin 1950). Lipids stored in intracellular lipid bodies in oleaginous yeasts are mainly of diacyl and triacylglycerols (TAGs). Fatty acid composition of lipids accumulated (palmitic acid, stearic acid, myristic acid, oleic acid, linolenic acid, and linoleic acid) makes oleaginous yeasts, the most preferred microorganism for the production of triglyceride feedstock (Sagia et al. 2020; Wang et al. 2019; Fakas et al. 2009).

Oleaginous yeasts can accumulate lipids in the range of 40–70% of their dry weight with the capability to grow on a multitude of carbon sources (glucose, xylose, glycerol, arabinose, mannose, etc.). The typical oleaginous yeast genera include *Rhodosporidium, Rhodotorula, Candida, Lipomyces, Trichosporon, Yarrowia*, and

Cryptococcus. Oleaginous yeasts for SCO production possess advantages over other microorganisms like filamentous fungi and microalgae viz. shorter duplication period, higher growth rates, higher lipid content, easier scale-up, no requirement of light, and better control of bacterial contamination due to lower pH requirements. Lipid accumulation in oleaginous yeasts is categorized into two types of mechanism—de-novo and ex-novo lipid accumulation. De-novo lipid accumulation occurs with hydrophilic substrates under nitrogen limited conditions, whereas ex-novo lipid accumulation occurs when hydrophobic resources are used as a substrate.

The composition and proportion of fatty acids in the single cell oil (SCO) vary depending on the type of cultivation process and the substrate (Tanimura et al. 2014). Yeasts can utilize several types of carbon source for biomass and lipid production including glucose, xylose, cellobiose (Yu et al. 2014b), acetate (Gong et al. 2015), molasses (Karatay and Dönmez 2010), glycerol (Polburee et al. 2015), hydrolysate of cassava starch (Wang et al. 2012), industrial and municipal organic wastes (Zhou et al. 2013), and lignocellulose hydrolysates such as rice straw (Huang et al. 2009), corncob (Gao et al. 2014), sugarcane bagasse (Huang et al. 2012), wheat straw (Yu et al. 2011), and fruit pulp (Patel et al. 2015). Accumulation of lipid in oleaginous yeasts occurs under the limitation of nitrogen or other nutrient sources except for carbon (Zhao et al. 2008; Wu et al. 2010), and the lipid accumulation is found to be optimal at molar C:N ratio of 65–100 (Calvey et al. 2016). Hydrolytic properties in addition to lipogenic properties possessed by certain yeasts prove to be advantageous in using low-cost substrates for oil production. *Trichosporon asahii* was reported with endoglucanase (CMCase) and β-glucosidase activity of about 0.11 IU/mL and 0.55 IU/mL, respectively. Lipase activity of about 50 IU/mL and 64% w/w lipid production with soap stock of pomace olive oil refining was reported in *Yarrowia lipolytica* (Ayadi et al. 2018).

A pilot-scale study was undertaken for biodiesel production with single cell oil from *Rhodosporidium toruloides* with sugarcane juice as the carbon source (Soccol et al. 2017). Lipid productivity of about 0.44 g/L/h was obtained in the study. Diesel engine test with the obtained lipids showed 220% reduction in CO_2 emission, sevenfold reduction in CO emission, and 50% reduction in NO_x emission when compared with first-generation biodiesel from soybean oil. Although the main storage form of yeast lipids is triacylglycerol (TAG), they also contain a relative amount of C_{16} and C_{18} fatty acids. Apart from being a potential feedstock for biodiesel, lipids from yeasts can also be used as cocoa butter substitute (CBS). Cocoa butter is chiefly composed of three types of triacylglycerols—1,3-dipalmitil-2-oleoil glycerol (POP) ($C_{16:0}$–$C_{18:1}$–$C_{16:0}$), 1(3)-palmitil-3(1)-estearil-2- glycerol (POS) ($C_{16:0}$–$C_{18:1}$–$C_{18:0}$), and 1,3-diestearil-2-oleoil glycerol (SOS) ($C_{18:0}$–$C_{18:1}$–$C_{18:0}$) (Tanimura et al. 2014). *Trichosporon oleaginosus* was reported to produce 28% POP and POS with a total TAG of about 0.3 g/g dry cell weight (Dionisi et al. 2004; Wei et al. 2017).

13.3 Biochemistry of Lipid Accumulation in Oleaginous Yeasts

De-novo accumulation of lipid in oleaginous yeasts occurs through quasi-inverted β-oxidation process with acetyl Co-A from intermediate cellular metabolism as the basic unit. The fatty acids being formed will be esterified with glycerol forming structural and storage lipids (TAG-triacylglycerol). Glucose and xylose are the abundant simple sugar compounds found in the lignocellulosic biomass. Glucose metabolism and xylose metabolism yield about 1.1 and 1.2 moles of acetyl Co-A per 100 g of glucose (~0.56 moles) and 100 g of xylose (0.66 moles), respectively. If all acetyl Co-A produced from glucose and xylose metabolism is channelized to lipid biosynthesis, theoretical lipid yield will be 0.32 g g^{-1} and 0.34 g g^{-1} for glucose and xylose, respectively (Ratledge 1988).

Pyruvic acid, the net product of glycolysis, will be decarboxylated by pyruvate dehydrogenase to acetyl Co-A. This acetyl Co-A either enters the Krebs cycle or into the pathway for lipid biosynthesis. However, in oleaginous microorganisms, acetyl Co-A for lipid accumulation comes from the TCA cycle intermediate, citric acid. Under the limitations of nitrogen in the culture media, AMP deaminase (adenosine monophosphate deaminase) of oleaginous yeasts converts AMP into IMP (Inosine monophosphate) and $NH4^+$. This $NH4^+$ serves as an intracellular nitrogen source for cell material synthesis under nitrogen exhaustion conditions. The events result in a decrease in the concentration of intracellular AMP which in turn alters the TCA cycle. Isocitrate dehydrogenase (responsible for isocitrate to α-ketoglutarate transformation) which is allosterically activated by AMP loses its activity. Loss in the activity of isocitrate dehydrogenase leads to the accumulation of isocitric acid inside the mitochondria (at concentration equilibrium with citric acid). Citric acid enters the cytoplasm in exchange with malic acid when the citric acid concentration inside the mitochondria reaches the critical value. Citric acid will be broken down into oxaloacetate and acetyl Co-A by ATP-citrate lyase (ATP-CL), a key enzyme of lipid bio-synthesis in oleaginous microorganisms. This acetyl Co-A will be used for fatty acids synthesis by the quasi-inverted β-oxidation pathway. ATP-citrate lyase (ATP-CL) is found to be absent in non-oleaginous yeasts. In non-oleaginous yeasts, the citric acid accumulated as a result of nitrogen exhausted is secreted into the extracellular environment or accumulated as intracellular polysaccharides on inhibiting 6-phosphofructokinase (Boulton and Ratledge 1980; Boulton and Ratledge 1981; Wynn et al. 2001).

In ex-novo lipid accumulation, free fatty acids which are produced by the hydrolysis of the hydrophobic substrates with extracellular lipase are first incorporated inside the microbial cells. These fatty acids may either be dissimilated for cellular growth or subjected to bio-transformations, wherein lipid profile with new fatty acid profile indifferent to the initial hydrophobic substrate will be synthesized. Ex-novo lipid accumulation is a growth-coupled process, i.e., lipid accumulation takes place concurrently with cell growth and is independent of nitrogen limitation in the culture media (Papanikolaou et al. 2001; Papanikolaou et al. 2002) (Fig. 13.1).

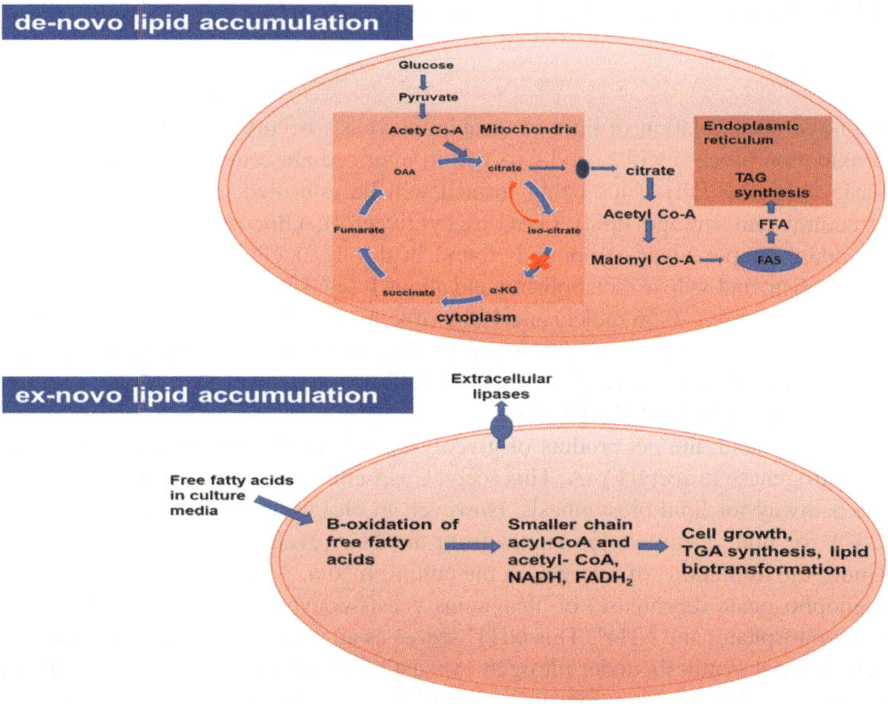

Fig. 13.1 Biochemistry of lipid accumulation in oleaginous yeast

13.4 Low-Cost Substrates for SCO Production

Being a developing technology, the cost of microbial oil production is higher than that of plant oils due to the high cost of culture media. Therefore, exploration and use of low-cost substrates together with efficient oleaginous yeasts to utilize low-cost renewable substrates are essential. The substrates used for lipid accumulation can be categorized into two major groups—hydrophilic and hydrophobic substrates—based on the type of lipid accumulation (de-novo and ex-novo). Various hydrophilic substrates like cane and beet molasses, glycerol, acetate, wastewater, lignocellulosic hydrolysate, cellobiose, brine, starch hydrolysate, propionic acid, and butyric acid have been used for de novo lipid production. Hydrophobic materials like fatty esters, vegetable oils, soap stocks, pure free fatty acids, and fish oils are used for ex-novo lipid production (Qin et al. 2017) (Fig. 13.2). The fatty acid profile varies with the yeast strain and substrate used (Table 13.1).

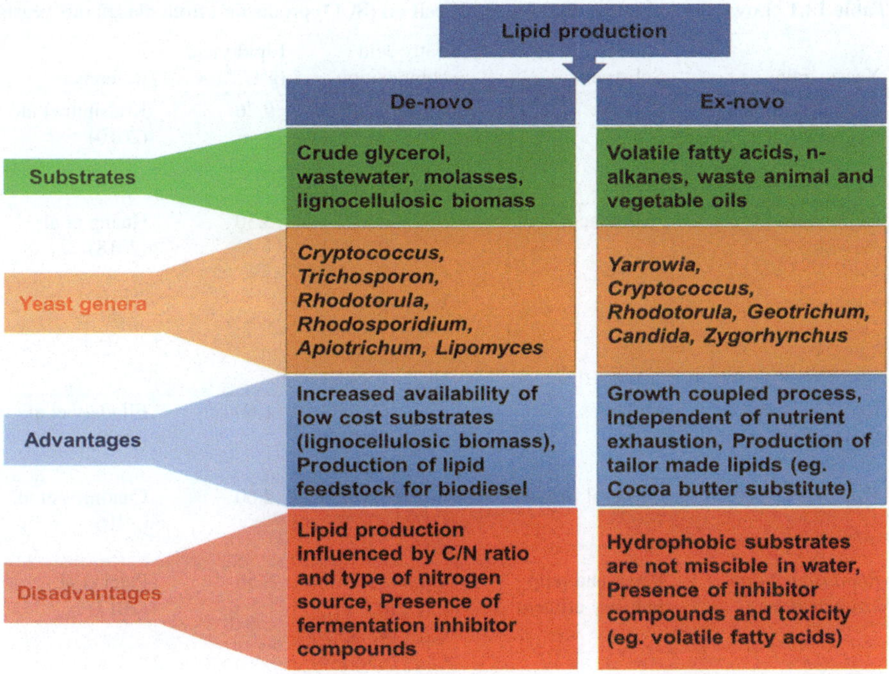

Fig. 13.2 Low-cost substrate used for SCO production from oleaginous yeast

13.4.1 Hydrophilic Substrates

13.4.1.1 Molasses

Molasses is a brown viscous liquid byproduct of sugar manufactured from sugarcane or sugar beet. Major sugars present in the molasses include sucrose, fructose, and glucose and therefore are widely used for industrial fermentation of ethanol, levan, biosurfactant, and lactic acid. Even though oleaginous microorganisms grow well in molasses, lipid accumulation is limited due to the low C/N ratio (12.5) of molasses (Jiru et al. 2018).

13.4.1.2 Crude Glycerol

Crude glycerol (80% glycerol) is a major byproduct of biodiesel production. The manufacture of 10 kg biodiesel generates about 1 kg glycerol as a byproduct (Bauer and Hulteberg 2013). This crude glycerol is treated as a waste since its purification is expensive and cumbersome. A wide variety of oleaginous yeasts were reported to convert glycerol into single cell oil. With a greater degree of reduction, low cost, and less competition with food production, crude glycerol can be a potential carbon

Table 13.1 Low-cost substrates used for single cell oil (SCO) production from oleaginous yeasts

Yeast strain	Substrate	Fatty acid composition	Lipid yield (g L^{-1})	Reference
Rhodosporidium toruloides	Glucose	$C_{16:0}$ (22.49%) $C_{18:0}$ (14.56%) $C_{18:1}$ (41.54%) $C_{18:2}$ (15.12%)	9.26	Kraisintu et al. (2010)
Cryptococcus curvatus	Acetic acid	$C_{16:0}$ (8.38%) $C_{17:0}$ (0.40%) $C_{18:0}$ (29.75%) $C_{18:1}$ (50.37%) $C_{18:2}$ (6.44%) $C_{18:3}$ (1.50%) $C_{20:1}$ (1.53%)	5.30	Huang et al. (2018)
Cryptococcus curvatus	Molasses	$C_{16:0}$ (16.74%) $C_{18:1}$ (22.66%) $C_{18:2}$ (30.68%)	1.60	Elfadaly et al. (2009)
Yarrowia lipolytica	Glycerol	C16:0 (21%) C16:1 (21%) C18:1 (36%)	2.60	Canonico et al. (2016)
Rhodosporidium kratochvilovae	Paper and pulp industry effluent	$C_{16:0}$ (21.86%) $C_{18:0}$ (0.5%) $C_{18:1}$ (45.43%) $C_{18:2}$ (15.91%)	8.56	Patel et al. (2017)
Lipomyces starkeyi	Sewage sludge	$C_{16:0}$ (55.93%) $C_{18:0}$ (13.8%) $C_{18:1}$ (25.89%) $C_{18:3}$ (0.12%)	1.00	Angerbauer et al. (2008)
Rhodosporidium *toruloides*	Bioethanol wastewater	$C_{16:1}$(11.2%) $C_{18:0}$ (16.9%) $C_{18:1}$ (49.9%) $C_{18:2}$ (13.6%)	3.8	Zhou et al. (2013)
Cryptococcus curvatus	Municipal wastewater	–	29.9	Chi et al. (2011)
Rhodotorula glutinis	Wheat straw	$C_{16:1}$(31.4%) $C_{18:0}$ (9.0%) $C_{18:1}$ (31.6%) $C_{18:2}$ (19.3%)	1.4	Mast et al. (2014)
Cryptococcus curvatus	Volatile fatty acids + glucose	$C_{16:0}$ (24%) $C_{18:0}$ (13%) $C_{18:1}$ (33%) $C_{18:2}$ (18%)	14.5	Christophe et al. (2012)

source for SCO production from oleaginous yeasts. The re-use of crude glycerol from biodiesel production for oil production by oleaginous yeasts not only reduces the cost of production but also serves to recycle the waste glycerol. Moreover, crude glycerol also possesses other macro elements like calcium, potassium, and magnesium which support the growth of yeast. Yeast genera, namely *Rhodosporidium, Rhodotorula, Candida, Trichosporonoides, Lipomyces, Yarrowia, Cryptococcus,* and *Schizosaccharomyces*, are reported to utilize crude glycerol for SCO production (Guerfali et al. 2020).

13.4.1.3 Wastewaters

Wastewaters from food processing and agro-industries are difficult and expensive to treat since it has very high concentrations of organic matter. Such wastewaters with abundant organic materials and free sources of nutrients can be used as a carbon source for yeast lipid production. The use of wastewater as a raw material for lipogenesis can also reduce the energy spent on treating wastewaters. Several wastewater sources like industrial wastewaters, olive mill wastewaters, sewage sludge, monosodium glutamate wastewater, butanol wastewaters, and livestock wastewaters have been used for SCO production. Several yeast species have been found capable of producing versatile extracellular enzymes like protease, lipase, and lignin peroxidase for better utilization of nutrients in wastewater (Yang et al. 2013). However, there is a necessity to screen and improve the yeast strains to be capable of lipogenesis in high organic concentrations in wastewater for efficient and cost-effective lipid production. Since oleaginous yeasts can flourish at a low pH, this criterion can be exploited for scaling up of SCO production with unsterilized wastewater as a substrate. Acidic pH can be used to generate a yeast-dominated microflora under non-sterile conditions since it is cumbersome to sterilize large volumes of wastewater.

13.4.1.4 Lignocellulosic Biomass Hydrolysate

Lignocellulosic biomass serves as the most abundant and promising feedstock for future renewable biofuels. Among the various low-cost substrates, non-edible biomass like lignocellulosic biomass seems to be efficient which would be converted into fuel. Almost in most of the developing countries, lignocellulosic biomass is subjected to direct combustion for heat generation, cooking, and waste elimination in agricultural fields. This direct combustion leads to various problems including environmental pollution. Instead, lignocellulosic biomass can be valorized into high-quality products like bioethanol and lipids using microbes (Cherubini and Ulgiati 2010).

13.4.2 Hydrophobic Waste Resources

Hydrophobic wastes like volatile fatty acids, n-alkanes can be used as a feedstock for ex-novo lipid production. This ex-novo lipid accumulation results in the intracellular biomodification of lipid substrates by oleaginous yeast leading to the production of new fatty acid profiles. Therefore, waste fat resources can be upgraded to lipid products of higher value like cocoa butter substitutes (Vajpeyi and Chandran 2015). Lipid yield in oleaginous yeast was found to increase with supplementation of volatile fatty acids with an additional simple carbon source. Volatile fatty acids (VFAs) which are derived from wastes also contain organic acids like propionic acid, isobutyric acid, acetic acid, n-butyric acid, and isovaleric acid (Huang et al. 2016). Conversion of these organic acids into SCO by oleaginous yeasts can therefore help in sustainable waste management (Bialy et al. 2011).

13.5 Lignocellulosic Biomass as a Substrate for SCO Production from Oleaginous Yeasts

13.5.1 Structure, Composition, and Recalcitrance of Lignocellulosic Biomass

Lignocellulose is composed of several polymers including cellulose, hemicellulose, lignin, pectin, starch, and ashes in lesser quantities. Cellulose is the most abundant biopolymer and is made of glucopyranose units (500–1400) linked by β 1–4 glycosidic linkage. Cellobiose is the fundamental repeating unit of cellulose (Robak and Balcerek 2018). Cellulose concentration ranges from 34 to 50% in softwood species, 41–50% in hardwood species, and 15–45% in most of the agricultural crop species, and it increases with maturity of the plant (Monlau et al. 2014). The degree of polymerization and crystallinity plays a critical role in the recalcitrance of lignocellulosic biomass. Lignocellulosic biomass with shorter cellulose chains and fewer hydrogen bonds is easier to hydrolyze than long cellulose chains with more hydrogen bonds. Crystalline cellulose fibers are slower to hydrolyze than amorphous cellulose. Hemicelluloses constitute 20–35% of lignocellulosic biomass and may be homoglycans or heteroglycans (Chandel et al. 2018). Hemicellulose is a branched heteropolysaccharide and composed of hexoses like mannose, glucose, and galactose; uronic acids like glucuronic and galacturonic acids; and pentoses like xylose and arabinose. Xylan is the predominant hemicellulose ranging from 12 to 37% of the lignocellulosic biomass in the case of agricultural residues (Monlau et al. 2014). Acetylation of hemicellulose units limits the cellulose accessibility to enzymatic hydrolysis (Pan et al. 2006).

Lignin is a polymer of aromatic nuclei made up of a single repeating unit or several similar components. It is a very complex heteropolymer with phenyl propane (sinapyl, coumaryl, and coniferyl alcohol) as the basic unit of lignin. Lignin constitutes about 15–35% of the lignocellulosic biomass of wood species and 3.5–30% of grass species (Monlau et al. 2014). Cellulose and lignin content is higher in hardwood and softwood plant species than in agricultural crops. Lignin is insoluble in neutral organic solvents and hot water. Cellulose and hemicellulose remain associated with hydrogen bonds, and lignin is linked covalently to hemicellulose thus forming a lignin-carbohydrate complex (LCC). Lignin thus presents a physiochemical barrier to enzymatic degradation of lignocellulosic biomass by forming close inter-linkage with cellulose and hemicelluloses. Hydrophobic structural characteristics of lignin also adsorb hydrolytic enzymes irreversibly (Tarasov et al. 2018; Valdés et al. 2020). Therefore, an effective pretreatment method is required for the disruption of lignin-cellulose matrix and reducing enzyme adsorption to lignin complex for the subsequent valorization of lignocellulosic biomass to valuable products like biofuels.

13.5.2 Pretreatment of Lignocellulosic Biomass

The recalcitrance of lignocellulosic biomass makes its hydrolysis ineffective. Therefore, pretreatment is needed to degrade the crystalline structure and to separate these polymers, thus making each of the polymers available for enzymatic hydrolysis. Particle size reduction by milling, grinding, and extrusion before pretreatment efficiently deconstructs lignocellulosic biomass with an increased rate of hydrolysis (size threshold depends on the lignocellulosic feedstocks). Various methods of pretreatments like physical (mechanical comminution, extrusion, pyrolysis, and pulsed electric field), chemical (alkali pretreatment, acid pretreatment, ozonolysis, and organosolv process), physiochemical process (steam explosion, ultrasound treatment, CO_2 explosion, liquid hot water treatment, ammonia fiber expansion, oxidative pretreatment, and wet oxidation), thermochemical process, and biological pretreatment (fungal or bacterial) are used. An effective pretreatment should be cost-effective, should remove lignin portion without degrading cellulose and hemicelluloses layers, should produce minimum inhibitory compounds, should be low energy demanding, and should be ecofriendly and safe to use. Chemical pretreatments are the most commonly used technique for lignocellulosic deconstruction.

13.5.2.1 Physical Methods

Mechanical comminution and pyrolysis are the commonly used physical methods of pretreating lignocellulosic biomass. In mechanical comminution, the size and crystallinity of biomass are reduced by the combination of chipping, milling, and grinding (Cadoche and López 1989). In pyrolysis, cellulose is decomposed by exposing the biomass to high temperature (>300 °C) (Kilzer and Broido 1965).

13.5.2.2 Chemical Methods

Acid Pretreatment

Acid pretreatment (with hydrochloric acid, sulfuric acid, phosphoric acid, and acetic acid) solubilizes hemicellulose and reduce cellulose through disruption of hydrogen and covalent bonds in the lignocellulosic complex. This process hydrolyzes hemicellulose especially xylan into its monomers. Acid pretreatment is found to be suitable for the disruption of lignocellulose complex in agriculture residues and hardwood species. Pretreatment with concentrated acids results in effective hydrolysis releasing a high concentration of simple sugars. However, the use of concentrated acids partially degrades hemicellulose into furfural, hydroxymethyl furfural, and other organic acids whose presence in the hydrolysate inhibits the fermentation process (McMillan et al. 1994). Therefore, a two-stage acid pretreatment method has been proposed, wherein the extraction of hemicellulose is done with less concentrated acid in the first phase followed by high concentrated acid in the second phase for cellulose destruction. But using concentrated acids necessitates the use of corrosion-resistant equipment, thus increasing the cost and also possesses safety issues. Pretreatment with dilute acids, on the other hand, is less aggressive, economical, more environmentally friendly, and generates fewer inhibitory compounds (furfural and HMF). However, pretreatment with dilute acid requires a higher temperature than that required for concentrated acids (Solarte-Toro et al. 2019; Singh et al. 2015). The dry dilute acid method has been used as an alternative to the wet acid method (concentrated and dilute acids), wherein both the biomass and product are solid. Dry biomass is impregnated with acid for efficient adsorption. Comparable assimilable sugar yields with reduced amounts of inhibitory compounds can be obtained with the dry dilute acid method (He et al. 2014).

Alkali Method

This method of pretreatment is carried out by the addition of bases like NaOH, KOH, and $Ca(OH)_2$ to the lignocellulosic biomass. In this process, the internal surface area of the biomass is increased by swelling the biomass, reduces the degree of polymerization, crystallinity, and breaks the lignin carbohydrate complex. Ester and ionic bonds interlinking hemicellulose and other components are saponified thereby increasing the porosity of the lignocellulosic complex (Tarkow and Feist 1969). Alkali pretreatment is found to be effective with biomass with low lignin content compared to those with high lignin content (Singh et al. 2015; Xu and Sun 2016). Alkali pretreatment resulted in the highest recovery of cellulose (59.66) and hemicelluloses (28.34) from corn cob (Sharma et al. 2017), 92.5% delignification, and 81.5% w/w cellulose yield in paddy straw (Kobkam et al. 2018). The main advantage is its low cost and mild operation conditions. But the process results in the formation of salts that are difficult to remove, and the reaction time is also longer.

Ozonolysis

Ozonolysis is used for the efficient degradation of lignin at room temperature and pressure. Yield from enzymatic hydrolysis has been shown to increase after pretreating the biomass with ozone. While lignin is efficiently removed, hemicellulose is only partially degraded in this process. Ozonolysis does not lead to the production of any inhibitor compounds during the pretreatment. However, this method requires large quantities of ozone, thus making ozonolysis expensive (Vidal and Molinier 1988).

Oxidative Delignification

In this method, peroxidase enzyme is used to decompose lignin in the presence of hydrogen peroxide. Solubilization of 50% lignin and hemicellulose was achieved with 2% hydrogen peroxide within 8 h at 30 °C. Glucose yield with enzymatic hydrolysis of cellulose after pretreatment was also found to be increased with oxidative delignification (Azzam 1989).

Organosolv Pretreatment

In this method, organic solvent (ethanol, methanol, glycerol, phenol, acetone, formic acid, and acetic acid) is added to the lignocellulosic biomass to separate lignin from cellulose. Solid phase with cellulose, hemicellulose, and liquid fraction with lignin is obtained at the end of pretreatment (Borand and Karaosmanoğlu 2018). Excision of O-aryl bonds of lignin with carbohydrates occurs during solvent treatment resulting in the dissolution of lignin along with organic solvent. This process also generates acetyl compounds which help in the autohydrolysis of hemicellulose and cellulose. Organic solvents can be recovered and reused. The main disadvantage of this process is the generation of several inhibitor compounds (guaiacol, vanillin, vanillic acid, syringaldehyde, syringic acid, and ferulic acid) (Zhao et al. 2009).

13.5.2.3 Physicochemical Method

Steam Explosion Pretreatment

Steam explosion, a physicochemical method, is a very energy efficient method (Conde-Mejía et al. 2012). In this process, the biomass is first exposed to saturated steam at high temperature (162–260 °C) and pressure (5–50 atm) for a short interval of time. The steam expands into the lignocellulose matrix as the pressure is reduced gradually, thus separating the cellulose fibers and thus disrupting the cell wall. This process also generates acetyl compounds that auto-hydrolyze hemicellulose (Grous et al. 1986). This method generates less inhibitory compounds than acid and alkali methods. Steam explosion is less effective in softwood species with less content of acetyl groups. Acid compounds can be used as a catalyst in such cases to improve cell wall deconstruction, but it leads to the generation of fermentation inhibitor compounds (Singh et al. 2015).

Ammonia Fiber Explosion (AFEX)

In this method, the biomass is treated with liquid ammonia (1.2 kg of liquid ammonia for 1 kg of lignocellulosic biomass) at high temperature (90 °C) and for a short period (30 min), after which the pressure is reduced rapidly. The AFEX method is not efficient for biomass with high lignin, and hemicellulose is not solubilized significantly with this method. However, this method does not require a small particle size and does not produce inhibitor compounds (Holtzapple et al. 1991).

CO_2 Explosion

This method works by increasing the hydrolysis rate with carbonic acid generated from carbon dioxide. But the yields of hydrolysis are low when compared to that achieved with ammonia fiber explosion methods and steam explosion. However, pretreatment of lignocellulosic biomass with CO_2 explosion does not result in the formation of fermentation inhibitor compounds (Zheng et al. 1998).

13.5.2.4 Biological Pretreatment

Microorganisms or biocatalysts from microbes are used to open the cell wall matrix in this method of pretreatment. White-rot, brown-rot, and soft-rot fungi and actinomycetes can be used to degrade lignin and hemicellulose components of waste plant biomass. White-rot fungi target lignin, whereas brown-rot fungi degrade cellulose (Hatakka 1983). White-rot fungi have been reported with enzymes degrading lignin, cellulose, and hemicellulose polymers. *Phanerochaete chrysosporium* is the well-studied white-rot fungi with lignin-degrading property. *P. chrysosporium* has been shown to produce lignin-degrading enzymes like manganese-dependent peroxidases and lignin peroxidases (Boominathan and Reddy 1992). Actinomycetes were also investigated for their ability to degrade lignin. Small laccases similar to fungal laccases were shown to affect lignin degradation in *Streptomyces coelicolor*, *Streptomyces lividans*, *Streptomyces viridosporus,* and *Amycolatopsis* sp. (Saritha et al. 2013; Majumdar et al. 2014). The requirement of longer duration, specific growth conditions, and aseptic environment makes it less preferable for industrial-scale operation. A combination of biological methods with a common pretreatment method can be advantageous in the deconstruction of the lignocellulosic cell wall matrix.

13.5.3 Conversion of Biomass into Fermentable Sugars

Since most of the oleaginous yeasts lack cellulolytic activity, the pretreated lignocellulosic biomass should be hydrolyzed before the fermentation process. This process depolymerizes cellulose and hemicellulose into hexose and pentose monomers which can be assimilated by oleaginous yeasts into biolipids/single cell oil. Enzymatic hydrolysis is advantageous at mild process conditions compared to acid and alkaline hydrolysis. Microorganisms like bacteria and fungi are endowed with the abilities to produce hydrolytic enzymes for depolymerizing lignocellulosic biomass. Bacterial genera, *Bacillus, Clostridium, Ruminococcus, Microbispora, Cellulomonas, Erwinia, Bacteriodes, Thermomonospora, Acetovibrio,* and *Streptomyces*, have been reported to produce cellulolytic enzymes. Among the cellulolytic fungi (*Phanerochaete chrysosporium, Aspergillus, Sclerotium rolfsii, Schizophyllum, Trichoderma,* and *Penicillium*), *Trichoderma* has been widely used for the cellulase production (Duff and Murray 1996). Cellulases are a group of enzymes that includes endoglucanase, exoglucanase/cellobiohydrolase, and β-glucosidase. Endoglucanase is active against amorphous regions/less crystalline regions and creates new free chain ends which are then attacked by other enzymes. Exoglucanase attacks crystalline cellulose and generates glucose/cellobiose units. Finally, β-glucosidase hydrolyzes cellobiose to monomer sugars like glucose. Other accessory enzymes attacking hemicellulose, glucuronidase, acetylesterase, β-xylosidase, glucomannanase, galactomannanase, and xylanase act synergistically with cellulase enzymes and convert cellulose-hemicellulose into assailable free sugars (Singh et al. 2015; Fan et al. 2012; Sternberg 1976). The rate of hydrolysis can be improved by increasing the concentration of cellulase. Usually, in laboratory, cellulase is used at the dose of 10 FPU/g cellulose for high monomer (glucose) yield in 48–72 h of reaction time (Gregg and Saddler 1996). Irreversible adsorption of cellulase to cellulose and lignin deactivates the enzyme which can be minimized by the use of surfactants (Tween 20, 80, cationic Q-86 W, antihole 20BS, polyoxyethylene glycol, Emulgen 147, and anionic Neopelex F-25) in enzymatic hydrolysis (Wu and Ju 1998; Park et al. 1992; Ooshima et al. 1986; Helle et al. 1993). Cellobiose and glucose, which are the end products of hydrolysis, inhibit the activity of cellulase. High loading of enzymes, removal of hydrolysis products formed during hydrolysis, and supplement of β-glucosidases during the reaction can minimize the inhibition of cellulase by end products. From the reaction mixture, cellulases can be recovered and reused for the next batch of hydrolysis (Kumar et al. 2017). But the efficiency of hydrolysis decreases gradually with each step of recycling (Ramos et al. 1993).

13.5.4 SCO Production with Oleaginous Yeasts from Lignocellulosic Hydrolysate

Various waste products and lignocellulosic hydrolysates have been used by researchers for single cell oil production from oleaginous yeasts (Table 13.2). SCO was produced with 45% lipid yield from *Endomycopsis vernalis* with sulfite waste liquor as a carbon source by Lindner in 1922. Lignocellulosic hydrolysate

Table 13.2 SCO production by oleaginous yeasts from lignocellulosic biomass

Substrate	Pretreatment strategy	Oleaginous yeast	Lipid yield (g L^{-1})		Reference
Wheat straw	Dilute sulfuric acid	*Yarrowia lipolytica*	Detoxified hydrolysate	Non-detoxified hydrolysate	Tanimura et al. (2014)
			0.30	0.40	
		Rhodotorula glutinis	2.40	3.50	
		Lipomyces starkeyi	3.70	4.50	
		Cryptococcus curvatus	4.20	5.80	
Paper mill sludge	Ultrasonication	*Cryptococcus vishniaccii*	7.80		Deeba et al. (2016)
Corn cob residues	Enzymatic hydrolysis	*Trichosporon cutaneum*	12.30		Gao et al. (2014)
Wheat straw	Acid hydrolysis	*Rhodotorula glutinis*	1.40		Mast et al. (2014)
Sugarcane bagasse	Sulfuric acid	*Trichosporon fermentans*	15.80		Huang et al. (2012)
Paddy straw	Sulfuric acid	*Trichosporon fermentans*	7.70		Huang et al. (2009)
Rice bran	Defatting and acid hydrolysis	*Yarrowia lipolytica*	48.02% of dry cell weight		Tsigie et al. (2012)
Wheat straw	Dilute sulfuric acid	*Rhodotorula mucillaginosa.*	9.70		Enshaeieh et al. (2015)
Corn stover	Alkaline hydrolysis	*Cryptococcus humicola*	15.5		Sitepu et al. (2014)
Sugarcane bagasse	Acid hydrolysis	*Lipomyces starkeyi*	0.14		Xavier et al. (2017)
Waste sweet potato vines	Enzymatic hydrolysis	*Trichosporon fermentens*	9.6		Zhan et al. (2013)
Paddy straw	Alkaline hydrolysis	*Trichosporon mycotoxinivorans*	5.17		Sagia et al. (2020)

contains hexoses like glucose, galactose, and mannose and pentoses like xylose and arabinose. Hexoses can be readily utilized by all microorganisms, but the utilization of pentoses which constitute a significant portion of hydrolysate is essential for the complete valorization of lignocellulosic biomass into single cell oil. Various oleaginous yeast strains have been reported with the ability to use both these hexoses and pentoses as carbon sources for lipid accumulation. Therefore, xylose- and pentose-utilizing oleaginous yeast isolates possess an advantage for the economical production of SCO from lignocellulosic biomass.

High biomass loading during saccharification leads to a higher level of inhibitor compounds having an inhibitory effect on fermenting microbes and subsequent reduction in lipid yield. Greater biomass loading also results in lower sugar yield due to feedback inhibition in enzymatic hydrolysis and subsequent reduction in lipid yield.

Lipid production by oleaginous yeasts from lignocellulosic biomass can be proceeded after pretreatment by three processes—SHLP (separate hydrolysis and lipid production), SSLP (simultaneous saccharification and lipid production), and CBP (consolidated bioprocessing). Separate hydrolysis and lipid production is commonly used, wherein lipid production is carried out in hydrolysate after saccharification by hydrolytic enzymes. The main disadvantage of SHLP is the feedback inhibition of hydrolytic enzymes by the end products. Simultaneous saccharification and lipid production can resolve the problem of feedback inhibition, wherein saccharification and lipid production is carried out simultaneously. However, it requires the use of thermotolerant microorganisms for fermentation since enzymatic hydrolysis requires the optimal temperature of 50 °C, but the optimal temperature for most oleaginous yeasts for fermentation is $\leq$30 °C. SSLP was demonstrated with *Cryptococcus curvatus* at 37 °C and regenerated corn stover as the substrate. The lipid yield of 6 g L^{-1} was obtained with 5% substrate loading after 48 h. Consolidated bioprocessing is extensively used in bioethanol production from lignocellulosic biomass, wherein enzyme production, carbohydrate hydrolysis, and fermentation/lipid production are integrated into one process. Isolation of natural populations of cellulolytic oleaginous yeast strains or genetic engineering of oleaginous yeasts for production of extracellular cellulolytic enzymes will pave the way for inexpensive and rapid lipid production from lignocellulosic biomass (Gong et al. 2013). Consolidated bioprocessing with genetically engineered cellulolytic yeast *Yarrowia lipolytica* was attempted with 12 gL^{-1} cellulose consumption and 14% lipid accumulation (Guo et al. 2018).

Based on the method of pretreatment and hydrolysis, toxic lignocellulosic degradation byproducts may be produced. These include acetic acid, furfural, hydroxymethyl furfural, formic acid, and vanillin. These degradation compounds can inhibit cell growth and subsequent fermentation. Furfural was found to be the most toxic among the other degradation compounds. A decrease in the yeast biomass weight and lipid yield of *Cryptococcus curvatus* by 78.4% and 61% for glucose and 72% and 59.3% for xylose, respectively, was reported in the presence of furfural (1.0 g L^{-1}) (Yu et al. 2014a). The generation of inhibitory compounds necessitates detoxification of hydrolysate which may further increase the cost of the whole production process. The selection of oleaginous yeast with high tolerance to inhibitory compounds or the ones capable of utilizing lignin degradation compounds as a carbon source is therefore a good tactic for lipid production from lignocellulosic hydrolysates. A fed-batch lipid production from *Trichosporon cutaneum* was performed with 4-hydroxybenzaldehyde as the sole carbon substrate. The lipid yield obtained was 0.85 g L^{-1} (0.039 g/g of 4-hydroxybenzaldehyde) (Hu et al. 2018). An inhibitor degradation study was undertaken, wherein the biodegradation of inhibitors was examined by providing each inhibitor as the solitary carbon source. It was found that furfural, hydroxymethyl furfural, 4-hydroxybenzaldehyde, vanillin, and syringaldehyde were converted to corresponding nontoxic acid—furoic acid, HMF acid, 4-hydroxybenzoate, vanillate, and syringate—by *Trichosporon cutaneum*. The enzymes involved in the biodegradation of inhibitors were found to be alcohol dehydrogenases, aldehyde reductases, aldehyde dehydrogenases, salicylaldehyde dehydrogenase, D-lactaldehyde dehydrogenase, aminoadipate-semialdehyde dehydrogenase, betaine aldehyde dehydrogenase, semialdehyde dehydrogenase, alcohol oxidase, vanillyl alcohol oxidase, glucose oxidase, and choline oxidase (Wang et al. 2016). Phenolic aldehyde from lignin was also used as the only carbon source for SCO production by *Trichosporon cutaneum*. Resistance to inhibitors was demonstrated in *Rhodosporidium toruloides*. The study conducted showed that the presence of inhibitory compounds does not have a profound effect in the distribution of major fatty acids of *Rhodosporidium toruloides*—palmitic acid (C16:0), stearic acid (C18:0), and oleic acid (C18:1) (Hu et al. 2009). High recalcitrance of lignin makes it difficult to valorize it into valuable bioproducts. However, aromatic metabolic pathways for metabolizing lignin-related aromatic compounds have been found in certain oleaginous yeast species. Products of this aromatic metabolism include acetyl Co-A which is a precursor for fatty acid synthesis thus leading to the conversion of lignin-related aromatic compounds into lipids (Yaguchi et al. 2020).

13.6 Genetic Engineering for Enhanced SCO Production from Oleaginous Yeasts

Under the conditions of excess carbon and limited nitrogen conditions, non-oleaginous microbes accumulate excess carbon as polysaccharides like glycogen, whereas oleaginous yeast accumulates excess carbon as intracellular lipids.

Two different approaches are generally used for the improvement of the wild microbial strains for enhanced single cell oil production. The first is to improve the metabolic pathways of lipid biosynthesis in oleaginous yeasts, and the second is to recombine fatty acid synthesis genes from oleaginous yeasts into non-oleaginous yeasts or other microbes (*Saccharomyces cerevisiae* and *Escherichia coli*). Approaches used for metabolic engineering to enhance the lipid yield include overexpressing the enzymes involved in fatty acid and TAG (triacylglycerol) biosynthesis pathway, regulation of enzymes related to TAG biosynthesis, and inhibition of lipid catabolism (Fig. 13.3). Key lipid biosynthesis genes identified are ATP-citrate lyase (ACL), acetyl Co-A carboxylase (ACC), diacylglycerol acetyl transferase (DGAT), glycerol 3-phosphate dehydrogenase, glycerol 3-phosphate acyl transferase (GPAT), and acetyl Co-A synthetase (ACS) (Liang and Jiang

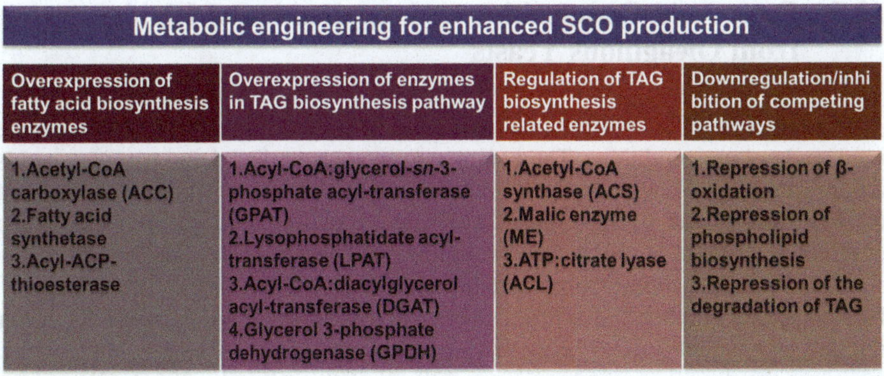

Fig. 13.3 Metabolic engineering strategies for enhancing the lipid yield from oleaginous microorganisms

2013). A two-fold increase in lipid content in *Yarrowia lipolytica* was achieved by the overexpression of the key enzyme ACC 1 (acetyl Co-A carboxylase). An increase in lipid content by 41% in *Yarrowia lipolytica* was achieved by simultaneous co-expression of ACC 1 and DGA 1 (acetyl Co-A carboxylase and diacylglycerol acetyl transferase) by combining both the genes in a gene construct (Tai and Stephanopoulos 2013). A three-fold increase in lipid yield has been achieved by redirecting the carbon flux towards TAG biosynthesis by deletion of GUT 1 (glycerol 3-phosphate dehydrogenase) (Beopoulos et al. 2008). Ester synthesizing genes can be introduced into non-oleaginous yeasts for fatty acid esters production from carbohydrates. This approach results in the direct biodiesel production from raw material rather than lipid production (Kalscheuer et al. 2006; Schmidt-Dannert and Holtzapple 2011). Enzymes not directly involved in lipid biosynthesis like malic enzyme and ATP:citrate lyase (ACL) also influence the lipid yield. Malic enzyme supplies NADH for fatty acid synthase (FAS) and desaturases. ACL catalyzes the citrate to acetyl-CoA conversion and is the key enzyme in oleaginous microorganisms. Besides genetic modification approach for high lipid yield, engineering for other desirable characteristics like simultaneous/co-utilization of various sugars (like glucose and xylose), lipid production at high temperature thus facilitating for simultaneous saccharification and fermentation after pretreatment of lignocellulosic biomass, and resistance to inhibitor compounds generated from various pretreatment approaches of lignocellulosic biomass could lead to economical and sustainable single cell oil production.

13.7 Fatty Acid Composition and Application of Lipids from Oleaginous Yeasts

The fatty acid profile of oleaginous yeast varies with the species, strain, substrate used, and culture conditions. The common fatty acids in diacyl and triacylglycerols accumulated by oleaginous yeast includes myristic acid (C14:0), palmitic acid (C16:0), stearic acid (C18:0), oleic acid (C18:1), and linoleic acid (C18:2) (Sagia et al. 2020; Patel et al. 2014; Patel et al. 2015). The fatty acid composition of the single cell oil determines the potentiality of biodiesel to be used in diesel engines. Higher octane number and a lower degree of unsaturation are essential for better ignition and stability. EU (European Union) has set a limit for iodine value, which measures the degree of unsaturation as 120 g I_2/100 g (Knothe 2006). Oleaginous yeasts like *Trichosporon fermentans and Rhodotorula glutinis* have been reported to produce single cell oil from lignocellulosic hydrolysate with iodine values within the threshold limit (Hoekman et al. 2012). High saturated fatty acids increase the shelf life of biodiesel, whereas unsaturated fatty acids determine the cold flow plugging property of biodiesel. An optimum ratio of saturated to unsaturated fatty acids is necessary for kinematic viscosity and oxidative stability of biodiesel. The cold flow plugging property (CFPP) determines the low-temperature operability of the biodiesel. CFPP is defined as the lowest temperature at which biodiesel (20 ml) flows through a wire mesh screen in 60 seconds under vacuum. Biodiesel solidifies and blocks the engine once the CFPP is reached. Several oleaginous yeasts have been reported with CFPP within the threshold limits of CFPP set by EU ($\leq$5/$\leq$−20). Oxidative stability of the biodiesel increases its shelf life. Oxidative stability is inversely proportional to the number of double bonds in the cis configuration. Linolenic acid (18:3) is highly prone to auto-oxidation, and therefore, a limit of 12% linolenic acid is set by EU (Knothe 2006; Patel et al. 2016). Oleaginous yeasts like *Yarrowia lipolytica, Trichosporon cutaneum, Rhodosporidium toruloidies,* and *Lipomyces starkeyi* have been shown to produce SCO from lignocellulosic hydrolysates with physical properties suitable for biodiesel application in diesel engines (Patel et al. 2016).

13.8 Possibilities for Improved Profitability from SCO Production

Oleaginous yeast can be co-cultured with microalgae for enhanced biomass and lipid yield. Microalgae provide oxygen for heterotrophic yeast, and yeast supplies carbon dioxide for autotrophic microalgae thus minimizing the requirement for mechanical aeration. Enhanced lipid yield and biomass were achieved with the synergistic association of microalgae (*Chlorella vulgaris*) with yeast (*Rhodotorula glutinis*) compared to pure cultures (Zhang et al. 2014).

In addition to lipid production, oleaginous yeast can also be used for the synthesis of other value-added chemicals like β-carotene, torularhodin, and torulene. Carotenoids exhibiting provitamin A was shown to be produced by the oleaginous yeast *Rhodotorula glutinis* (Chaturvedi et al. 2018). Oleaginous yeast is also reported to produce enzymes like phenylalanine ammonia lyase (for aspartame production), invertase, α-L-arabinofuranosidase, tannase, and pectinase (Cui et al. 2015; Kot et al. 2016; Taskin 2013).

13.9 Conclusion

Biodiesel has been currently produced with high-cost vegetable oils. Lipids from oleaginous yeast have been confirmed as a potential alternate feedstock to vegetable oil for biodiesel production. Therefore, the use of oleaginous yeasts for microbial lipid production is a promising way for biodiesel production and to combat the energy crisis. Abundant and nonedible lignocellulosic biomass can be used as a low-cost raw material for single cell oil production, thus making biodiesel production sustainable, economical, and renewable. However, the recalcitrance of lignocellulosic biomass necessitates pretreating the biomass which results in the production of various inhibitor compounds. Selection and improvement of oleaginous yeast isolates with high lipid yield, biomass yield, osmotolerance, inhibitor resistance, SCO production at high temperature, and low pH make the process highly advantageous over vegetable oil production. Optimization, scale-up, and technological advancements in the sustainable conversion of lignocellulosic biomass into SCO by oleaginous yeasts can help to meet the increasing energy demand by the increasing population.

References

Almeida JRM, Bertilsson M, Gorwa-Grauslund MF, Gorsich S, Lidén G (2009) Metabolic effects of furaldehydes and impacts on biotechnological processes. Appl Microbiol Biotechnol 82 (4):625–638. https://doi.org/10.1007/s00253-009-1875-1

Angerbauer C, Siebenhofer M, Mittelbach M, Guebitz GM (2008) Conversion of sewage sludge into lipids by *Lipomyces starkeyi* for biodiesel production. Bioresour Technol 99(8):3051–3056. https://doi.org/10.1016/j.biortech.2007.06.045

Ayadi I, Belghith H, Gargouri A, Guerfali M (2018) Screening of new oleaginous yeasts for single cell oil production, hydrolytic potential exploitation and agro-industrial by-products valorization. Process Saf Environ Prot 119:104–114. https://doi.org/10.1016/j.psep.2018.07.012

Azzam AM (1989) Pretreatment of cane bagasse with alkaline hydrogen peroxide for enzymatic hydrolysis of cellulose and ethanol fermentation. J Environ Sci Health Part B 24(4):421–433. https://doi.org/10.1080/03601238909372658

Bauer F, Hulteberg C (2013) Is there a future in glycerol as a feedstock in the production of biofuels and biochemicals? Biofuels Bioprod Biorefin 7(1):43–51. https://doi.org/10.1002/bbb.1370

Beopoulos A, Mrozova Z, Thevenieau F, Le Dall M-T, Hapala I, Papanikolaou S, Chardot T, Nicaud J-M (2008) Control of lipid accumulation in the yeast Yarrowia lipolytica. Appl Environ Microbiol 74(24):7779. https://doi.org/10.1128/aem.01412-08

Bialy HE, Gomaa OM, Azab KS (2011) Conversion of oil waste to valuable fatty acids using oleaginous yeast. World J Microbiol Biotechnol 27(12):2791–2798. https://doi.org/10.1007/s11274-011-0755-x

Boominathan K, Reddy CA (1992) cAMP-mediated differential regulation of lignin peroxidase and manganese-dependent peroxidase production in the white-rot basidiomycete *Phanerochaete chrysosporium*. Proc Natl Acad Sci U S A 89(12):5586. https://doi.org/10.1073/pnas.89.12.5586

Borand MN, Karaosmanoğlu F (2018) Effects of organosolv pretreatment conditions for lignocellulosic biomass in biorefinery applications: a review. J Renew Sustain Energy 10(3):033104. https://doi.org/10.1063/1.5025876

Boulton CA, Ratledge C (1980) Regulatory studies on citrate synthase in Candida 107, an oleaginous yeast. Microbiology 121(2):441–447. https://doi.org/10.1099/00221287-121-2-441

Boulton CA, Ratledge C (1981) Correlation of lipid accumulation in yeasts with possession of ATP: citrate lyase. Microbiology 127(1):169–176. https://doi.org/10.1099/00221287-127-1-169

Cadoche L, López GD (1989) Assessment of size reduction as a preliminary step in the production of ethanol from lignocellulosic wastes. Biol Wastes 30(2):153–157. https://doi.org/10.1016/0269-7483(89)90069-4

Calvey CH, Su Y-K, Willis LB, McGee M, Jeffries TW (2016) Nitrogen limitation, oxygen limitation, and lipid accumulation in *Lipomyces starkeyi*. Bioresour Technol 200:780–788. https://doi.org/10.1016/j.biortech.2015.10.104

Canonico L, Ashoor S, Taccari M, Comitini F, Antonucci M, Truzzi C, Scarponi G, Ciani M (2016) Conversion of raw glycerol to microbial lipids by new Metschnikowia and *Yarrowia lipolytica* strains. Ann Microbiol 66(4):1409–1418. https://doi.org/10.1007/s13213-016-1228-0

Chandel AK, Garlapati VK, Singh AK, Antunes FAF, da Silva SS (2018) The path forward for lignocellulose biorefineries: bottlenecks, solutions, and perspective on commercialization. Bioresour Technol 264:370–381. https://doi.org/10.1016/j.biortech.2018.06.004

Chaturvedi S, Bhattacharya A, Khare SK (2018) Trends in oil production from oleaginous yeast using biomass: biotechnological potential and constraints. Appl Biochem Microbiol 54(4):361–369. https://doi.org/10.1134/s000368381804004x

Cherubini F, Ulgiati S (2010) Crop residues as raw materials for biorefinery systems—a LCA case study. Appl Energy 87(1):47–57. https://doi.org/10.1016/j.apenergy.2009.08.024

Chi Z, Zheng Y, Jiang A, Chen S (2011) Lipid production by culturing oleaginous yeast and algae with food waste and municipal wastewater in an integrated process. Appl Biochem Biotechnol 165(2):442–453. https://doi.org/10.1007/s12010-011-9263-6

Christophe G, Deo JL, Kumar V, Nouaille R, Fontanille P, Larroche C (2012) Production of oils from acetic acid by the oleaginous yeast *Cryptococcus curvatus*. Appl Biochem Biotechnol 167(5):1270–1279. https://doi.org/10.1007/s12010-011-9507-5

Conde-Mejía C, Jiménez-Gutiérrez A, El-Halwagi M (2012) A comparison of pretreatment methods for bioethanol production from lignocellulosic materials. Process Saf Environ Prot 90(3):189–202. https://doi.org/10.1016/j.psep.2011.08.004

Cui J, Liang L, Han C, Lin Liu R (2015) Stabilization of phenylalanine ammonia lyase from *Rhodotorula glutinis* by encapsulation in polyethyleneimine-mediated biomimetic silica. Appl Biochem Biotechnol 176(4):999–1011. https://doi.org/10.1007/s12010-015-1624-0

Deeba F, Pruthi V, Negi YS (2016) Converting paper mill sludge into neutral lipids by oleaginous yeast *Cryptococcus vishniaccii* for biodiesel production. Bioresour Technol 213:96–102. https://doi.org/10.1016/j.biortech.2016.02.105

Dionisi F, Golay P-A, Hug B, Baumgartner M, Callier P, Destaillats F (2004) Triacylglycerol analysis for the quantification of cocoa butter equivalents (CBE) in chocolate: feasibility study and validation. J Agric Food Chem 52(7):1835–1841. https://doi.org/10.1021/jf035391q

Duff SJB, Murray WD (1996) Bioconversion of forest products industry waste cellulosics to fuel ethanol: a review. Bioresour Technol 55(1):1–33. https://doi.org/10.1016/0960-8524(95)00122-0

Elfadaly H, El-Naggar N, Marwan E-S (2009) Single cell oil production by an oleaginous yeast strain in a low cost cultivation medium. Res J Microbiol 4:301–313. https://doi.org/10.3923/jm.2009.301.313

Enshaeieh M, Abdoli A, Madani M, Bayat M (2015) Recycling of lignocellulosic waste materials to produce high-value products: single cell oil and xylitol. Int J Environ Sci Technol 12(3):837–846. https://doi.org/10.1007/s13762-014-0687-8

Fakas S, Papanikolaou S, Batsos A, Galiotou-Panayotou M, Mallouchos A, Aggelis G (2009) Evaluating renewable carbon sources as substrates for single cell oil production by *Cunninghamella echinulata* and *Mortierella isabellina*. Biomass Bioenergy 33(4):573–580. https://doi.org/10.1016/j.biombioe.2008.09.006

Fan LT, Gharpuray MM, Lee YH (2012) Cellulose hydrolysis. Springer Science and Business Media (3)

Gao Q, Cui Z, Zhang J, Bao J (2014) Lipid fermentation of corncob residues hydrolysate by oleaginous yeast *Trichosporon cutaneum*. Bioresour Technol 152:552–556. https://doi.org/10.1016/j.biortech.2013.11.044

Gong Z, Shen H, Wang Q, Yang X, Xie H, Zhao ZK (2013) Efficient conversion of biomass into lipids by using the simultaneous saccharification and enhanced lipid production process. Biotechnol Biofuels 6(1):36. https://doi.org/10.1186/1754-6834-6-36

Gong Z, Shen H, Zhou W, Wang Y, Yang X, Zhao ZK (2015) Efficient conversion of acetate into lipids by the oleaginous yeast *Cryptococcus curvatus*. Biotechnol Biofuels 8(1):189. https://doi.org/10.1186/s13068-015-0371-3

Gregg DJ, Saddler JN (1996) Factors affecting cellulose hydrolysis and the potential of enzyme recycle to enhance the efficiency of an integrated wood to ethanol process. Biotechnol Bioeng 51(4):375–383. https://doi.org/10.1002/(SICI)1097-0290(19960820)51:4<375::AID-BIT1>3.0.CO;2-F

Grous WR, Converse AO, Grethlein HE (1986) Effect of steam explosion pretreatment on pore size and enzymatic hydrolysis of poplar. Enzym Microb Technol 8(5):274–280. https://doi.org/10.1016/0141-0229(86)90021-9

Guerfali M, Ayadi I, Sassi H-E, Belhassen A, Gargouri A, Belghith H (2020) Biodiesel-derived crude glycerol as alternative feedstock for single cell oil production by the oleaginous yeast *Candida viswanathii* Y-E4. Ind Crop Prod 145:112103. https://doi.org/10.1016/j.indcrop.2020.112103

Guo Z-p, Robin J, Duquesne S, O'Donohue MJ, Marty A, Bordes F (2018) Developing cellulolytic *Yarrowia lipolytica* as a platform for the production of valuable products in consolidated bioprocessing of cellulose. Biotechnol Biofuels 11(1):141. https://doi.org/10.1186/s13068-018-1144-6

Hatakka AI (1983) Pretreatment of wheat straw by white-rot fungi for enzymic saccharification of cellulose. Eur J Appl Microbiol Biotechnol 18(6):350–357. https://doi.org/10.1007/bf00504744

He Y, Zhang J, Bao J (2014) Dry dilute acid pretreatment by co-currently feeding of corn stover feedstock and dilute acid solution without impregnation. Bioresour Technol 158:360–364. https://doi.org/10.1016/j.biortech.2014.02.074

Helle SS, Duff SJB, Cooper DG (1993) Effect of surfactants on cellulose hydrolysis. Biotechnol Bioeng 42(5):611–617. https://doi.org/10.1002/bit.260420509

Hoekman SK, Broch A, Robbins C, Ceniceros E, Natarajan M (2012) Review of biodiesel composition, properties, and specifications. Renew Sust Energ Rev 16(1):143–169. https://doi.org/10.1016/j.rser.2011.07.143

Holtzapple MT, Jun J-H, Ashok G, Patibandla SL, Dale BE (1991) The ammonia freeze explosion (AFEX) process. Appl Biochem Biotechnol 28(1):59–74. https://doi.org/10.1007/bf02922589

Hu C, Zhao X, Zhao J, Wu S, Zhao ZK (2009) Effects of biomass hydrolysis by-products on oleaginous yeast *Rhodosporidium toruloides*. Bioresour Technol 100(20):4843–4847. https://doi.org/10.1016/j.biortech.2009.04.041

Hu M, Wang J, Gao Q, Bao J (2018) Converting lignin derived phenolic aldehydes into microbial lipid by *Trichosporon cutaneum*. J Biotechnol 281:81–86. https://doi.org/10.1016/j.jbiotec.2018.06.341

Huang C, Zong M-H, Wu H, Liu Q-P (2009) Microbial oil production from rice straw hydrolysate by *Trichosporon fermentans*. Bioresour Technol 100(19):4535–4538. https://doi.org/10.1016/j.biortech.2009.04.022

Huang C, Wu H, R-f L, M-h Z (2012) Improving lipid production from bagasse hydrolysate with *Trichosporon fermentans* by response surface methodology. New Biotechnol 29(3):372–378. https://doi.org/10.1016/j.nbt.2011.03.008

Huang X-F, Liu J-N, Lu L-J, Peng K-M, Yang G-X, Liu J (2016) Culture strategies for lipid production using acetic acid as sole carbon source by *Rhodosporidium toruloides*. Bioresour Technol 206:141–149. https://doi.org/10.1016/j.biortech.2016.01.073

Huang X-F, Shen Y, Luo H-J, Liu J-N, Liu J (2018) Enhancement of extracellular lipid production by oleaginous yeast through preculture and sequencing batch culture strategy with acetic acid. Bioresour Technol 247:395–401. https://doi.org/10.1016/j.biortech.2017.09.096

Jiru TM, Steyn L, Pohl C, Abate D (2018) Production of single cell oil from cane molasses by *Rhodotorula kratochvilovae* (syn, Rhodosporidium kratochvilovae) SY89 as a biodiesel feedstock. Chem Cent J 12(1):91. https://doi.org/10.1186/s13065-018-0457-7

Kalscheuer R, Stölting T, Steinbüchel A (2006) Microdiesel: *Escherichia coli* engineered for fuel production. Microbiology 152(9):2529–2536. https://doi.org/10.1099/mic.0.29028-0

Karatay SE, Dönmez G (2010) Improving the lipid accumulation properties of the yeast cells for biodiesel production using molasses. Bioresour Technol 101(20):7988–7990. https://doi.org/10.1016/j.biortech.2010.05.054

Kilzer FJ, Broido A (1965) Speculations on the nature of cellulose pyrolysis. Pyrodynamics 2:151–163

Knothe G (2006) Analyzing biodiesel: standards and other methods. J Am Oil Chem Soc 83(10):823–833. https://doi.org/10.1007/s11746-006-5033-y

Kobkam C, Tinoi J, Kittiwachana S (2018) Alkali pretreatment and enzyme hydrolysis to enhance the digestibility of rice straw cellulose for microbial oil production. Int J Appl Sci Technol. https://doi.org/10.14416/j.ijast.2018.07.003

Kot AM, Błażejak S, Kurcz A, Gientka I, Kieliszek M (2016) Rhodotorula glutinis—potential source of lipids, carotenoids, and enzymes for use in industries. Appl Microbiol Biotechnol 100(14):6103–6117. https://doi.org/10.1007/s00253-016-7611-8

Kraisintu P, Yongmanitchai W, Limtong S (2010) Selection and optimization for lipid production of a newly isolated oleaginous yeast, *Rhodosporidium toruloides* DMKU3-TK16. Kasetsart J Nat Sci 44:436–445

Kumar A, Singh S, Tiwari R, Goel R, Nain L (2017) Immobilization of indigenous holocellulase on iron oxide (Fe2O3) nanoparticles enhanced hydrolysis of alkali pretreated paddy straw. Int J Biol Macromol 96:538–549

Li Q, Du W, Liu D (2008) Perspectives of microbial oils for biodiesel production. Appl Microbiol Biotechnol 80(5):749–756. https://doi.org/10.1007/s00253-008-1625-9

Liang M-H, Jiang J-G (2013) Advancing oleaginous microorganisms to produce lipid via metabolic engineering technology. Prog Lipid Res 52(4):395–408. https://doi.org/10.1016/j.plipres.2013.05.002

Lotero E, Goodwin JG, Bruce DA, Suwannakarn K, Liu Y, Lopez DE (2006) The catalysis of biodiesel synthesis. Catalogue 19(1):41–83

Lundin H (1950) Fat synthesis by micro-organisms and its possible applications in industry. J Inst Brew 56(1):17–28. https://doi.org/10.1002/j.2050-0416.1950.tb01516.x

Majumdar S, Lukk T, Solbiati JO, Bauer S, Nair SK, Cronan JE, Gerlt JA (2014) Roles of small laccases from Streptomyces in lignin degradation. Biochemist 53(24):4047–4058. https://doi.org/10.1021/bi500285t

Mast B, Zöhrens N, Schmidl F, Hernandez R, French WT, Merkt N, Claupein W, Graeff-Hönninger S (2014) Lipid production for microbial biodiesel by the oleagenious yeast *Rhodotorula glutinis* using hydrolysates of wheat straw and miscanthus as carbon sources. Waste Biomass Valori 5 (6):955–962. https://doi.org/10.1007/s12649-014-9312-9

McMillan JD, Baker JO, Overend RP (1994) Enzymatic conversion of biomass for fuels production. American Chemical Society, Washington, DC, pp 292–324

Monlau F, Sambusiti C, Barakat A, Quéméneur M, Trably E, Steyer JP, Carrère H (2014) Do furanic and phenolic compounds of lignocellulosic and algae biomass hydrolyzate inhibit anaerobic mixed cultures? A comprehensive review. Biotechnol Adv 32(5):934–951. https://doi.org/10.1016/j.biotechadv.2014.04.007

Ooshima H, Sakata M, Harano Y (1986) Enhancement of enzymatic hydrolysis of cellulose by surfactant. Biotechnol Bioeng 28(11):1727–1734. https://doi.org/10.1002/bit.260281117

Pan X, Gilkes N, Saddler JN (2006) Effect of acetyl groups on enzymatic hydrolysis of cellulosic substrates. Holzforschung 60(4):398–401. https://doi.org/10.1515/HF.2006.062

Papanikolaou S, Chevalot I, Komaitis M, Aggelis G, Marc I (2001) Kinetic profile of the cellular lipid composition in an oleaginous *Yarrowia lipolytica* capable of producing a cocoa-butter substitute from industrial fats. Antonie Leeu 80(3):215–224. https://doi.org/10.1023/a:1013083211405

Papanikolaou S, Chevalot I, Komaitis M, Marc I, Aggelis G (2002) Single cell oil production by *Yarrowia lipolytica* growing on an industrial derivative of animal fat in batch cultures. Appl Microbiol Biotechnol 58(3):308–312. https://doi.org/10.1007/s00253-001-0897-0

Park JW, Takahata Y, Kajiuchi T, Akehata T (1992) Effects of nonionic surfactant on enzymatic hydrolysis of used newspaper. Biotechnol Bioeng 39(1):117–120. https://doi.org/10.1002/bit.260390117

Patel A, Pravez M, Deeba F, Pruthi V, Singh RP, Pruthi PA (2014) Boosting accumulation of neutral lipids in *Rhodosporidium kratochvilovae* HIMPA1 grown on hemp (*Cannabis sativa* Linn) seed aqueous extract as feedstock for biodiesel production. Bioresour Technol 165:214–222. https://doi.org/10.1016/j.biortech.2014.03.142

Patel A, Sindhu DK, Arora N, Singh RP, Pruthi V, Pruthi PA (2015) Biodiesel production from non-edible lignocellulosic biomass of *Cassia fistula* L. fruit pulp using oleaginous yeast *Rhodosporidium kratochvilovae* HIMPA1. Bioresour Technol 197:91–98. https://doi.org/10.1016/j.biortech.2015.08.039

Patel A, Arora N, Sartaj K, Pruthi V, Pruthi PA (2016) Sustainable biodiesel production from oleaginous yeasts utilizing hydrolysates of various non-edible lignocellulosic biomasses. Renew Sust Energ Rev 62:836–855. https://doi.org/10.1016/j.rser.2016.05.014

Patel A, Arora N, Pruthi V, Pruthi PA (2017) Biological treatment of pulp and paper industry effluent by oleaginous yeast integrated with production of biodiesel as sustainable transportation fuel. J Clean Prod 142:2858–2864. https://doi.org/10.1016/j.jclepro.2016.10.184

Polburee P, Yongmanitchai W, Lertwattanasakul N, Ohashi T, Fujiyama K, Limtong S (2015) Characterization of oleaginous yeasts accumulating high levels of lipid when cultivated in glycerol and their potential for lipid production from biodiesel-derived crude glycerol. Fungal Biol 119(12):1194–1204. https://doi.org/10.1016/j.funbio.2015.09.002

Qin L, Liu L, Zeng A-P, Wei D (2017) From low-cost substrates to single cell oils synthesized by oleaginous yeasts. Bioresour Technol 245:1507–1519. https://doi.org/10.1016/j.biortech.2017.05.163

Ramos LP, Breuil C, Saddler JN (1993) The use of enzyme recycling and the influence of sugar accumulation on cellulose hydrolysis by *Trichoderma* cellulases. Enzym Microb Technol 15 (1):19–25. https://doi.org/10.1016/0141-0229(93)90111-E

Ratledge C (1988) Biochemistry, stoichiometry, substrates and economics. In: Moreton RS (ed) Single cell oil. Longman Scientific & Technical, Harlow, pp 33–70

Robak K, Balcerek M (2018) Review of second generation bioethanol production from residual biomass. Food Technol Biotechnol 56(2):174–187. https://doi.org/10.17113/ftb.56.02.18.5428

Sagia S, Sharma A, Singh S, Chaturvedi S, Nain PKS, Nain L (2020) Single cell oil production by a novel yeast *Trichosporon mycotoxinivorans* for complete and ecofriendly valorization of paddy straw. Electron J Biotechnol 44:60–68. https://doi.org/10.1016/j.ejbt.2020.01.009

Saritha M, Arora A, Lata (2012) Biological pretreatment of lignocellulosic substrates for enhanced delignification and enzymatic digestibility. Indian J Microbiol 52(2):122–130. https://doi.org/10.1007/s12088-011-0199-x

Saritha M, Arora A, Singh, S and Nain L (2013) *Streptomyces griseorubens* mediated delignification of paddy straw for improved enzymatic saccharification yields. Bioresour Technol 135:12–17. https://doi.org/10.1016/j.biortech.2012.11.040

Schmidt-Dannert C, Holtzapple EK (2011) U.S. Patent No. 7,897,369. Washington, DC: U.S. Patent and Trademark Office

Sharma A, Nain V, Tiwari R, Singh S, Adak A, Nain PKS, Nain L (2017) Simultaneous saccharification and fermentation of alkali-pretreated corncob under optimized conditions using cold-tolerant indigenous holocellulase. Korean J Chem Eng 34(3):773–780. https://doi.org/10.1007/s11814-016-0334-9

Singh J, Suhag M, Dhaka A (2015) Augmented digestion of lignocellulose by steam explosion, acid and alkaline pretreatment methods: a review. Carbohydr Polym 117:624–631. https://doi.org/10.1016/j.carbpol.2014.10.012

Sitepu IR, Jin M, Fernandez JE, da Costa SL, Balan V, Boundy-Mills KL (2014) Identification of oleaginous yeast strains able to accumulate high intracellular lipids when cultivated in alkaline pretreated corn stover. Appl Microbiol Biotechnol 98(17):7645–7657. https://doi.org/10.1007/s00253-014-5944-8

Soccol CR, Dalmas Neto CJ, Soccol VT, Sydney EB, da Costa ESF, Medeiros ABP, de Souza Vandenberghe LP (2017) Pilot scale biodiesel production from microbial oil of *Rhodosporidium toruloides* DEBB 5533 using sugarcane juice: performance in diesel engine and preliminary economic study. Bioresour Technol 223:259–268. https://doi.org/10.1016/j.biortech.2016.10.055

Solarte-Toro JC, Romero-García JM, Martínez-Patiño JC, Ruiz-Ramos E, Castro-Galiano E, Cardona-Alzate CA (2019) Acid pretreatment of lignocellulosic biomass for energy vectors production: a review focused on operational conditions and techno-economic assessment for bioethanol production. Renew Sust Energ Rev 107:587–601. https://doi.org/10.1016/j.rser.2019.02.024

Sternberg D (1976) Production of cellulase by Trichoderma. In: Biotechnology and bioengineering symposium, vol 6, p 35

Tai M, Stephanopoulos G (2013) Engineering the push and pull of lipid biosynthesis in oleaginous yeast *Yarrowia lipolytica* for biofuel production. Metab Eng 15:1–9. https://doi.org/10.1016/j.ymben.2012.08.007

Tanimura A, Takashima M, Sugita T, Endoh R, Kikukawa M, Yamaguchi S, Sakuradani E, Ogawa J, Shima J (2014) Selection of oleaginous yeasts with high lipid productivity for practical biodiesel production. Bioresour Technol 153:230–235. https://doi.org/10.1016/j.biortech.2013.11.086

Tarasov D, Leitch M, Fatehi P (2018) Lignin–carbohydrate complexes: properties, applications, analyses, and methods of extraction: a review. Biotechnol Biofuels 11(1):269. https://doi.org/10.1186/s13068-018-1262-1

Tarkow H, Feist WC (1969) A mechanism for improving the digestibility of lignocellulosic materials with dilute alkali and liquid ammonia. In: Cellulases and their applications, vol 95. Advances in chemistry, pp 197–218. https://doi.org/10.1021/ba-1969-0095.ch01210.1021/ba-1969-0095.ch012

Taskin M (2013) Co-production of tannase and pectinase by free and immobilized cells of the yeast *Rhodotorula glutinis* MP-10 isolated from tannin-rich persimmon (Diospyros kaki L.) fruits. Bioprocess Biosyst Eng 36(2):165–172. https://doi.org/10.1007/s00449-012-0771-8

Tsigie YA, Wang C-Y, Kasim NS, Diem Q-D, Huynh L-H, Ho Q-P, Truong C-T, Ju Y-H (2012) Oil production from *Yarrowia lipolytica* Po1g using rice bran hydrolysate. J Biomed Biotechnol 2012:378384. https://doi.org/10.1155/2012/378384

Vajpeyi S, Chandran K (2015) Microbial conversion of synthetic and food waste-derived volatile fatty acids to lipids. Bioresour Technol 188:49–55. https://doi.org/10.1016/j.biortech.2015.01.099

Valdés G, Mendonça RT, Aggelis G (2020) Lignocellulosic biomass as a substrate for oleaginous microorganisms: a review. Preprints.org. https://doi.org/10.20944/preprints202009.0449.v1

Vidal PF, Molinier J (1988) Ozonolysis of lignin—improvement of in vitro digestibility of poplar sawdust. Biomass 16(1):1–17. https://doi.org/10.1016/0144-4565(88)90012-1

Wang Q, Guo F-J, Rong Y-J, Chi Z-M (2012) Lipid production from hydrolysate of cassava starch by *Rhodosporidium toruloides* 21167 for biodiesel making. Renew Energy 46:164–168. https://doi.org/10.1016/j.renene.2012.03.002

Wang J, Gao Q, Zhang H, Bao J (2016) Inhibitor degradation and lipid accumulation potentials of oleaginous yeast *Trichosporon cutaneum* using lignocellulose feedstock. Bioresour Technol 218:892–901. https://doi.org/10.1016/j.biortech.2016.06.130

Wang X, Fosse HK, Li K, Chauton MS, Vadstein O, Reitan KI (2019) Influence of nitrogen limitation on lipid accumulation and EPA and DHA content in four marine microalgae for possible use in aquafeed. Front Mar Sci. 6(95). https://doi.org/10.3389/fmars.2019.00095

Wei Y, Siewers V, Nielsen J (2017) Cocoa butter-like lipid production ability of non-oleaginous and oleaginous yeasts under nitrogen-limited culture conditions. Appl Microbiol Biotechnol 101(9):3577–3585. https://doi.org/10.1007/s00253-017-8126-7

Woodbine M (1959) Microbial fat: microorganisms as potential fat producers. Prog Indus Microbiol 1:181–245

Wu J, Ju L-K (1998) Enhancing enzymatic saccharification of waste newsprint by surfactant addition. Biotechnol Prog 14(4):649–652. https://doi.org/10.1021/bp980040v

Wu S, Hu C, Jin G, Zhao X, Zhao ZK (2010) Phosphate-limitation mediated lipid production by *Rhodosporidium toruloides*. Bioresour Technol 101(15):6124–6129. https://doi.org/10.1016/j.biortech.2010.02.111

Wynn JP, Hamid AA, Li Y, Ratledge C (2001) Biochemical events leading to the diversion of carbon into storage lipids in the oleaginous fungi *Mucor circinelloi*des and *Mortierella alpina*. Microbiology 147(10):2857–2864. https://doi.org/10.1099/00221287-147-10-2857

Xavier MCA, Coradini ALV, Deckmann AC, Franco TT (2017) Lipid production from hemicellulose hydrolysate and acetic acid by *Lipomyces starkeyi* and the ability of yeast to metabolize inhibitors. Biochem Eng J 118:11–19. https://doi.org/10.1016/j.bej.2016.11.007

Xu JK, Sun RC (2016) Chapter 19—recent advances in alkaline pretreatment of lignocellulosic biomass. In: Mussatto SI (ed) Biomass fractionation technologies for a lignocellulosic feedstock based biorefinery. Elsevier, Amsterdam, pp 431–459. https://doi.org/10.1016/B978-0-12-802323-5.00019-0

Yaguchi A, Franaszek N, O'Neill K, Lee S, Sitepu I, Boundy-Mills K, Blenner M (2020) Identification of oleaginous yeasts that metabolize aromatic compounds. J Ind Microbiol Biotechnol 47(9–10):801–813. https://doi.org/10.1007/s10295-020-02269-5

Yang Q, Zhang H, Li X, Wang Z, Xu Y, Ren S, Chen X, Xu Y, Hao H, Wang H (2013) Extracellular enzyme production and phylogenetic distribution of yeasts in wastewater treatment systems. Bioresour Technol 129:264–273. https://doi.org/10.1016/j.biortech.2012.11.101

Yu X, Zheng Y, Dorgan KM, Chen S (2011) Oil production by oleaginous yeasts using the hydrolysate from pretreatment of wheat straw with dilute sulfuric acid. Bioresour Technol 102(10):6134–6140. https://doi.org/10.1016/j.biortech.2011.02.081

Yu X, Zeng J, Zheng Y, Chen S (2014a) Effect of lignocellulose degradation products on microbial biomass and lipid production by the oleaginous yeast *Cryptococcus curvatus*. Process Biochem 49(3):457–465. https://doi.org/10.1016/j.procbio.2013.10.016

Yu X, Zheng Y, Xiong X, Chen S (2014b) Co-utilization of glucose, xylose and cellobiose by the oleaginous yeast *Cryptococcus curvatus*. Biomass Bioenergy 71:340–349. https://doi.org/10.1016/j.biombioe.2014.09.023

Zhan J, Lin H, Shen Q, Zhou Q, Zhao Y (2013) Potential utilization of waste sweetpotato vines hydrolysate as a new source for single cell oils production by *Trichosporon fermentans*. Bioresour Technol 135:622–629. https://doi.org/10.1016/j.biortech.2012.08.068

Zhang Z, Ji H, Gong G, Zhang X, Tan T (2014) Synergistic effects of oleaginous yeast *Rhodotorula glutinis* and microalga *Chlorella vulgaris* for enhancement of biomass and lipid yields. Bioresour Technol 164:93–99. https://doi.org/10.1016/j.biortech.2014.04.039

Zhao X, Kong X, Hua Y, Feng B, Zhao Z (2008) Medium optimization for lipid production through co-fermentation of glucose and xylose by the oleaginous yeast *Lipomyces starkeyi*. Eur J Lipid Sci Technol 110(5):405–412. https://doi.org/10.1002/ejlt.200700224

Zhao X, Cheng K, Liu D (2009) Organosolv pretreatment of lignocellulosic biomass for enzymatic hydrolysis. Appl Microbiol Biotechnol 82(5):815–827. https://doi.org/10.1007/s00253-009-1883-1

Zheng Y, Lin H-M, Tsao GT (1998) Pretreatment for cellulose hydrolysis by carbon dioxide explosion. Biotechnol Prog 14(6):890–896. https://doi.org/10.1021/bp980087g

Zhou W, Wang W, Li Y, Zhang Y (2013) Lipid production by *Rhodosporidium toruloides* Y2 in bioethanol wastewater and evaluation of biomass energetic yield. Bioresour Technol 127:435–440. https://doi.org/10.1016/j.biortech.2012.09.067

Chapter 14
Techno-economic and Life Cycle Assessments of Microbial Process in Renewable Energy Production

Na Wu, Shunchang Yang, Pratap Pullammanappallil, and Ghasideh Pourhashem

Abstract Energy security, environmental concerns, and the increasing demand of a growing population present opportunities for adopting alternative pathways for energy and chemicals. Microbial biotechnologies have been making progress in the context of renewable energy production towards creating more sustainable societies. While the state-of-the-art production of bio-based energy, chemicals, and materials promises competitive functionality and quality, evaluation of their sustainability is crucial, particularly for emerging biotechnologies. Analytical methods such as techno-economic analysis (TEA) and life cycle assessment (LCA) are standardized techniques that are used to quantify economic viability and environmental sustainability of processes and products and offer decision-making information on their research, development, and deployment. However, challenges still exist for TEA and LCA studies to support biotechnology transition to a more sustainable future. Examples of such challenges include data availability and accessibility considering technology readiness levels in TEA studies, broadening the impact assessment to categories other than a single impact indicator (e.g., global warming potential), and estimating full life cycle performance in LCA studies. To address these challenges and to promote a sustainable bio-based economy, this chapter provides a systematic overview of the status of renewable bioenergy and

N. Wu
Department of Coatings and Polymeric Materials, North Dakota State University, Fargo, ND, USA

Department of Agricultural and Biological Engineering, University of Florida, Gainesville, FL, USA

S. Yang · P. Pullammanappallil
Department of Agricultural and Biological Engineering, University of Florida, Gainesville, FL, USA

G. Pourhashem (✉)
Department of Coatings and Polymeric Materials, North Dakota State University, Fargo, ND, USA
e-mail: ghasideh.pourhashem@ndsu.edu

J. K. Saini, R. K. Sani (eds.), *Microbial Biotechnology for Renewable and Sustainable Energy*, Clean Energy Production Technologies,
https://doi.org/10.1007/978-981-16-3852-7_14

biochemicals commercialization, markets, and policies. Additionally, the chapter discusses possible knowledge-based process design approaches, identifying the interrelations between the challenges and development regarding resource efficiency and waste minimization, and bridging the gap between research and commercialization. Case studies of biobutanol production pathways are also discussed for learning and optimization potential for sustainability gains. Finally, the chapter emphasizes the engagement of multiplayers for interdisciplinary work to bring renewable energy into reality.

Keywords Techno-economic analysis · Life cycle assessment · Sustainability · Microbial process · Biobutanol

14.1 Introduction

Techno-economic analysis (TEA) and life cycle assessment (LCA) are prospective methods in the assessment of green/sustainable technologies. Green products or technologies, by definition, are to be environment friendly. Research, development, and deployment (RD&D) of green products or technologies are actions taken to achieve different sustainable development goals (SDGs), such as the ones adopted by United Nations (UN) in 2015, including affordable and clean energy, climate action, economic growth, and clean water, for creating a more sustainable future. While technical feasibility and functionality of biotechnologies is a knockout criterion for researchers and interested industry players to identify opportunities, further assessments such as economic viability, ecologic sustainability, and social acceptance are also essential. For example, "yields," "energy efficiency," and "reaction rate" are typical technical indicators; however, high scores in these indicators may be accompanied by downsides such as expensive equipment, high global warming potentials, or high eutrophication risks. Extreme cases are green technologies that are neither "green" nor affordable but run against the SDGs. Environmental and economic impacts are two essential criteria for these green chemical technologies, not only to their qualification for lower environmental impact than their conventional counterparts but also their ability to replace conventional technologies commercially.

The fundamental idea of renewable energy production through microbial processes is the processing of biobased resources into energy/chemicals/materials. The challenge for such process development is the scarcity of resources in terms of natural capital and money (Buchner et al. 2018). Thus, the process development needs to achieve three goals: (1) maximizing utilization of all biomass components and minimize waste, (2) evaluating the tradeoffs resulting from the interactions between technical advances and sustainability parameters, and (3) building the decision-making platform of resource allocation for raw material suppliers, producers and stakeholders (Wu et al. 2019). Based on these objectives, this chapter presents the structure and content of TEA and LCA methodologies to understand the framework of sustainability analysis; reviews and discusses challenges and opportunities for microbial processes in renewable energy production to address the importance of biotechnology for a biobased economy; and finally illustrates TEA and LCA applications through case studies. In the end, key factors and concepts are discussed for the roadmap to bring the microbial process in renewable energy production into reality.

14.2 TEA and LCA Methodologies

Developing appropriate tools and methods for measuring sustainability is necessary for inducing new technologies, especially those in a position of making a difference in developing our sustainable future. This chapter focuses on the specific context of two popular ones: TEA and LCA.

LCA is a systematic technique to assess the environmental impacts associated with all the stages (production, distribution, use, and end-of-life phases) of a product's or service's life. During an LCA, the upstream and downstream processes throughout the entire life cycle of a product, process, or service are included. For example, in the LCA of bioenergy, the environmental impacts cover biomass cultivation with all relevant inputs and outputs from the environment (e.g., carbon dioxide emission or sequestration and water consumption) as well as emissions from incineration into the air, water, and soil.

Techno-economic analysis (TEA) measures the technical and economic performance of a process, product, and service. To evaluate a specific technology (e.g., compare different options, analyze commercialization feasibility), TEA is an integral tool that usually combines process design and simulation/model with establishing capital and operating cost profiles. For profit-oriented stakeholders, TEA is the most important basis for decisions about research, development, and deployment (RD&D). Specifically, TEA connects research, engineering, and business. Having the capability of being conducted at different technological stages and production scales, TEA can be used as a basis for making a variety of decisions. For example, researchers can use TEA to identify process hotspots of production cost at bench scale, engineers can compare process conditions and configurations for financial impact during process design and development, and investors can determine the potential economic viability of a project by averting unnecessary expenditures.

TEA and LCA share similar logic for contents. They are assessments of a product or process that provide essential decision-making information. As defined by ISO standards, LCA consists of four phases: goal and scope definition, life cycle inventory (LCI) analysis, life cycle impact assessment (LCIA), and interpretation of results (Fig. 14.1). Cost estimation and market investment are important components of TEA, where cost and revenue are calculated for profitability analysis. Similar to TEA, life cycle costing (LCC) is a cost assessment tool over the life of a project. Therefore, LCC and LCA have analogous procedures with a consistent definition of the product system and measures the financial impacts. Considering the scope and

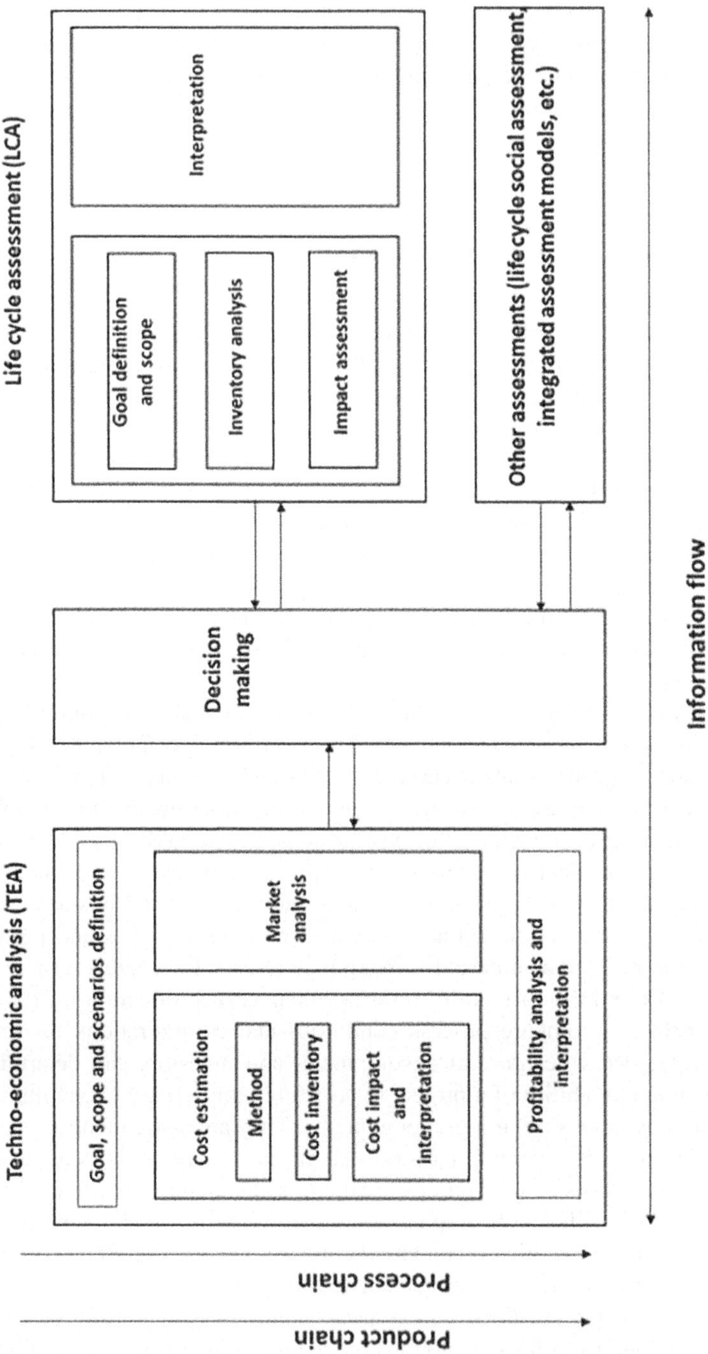

Fig. 14.1 Structure of TEA and LCA

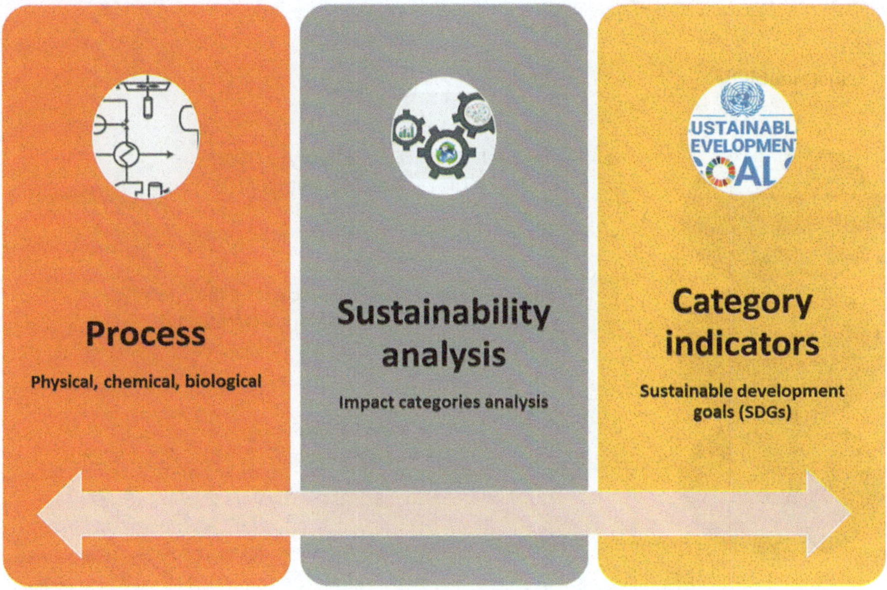

Fig. 14.2 Overall framework interlinking methods and goals for sustainability

boundaries, TEA can be applied as the basis for life cycle costs inside the plant gate. In a broader concept, LCC and LCA together with social life cycle assessment are three pillars of life cycle sustainability assessment (LCSA). LCSA is an integrated framework for the application of life cycle thinking. For example, in modern business practice, life cycle management is a comprehensive decision process, which addresses the three pillars and assesses the cost and performance tradeoffs.

When carried out in parallel, TEA and LCA have usually the same goal and scope, as well as they overlap in inventories such as mass and energy balances in terms of physical, chemical, and biological flows (Fig. 14.2). The assessment outcomes can be reflected in different category indicators/indices/metrics. For example, the sustainability metrics cover a wide range of aspects including economic, environmental, and social factors. Currently, no universal metrics are recognized for evaluating the sustainability of a product or process; however, many studies (Bare et al. 2006; Horváth et al. 2017; Tabone et al. 2010) employ widely used approaches such as green chemistry metrics and life cycle assessment. A positive correlation has been found between adherence to green design principles and a reduction of the environmental impacts of a process (Tabone et al. 2010). The principles for green chemistry and green engineering (green metrics) are well known for the design of chemical products and processes that utilize resources (e.g., raw materials, energy) efficiently and reduce waste and toxic/hazardous chemicals use. Mass-based metrics such as atom economy and E-factor (environmental factor), which are core parts of green process design, need to be augmented by metrics of measuring the environmental impact and assessing economic viability (Sheldon 2018). With limited

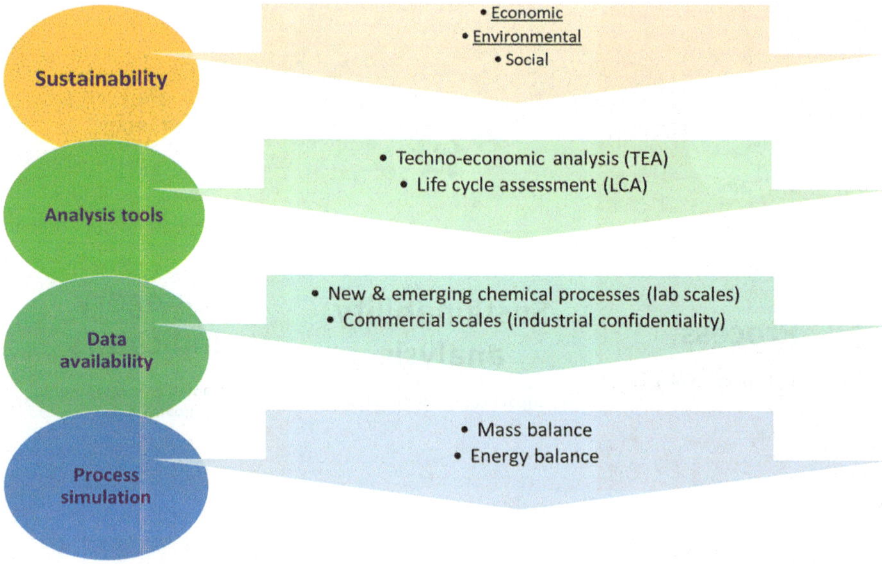

Fig. 14.3 The logic of process modeling as a simplified approach for TEA and LCA

information and resources, identifying the essential metrics could be important for decision-making in chemical manufacturing processes.

Data access and quality are common issues faced by both TEA and LCA methods, which can complicate conducting such analyses. One crucial concept, here, is the technology readiness level (TRL). TRL rates the technological maturity of R&D projects, which indicates the data availability and the corresponding accuracy of the results. For LCA practitioners, the method of building life cycle inventory is also closely related to the database, which could be challenging considering both time requirement and model accuracy. Building and running TEA and LCA models require extensive data collection and analysis for inventories. For example, cost inventory requires operational and capital expenditure and life cycle inventory requires material and energy flow data sets, not only for the studied process but also for upstream chemicals/materials processes. To address the missing data issue, process simulation is a valuable method for inventory data estimation, especially either for chemicals that are not currently commercially produced or for which the primary industrial data are not accessible. An example of such is that the process to produce some chemicals at commercial scales is kept confidential. Figure 14.3 shows the logic of process modeling as a simplified approach for TEA and LCA. The process simulation, however, still requires sufficient data such as concept proof/validation in the laboratory, knowledge of detailed process design parameters and operating conditions, which could have an impact on the quality of inventory data obtained from the simulation. Therefore, the more details achieved in the process model, the more accurate results are obtained from those assessments. In this chapter, we will also introduce a computer simulation platform integrating sustainability assessment tools along with technical performance to compare biotechnology alternatives.

14.3 Microbial Process in Renewable Energy Production: Challenges and Opportunities

Biotechnology uses microorganisms and enzymes for renewable energy production. Microorganisms are "unseen majority"—abundant and diversified, which have the potential to help solve the global energy and climate change challenge (Cavicchioli et al. 2019). Specifically, microbial technologies provide mitigation solutions such as biofuels and CO_2 fixation from the contribution of marine and territorial biome. Biofuels, as renewable energies, could be a large-scale approach. Microorganisms and their environment interact and affect each other. On the one hand, microorganisms convert nutrients to various potentially useful by-products (e.g., biofuel) with evolved metabolic strategies under changing environments; on the other hand, the environment is influenced by the products (e.g., methane) generated by the microorganisms. In addition to a systematic understanding of the biological mechanism of energy and carbon transformation, in many cases, the development of microbial processes requires economically viable options, optimized process design, scale-up concepts, and ecological insights. With this in mind, this section presents the challenges for R&D, commercialization aspects, and success/failure cases and closely relates TEA and LCA in the following aspects: (1) early stages for directing research efforts, (2) commercial-scale production for developing a framework, and (3) promote biotechnology contributions to solving environmental sustainability problems.

14.3.1 Status of Renewable Bioenergy/Biochemicals Commercialization, Markets, and Policies

Until now, various biofuel types (alcohols, biogas, hydrogen, biodiesel, hydrocarbons) using a variety of feedstocks (e.g., lignocellulosic, algal biomass, industrial waste) and strains of microorganisms have been researched and developed at different levels from laboratory scales to industrial scales. However, substantial commercial production of biofuels such as cellulosic biofuels is still limited if any available. An example is cellulosic ethanol, an important biofuel whose production has been scaled up for commercialization. Among the three major commercial startup projects in 2014, namely DowDuPont, POET-DSM, and Abengoa SA, none produces cellulosic ethanol commercially at present. Abengoa sold its U. S. ethanol plants in 2016, DowDuPont sold its plant to the company Verbio in 2018, which produces renewable natural gas instead of ethanol, and POET- DSM's plant is converted to an R&D facility while their ethanol production ceased in 2019. The commercialization of cellulosic ethanol has been tried without success. The reason for this could be three-fold: operational, feedstock-related, and socio-economic aspects. Technically, operation difficulties such as temperature control, microbes contamination, solids handling, and equipment functioning have made processing conditions suboptimal, which prevent the laboratory results (yields and conversion efficiencies) from being realized and economically viable at industrial scales. Likewise, the supply of biomass, including the quantities, collection, transportation, and storage feedstock is still not reliable. Moreover, the socio-economic aspects such as market competition with traditional corn ethanol, the overall "blend-wall" for ethanol, social acceptance of vehicles with high ethanol blends (e.g., E85), and regulatory uncertainties (e.g., investment deterrence) have hampered the development of cellulosic ethanol.

Biodiesel has been commercialized, especially in Europe, as a key biofuel. Currently, the main feedstock for biodiesel conversion is still plant oil, which may be a controversial topic for the potential impact on food markets. Waste and microbial oil (e.g., microalgal lipids) show good future potential for biodiesel development. Renewable natural gas, which is from the biogas product of anaerobic digestion, has been increasingly addressed, for its flexibility in utilizing renewable waste materials (e.g., agricultural residue, municipal solids waste, urban wastewater, livestock manure). Biohydrogen produced through biological means promises merits as a clean fuel with high energy content, however, its commercialization needs to be further validated by improving yield, storage, and transportation logistics, and overcoming the difficulties in strains, fermentation (e.g., substrate), engineering aspects (e.g., bioreactors design). Algal biofuel has been a very active research field since 2005, for its promising features (e.g., high photosynthetic efficiency, using low non-arable land and low-quality water) over terrestrial feedstocks. Although algal biomass depicts a bright future of sustainable energy, additional effort including strain selection, cultivation conditions, and the downstream process is required to advance the practical utilization of algal biomass. Table 14.1 shows the commercialization status of different types of renewable energy through a variety of microbial processes, industrial plants, and future deployment considerations. Policies play important roles in the development of biofuels, both in the R&D and market stages. In general, policies related to bioenergy in the US include feed-in tariffs, carbon tax, biofuel standards for transportation, sustainability standards, and certification, and electricity and heat policies. Due to the complex interaction of various factors, policies have been a controversial topic for promoting bioenergy use and lowering emissions. For example, a carbon tax may affect some economic sectors such as the coal industry and interfere the social equity, while the policy itself may to some extent be limited in impacting climate change. However, good practices and considerations could be designed to adapt to target-specific policies (Smolinksi and Cox 2016). For instance, flexible rates and differentiating payments according to different scenarios (e.g., fuel type, project size, upstream producers/downstream customers), and integrating other policies such as water/land/agriculture could be more resilient and effective in facing the implementation challenges. Policy

Table 14.1 Commercialization status of different types of bioenergy

Type of bioenergy	Commercialization status	Examples	Future deployment considerations
Ethanol	Commercialized, mostly first-generation ethanol, a small share of cellulosic ethanol	DowDuPont, POET-DSM, and Abengoa	Feedstock supply, enzyme recycling, yields, efficiency
Renewable natural gas	Commercially viable in Europe under preconditions such as high subsidized market prices for electricity. In the US, biogas projects are operational/under construction/planned for pipeline injection or use as vehicle fuel	Ameresco's, Vanguard Renewables	Optimized digesters, steady market and subsidized prices, low transportation and Operation&Maintainance (O&M) cost, improve gas yield such as supplement addition
Biodiesel	Commercialized and takes 80% and 6% of the market for transport biofuels in Europe and the US, respectively	Advanced Biodiesel Inc., Agromond USA LLC, Allied Renewable Energy LLC	Non-food feedstock such as waste oil instead of oil crops
Hydrogen	Limited information in commercialization, ample findings in R&D	Verbio	Improve process performance such as yields and energy requirement, solve distribution and storage issues
Microalgae-based biofuels	Limited information in commercialization, remains in the R&D and demonstration stage	Algenol	Improve algal biomass cultivation strategy, scaling, harvesting, and dewatering techniques

innovation could also be a powerful tool in reducing risks, thereby encouraging investments in promising bioenergy technologies. One example is the suggestion of a reverse auction instead of government subsidies for corn stover biofuels to reduce the long-term risk for investors as presented in TEA research by Petter and Tyner (2014). Overall, for bioenergy markets and bioeconomies, proper policies could play a key role, especially in a phenomenon of fragile crude oil price and market fluctuations, to secure a structural transition to a downward trend in non-renewables.

14.3.2 TEA and LCA in Research, Development, and Technology Deployment

The scientific literature explicitly using TEA and LCA or an economic and life-cycle approach, to estimate the economic and environmental impacts of bioenergy production and use, as well as other sustainability dimensions has been increasing. These publications can be classified into three main categories: (1) technological system and its direct impacts (e.g., financial performance, emissions), (2) development of evaluation tools (e.g., TEA and LCA methods, sustainability metrics), and (3) sustainability trade-offs and indirect impacts (e.g., social benefits, land use, food security, biodiversity). Intensive research addresses the first category, due to the relatively low technology readiness level of the entire bioenergy industry. Specifically, for the microbial process, the fermentation step is emphasized (Crater et al. 2018). For example, effective microbial communities by systems biotechnology and enzyme/biocatalyst engineering in fermentation have improved capabilities in bioenergy conversion (e.g., higher yields, less inhibition) (Srivastava 2019). This is critical because the fermentation step has a direct impact not only on the economics but also on the technical performance of downstream processing. The importance of "begin with the end" should be also noted for understanding scale-up effects. In many studies, feedstock, pretreatment, and geographical information are starting points within the context of a conceptual design and early guidance for reliable scale-up results of end production. Cherubini and Strømman (2011) have reviewed evolving bioenergy LCA studies and found most research results show more favorable environmental impacts of bioenergy than that of fossil fuels in terms of GHG emission reductions and fossil energy consumption. However, the economic outcomes are more diversified depending on the assumptions (e.g., government incentives, feedstock compositions, geographic and seasonal factors) (Vasco-Correa et al. 2018). Nevertheless, these discussions about bioenergy encourage moving the research and technology deployment towards the direction of development in a more sustainable manner. The areas can be the following technical aspects.

14.3.2.1 Biorefinery Concept

The biorefinery concept has been increasingly focused by researchers, especially TEA and LCA practitioners. Strategies such as byproducts valorization and diversifying product portfolio could potentially reduce the economic risk of investing in a single product by maximizing resource utilization and minimizing "waste". Accordingly, methodological progress is needed such as allocation of LCA, which should be selected to represent the system with less uncertainty or avoided by using the proper functional unit and defining different system boundaries.

14.3.2.2 "Waste" Materials/Non-food Crops as Feedstock

Biomass as a feedstock for bioenergy production could be an expensive choice, which may also entail environmental burdens to some extent. For example, food-based feedstock cultivation could require substantial inputs such as fertilizer and water and has an indirect influence on the food price, which could result in an increase in both monetary values and mass inflows in the system. As discussed previously, using "waste" materials does not necessarily mean automatic cost-effectiveness or an eco-efficient process. For instance, the process of using lignocellulosic biomass to produce bioenergy or other bio-based products is still limited to the R&D stage, due to the technical difficulties and trade-offs across various sustainability sectors.

14.3.2.3 Process Enhancement

Despite the endeavors by researchers and engineers to use the biomass more efficiently in process hotspots such as pretreatments, microbial culturing and processing, and integrated downstream processes, there is no clear breakthrough technology that significantly makes changes to the energy conversion and the developed system that delivers gains in both bio-based production/processing and waste treatment. At least, fundamental issues such as microorganisms' potentials, plant phenotyping, reaction mechanism, and inter-and transdisciplinary research need to be more thoroughly understood.

Complexity and diversity of the bioenergy systems (e.g., system boundaries in LCA, production capacity in TEA) have made different studies non-comparable, which means there is still space to improve the methodology for knowledge-based decisions. For example, considering end-of-life scenarios and environmental portfolios including indirect effects (not just GHG emission), addressing data scarcity, and building the analysis framework. Moreover, the 2020 trade-offs should be addressed such as industry bearing and competing for scenery (e.g., regions revenue, market growth trends, manufacturers) (Escobar and Laibach 2020). The example of the first generation of bioethanol and cellulosic ethanol could illustrate the concept. The existence of first-generation bioethanol (such as sugar-based) with its market and suppliers, although criticized by many researchers for its long-term impacts on the environment and food security, could be the result of trade-offs in economic drivers, energy security, resource re-allocation, and the farmers' benefits. The rare success stories of cellulosic ethanol could be partially attributed to the competition with traditional ethanol, either first-generation or fossil-based, where the underlying approaches are the interactions of different groups of interest. For the chemical industry using biomass as the raw materials, sustainability metrics such as green chemistry metrics need to be taken into account when designing the process and evaluating its economic, environmental, and societal impacts. To bring bioeconomy into the reality, as harnessed by bioenergy, it requires the efforts of players from a wide range such as chemists, engineers, microbiologists, economists, governments, stakeholders, and the communities.

14.4 Case Studies: Lignocellulosic Butanol as an Advanced Biofuel

To illustrate TEA and LCA in evaluating the learning and optimization potential of bioenergy technologies, the biobutanol production processes were investigated and compared as a case study, which covered novel approaches, traditional fermentation methods, and the fossil-based benchmark. This section presents the background of biobutanol production, TEA and LCA modeling details, and key aspects to reinvigorate butanol for bioenergy applications.

14.4.1 TEA of Biobutanol Production Alternatives

The traditional fermentation method for butanol production is called Acetone–butanol–ethanol (ABE) fermentation. Although ABE fermentation has been industrially exploited in the US since the beginning of the last century, it was replaced by the petrochemical industry around the 1960s (Ezeji et al. 2007). The main problems included high feedstock cost, product inhibition, low ABE yield, low productivities, and inefficient recovery processes. However, butanol has increasingly attracted researchers' attention for its various advantages (high energy content, low water solubility, high blending ratio in gasoline, etc.). Specifically, utilizing cost-effective cellulosic feedstock has motivated the biosynthesis of butanol in the recent era (Kumar et al. 2012). Table 14.2 shows the status of leading biofuel companies producing bio-butanol.

Economic analysis of ABE fermentation has been performed by several researchers (Pfromm et al. 2010; Kumar et al. 2012; Tao et al. 2014; Qureshi et al. 2013) with regard to different feedstocks and process parameters (fermenter size, plant capacity, microbial strains, production yield, etc.). In these studies, the ABE fermentation butanol yields are 0.11–0.3 g/g biomass. Many of these studies were performed on the lab scale and multiple additional assumptions. The low yields were due to the low concentration of butanol in the fermentation broth (12–18 g/L) and the presence of a variety of inhibitory chemicals (furfural, hydroxymethylfurfural (HMF), etc.) generated before and during fermentation. The industrially confirmed yield of 0.11 g butanol/g of corn corresponds to 34 wt% conversions of solvents (Pfromm et al. 2010). Debates exist in energy yield comparison between ethanol fermentation and acetone–butanol–ethanol (ABE) fermentation (Wu et al. 2007; Swana et al. 2011; Tao et al. 2014). To improve the yields of bio-butanol production as an advanced biofuel, a new and promising scheme for the "hybrid conversion" process employs anaerobic bacteria to produce an alternative intermediate—butyric acid, which has a higher titer (more than 60 g/L) and then converting butyric acid to butanol through a catalytic process (more than 98% conversion rate) (Lee et al. 2014).

Table 14.2 The status of bio-butanol production in leading biofuel companies

Company	Product	Status	Note
Cobalt Technologies	n-butanol	Closed	One of the leading companies of commercializing the production of bio n-butanol for chemical and fuel
Gevo	Isobutanol	Conversion of corn ethanol plants for butanol production, process optimization	More plants for cellulosic isobutanol
Eastman	n-butanol	Producing n-butanol from petroleum	Commercialization of the bio-catalysis technology for producing bio-based butanol
Green Biologistics	n-butanol	Commercial facility operation ceased with possible reasons of cost disadvantage, small volume fermentation	Producing n-butanol from corn
Butamax	Isobutanol	Develop a commercial facility for the biobutanol production process	Previous work includes piloting and risk mitigation, beginning of isobutanol retrofit project
Butalco GmBH	Isobutanol	Focused on bioethanol fermentation	Develop integrated production processes to ferment xylose into isobutanol by yeast strain
Cathay Industrial Biotech	n-butanol	Shut down	Scaled-up biobutanol production from corn
ZeaChem	Butanol	–	Indirect production of butanol from ethanol

There is very limited research on the comparisons of traditional ABE fermentation and the butyric acid to butanol catalytic process from domestic lignocellulosic biomass such as corn stover and wheat straw, which are representative of their high cellulose content and biomass yield per unit area (Swana et al. 2011). Thus, this research will focus on bio-butanol production with lignocellulosic feedstock and concentrate on one of the major bottlenecks in the overall process—the difficulty in product purification from the fermentation broth. To address the challenge, different biorefinery scenarios (conversion and product recovery) are discussed to separate butyric acid/butanol from other byproducts, mainly acetic acid/ethanol in both perspectives of energy and economic analysis.

14.4.1.1 TEA Method

This research is focused on the catalytic process for converting butyric acid to butanol of the "hybrid" conversion process. Here, butyric acid is used as direct input in the fermentation broth. Information such as the butyric acid concentration and yields fermentation process is based on literature (Sjöblom et al. 2015). Since the

Table 14.3 Thermodynamic properties of acetic acid and butyric acid

Component	Formula	Molar mass	Boiling point
Acetic acid	CH_3COOH	60 g/mol	244.6 °F (118.1 °C)
Butyric acid	$C_4H_8O_2$	88 g/mol	326.3 °F (163.5 °C)

fermentation broth contains butyric acid and other coproducts (mainly acetic acid), two scenarios were investigated:

Scenario 1: First catalytically convert the acids (butyric acid and acetic acid) in the mixture from fermentation broth to alcohols and then separate the alcohols to around 95% mass purity.

Scenario 2: First separate the two acids in the mixture, catalytically convert each of them to their corresponding alcohol, and finally purify the alcohol to 95% mass purity.

The thermodynamic properties of butyric acid and acetic acid are shown in Table 14.3. Considering a plant capacity of 30 million gallons/year of butanol, assumptions made in this study are as the following:

- Acetic acid and butyric acid could be catalyzed by the same catalyst (ZnO-supported Ru-Sn bimetallic catalyst).
- The catalysts have the same selectivity (99.9%) and conversion rates (98.6%) on both acetic acids and butyric acid.
- The concentration of acetic acids and butyric acid does not affect the catalyst's selectivity and conversion rates.
- The catalytic process was operated on the same condition: 265 °C and 25 atm.
- The concentration of butyric acid and acetic acid in the fermentation is 58.8 g/L and 11.46 g/L, respectively.
- The capital cost is borrowed at an interesting rate of 10% for 20 years.

The catalytic process is through the conversion of hydrogenation of acids in the vapor phase by a stable and selective catalyst. Metal catalysts such as $Cu/ZnO/Al_2O_3$ and ZnO-supported Ru-Sn bimetallic catalysts could have more than 98% yield of butanol from biomass-derived butyric acid. The selectivity (ratio of substrate converted to desired product to total substrate converted, addressing unwanted reactions) and conversion rates are important criteria in selecting the catalysts. Here, the main reactions are:

$$\text{Acetic acid(CH3COOH)} + 2H_2 \rightarrow \text{Ethanol} + \text{Water}$$

$$\text{Butyric acid} + 2H_2 \rightarrow \text{Butanol} + \text{Water}$$

Then Aspen plus V8.8 was used to simulate the processes of the two scenarios and the economic performance is evaluated.

Scenario 1: The process flow diagram (PFD) of scenario 1 is shown in Fig. 14.4. The feed broth and hydrogen are introduced into the catalytic reactor R1, where acetic acid and butyric acid are converted to ethanol and butanol through

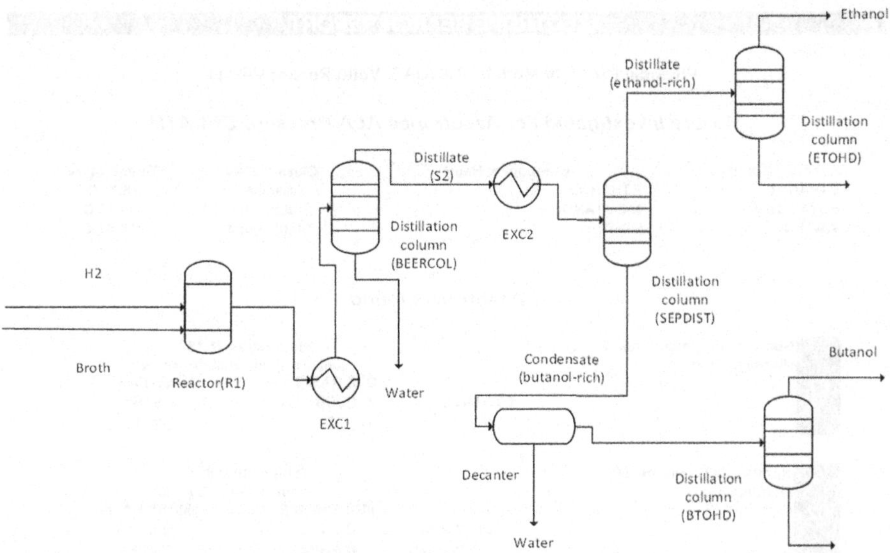

Fig. 14.4 Process flow diagram (PFD) of Scenario 1

hydrogenation reaction, respectively. The effluent from the reactor goes into a distillation column (BEERCOL). Here, two azeotropes are formed ethanol and water, butanol, and water (as analyzed by ASPEN, shown in Fig. 14.5). The distillate (S2) contains most ethanol and butanol as well as a portion of water. The S2 is sent for further distillation SEPDIST, where ethanol and butanol are separated for individual distillation for a 95% mass purity. The distillation column ETOHD produces the target ethanol and column BTOHD produces the target butanol. For butanol purification, a decanter is used for two liquid phase separation for removing water. The n-butanol/water azeotrope is heterogeneous, which is different from the ethanol/water system (homogeneous), and therefore the constituents of the mixture are not completely miscible in the decanter (two liquid phases). This process refers to the double effect distillation to obtain ABE as final products (Naleli 2016). Here, the property method chosen is UNIQUAC (universal quasichemical). Vapor-liquid equilibrium for ethanol and butanol is shown in Figs. 14.6 and 14.7. The ternary diagram for butanol ethanol and water is shown in Fig. 14.8.

Scenario 2: The flowsheet of the process of scenario 1 is shown in Fig. 14.9. Different from scenario 1, in this scenario, the mixture of butyric acid and acetic acid is sent to the distillation column DIST01 for separation. Here, the acetic acid solution AA is obtained at the bottom of the distillation column, and the azeotrope of butyric acid and water is obtained as distillate, as analyzed by the azeotrope search report (Fig. 14.10) in ASPEN. Then, acetic acid and butyric acid are sent to the catalytic process separately. In reactors RAA and RBB, each acid is converted to its alcohol. The ethanol and butanol solutions obtained are sent for purification by distillation.

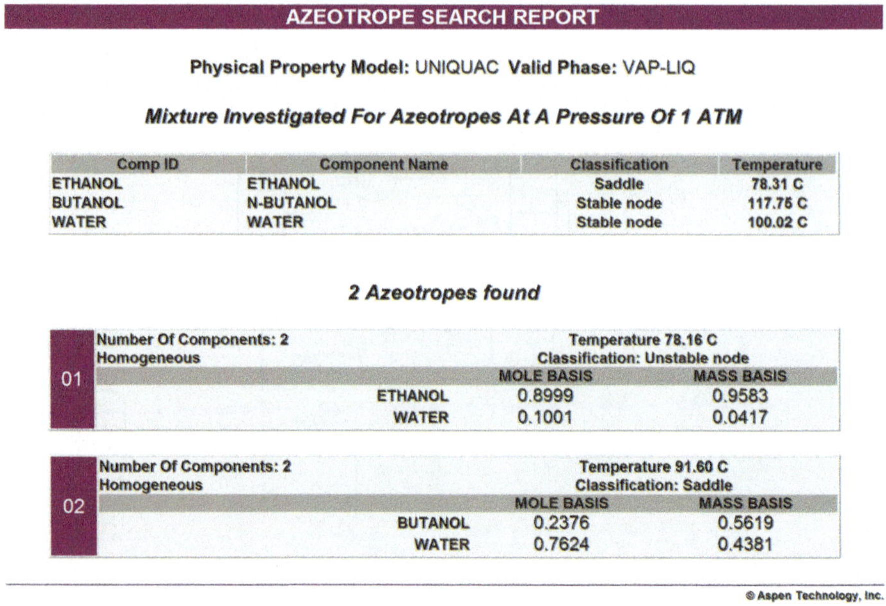

AZEOTROPE SEARCH REPORT

Physical Property Model: UNIQUAC **Valid Phase:** VAP-LIQ

Mixture Investigated For Azeotropes At A Pressure Of 1 ATM

Comp ID	Component Name	Classification	Temperature
ETHANOL	ETHANOL	Saddle	78.31 C
BUTANOL	N-BUTANOL	Stable node	117.75 C
WATER	WATER	Stable node	100.02 C

2 Azeotropes found

01 — Number Of Components: 2; Homogeneous; Temperature 78.16 C; Classification: Unstable node

	MOLE BASIS	MASS BASIS
ETHANOL	0.8999	0.9583
WATER	0.1001	0.0417

02 — Number Of Components: 2; Homogeneous; Temperature 91.60 C; Classification: Saddle

	MOLE BASIS	MASS BASIS
BUTANOL	0.2376	0.5619
WATER	0.7624	0.4381

Fig. 14.5 Azeotropes in Scenario 1

Then, over 95% mass purity alcohols are obtained. The butanol purification process is similar to that of scenario 1.

14.4.1.2 Results and Discussion

The capital cost and operation cost were obtained by ASPEN Process Economic Analyzer with its built-in evaluation method of sizing based on the mass and energy balance. The economic analysis summary is shown in Table 14.4. The cost of the main equipment is shown in Table 14.5. The utilities include electricity, steam, refrigerant, and cooling water. The overall economic performance of scenario 1 is better than that of scenario 2 due to the significant savings in operating costs. The high capital and operating costs of Scenario 2 are mainly caused by the distillation difficulties in separating butyric acid and acetic acid and huge utility requirements. Here, without considering the butyric acid fermentation cost, the unit cost for scenario 1 is 0.21 $/L butanol, while scenario 2 has a unit cost of 0.84 $/L butanol. Thus, the process in which the butyric acid fermentation broth was catalyzed before products recovery has better economic performance.

The butyric acid fermentation process is similar to the bioethanol fermentation process. The major difference is the microbes involved in the fermentation. Considering the butyric acid concentration of 58.8 g/L (Sjöblom et al. 2015), ethanol fermentation has a similar titer. The lignocellulosic ethanol fermentation process

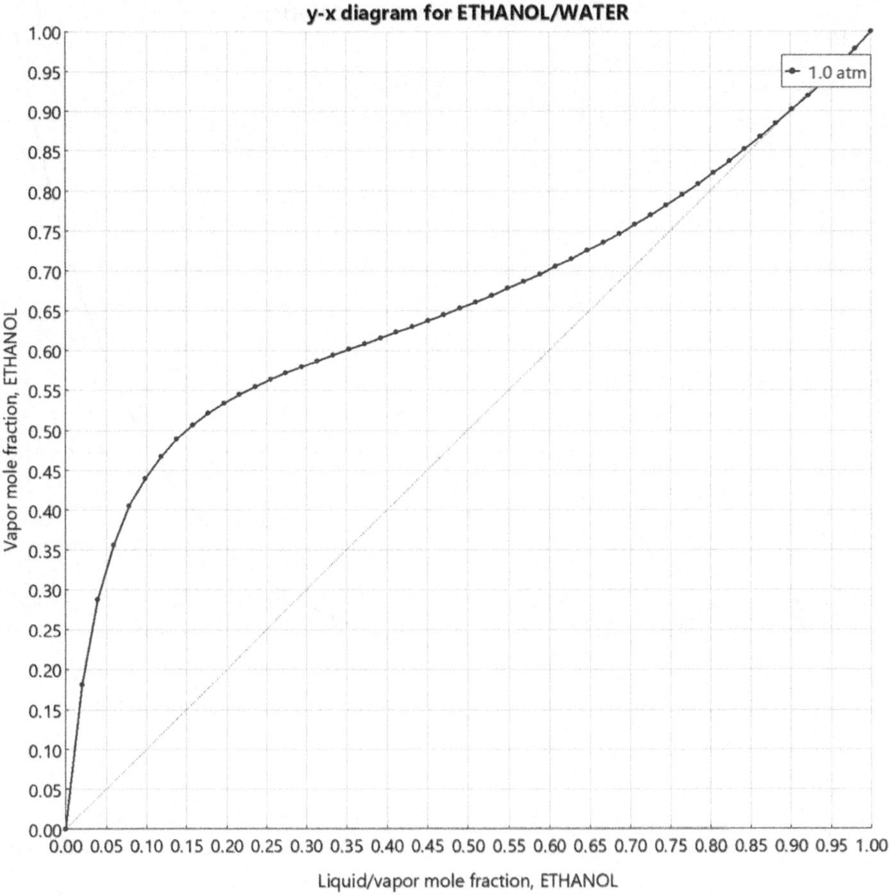

Fig. 14.6 Vapor-liquid equilibrium of the mixture of ethanol and water (1 atm)

(Wu 2018) was used as a reference for the economic analysis of butyric acid production cost. The butyric acid production cost is estimated to be 0.71 US$/L. To produce 1 kg of butanol, 1.19 kg of butyric acid is required. The butanol production cost is estimated to be 0.87 U$/L in Scenario 1. Due to limited studies available in the literature about the production cost of butyric acid, future work of evaluating the production cost of butyric acid for the specific fermentation methods is necessary.

Baral and Shah (2016) estimated the butanol production cost from traditional ABE fermentation to be 1.8 $/L. Qureshi et al. (2013) also presented a techno-economic analysis of ABE fermentation with a production cost of 1US$/L. However, different assumptions were made regarding the plant capacity, biorefinery concepts, and recovery methods. Therefore, it is difficult to make comparisons in many aspects.

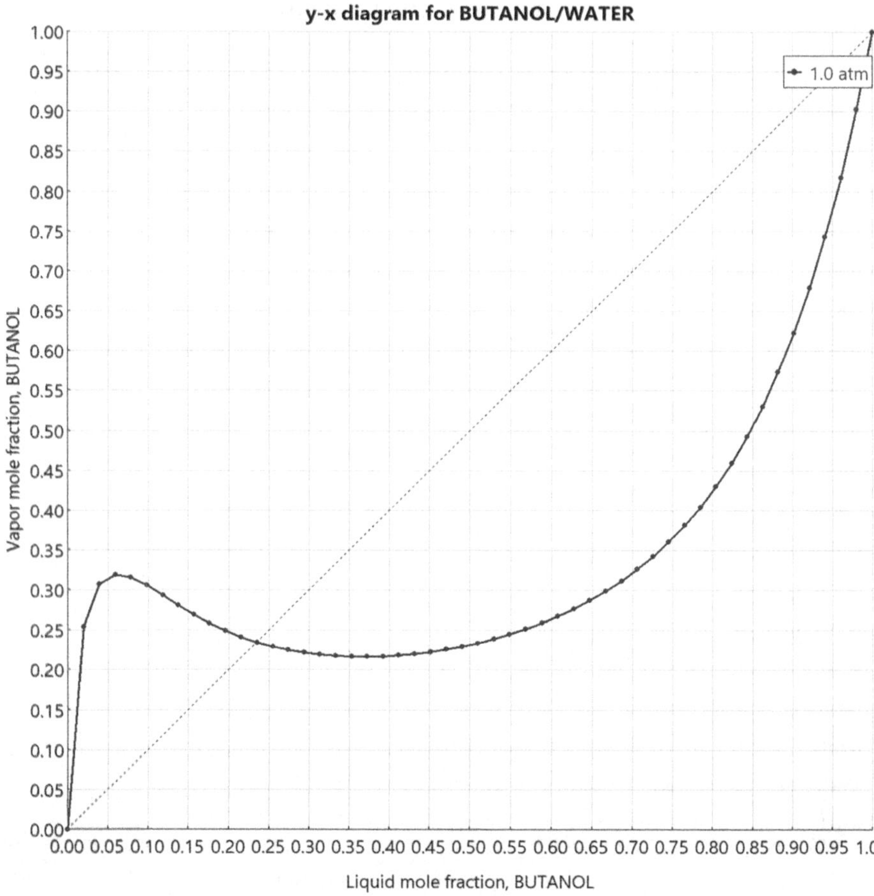

Fig. 14.7 Vapor-liquid equilibrium of the mixture of butanol and water (1 atm)

The butanol purification process could be further optimized as the following: Butanol-water system will form two liquid phases once condensed. This is a steady-state simulation of an azeotrope mixture of system butanol and water in which case two columns were used with a decanter located in between (Luyben 2008). Decanter separated two liquid phases and returned on the aqueous phase and organic (butanol rich) phase to a column as a reflux stream. Recycling and recovering steps for the remaining product in the waste stream are needed but not discussed in this study, which could be further investigated in future work.

The TEA work studied different scenarios about the butyric acid to butanol catalytic process to obtain the final product—butanol. Catalytically converting the acids (butyric acid and acetic acid) in the fermentation broth to alcohols before separating the alcohols shows promising economic advantages. With the advantage of a higher titer than ABE fermentation, butyric acid fermentation still needs a more detailed techno-economic analysis to investigate whether it achieves a competitive

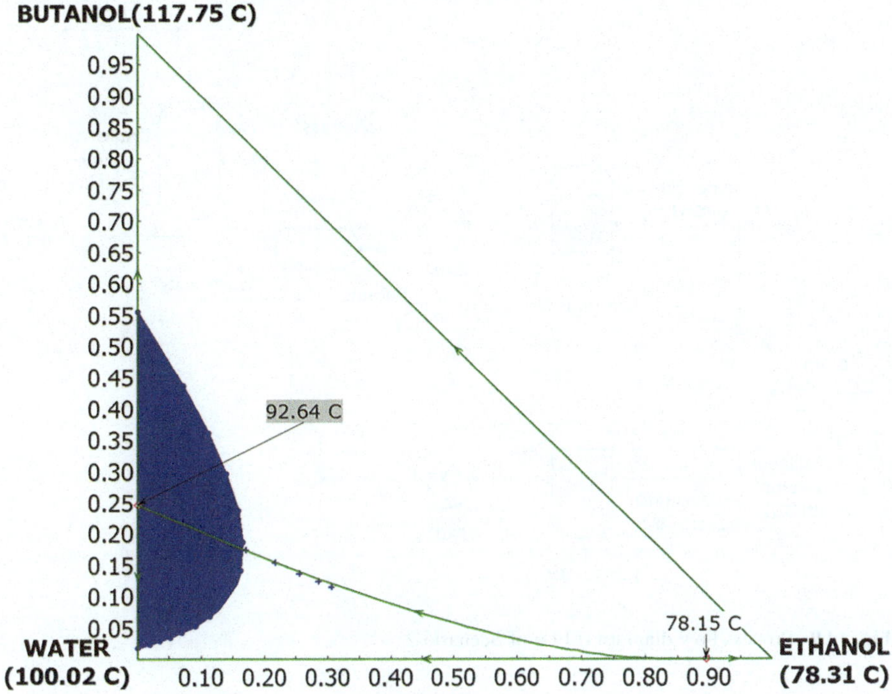

Fig. 14.8 Ternary diagram for butanol ethanol and water

cost or not. Besides, the waste stream from the whole process is another area for future research with the purpose of recovering energy and improving economic performance.

14.4.2 Environmental Impacts Considerations: LCA of Butanol Production Alternatives

An LCA study was carried out to evaluate environmental impacts along with process design for the implementation of bio-butanol technologies, support the strategic decision-making process, and analyze and compare different production alternatives of butanol. A wide variety of processes for butanol production have been studied through LCA such as the effect of different pretreatment methods (Baral et al. 2018), conversion methods such as oxo synthesis (Brito and Martins 2017), and ABE fermentation (Pereira et al. 2015), different feedstocks (e.g., corn and wheat straw) (Wu et al. 2007), different microbial strains (e.g., clostridia, cyanobacteria) (Nilsson et al. 2020), and different separation processes (Mahmud and Rosentrater 2020). The assessment applied in this case study used the TEA results from the previous section,

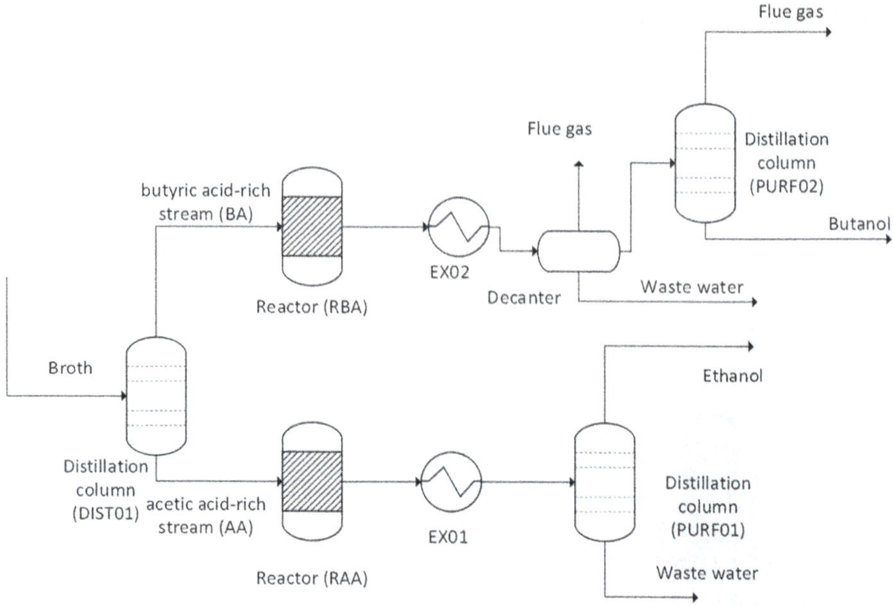

Fig. 14.9 Process flow diagram (PFD) of Scenario 2

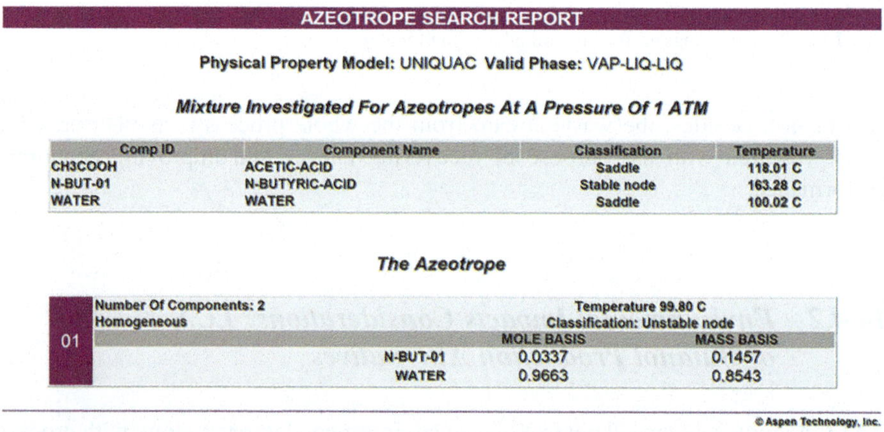
AZEOTROPE SEARCH REPORT

Physical Property Model: UNIQUAC **Valid Phase:** VAP-LIQ-LIQ

Mixture Investigated For Azeotropes At A Pressure Of 1 ATM

Comp ID	Component Name	Classification	Temperature
CH3COOH	ACETIC-ACID	Saddle	118.01 C
N-BUT-01	N-BUTYRIC-ACID	Stable node	163.28 C
WATER	WATER	Saddle	100.02 C

The Azeotrope

01 — Number Of Components: 2; Homogeneous; Temperature 99.80 C; Classification: Unstable node

	MOLE BASIS	MASS BASIS
N-BUT-01	0.0337	0.1457
WATER	0.9663	0.8543

© Aspen Technology, Inc.

Fig. 14.10 Azeotropes in Scenario 2

Table 14.4 Economic summary of butyric acid to butanol catalytic process

	Scenario 1	Scenario 2
Total capital cost (million $)	15.5	27.5
Capital charges (million $)	1.8	3.2
Total operating cost (million $)	21.7	92.7
Total utility cost (million $)	18.3	83.5

Table 14.5 Major unit operation equipment cost and installation cost

	Name	Equipment cost (million $)	Installed cost (million $)
Scenario 1	Hydrogenation Reactor (R1)	0.27	0.46
	Distillation column (SEPDIST)	0.38	0.88
	Heat exchanger (EXC1)	0.44	1.05
	Heat exchanger (EXC2)	0.08	0.25
	Decanter	0.02	0.13
	Distillation column (ETOHD)	0.15	0.55
	Distillation column (BTOHD)	0.11	0.48
	Distillation column (BEERCOL)	1.35	2.75
Scenario 2	Distillation column (DIST01)	8.57	14.97
	Heat exchanger (EX01)	0.02	0.09
	Heat exchanger (EX02)	0.63	0.99
	Decanter	0.02	0.12
	Distillation column (PURF01)	0.21	0.66
	Distillation column (PURF02)	0.15	0.51
	Reactor (RAA)	0.08	0.23
	Reactor (RBA)	0.14	0.32

which compare different alternatives of the butyric acid catalytic process to show the economic potential.

14.4.2.1 LCA Method

We used the LCA method as a tool to environmentally assess, identify hotspots and recommend strategies to improve the butanol production process. The LCA model was built according to the international standards ISO 14040 and ISO 14044. To conduct the LCA, the SimaPro 9.0 and Traci 2.1 V1.05/US 2008 methods were used.

14.4.2.2 Goal and Scope Definition, Functional Units, and System Boundary

The main purpose of the assessment was to compare different proposed process configurations for butanol production as a fuel. Here, three cases were investigated (Fig. 14.11): (1) butanol production through the catalytic process of butyric acid (two scenarios) from lignocellulosic (wheat straw), (2) butanol production through direct fermentation: ABE by clostridia (wheat straw) and fermentation by *E.coli* of lignocellulosic (corn stover) biomass, and (3) Butanol production through the petrochemical pathway. The functional unit is defined as 1 MJ of butanol product.

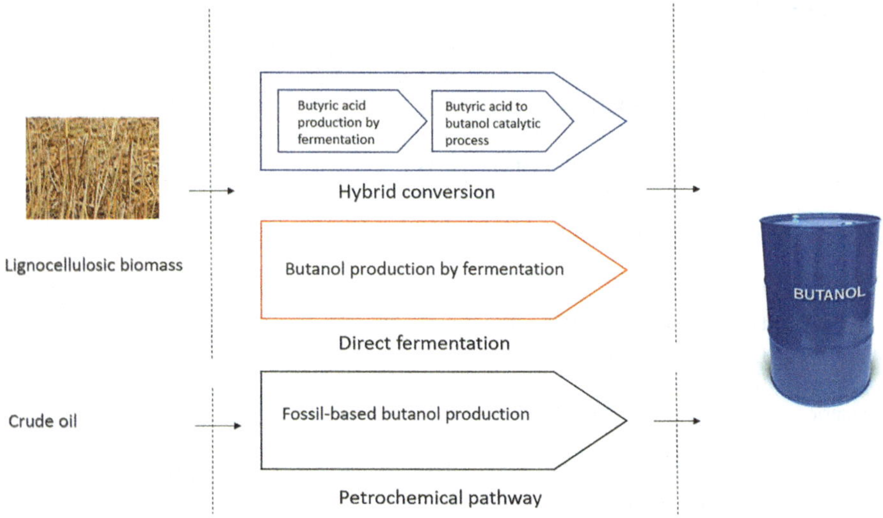

Fig. 14.11 Butanol production scenarios investigated

A "well to wheel (WTW)" system boundary is considered in this case study, which uses butanol as the final product of the industrial production facility.

14.4.2.3 Life Cycle Inventory

The materials and energy inputs of the inventories for the production of butanol using different processes are obtained from a combination of sources: process simulation results were mainly used and literature and Ecoinvent database v.3 were used for data gaps when needed. The inventory for the petrochemical pathway was directly provided by the Ecoinvent database v.3. Specifically, the petrochemical process includes propylene hydroformylation (oxo synthesis) with subsequent hydrogenation of the aldehydes formed. The hybrid conversion process includes two main processes: butyric acid fermentation, and butyric acid to butanol catalytic process. The butyric acid fermentation process data were based on the process model developed by Baroi et al. (2017) (Table 14.6), where the yield and concentration of butyric acid are in the same range as the TEA model presented in the previous section. Considering the substitutive catalytic process, minor modifications for the inventory data were made to exclude extraction and purification steps. Mass allocation was considered for the two main products: butyric acid and acetic acid. The energy consumption for the following butanol catalytic process was estimated by the previous process simulation section for the industrial scenarios. As a benchmark for the hybrid conversion process, two butanol production through direct fermentation processes were also evaluated: ABE process data from Brito and Martins (2017) and butanol conversion process using corn stover hydrolyzed sugars from the GREET

Table 14.6 Life Cycle Inventory (LCI) for butyric acid production

Input (materials and energy)	Unit	Value
Enzyme mix	tonnes/year	700.52
Wheat straw	tonnes/year	46150.21
KOH	tonnes/year	10.58
K2HPO4	tonnes/year	157.13
NaOH	tonnes/year	622.22
H2SO4	tonnes/year	857.14
Urea	tonnes/year	1078.58
Water	tonnes/year	1,10,120.89
Output (product)		
Butyric acid	kg/year	8,900,000
Acetic acid	kg/year	1,100,000

Table 14.7 Summary of data sources for processes of different cases involved in this study

Case	Process pathways	Data source
1. Hybrid conversion	Step 1. Butyric acid fermentation	Baroi et al. (2017)
	Step 2. Butyric acid to butanol catalytic process (two scenarios as in the TEA)	Process simulation
2. Direct fermentation	Pathway 1. ABE fermentation by clostridia (wheat straw)	Brito and Martins (2017)
	Pathway 2. Butanol fermentation by E.coli (corn stover)	Dunn et al. (2015)
3. Petrochemical conversion	Hydroformylation of propylene	Ecoinvent database v.3

model by Argonne National Laboratory (Dunn et al. 2015). The summary of data sources for processes of different cases considered in the LCA is shown in Table 14.7.

14.4.2.4 Results and Discussion

This case study of LCA covers a wide range of mostly hypothetical processes, which also means the process is still in the R&D stage and currently does not exist at a commercial scale. The overall objective of this study was to explore the potential of butanol production through a novel catalytic process (the modeled hybrid conversion system) by comparing it with the traditional configurations. Our results could be helpful for researchers to focus on areas for sustainability in the future.

14.4.2.5 Environmental Impact Assessment

The main difference observed in the environmental impacts between the process pathways is related to energy consumption in terms of electricity, heat, and cooling energy. Figure 14.12 presents the comparison of impacts for 1 MJ of butanol through

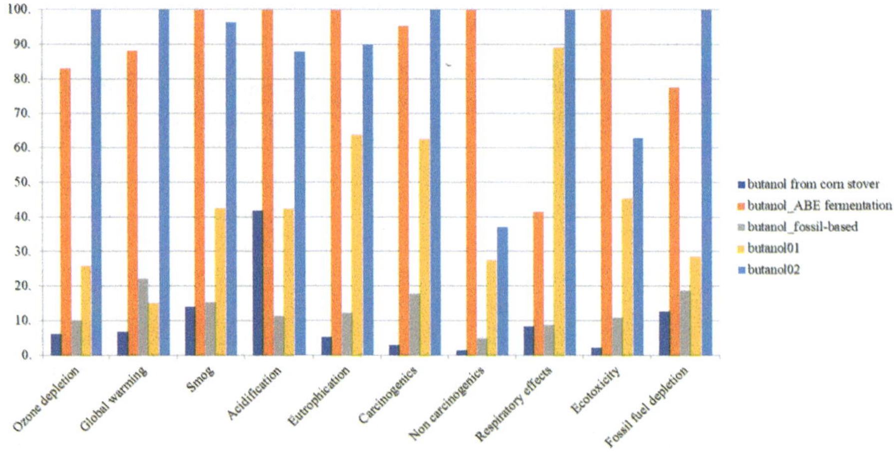

Fig. 14.12 Comparison of butanol production (1 MJ) alternatives

all the cases investigated. The hybrid conversion processes include two scenarios of process design, where scenario 1 (butanol 01) performs better in all environmental categories than scenario 2 (butanol 02). It can be inferred that the separation of final products after the catalytic reaction shows benefits in the environmental impacts, which is also in agreement with the economic results (lower operational cost due to lower utility usage). Thus, the results highlight the importance of catalytic process development in the aqueous phase for both economic and environmental advantages. The comparison also points out the environmental desirability of the hybrid conversion process for butanol, compared to the traditional ABE fermentation, since both scenarios of the hybrid conversion process show less environmental burden in most TRACI 2.1 categories, especially acidification, non-carcinogenic, and ecotoxicity. It should be noted that although butanol production from corn stover through fermentation of *E.coli* shows promising results in all the categories, it was modeled on additional assumptions such as fermentation temperature and retention time, as described in the GREET model, where further process refinement is required. Surprisingly, most of the bio-based routes for butanol production, except butanol from corn stover, have more environmental burden than the fossil-based route. This may partially be due to a lack of optimization of energy networks for the bio-based systems, whereas the fossil-based route is a mature industrial technology. Thus, we can conclude that at the current stage of biobutanol production, it may be difficult to compete with fossil-based butanol, both economically and environmentally.

For the process improvement, Fig. 14.13 shows the impact analysis of butanol hybrid conversion for both scenarios 1 and 2. Heat energy consumption and butyric acid are the main contributors to the environmental burdens, regardless, scenario 2 is more energy-intensive in terms of product purification steps. Figure 14.14 shows the environmental impacts in ten categories for 1 kg of butyric acid production. Electricity is a key factor for the technology, where it was mainly used for removing and recovering organic acids (butyric acid and acetic acid) using membranes. The

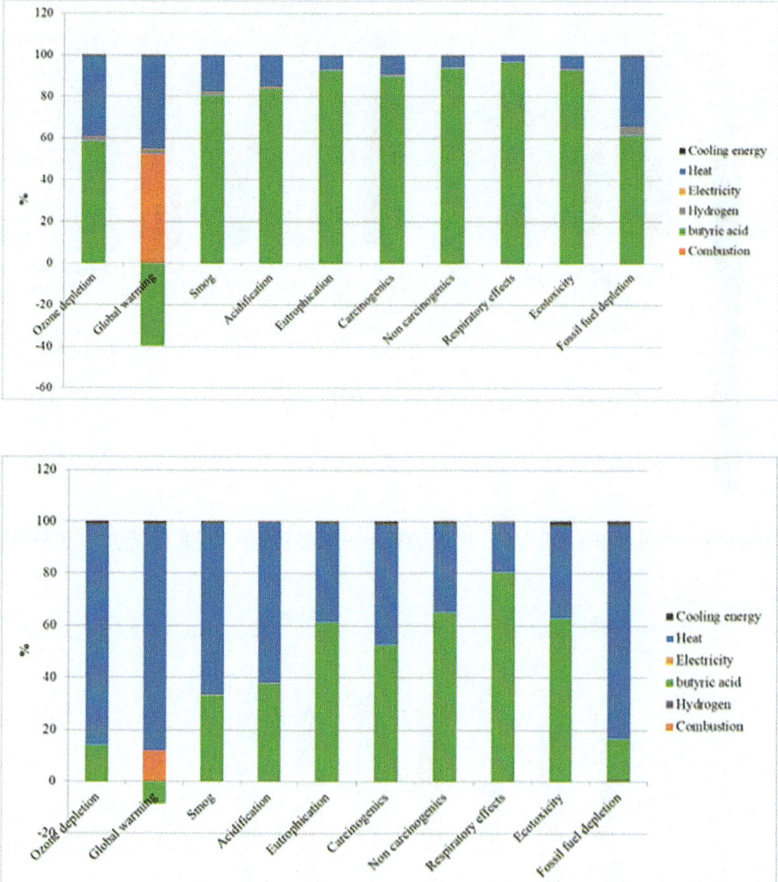

(Top)

(Bottom)

Fig. 14.13 Environmental impact analysis of butanol (1 MJ) hybrid conversion scenario 1 (top) and 2 (bottom)

separation of organic acids is essential for the downstream processes, either catalytic process or direct distillation. The connections between the steps determine the success of the biotechnologies' development to some extent. Similar to butanol hybrid conversion scenario 2, the ABE fermentation process requires significant amounts of steam for butanol purification (Fig. 14.15). Nevertheless, the bottleneck for butanol production could be the energy-intensive downstream product purification process. The high-energy demand will also be reflected in a higher production cost. For the fermentation step in both the butyric acid and ABE process, enzymes inputs are the second major contributors to the environmental impacts, especially in categories such as eutrophication, non-carcinogenic, and ecotoxicity. It should also be noted that the waste stream treatment process is not considered for all the cases, which could add credits (energy and nutrients recovery) to these emerging biotechnologies.

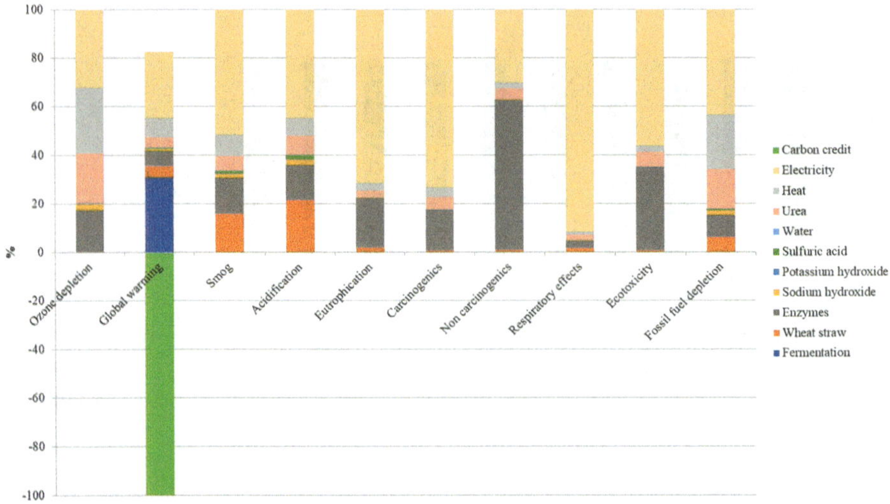

Fig. 14.14 Environmental impact analysis of butyric acid production (1 kg as the intermediate product)

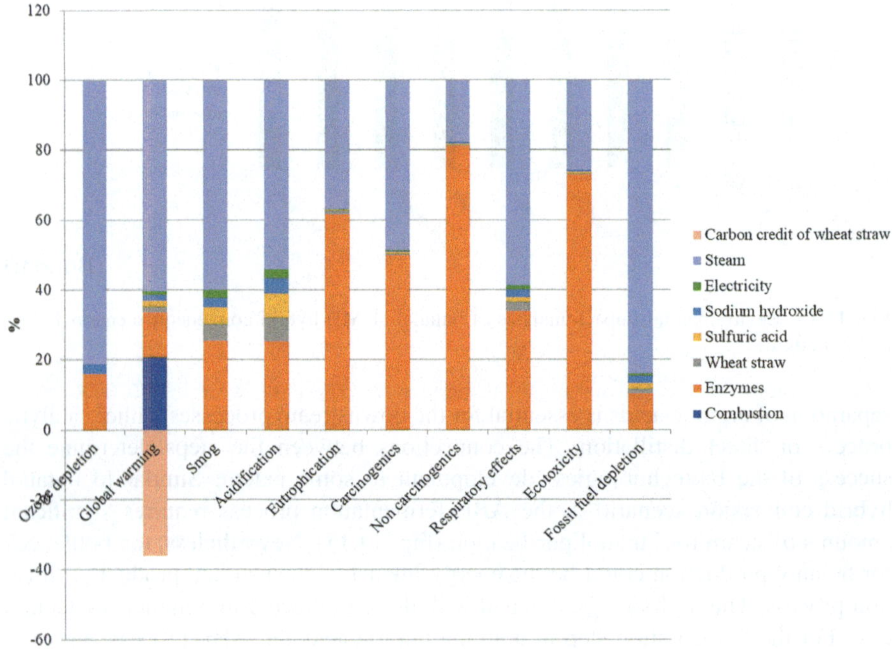

Fig. 14.15 Environmental impact analysis of butanol production (1 MJ) through ABE fermentation

14.4.3 *Uncertainty*

Variability in system parameters including inputs and design leads to the inherent uncertainty in the outcomes. Similar to other technology assessment methods, LCA and TEA could be associated with uncertainty risks when used for informing decisions. In the presented case studies, variability in feedstock composition, production and processing, allocation decision, geographic factors, and data estimation methods could cause variation in the environmental impacts and therefore uncertainty in the results. The sustainability impact scores may change substantially if variations in the input parameters are taken into account. Identifying the sources of uncertainty in the model and addressing them through a comprehensive uncertainty analysis will help increase the robustness of the results and reliability of recommendations for a wide range of potential process and market conditions. This highlighted data-driven research as a powerful tool in minimizing risks and maximizing benefits.

14.5 Conclusions, Guidelines, and Roadmap for the Future

This chapter provides state-of-the-art information and presents knowledge of analytical methods such as techno-economic analysis (TEA) and life cycle assessment (LCA) as standardized techniques that are used to quantify economic viability and environmental sustainability of microbial processes and products and assist with decision-making. R&D challenges while incorporating sustainability analysis to support biotechnology transition to a more sustainable future were discussed, especially in the area of data availability and accessibility considering technology readiness levels. Both challenges and opportunities for microbial process in renewable energy production were presented by a systematic review of commercialization status of renewable bioenergy/biochemicals, where success/failure stories were discussed and the main TEA and LCA findings in R&D and Technology Deployment were summarized.

The application of TEA and LCA tools and their main features in assessing the economic viability and environmental performance of microbial processes were further demonstrated through a case study of biobutanol production. The study evaluated the economic and environmental implications of different biobutanol production pathways representing the development of technologies, as well as improved configurations in the future. In general, an integrated or combined butanol production pathway (microbial and chemical) can be beneficial in terms of sustainability performance such as exhibiting lower environmental impacts as well as promising outcomes in the financial assessment. Although compared to a fossil-based route, the results of the case study may depict an unfavorable situation for the biobased route under current technological conditions, key sustainability improvements can be obtained by considering technological advances, waste treatment, and optimized energy networks. Compared to fossil-based energy, the microbial process in energy application still needs to overcome many shortcomings such as the input enzyme production, product recovery, and the interactions between the upstream and downstream steps. To bridge the gap between research and commercialization, this chapter emphasizes the role of interdisciplinary work, namely, analytics, science, engineering, politics, business, and society in building a harmonized and realistic roadmap for future sustainable biotechnology.

Acknowledgments This work was supported by funding from (1) North Dakota State University, Department of Coatings and Polymeric Materials and Center for Sustainable Materials Science (CSMS), and (2) the U.S. Department of Energy's Office of Energy Efficiency and Renewable Energy, Bioenergy Technologies Office and sponsored by the U.S. Department of Energy's International Affairs under award number, DEPI0000031.

References

Baral NR, Shah A (2016) Techno-economic analysis of cellulosic butanol production from corn stover through acetone–butanol–ethanol fermentation. Energy Fuel 30(7):5779–5790

Baral NR, Quiroz-Arita C, Bradley TH (2018) Probabilistic lifecycle assessment of butanol production from corn stover using different pretreatment methods. Environ Sci Technol 52(24):14528–14537. https://doi.org/10.1021/acs.est.8b05176

Bare J, Gloria T, Norris G (2006) Development of the method and US normalization database for life cycle impact assessment and sustainability metrics. Environ Sci Technol 40(16):5108–5115. https://doi.org/10.1021/es052494b

Baroi GN, Gavala HN, Westermann P, Skiadas IV (2017) Fermentative production of butyric acid from wheat straw: economic evaluation. Ind Crop Prod 104:68–80

Brito M, Martins F (2017) Life cycle assessment of butanol production. Fuel 208:476–482. https://doi.org/10.1016/j.fuel.2017.07.050

Buchner GA, Zimmermann AW, Hohgräve AE, Schomäcker R (2018) Techno-economic assessment framework for the chemical industry—based on technology readiness levels. Ind Eng Chem Res 57(25):8502–8517. https://doi.org/10.1021/acs.iecr.8b01248

Cavicchioli R, Ripple WJ, Timmis KN, Azam F, Bakken LR, Baylis M et al (2019) Scientists' warning to humanity: microorganisms and climate change. Nat Rev Microbiol 17(9):569–586. https://doi.org/10.1038/s41579-019-0222-5

Cherubini F, Strømman AH (2011) Life cycle assessment of bioenergy systems: state of the art and future challenges. Bioresour Technol 102(2):437–451. https://doi.org/10.1016/j.biortech.2010.08.010

Crater JS, Lievense JC (2018) Scale-up of industrial microbial processes. FEMS Microbiol Lett 365(13):fny138

Dunn JB, Adom F, Sather N, Han J, Snyder S, He C, ... & You F (2015) Life-cycle analysis of bioproducts and their conventional counterparts in GREET (No. ANL/ESD-14/9 Rev.). Argonne National Lab.(ANL), Argonne

Escobar N, Laibach N (2020) Sustainability check for bio-based technologies: a review of process-based and life cycle approaches. Renew Sustain Energy Rev 135:110213. https://doi.org/10.1016/j.rser.2020.110213

Ezeji T, Qureshi N, Blaschek HP (2007) Butanol production from agricultural residues: impact of degradation products on Clostridium beijerinckii growth and butanol fermentation. Biotechnol Bioeng 97:1460–1469. https://doi.org/10.1002/bit.21373

Horváth IT, Cséfalvay E, Mika LT, Debreczeni M (2017) Sustainability metrics for biomass-based carbon chemicals. ACS Sustain Chem Eng 5(3):2734–2740. https://doi.org/10.1021/acssuschemeng.6b03074

Kumar M, Goyal Y, Sarkar A, Gayen K (2012) Comparative economic assessment of ABE fermentation based on cellulosic and non-cellulosic feedstocks. Appl Energy 93:193–204. https://doi.org/10.1016/j.apenergy.2011.12.079

Lee JM, Upare PP, Chang JS, Hwang YK, Lee JH, Hwang DW et al (2014) Direct hydrogenation of biomass-derived butyric acid to n-butanol over a ruthenium–tin bimetallic catalyst. ChemSusChem 7(11):2998–3001. https://doi.org/10.1002/cssc.201402311

Luyben WL (2008) Control of the heterogeneous azeotropic n-butanol/water distillation system. Energy Fuel 22:4249–4258. https://doi.org/10.1021/ef8004064

Mahmud N, Rosentrater KA (2020) Life-cycle assessment (LCA) of different pretreatment and product separation technologies for butanol bioprocessing from oil palm frond. Energies 13(1):155. https://doi.org/10.3390/en13010155

Naleli K (2016) Process modelling in production of biobutanol from lignocellulosic biomass via ABE fermentation. (Master thesis). Stellenbosch University. http://hdl.handle.net/10019.1/98620

Nilsson A, Shabestary K, Brandão M, Hudson EP (2020) Environmental impacts and limitations of third-generation biobutanol: life cycle assessment of n-butanol produced by genetically engineered cyanobacteria. J Ind Ecol 24(1):205–216. https://doi.org/10.1111/jiec.12843

Pereira LG, Chagas MF, Dias MO, Cavalett O, Bonomi A (2015) Life cycle assessment of butanol production in sugarcane biorefineries in Brazil. J Clean Prod 96:557–568. https://doi.org/10.1016/j.jclepro.2014.01.059

Petter R, Tyner WE (2014) Technoeconomic and policy analysis for corn stover biofuels. International Scholarly Research Notices, 2014

Pfromm PH, Amanor-Boadu V, Nelson R, Vadlani P, Madl R (2010) Bio-butanol vs. bio-ethanol: a technical and economic assessment for corn and switchgrass fermented by yeast or Clostridium acetobutylicum. Biomass Bioenergy 34(4):515–524. https://doi.org/10.1016/j.biombioe.2009.12.017

Qureshi N, Saha BC, Cotta MA, Singh V (2013) An economic evaluation of biological conversion of wheat straw to butanol: a biofuel. Energy Convers Manag 65:456–462. https://doi.org/10.1016/j.enconman.2012.09.015

Sheldon RA (2018) Metrics of green chemistry and sustainability: past, present, and future. ACS Sustain Chem Eng 6(1):32–48. https://doi.org/10.1021/acssuschemeng.7b03505

Sjöblom M, Matsakas L, Christakopoulos P, Rova U (2015) Production of butyric acid by Clostridium tyrobutyricum (ATCC25755) using sweet sorghum stalks and beet molasses. Ind Crop Prod 74:535–544. https://doi.org/10.1016/j.indcrop.2015.05.041

Smolinksi S, Cox S (2016) Policies to enable bioenergy deployment: key considerations and good practices (No. NREL/TP-7A40-66322). National Renewable Energy Lab.(NREL), Golden, CO (United States). https://doi.org/10.2172/1253707

Srivastava RK (2019) Bio-energy production by contribution of effective and suitable microbial system. Mater Sci Energy Technol 2(2):308–318. https://doi.org/10.1016/j.mset.2018.12.007

Swana J, Yang Y, Behnam M, Thompson R (2011) An analysis of net energy production and feedstock availability for biobutanol and bioethanol. Bioresour Technol 102(2):2112–2117. https://doi.org/10.1016/j.biortech.2010.08.051

Tabone MD, Cregg JJ, Beckman EJ, Landis AE (2010) Sustainability metrics: life cycle assessment and green design in polymers. Environ Sci Technol 44(21):8264–8269. https://doi.org/10.1021/es101640n

Tao L, Tan EC, McCormick R, Zhang M, Aden A, He X, Zigler BT (2014) Techno-economic analysis and life-cycle assessment of cellulosic isobutanol and comparison with cellulosic ethanol and n-butanol. Biofuels Bioprod Biorefin 8(1):30–48. https://doi.org/10.1002/bbb.1431

United Nations (2015) Transforming our world: the 2030 agenda for sustainable development. http://www.un.org/ga/search/view_doc.asp?symbol=A/RES/70/1&Lang=E

Vasco-Correa J, Khanal S, Manandhar A, Shah A (2018) Anaerobic digestion for bioenergy production: global status, environmental and techno-economic implications, and government policies. Bioresour Technol 247:1015–1026. https://doi.org/10.1016/j.biortech.2017.09.004

Wu N (2018) Cultivation, techno-economic analysis of biofuels production. Doctoral dissertation, University of Florida

Wu M, Wang M, Liu J, Huo H (2007) Life-cycle assessment of corn-based butanol as a potential transportation fuel (No. ANL/ESD/07-10). Argonne National Lab.(ANL), Argonne, IL (United States). https://doi.org/10.2172/925379

Wu N, Moreira C, Zhang Y, Doan N, Yang S, Phlips E, ..., Pullammanappallil P (2019) Techno-economic analysis of biogas production from microalgae through anaerobic digestion. In: Biogas. IntechOpen. https://doi.org/10.5772/intechopen.86090

The manufacturer's authorised representative in the EU is Springer
Nature Customer Service Centre GmbH, Europaplatz 3, 69115 Heidelberg,
Germany. If you have any concerns regarding our products, please
contact ProductSafety@springernature.com

Printed and bound by CPI Group (UK) Ltd, Croydon, CR0 4YY
30/09/2026
02225329-0015